# Preliminary Chemical Engineering Plant Design

# Preliminary Chemical Engineering Plant Design

## William D. Baasel

Professor of Chemical Engineering
Ohio University

**ELSEVIER**

New York/Oxford/Amsterdam

AMERICAN ELSEVIER PUBLISHING COMPANY, INC.
52 Vanderbilt Avenue, New York, N.Y. 10017

ELSEVIER SCIENTIFIC PUBLISHING COMPANY
335 Jan Van Galenstraat, P.O. Box 211
Amsterdam, The Netherlands

**Library of Congress Cataloging in Publication Data**

Baasel, William D
Preliminary chemical engineering plant design.

Bibliography: p.
Includes index.
1. Chemical plants—Design and construction.
I. Title.
TH4524.B3    660.2'8    74-19453
ISBN 0-444-00152-2

*To my wife Patricia,*
*without whose support and assistance this book*
*could never have been completed*

# Contents

# Preface

The idea for this book was conceived while I was on a Ford Foundation residency at the Dow Chemical Company in Midland, Michigan. I was assigned to the process engineering department, where I was exposed to all areas of process engineering, project engineering, and plant construction. My previous industrial experiences had been in pilot plants and research laboratories. Much to my surprise, I found that what was emphasized in the standard plant design texts was only a part of preliminary process design. Such areas as writing a scope, site selection, equipment lists, layout, instrumentation, and cost engineering were quickly glossed over. After I returned to Ohio University and began to teach plant design, I decided a book that emphasized preliminary process engineering was needed. This is the result. It takes the reader step by step through the process engineering of a chemical plant, from the choosing of a site through the preliminary economic evaluation.

So that the reader may fully understand the design process, chapters dealing with planning techniques, optimization, and sophisticated computer programs are included. These are meant merely to give the reader an introduction to the topics. To discuss them thoroughly would require more space than is warranted in an introductory design text. They (and other sophisticated techniques, like linear programming) are not emphasized more because before these techniques can be applied a large amount of information about the process must be known. When it is not available, as is often the case, the engineer must go through the preliminary process design manually before these newer techniques can be used. It is to this initial phase of design that this book is directed.

Three types of design problems fit this situation. One is the design of a plant for a totally new product. The second is the design of a new process for a product that currently is being produced. The last is the preliminary design of a competitor's plant, to determine what his costs are. In each of these, little is known about the process, so that a large amount of educated guessing must occur.

As time goes on, more and more people are being involved in these types of plant design. Most chemical companies estimate that 50% of their profits 10 years hence will come from products not currently known to their research laboratories. Since these will compete with other products now on the market, there will be a great need for improving present processes and estimating a rival's financial status.

This book deals mainly with chemical plant design, as distinct from the design of petroleum refineries. For the latter, large amounts of data have been accumulated, and the procedures are very sophisticated. It is assumed that the reader has some familiarity with material and energy balances. A background in unit operations and thermodynamics would also be helpful, although it is not necessary. No attempt is made to repeat the material presented in these courses.

This book applies a systems philosophy to the preliminary process design and cost estimation of a plant. In doing so, it tries to keep in perspective all aspects of the design. There is always a tendency on the part of designers to get involved in specific details, and forget that their job is to produce a product of the desired quality and quantity, at the lowest price, in a safe facility. What is not needed is a technological masterpiece that is difficult to operate or costly to build.

For those using this book as a text, I suggest that a specific process be chosen. Then, each week, one chapter should be read, and the principles applied to the specific process selected. The energy balance and economic chapters may each require two weeks. The pollution abatement chapter may be included after Chapter 8, or it can be studied as a separate topic unrelated to the over-all plant design.

Each student or group of students may work on a different process, or the whole class may work on the same process. The advantage of the latter method is that the whole class can meet weekly to discuss their results. This has worked very success-fully at Ohio University. In the discussion sections, the various groups present their conclusions, and everyone, especially the instructor, benefits from the multitude of varied and imaginative ideas.

Initially, this procedure poses a problem, since in most college courses there is a right and a wrong answer, and the professor recognizes and rewards a correct response. In designing a plant, many different answers may each be right. Which is best often can be determined only by physically building more than one plant, and evaluating each of them. Of course, no company would ever do this. It would build the plant that appears to contain fewer risks, the one that seems to be best economically, or some combination of these.

Since the student will build neither, and since the professor probably cannot answer certain questions because of secrecy agreements or lack of knowledge, the student must learn to live with uncertainty. He will also learn how to defend his own views, and how to present material so as to obtain a favorable response from others. These learning experiences, coupled with exposure to the process of design as distinct from that of analysis and synthesis, are the major purposes of an introduc-tory design course.

Besides students, this book should be useful to those in industry who are not intimately familiar with process engineering. Researchers should be interested in process design because their projects are often killed on the basis of a process engineering study. Administrators need to have an understanding of this because they must decide whether to build a multi-million-dollar plant designed by a process engineering team. Operating personnel should know this because they must run plants designed by process engineers. Similarly, project engineers and contractors need to understand process engineering because they must take the resultant plans and implement them. Finally, pilot plant and semi-plant managers and operators need to know the problems that can arise during process design because they often must determine whether the various schemes devised by process designers are feasible.

The importance of preliminary design cannot be underestimated. For every plant built, 10 partially engineered plants are rejected. For some of these, over $100,000 worth of engineering will have been completed before the plant is rejected. Often this loss could have been avoided if there had been a greater understanding of preliminary chemical engineering process design by all concerned.

I wish to express my deep thanks to the Dow Chemical Company, particularly to my preceptors Dr. Harold Graves and James Scovic, and everyone in the Process Engineering Department. They were completely open with me, and showed me how chemical engineering plant design is done. Also, I would like to thank all those others at Dow who spent a lot of time educating me.

I would also like to acknowledge the support of the Chemical Engineering Department at Ohio University, and especially its chairman, Dr. Calvin Baloun. However, the group that had the greatest influence on the final form of this book was the Ohio University Chemical Engineering seniors of 1970, 1971, 1972, 1973, and 1974. They evaluated the material and suggested many improvements that were incorporated into this book. To them I am deeply indebted. I would also like to thank the following people who assisted me in the preparation of the manuscript: Linda Miller, Carolyn Bartels, Audrey Hart, Joan Losh, Cindy Maggied, and Judy Covert.

William D. Baasel

March 18, 1974

# CHAPTER 1

# Introduction to Process Design

Design is a creative process whereby an innovative solution for a problem is conceived. A fashion designer creates clothes that will enhance the appeal of an individual. An automobile designer creates a car model that will provide transportation and a certain appeal to the consumer. The car's appeal may be because of its power, beauty, convenience, economy, size, operability, low maintenance, uniqueness, or gimmicks. A process engineer designs a plant to produce a given chemical. In each of these instances a new thing is created or an old thing is created in a new way.

Design occurs when a possible answer for a present or projected need or desire by people or industry has been found. If a product were not expected to meet a need or desire, there would be no reason to produce it and hence no reason for design. A company or person is not going to manufacture something that cannot be sold at a profit.

The needs may be basic items like substances with which to clean ourselves, coverings to keep our bodies warm, dishes upon which to place our food, or cures for our diseases. The desires may be created by the advertising firms, as in the case of vaginal deodorants and large sexy cars.

Often the need or desire can be satisfied by a substance that is presently on the market, but it is projected that a new product will either do a better job, cost less, or require less time and effort. The toothpastes produced before 1960 did a respectable job of cleaning teeth, but the addition of fluoride made them better cavity preventatives, and those toothpastes that added fluorides became the best sellers. Orange juice could be shipped in its natural form to northern markets, but frozen concentrated orange juice occupies one-fourth the volume and costs less to the consumer. TV dinners and ready-to-eat breakfast cereals cost more than the same foods in their natural state, but they reduce the time spent in the kitchen. All of these items resulted from research followed by design.

Most companies in the consumer products industries realize that their products and processes must be continually changed to compete with other items that are attempting to replace them. Sometimes almost a complete replacement occurs within a short time and a company may be forced to close plants unless an alternate use of its products is found. As an example, consider the case of petroleum waxes. In the late 1950s the dairy industry consumed 220,000 tons per year of petroleum waxes for coating paperboard cartons and milk bottle tops. This was 35% of the total U.S. wax production. By 1966 this market had dropped to 14% of its former level (25,000 tons / yr) because polyethylene and other coatings had replaced it.[1]

One reason for conducting research is to prevent such a change from completely destroying a product's markets. This may be done by improving the product, finding new uses for it, or reducing its costs. Cost reduction is usually accomplished by improving the method of producing the product. Research is also conducted to find new substances to meet industry's and people's needs and desires.

Once a new product that looks salable or an appealing new way for making a present product is discovered, a preliminary process design for producing the item is developed. From it the cost of building and operating the plant is estimated. This preliminary process design is then compared with all possible alternatives. Only if it appears to be the best of all the alternatives, if it has potential for making a good profit, and if money is available, will the go-ahead for planning the construction of a facility be given.

Since the goal of a chemical company is to produce the products that will make the most money for its stockholders, each of these phases is important; each will be discussed in greater detail.

## RESEARCH

Most large chemical companies spend around 5% of their total gross sales on some type of research. In 1967 the Gulf Research and Development Company, a wholly owned subsidiary of the Gulf Oil Corp., spent $30,000,000 on research and development.[2] Of this, 58% was for processes and 42% was for products. This means most of their sizeable research budget went into developing new processes or improving old ones.

A company sells its products because either they are better than, or they cost less than, a competitive product. If a company does not keep reducing its processing costs and improving quality it can easily lose its markets. An example of how technological improvements in the production of fertilizers have forced many older plants out of business is given in Chapter 3.

If Gulf's research budget is broken down another way, basic research received 8% of $30,000,000, applied research got 41%, development projects received 22%, and technical service ended up with 29%.

*Basic research* consists of exploratory studies into things for which an end use cannot be specified. It might include a study to determine the effect of chlorine molecules on the diffusivity of hydrocarbons or a study of the dissolution of single spheres in a flowing stream. The prospective dollar value of this research cannot be estimated.

*Applied research* has a definite goal. One company might seek a new agricultural pesticide to replace DDT. Another might be testing a new approach to manufacturing polystyrene. *Development projects* are related to the improvement of current production methods or to determining the best way of producing a new product. They could involve anything from designing a new waste recovery system to studying the feasibility of replacing conventional controllers in an existing plant with direct digital control.

*Technical service* is devoted to making the company's products more acceptable to the user. Its people try to convince prospective users of the advantages of using their company's chemicals. This cannot be done in the manner of a television commercial by using gimmicks or sex appeal, but must rely on cold, hard facts. Why should a manufacturer switch from a familiar, adequate product to a new one? Since no chemical is completely pure and since each manufacturer uses at least a slightly different process and often different raw materials, the impurities present in products from several suppliers will be different. How these impurities will affect products, processes, catalysts, and so on is often unknown. It is the job of technical service representatives to find out. For instance, caustic soda produced as a by-product of chlorine production in a mercury cell cannot be used in the food or photographic industries because trace amounts of mercury might be present.

One case where technical service representatives were called in occurred when a large chemical company which found it could easily increase its product purity without changing prices, did just that. About three months later it got a desperate call from a customer that produced fire extinguishers. All of their new fire extinguishers were rusting out very rapidly and they could not understand why. An investigation found that what had been removed from the upgraded product was a chemical that acted as a rust inhibitor. Neither of the companies had previously realized that this contaminant was actually indispensable to the producer of fire extinguishers.

Experiences like this make production men very hesitant to make changes. This can be very frustrating to a process engineer whose job is to improve the present process. One superintendent was able to increase the throughput in his plant by 60%. Six months later he insisted that the design of a new plant should be based on the old rate. He reasoned that not all the customers had tried the "new product" and there might be some objections to it. Yet he had not informed any of the users of the processing change.

## OTHER SOURCES OF INNOVATIONS

Research is not the only source of new ideas. They may occur to anyone, and most companies encourage all their employees to keep their eyes open for them. A salesman, in talking to a customer, may find that this customer has a given need that he has been unable to satisfy. A engineer at a convention may find out that someone has difficulty operating a specific unit because some needed additive has a deleterious side effect. The engineer and salesman report the details of these findings in the hope that some researcher within their own company may have discovered a product that can meet these needs. Another may hear or read about a new way of doing something, in some other country or in some other industry, that can be adapted to his company's projects. This is the way Dow found out about the Ziplock® feature of their food storage bags. In this instance, after further investigation they negotiated a contract with the Japanese inventors for the sole use of the device for consumer products sold within the United States.

Another source of design ideas is the production plant. There the operators and engineers must surmount the problems that arise daily in producing an adequate supply of a quality product. Sometimes accidentally, sometimes by hard work, new processing conditions are found that eliminate the need for some purification steps or that greatly increase the plant capacity. People who have transferred from another production operation are often able to come up with suggestions that worked in other circumstances and may profitably be applied to the process with which they are now involved.

## PROCESS ENGINEERING

*Process engineering* is the procedure whereby a means for producing a given substance is created or modified. To understand what is involved one must be familiar with chemical plants.

Chemical plants are a series of operations that take raw materials and convert them into desired products, salable by-products, and unwanted wastes. Fats and oils obtained from animals and plants are hydrolyzed (reacted with water) and then reacted with soda ash or sodium hydroxide to make soaps and glycerine. Bromine and iodine are recovered from sea water and salt brines. Nitrogen and hydrogen are reacted together under pressure in the presence of a catalyst to produce ammonia, the basic ingredient used in the production of synthetic fertilizers.

To perform these changes some or all of the following steps are needed.

1. Feed storage:             Incoming materials are placed in storage prior to use.

2. Feed preparation:         The raw materials are physically changed and purified.

3. Reaction:                 The raw materials are brought together under controlled conditions so that the desired products are formed.

4. Product purification:     The desired products are separated from each other and from the other substances present.

5. Product packaging and storage:   The products are packaged and stored prior to shipment.

6. Recycle, recovery, and storage:   Undesirable substances are separated from the reusable materials, which are then stored.

7. Pollution control:                    The waste is prepared for disposal.

To illustrate these steps, consider the process flow sheet for Armour's continuous soap-making process given in Figure 1-1[3]. The feed, consisting of fats and oils, is prepared by centrifuging it to remove proteins and other solid impurities, deaerating it to remove oxygen, which could degrade the product, and finally heating it. After this preparation the triglycerides, which comprise a majority of the fats and oils, are reacted with water to form fatty acids and glycerine. One such reaction is:

$$(C_{17}H_{35}COO)_3C_3H_5 + 3H_2O \longrightarrow 3C_{17}H_{35}COOH + C_3H_5(OH)_3$$

In this process both the reaction and the separation of the by-product, glycerine (sweet water), from fatty acids occur in splitters. The remaining steps in the sweet-water processing are all concerned with removal of the impurities to produce a clear glycerine. The settling tank allows time for any remaining acids to separate from the glycerine. These acids are sent to the fatty acid storage. Organic impurities that were not removed by the feed preparation steps are separated out by adding coagulants to which they will adhere, and then filtering them out. The water is removed by evaporation, followed by distillation, and any undesirable organics remaining are adsorbed on activated carbon and removed by filtration. The final product is then put in containers and stored before shipment to the customers.

Meanwhile, the fatty acids are purified before they are reacted with caustics to produce soaps. The steps involve a flash evaporation to remove water, and a vacuum distillation that removes some more water, any gases, and a fatty residue, which is recycled through the splitter. The vacuum still also separates the acids into two different streams. One of these is used to make toilet soaps and the other, industrial soaps. The process for making the industrial soap is not shown, but it is similar to that shown for toilet soaps. The soap is made in the saponifier. A typical reaction is

$$C_{17}C_{35}COOH + NaOH \longrightarrow C_{17}H_{35}COONa + H_2O.$$

The product is purified by removing water in a spray dryer. It is then extruded and cut into bars of soap, which are packaged for shipping.

A number of things are not shown on these process flow sheets. One is the storage facilities for the feed, product, and by-products. The second is the waste treatment facilities. All water leaving the process must be sent through treatment facilities before it can be discharged into lakes or rivers, and some means must be devised to get rid of the solid wastes from the filters and the centrifuge (see Chapter 16).

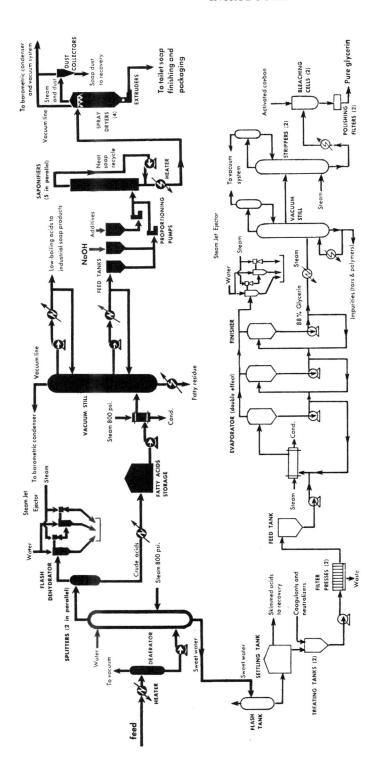

Figure 1-1  Flow Diagram for Armour's Soap Plant. Courtesy of Ladyn, H. W., "Fat Splitting and Soap Making Go Continuous," *Chemical Engineering;* Aug. 17, 1964, p. 106.

## PROFESSIONAL RESPONSIBILITIES

The process engineer is the person who constructs the process flow sheet. He decides what constitutes each of the seven steps listed at the beginning of the last section, and how they are to be interconnected. He is in charge of the process, and must understand how all the pieces fit together. The process engineer's task is to find the best way to produce a given quality product safely—"best," at least in part, being synonymous with "most economical."

The engineer assumes that the people, through their purchasing power in the market place, select what they deem best. He may devise a method of reducing pollution, but if it causes the price of the product to increase, it generally will not be installed unless required by the government. Other corporations and the public will not pay the increased price if they can get an equivalent product for less. This is true even if they would benefit directly from the reduced pollution. The engineer and his societies in the past have seldom crusaded for changes that would improve the environment and benefit the general public. The typical engineer just sat back and said, "If that's what they want, let them have it." Engineers have typically abrogated their social responsibilities and let the Rachel Carsons and Ralph Naders fight for the common good when engineers could have been manning the barricades.

Until the past few years, whenever the engineer spoke of ethics he meant loyalty to company. Now some are speaking about what is good for mankind. This trend could add a new dimension to process engineering just as great as the changes that occurred around 1958.

Between 1938 and 1958, the chemical and petrochemical industry could do nothing wrong. These were years of rapid expansion when the demand quickly exceeded the supply. The philosophy of the era was to build a plant that the engineer was sure would run at the design capacity. If it ran at 20, 30, or even 50% over the nominal capacity this was a feather in the superintendent's cap. There were proud boasts of a plant running at 180% of capacity. Anybody who could produce this was obviously in line for a vice-presidency. He was a manager's manager.

These were the years when whatever could be made could be sold at a profit. The United States was involved in a world war followed by a postwar business boom and the Korean War. Then came 1958. The Korean War had been over for five years. The United States was in the midst of a major recession.. The chemical industry that previously could do no wrong found that all of a sudden its profits were declining rapidly. A blow to the pocketbook causes a speedy reaction. A couple of major chemical concerns responded by firing 10% of salaried employees. This was the end of an era.

The Midas touch that had been associated with the chemical industry was gone, and firing all those men did not bring it back. A self-appraisal of company policy was begun; the process engineer's stature began to rise, but so did the demands that were placed upon him. The boards of directors of many companies decided they were the ones to pick the plant size. They began to request that the design capacity be within 10% of the actual capacity. They also asked that early design estimates be within 10% of the final cost. Competition was now so keen that no "fat" could be afforded in a process. Many plants were being run below design capacity because of a lack of sales. These companies realized that the excess capacity built into their plants was a liability rather than an asset. First, the larger the equipment the more expensive it is. This means the plant initially cost more than should have been spent. Second, a properly designed plant runs most efficiently at the design capacity. For instance, a pump will be chosen so that when it is operating at the design capacity it produces the desired flow rate and pressure at the lowest cost per pound of throughput. When it is operating at other rates the cost per pound increases. Thus, the cost of running a plant is at a minimum at the design capacity. An oversize plant could of course be run at design capacity until the product storage was full and then shut down until nearly all the product has been shipped to customers. However, the problems involved in starting up a plant usually rule this out as a practical solution.

This tightening-up trend will not be stopped, and more and more the process engineers will be expected to design a plant for the estimated cost that will safely produce the desired product at the chosen rate.

## COMPETING PROCESSES

There are various ways of producing a quality product. This can be seen by investigating how any given chemical is produced by competing companies. Consider the production of phenol. The most popular process is to obtain phenol from cumene. The four companies that offer process licenses are Allied Chemical Corporation, Hercules Inc., Rhone-Poulenc, and Universal Oil Products Co. These processes differ in the way the yield of phenol is maintained and how cumene hydroperoxide, a highly explosive material, is handled. The original method used to produce phenol was the sulfonation process. Only one of seven companies that announced plans to increase capacity in 1969 was planning to use this process. Currently Dow produces phenol at Midland by hydrolyzing monochlorobenzene with aqueous caustic soda, but it has been planning to phase out or scale down this operation. Phenol can also be produced by the direct oxidation of cyclohexane or by using the Rashig process.[4]

The facts that different companies using different processes can each make money and that even within the same company a product may be produced by two entirely different processes illustrate the challenge and headaches connected with process design. Design demands a large amount of creativity. It differs from the usual mathematics problem in that there is more than one acceptable answer. Theoretically there may be a best answer, but rarely are there enough data to show conclusively what this is. Even if it could be identified, this best design would vary with time, place, and company.

Advances in technology may make the best-appearing process obsolete before the plant can be put into operation. This happened to a multimillion dollar plant that Armour & Company built in the early 1950s for producing ACTH (adreno-corticotropic hormone). This is a hormone originally extracted from the pituitary glands of hogs. It provides relief from the painful inflammation of arthritis. Before the plant was completed a synthetic method of producing ACTH was proven. The plant designed to use hog gland extracts was never run. The old process could not compete economically with the new one. The process designers at Armour should not be condemned for what happened. They had no way of knowing that a newer process would make theirs obsolete.

The history of penicillin, which is produced from molds, is different. Penicillin is a powerful antibacterial substance that came into extensive use during World War II. There still is no known synthetic way of producing penicillin economically. If the pharmaceutical companies had refused to mass-produce this drug by fermentation because they feared it would soon be synthesized, then millions of people would have been deprived of its healing powers, and those who could have obtained it would have spent ten to one hundred times more for it.

Since each company keeps secret what it is researching, and how that research is progressing, process design risks must be taken based on the best information available.

## TYPICAL PROBLEMS A PROCESS ENGINEER TACKLES

The type of problem the process engineer is confronted with and the amount of information available vary widely. Four examples follow:

### A New Product

The applied research laboratory has developed a new substance that they feel has great potential as a gasoline additive. It improves the antiknock characteristics of gasoline, and does not noticeably increase the amount of air pollution. The marketing department estimates that within 5 years the market could reach 10,000,000 lb / yr. The process engineer is asked to design a plant to produce 10,000,000 lb / yr (4,500,000 kg / yr).

Since this is a new chemical, all that is known is the chemical process for making it, its normal boiling point, and its chemical formula. The only source of information is the chemist who discovered it. The process engineering study will determine the production costs, identify the most costly steps involved, and decide what further data must be obtained to ensure that the proposed process will work. The production costs are needed to determine if the new product can compete monetarily with tetraethyl lead and other additives.

It is important to identify the expensive steps, because it is here that research and development efforts should be concentrated. If the solvent recovery system is inexpensive, the prospective savings to be obtained by thoroughly studying it are small, and the cost of research may exceed any hoped-for saving. Conversely, should the reaction step be expensive, determining the kinetics of the reaction might result in the design of a recycle system that would reduce the number of reactors and save over a million dollars in one plant alone.

It is important to begin producing this additive as soon as possible. This is because the discovery of something new is frequently made by two or more independent investigators at about the same time, and the first producer sets the standards and gets the markets. In 1969 four chemical companies, Standard Oil Co. (Indiana), DuPont, Phillips Petroleum, and Montecatini-Edison, were all claiming to be the inventor of polypropylene. At that time, the U.S. Patent Office had still not decided who would get the U.S. patent, even though the work had been done over 10 years before.[5] Finally in December 1971 Montecatini-Edison received the patent.

A company usually sets product standards in such a way as to minimize the purification expenses. These standards are often empirical tests to ensure that the buyer will get the same product in each shipment. Examples would be the melt index of a polymer, the boiling-point range of the product, and the maximum amount of certain impurities. Another manufacturer using a different process would want to set different standards. His method of production will be different, and so the amount and kind of impurities will be different. Sometimes this means expensive purification steps must be installed to meet the specifications set by the initial manufacturer. If this competitor could have been the initial standard-setter then these steps would not be necessary.

The buyer adapts his process so he can use the first producer's products. He is not prone to switch unless the technical service department of the new manufacturer can convince him that it will save him money and that there are no risks involved. The fire extinguisher example given previously illustrates why the buyer is not eager to change.

The first company to produce a product also has the opportunity to set prices. Then when another producer enters or threatens to enter the lists, these prices can be dropped. The net result can be substantial profits for a company.

The importance of time means that only the critical questions raised by the process engineer's study can be answered before construction. Even some of these will not be fully answered until the plant starts up. This can pose problems. For example, suppose the process engineer assumes that the solvent can be separated from the product by a simple distillation. If an azeotrope is formed, this is impossible, and a much more costly separation step may be necessary. Should a plant be built before this is discovered, its product may be unsalable until a new separation step is designed and constructed. This could take 18 months — 18 months in which millions of dollars of equipment is sitting idle. To avoid this and still not delay construction, it may be necessary to continue investigating unverified steps while the plant is being designed and constructed. Then if it is found that certain steps do not work the necessary changes can be determined and the extra equipment ordered before the plant is completed. This procedure would only delay the startup 2 or 3 months.

**Changing a Process**


Polyvinyl chloride (PVC) is produced by a batch process. Since it is usually cheaper to produce chemicals if a flow process is used, the development department proposes a new process and has a process engineer assigned to design it and estimate its cost. If it is only slightly less expensive than the batch process, the new method will be dropped. If it appears that substantial savings can be realized by using the continuous process, further research and pilot-plant studies will be instituted to make certain it will work before the board of directors is asked to authorize the construction of the plant.

This situation differs from the previous one since usually much more is known about the product, and probably some of the proposed steps will involve operations currently being used in the batch process. There are also many people who are familiar with the product and have ideas about whether the proposed changes are feasible. This experience can be very helpful but can also lead to erroneous conclusions. Production engineers are continuously resolving on-line problems by analyzing what went wrong and hypothesizing why. Once the problem is resolved, the hypothesis as to why it happened is assumed correct, without being tested for proof. It often is wrong. The process engineer must be careful about accepting unproven hypotheses. He must also be wary about rejecting ideas that did not work previously. Just as processes have been continually improved, better equipment and processing techniques have made things possible which were impossible ten years ago.

However, the man who ignores advice that later proves to be correct look like a fool. Everybody loves to say "I told you so." The process engineer must use all the information he can get from the operating plant. He should talk not only to the bosses but also to the operators. They often know things that the superintendent doesn't. When a mistake occurs, human nature dictates that the operator attempt to rectify it before someone finds out. Often these operators know from experience that a higher pressure or temperature will not hurt the product. In one plant the operators, by just such observation of mishaps, found out how the reaction time could be reduced by one-third. They said nothing until the sales reached a high enough level that an additional shift was required. The workers did not want to work the night shift, so they told the superintendent what could be done to increase the throughput. The superintendent scoffed at them. It was not until years later, when another increase in capacity was needed, that the research and development department discovered the same thing.

The engineer should visit the plant and spend time observing the process. Often a process engineer will see where some innovation used in another plant can be applied here. He can also note where the trouble spots are.

**Increasing Capacity**

The Production of a surfactant is to be increased from 15,000,000 to 20,000,000 lb / yr.* With many new processes and some older ones, the operators and engineers find they can increase the throughput in certain units but are prevented from increasing production because other steps are running at the highest possible rates. The latter steps are called the bottlenecks. The process engineer must determine how to remove the bottlenecks from the process.

Again the process engineer must spend a large amount of time observing the operations in the plant and talking with supervisors and operators. Besides verifying which steps are the bottlenecks, he must determine if some of the other units must also be modified. For instance, a filter may be able to process 20% more material, but still be inadequate for the proposed new rates. If only the primary bottlenecks were removed, then the plant could still produce only 18,000,000 lb / yr, since this is the maximum amount that can be put through the filter.

Determining the capacity of the noncritical steps (those steps that are not bottlenecks) may require some testing. If a step is not critical there is no reason for the operators or engineers to determine its maximum throughput. Yet, as has been illustrated, this must be known to properly expand or to design a new plant.

For each unit that cannot produce at 20,000,000 lb / yr it must be decided whether the unit should be replaced with a larger one, whether a parallel unit should be installed, or whether to change operating conditions (which may require other modifications) and not make any changes in the equipment. An example of the latter would be to decrease the time each batch spends in the reactor. This would decrease the yield but increase the throughput.

Suppose Table 1-1 represents the yield obtained vs. time for each reactor cycle. If the reactor cycle is 8 hours and produces 15,000 lb of product per batch, then if the cycle time were cut to 5 hours the yield would be 13,250 lb per batch. The rates of production would be 1,875 lb / hr for the former and 2,650 lb / hr for the latter. For a plant operated 8,000 hours per year this would give a production rate of 15,000,000 lb / yr for the former and 21,200,000 lb / yr for the latter. A change of this sort would necessitate no increase in reactor capacity, but it would require changes in the recovery and recycle systems other than those solely due to the increase in capacity.

Table 1-1

Reactor Cycle Time vs. Yield

| Reactor Cycle Time (hours) | 2 | 3 | 4 | 5 | 6 | 7 | 8 |
|---|---|---|---|---|---|---|---|
| Yield, % | 35 | 60 | 75 | 83 | 88 | 92 | 94 |

*Those familiar with the metric system should substitute kilograms for pounds in this example.

### Determining Competitors' Costs

The research department has developed a new process for producing chlorobenzene and wants to pursue it further. The company has never produced chlorobenzene, but feels that if the price is right it would be willing to build a plant for its production. Before doing this, not only must it estimate what the proposed plant will cost but it must determine what costs the current manufacturers have. The proposed process will be dropped unless it has an economic advantage over the present process.

The process engineer must design a plant for the current process solely on the basis of published information. After he has completed his study no one will perform experiments to verify his assumptions, since the company does not plan to use that process. He is on his own. This type of problem is excellent for chemical engineering design classes. Some of best sources of material for such exercises are given at the end of this chapter.

### Factors in Problem-Solving

With each of the aforementioned problems, the process engineer begins by gathering all the information he can about the process. He talks with those in research, development, engineering, and production who might help him, and takes copious notes. He reads all the available literature and records anything that may be of future value. While doing this he develops a fact sheet on each of the substances he will be dealing with. This fact sheet should include all the chemical and physical information he can find. An example is given in Appendix C. During the process of design he will need to calculate heat and mass transfer coefficients, flow rates, efficiencies, and the like, and having this information at his finger tips will save him a lot of time. Since this information is general, many companies file it for future reference.

To become intimately familiar with a process takes time. For a process engineer this may take two weeks or more, depending on the complexity of the system and the engineer's previous experience. This time is not reduced substantially by the presence of large computers. It is a period for assimilating and categorizing a large amount of accumulated information.

The initial goal of the preliminary process study is to obtain an economic evaluation of the process, with the minimum expenditure of time and money. During this stage, all information necessary to obtain a reasonably accurate cost estimate for building and operating the plant is determined. It is expected that these costs will be within 10% of the actual costs.

The next 10 chapters are arranged in the order that a process engineer might follow in the design and evaluation of a process. These are the selection of a site, the writing of the scope (definition of project), the choosing of the process steps, the calculation of material balances, the listing of all major equipment with its specifications, the development of the physical layout of the plant, the instrumentation of the plant, the calculation of energy balances, the development of a cost estimate, and finally the economic evaluation of the process.

## COMPARISON WITH ALTERNATIVES

If the results of the economic evaluation appear promising, then this process must be compared with all other alternatives to determine whether taking the proposed action is really the best course to follow. As an example of possible alternatives that must be evaluated by upper management, consider the problems faced by the detergent industry in 1970. Nearly all detergents produced then contained a builder that assisted the surfactant in cleaning by sequestering calcium and magnesium ions.[6] Most of the large producers used sodium tripolyphosphate (STPP). This comprised about 40% of the detergent on the average, but in some cases was as much as 65%[7]. The phosphate was the nub of the problem. People were demanding that it be removed from detergents because it was accused of damaging the ecology of many lakes.

Phosphorous is a necessary plant nutrient, and in at least some lakes, prior to the Korean War, there was only a small amount of that element present. The amount was so small that some scientists speculate that its absence limited the growth of algae. Then detergents containing phosphates were introduced. Since phosphates are not removed by the usual primary and secondary sewage treatment plants, they were discharged into nearby rivers and lakes. The result was an increase in the phosphorous content of the waters, and an increase in the growth of algae. The growth was so rapid in some places that it depleted the oxygen supply in the water, causing the fish present to die. This angered both commercial and sports fishermen. It disturbed swimming enthusiasts when large numbers of algae and dead fish washed into swimming areas. It alarmed conservationists who are concerned about any upsets to the balance of nature.

The detergent industry had faced a similar crisis just 10 years before. Then the culprit was a surfactant, alkyl benzene sulfonate. Its purpose was to remove dirt, but it also foamed. This was fine for dishwashing, but very undesirable when it was discharged into rivers and lakes. It, like the phosphates, was not removed by the sewage treatment plants. This problem was solved by developing a group of new surfactants, linear alkylate sulfonates, which were biodegradable. This means that the secondary treatment facilities could remove them from the water. By 1965 these new compounds had completely replaced the former surfactants. The cost of obtaining this solution was over $150,000,000.[8]

The detergent industry hoped that this story would be repeated. It spent a lot of money on research and found a partial substitute for the phosphate, sodium nitrilo-triacetate (NTA). The chemical industry began to build plants for its production. Monsanto, which had built a plant to produce 75,000,000 lb / yr (35,000,000 kg / yr), planned to double that plant's capacity and to add another one to produce 200,000,000 lb / yr(90,000,000 kg / yr). W. R. Grace & Co. had facilities to produce 60,000,000 lb / yr(27,000,000 kg / yr), and the Ethyl Corporation planned to build a

250,000,000 lb / yr(115,000,000 kg / yr) plant. Everything looked rosy until a week before Christmas in 1970, when the Surgeon General of the United States and the head of the Environmental Protection Agency asked the detergent industry to refrain from using NTA. Tests run on rats indicated teratogenetic(fetal abnormalities)effects when NTA was administered in the presence of mercury or cadmium.[9] They were afraid similar effects might occur in men.

Other substitutes were available, but a report by the New York State Environmental Protection Agency stated these were inferior to phosphate, posed an alkalinity hazard, and reduced the effectiveness of flame-retardant materials.[10] Nevertheless, this did not stop the states of Indiana and New York from banning the sale of nearly all detergents containing phosphates in 1973. A number of other states, counties, and municipalities did not go this far, but limited the amount of STPP to 35%.[11]

What alternatives did a detergent producer have under these circumstances? The most obvious ones are to sell only in those areas with no bans, to produce a new product, or to stop producing cleaning compounds. Another possibility would be to try to convince the public that phosphates should not be banned. They might be able to convince cities to add to their waste treatment facilities the capability for removing phosphates. This could be financed by a detergent tax. Alternately, they might be able to show that the removal of phosphates from detergents is unimportant, since the detergent industry uses less then 15% of the phosphorus manufactured.[12] They might be able to convince the public that the major increase in environmental phosphates was due to the increased use of fertilizers and the runoff from feed lots.

Other alternatives would be to find either a use for the algae, or some predator or growth inhibitor that would prevent their rapid growth. If a use could be found, the detergent industry could promote harvesting of the algae as a commercial product. Then, by properly managing its production, maybe the algae, fish, and swimmers could live happily together. There are a large number of "ifs" in this solution, but probably no more than those involved in finding an algae inhibitor, killer, or predator. Here the problem would be finding a substance or organism that harms only algae and none of the other plants or animals present in lakes and rivers, nor anything that feeds directly or indirectly upon these plants or animals. This technique has been tried under many different circumstances, and has often failed. The English sparrow was introduced into the United States to kill the caterpillars of the snow-white Eugonia moths, which were defoliating shade trees. They performed their job well, but they also drove out other birds and this allowed the tussock moths, which they did not eat, to ravage the trees.[13] The starling was introduced into Australia and other countries to control certain insects. But after its introduction it changed its habits and is now a pest. In Jamaica the mongoose was introduced to fight a plague of rats. It increased rapidly and killed not only the rats, but also birds, snakes, and lizards. The result was a tremendous increase in insect pests.[13]

Procter & Gamble, which has about 50% of the detergent market, chose initially to remove their products from those areas having a total ban and reduce the phosphate content to permissible levels in the other areas. The other two big producers, Lever Brothers and Colgate Palmolive, chose to produce zero-phosphate products. Lever Brothers substituted sodium citrate as a builder. Colgate did not use any builder. Another course of action they might have followed would be to start producing soap again to compete with the detergents. Since soaps work well with softened water, this would be a good alternative if individuals, and even cities, could be convinced to install water softeners.

Lever Brothers, since it had to decide how to obtain the sodium citrate, had another host of alternatives to consider. It could produce or buy the product. If it chose to produce the builder, it could purchase the process from another firm or it could develop its own process. It could make the product from basic raw materials or from intermediate compounds. If it decided to let some other firm be the producer, it could buy the material on the open market or enter into a long-term agreement with another company. It might even do both by forming a joint company, such as Dow Corning, that would manufacture the builder. It could even buy a company that was currently producing it. All these possibilities must be economically evaluated to determine the best course of action to take.

The board of directors faced with such a decision must not only consider all the alternatives given above, but they must compare the best of these with all the other opportunities they have to invest the company's money. There is only so much money available, and the board is charged with obtaining the highest profits for the lowest risks. This means the decision about building a new detergent facility might need to be compared with plans for building a caustic chlorine facility at the Great Salt Lake, or enlarging a polyethylene plant in Peru, or buying controlling interest in a Belgian pharmaceutical firm.

## COMPLETING THE PROJECT

If, after comparing alternatives, a project is approved, then a project manager is appointed. His job is to shepherd the project through to completion by a designated date. During the preliminary process engineering only a few people will have worked on the project. The next phase will involve over a hundred people, and the project manager must arrange things so that nothing prevents the project from being completed on schedule. To do this, he may use critical path method(CPM)or program evaluation and review technique (PERT). These are discussed in Chapter 13.

After approval, the project is returned to process engineering for the detailed process design. Now the process engineer must provide all the information necessary to the project engineering specialists, so that equipment can be designed and specified. Between these two groups everything that goes into a chemical plant, from the smallest bolt to a 250-foot-high distillation tower, must be specified. Chapter 12 discusses this, plus the construction and startup phases.

## UNITS

For any project it is important that a consistent set of units are used. Most companies, in fact, prescribe that a given set of units be used for all calculations. This allows an experienced designer to easily run a rough check to determine if all the flow rates, temperatures, and sizes are reasonable. It allows persons working on different portions of the process to readily determine if there are any discontinuities at the interfaces between the sections. It also saves time and reduces the possibility of errors by minimizing the number of times that the units must be converted.

Table 1-2

Units to Be Used

| | English | Metric |
|---|---|---|
| Length | foot (ft)<br>inch (in) | meter (m)<br>centimeter (cm) |
| Capacity | gallons (gal)<br>cu ft | $m^3$<br>$m^3$ |
| Mass | pound (lb) ($lb_m$) | kilogram (kg) |
| Time | hour (hr)<br>second (sec)<br>minute (min) | hour (hr)<br>second (sec)<br>minute (min) |
| Temperature | degrees Fahrenheit (°F)<br>degrees Rankine (°R) | degrees Celsius (°C)<br>degrees Kelvin (°K) |
| Velocity | ft/sec<br>cu ft/min (CFM)<br>gal/min (GPM) | m/sec<br>$cm^3$/hr<br>$cm^3$/hr |
| Energy | British Thermal Unit (BTU) | kilogram-calorie (kcal) |
| Heat Capacity | BTU/lb°F | kcal/kg°C |
| Pressure | atmospheres (atm)<br>pounds per square in (psi)<br>in $H_2O$<br>in Hg | metric atmospheres (m atm)<br>$kg/cm^2$<br>mm $H_2O$<br>mm Hg |
| Power | horsepower (hp)<br>kilowatt (kw) | metric horsepower (mhp)<br>kilowatt (kw) |
| Viscosity | centipoise (cp) | centipoise (cp) |

Table 1-2 gives the two sets of units that will be used throughout this book.

In general, English units will be used throughout this book, with metric units given in parentheses. However, where metric units are the accepted practice, only these will be given. Only English units will be used in the examples, since the metric system has not yet been adopted by the chemical industry in the United States or Canada.

## References

1. "Wax Sales on Upswing Again," *Chemical Week*, Oct. 7, 1967, p. 61.
2. "Computer Counts Them Out," *Chemical Week*, Dec. 9, 1967, p. 69.
3. Ladyn, H.W.:"Fat Splitting and Soapmaking Go Continuous," *Chemical Engineering*, Aug. 17, 1964, p. 106.
4. "Will Success Spoil Phenol Success?" *Chemical Week*, Sept. 20, 1969, p. 163.
5. "Technology Newsletter," *Chemical Week*, May 3, 1969, p. 51.
6. Silvis, S.J.: "The World of Synthetic Detergents," *Chemical Week*, Oct. 29, 1969, p. 79.
7. Gruchow, N.: "Detergents: Side Effects of the Washday Miracle," *Science, 167*; 151, Jan. , 1970.
8. "Detergents Are Miscast as Pollution Villain," *What's New in Home Economics*, Sept. 1970, p. 75.
9. Rosenzweig, M.D.: "Soapers Face A New Race," *Chemical Engineering*, Feb. 8, 1971, p. 24.
10. "Coming Out in the Wash," *Chemical Week*, Jan. 21, 1973, p. 20.
11. "Crowding Phosphates off the Shelf," *Chemical Week*, Oct. 25, 1973, p. 27.
12. *Detergents, Phosphates and Environmental Control*, a report by Economics Laboratory Inc., St. Paul, Sept. 12, 1970.
13. Henderson, J.: *The Practical Value of Birds*, Macmillan, New York, 1934, pp. 30-34.

## Books Having Process and Product Information

Standen, A.: *Kirk-Othmer Encyclopedia of Chemical Technology*, Ed. 2, Interscience, New York, 1963-1972; supplemental volumes issued.

Twenty-two volumes that give all aspects of chemical technology, including a comprehensive discussion of chemical processes and processing conditions and a listing of properties for all mass-produced chemicals. Excellent.

Bikales, N.M. (ed.): *Encyclopedia of Polymer Science and Technology-Plastics, Resins, Rubbers, Fibers*, Interscience, New York, 1964-1972; supplemental volumes issued.

Sixteen volumes that give processes, processing techniques, theoretical aspects, and the properties for all polymers. Excellent.

Shreve, N.R.: *Chemical Process Industries*, Ed. 3, McGraw-Hill, New York, 1967.

A presentation of the processes and processing conditions for producing most major chemicals.
Groggins, P.H.: *Unit Processes in Organic Synthesis*, Ed. 5, McGraw-Hill, New York, 1958.

A presentation of the processes and processing conditions for making organic chemicals.

Sittig, M.: *Organic Chemical Process Encyclopedia*, Noyes Development Corporation, Park Ridge, N.J. 1966.

587 process flow sheets.

Thoor, T.J.W.v. (ed): *Chemical Technology: An Encyclopedic Treatment,* Barnes and Noble, New York, 1968 onward (all volumes not yet published).

Eight volumes that give the sources of raw materials and products together with processes and processing data.

Strickland, J.R. (ed): *Chemical Economics Handbook,* Stanford Research Institute, Menlo Park.

Product studies of most major chemicals giving the present projected economic characteristics. It gives the manufacturers' production processes, prices, many data sheets, and current indicators. It is constantly being revised.

Kent, J.A. (ed): *Riegel's Handbook of Industrial Chemistry,* Ed. 7, Reinhold, New York, 1973.

A presentation of the processes, production, and uses for a large number of chemicals.

Faith, W.L., Keyes, D.B., Clark, R.L.: *Industrial Chemicals*, Ed. 3, Wiley, New York, 1965.

A presentation of the processes, economics, manufacturers, and plant sites for a large number of chemicals.

## Periodicals Useful to the Process Engineer

*Chemical Engineering*

A biweekly publication of McGraw-Hill. This is an excellent publication having articles and news covering all aspects of process and project engineering. There is an article on process technology which often includes a process flow sheet and a review of some aspect of chemical engineering in each issue. Special desk book issues are published on specific topics each year. Twice yearly it publishes a listing of all chemical plants that have been proposed or are under construction. It also twice-yearly lists a summary of all new processes and technology.

*Hydrocarbon Processing*

A monthly publication of the Gulf Publishing Company. Another excellent publication that devotes itself to the process and project engineering of refineries and petrochemical operations. It has a flowsheet in every issue; its yearly handbook issue gives over 100 processes flowsheets, and its "NG, LNG, SNG Handbook" issue gives more. It also has very good product studies and an excellent thermodynamics data series. Three times a year it lists all new worldwide construction in the petroleum and petrochemical industry.

*Chemical Week*

A weekly publication of McGraw-Hill. This excellent publication concentrates on the news and business aspects of the chemical industry. It periodically has excellent marketing and sales studies for various products. It also publishes annually a plant-site-selection issue and a *Buyers' Guide*. The *Buyers' Guide* lists the major producers and the source of supply for over 6,000 chemical products, and has a list of chemical trade names. It publishes, quarterly, the current financial condition of 300 companies involved in chemical processing.

*Chemical and Engineering News*

A weekly publication of the American Chemical Society. It is an excellent publication giving the news that affects the chemical industry, including marketing, sales and business information. In its annual "Facts and Figures for the Chemical Industry," it gives the financial condition for over 100 different chemical companies, the production rates for large-volume chemicals, employment figures, foreign trade statistics, and world production statistics for select chemicals. It also has an annual issue on the world chemical economy.

### Oil and Gas Journal

A weekly publication of the Petroleum Publishing Company. This is a business- and technology-oriented publication for the petroleum and natural gas industry. It gives marketing production and exploration studies and industrial statistics, as well as current news. There are articles on processes, processing techniques, and costs. It is oriented toward the businessman and the production engineer.

### Modern Plastics

A monthly publication of McGraw-Hill. It is oriented to marketing, business, and production. Annually it issues the *Modern Plastics Encyclopedia*. This gives a list of chemical suppliers, equipment manufacturers, polymer properties, equipment descriptions, and machinery selector charts.

### Plastics World

A monthly publication of Rogers Publishing Company. This business-oriented publication gives production statistics. Annually it publishes a list of companies providing chemicals, services, and equipment to the plastics industry, and gives the production capacity by compound for each company.

### Plastics Technology

A monthly publication of Bill Brothers. It has articles on process production and manufacturing engineering. In a fall issue on plastics technology, it gives detailed equipment specifications, polymer specifications, and lists of equipment and chemical suppliers.

### Society of Plastics Engineers Journal

A monthly publication of the Society of Plastics Engineers. It contains articles giving engineering and technical information on plastics, as well as news of the society. Annually it publishes a *Digest of Plastics Standards*.

### Chemical Engineering Progress

A monthly publication of the American Institute of Chemical Engineers. It contains articles giving engineering and technical information, as well as news of the society. It also publishes a symposium series of volumes on specific topics.

### Industrial and Engineering Chemistry – Process Development and Design

A quarterly publication of the American Chemical Society. It reports research studies on various processes and unit operations. It is mainly concerned with laboratory developments.

### Industrial and Engineering Chemistry – Product Research and Development

It reports research studies on processes and products. It is concerned mainly with laboratory studies.

### Journal of Chemical and Engineering Data

A quarterly publication of the American Chemical Society. It gives lots of specific data.

### Journal of Physical and Chemical Reference Data

A quarterly publication of the American Chemical Society. It contains physical and chemical property data.

## Cost Engineering

A quarterly publication of Industrial Research Service, Inc., Dover, N.H. It gives cost data for process engineers. Each year it publishes an index and abstract of cost literature.

## Environmental Science and Technology

A monthly publication of the American Chemical Society, this presents the current environmental news and general technologically oriented articles followed by theoretical research studies. Yearly it publishes a *Pollution Control Directory*, a listing of equipment and chemical manufacturers and of service and consulting companies.

# Abstracts and Indexes Useful to the Process Engineer

## Chemical Abstracts

The most comprehensive of the abstracts for theoretical and applied chemistry and chemical engineering. Indexes and abstracts of patents and of worldwide periodicals.

## Engineering Index

Indexes and abstracts of technical literature from around the world.

## Chemical Market Abstracts

Abstracts and indexes of English-language periodicals giving information on new plants, chemical producers, and chemical consumers.

## Monthly Catalog of Government Publications

This is an index to U.S. government publications.

## Water Resource Abstracts

Abstracts of water resources research and development.

## Pollution Abstracts

Indexes and abstracts of worldwide technical literature on the environment.

# CHAPTER 2

# Site Selection

If one is to design a chemical plant the site must be known. The cost of energy and raw materials, the type of transportation to be used, and the availability of labor all depend on the plant site. Some examples follow in which the plant site as chosen because of the presence of a specific raw material or energy source.

Some years ago Proctor & Gamble was considering building a plant in Massachusetts. After looking at a number of possible locations the team responsible for choosing the site noticed a vacant area next to a power plant. They immediately realized that if the power company could supply them with steam at a reasonable price, they would not need to build steam generators as had been planned. Proctor & Gamble bought the site and negotiated a long-term contract for steam and power that was beneficial to both companies. This site would not have been brought to the team's attention by their local contacts. It was found because some of the team members were quick to recognize an unmentioned potential saving.

An American Salt Company plant and the Dow Chemical Company's Midland plant also benefit directly from each other's presence. Dow found that after recovering bromine from brine it had more salt left than it desired. American Salt needed salt. By locating next to Dow's plant it was able to buy this salt stream for less than it would cost to mine it or pump it from natural underground reservoirs. In turn, Dow was able to sell an unwanted stream that it would otherwise have had to pump back into the ground. The American Salt plant is typical of many satellite plants. These are plants that either use a by-product or a waste stream from another plant or are built mainly to supply a needed chemical to an adjacent plant. The nearby presence of another plant determines their location.

The design of a petroleum refinery is dependent on the type of crude available, and this in turn is dependent on the site. The crude petroleum obtained from Pennsylvania is noted for its high percentage of saturated aliphatic hydrocarbons and low sulfur content. The midcontinent crudes from Kansas and Wyoming, on the other hand, contain large amounts of naphthenic hydrocarbons along with a high sulfur content. The Pennsylvania oil is particularly suited for the manufacture of lubricating oils, but without extensive processing the gasoline would have a low octane rating. This would cause most high-compression automobiles to knock. The midcontinent crudes provide a higher octane gasoline fraction. They, however, usually must be treated with acids or solvents to make good motor oils. They also require larger sulfur-removal units.

    The costs of switching from using a sweet Gulf Coast crude (less than 0.5% sulfur)
to a sour Arabian crude (about 1.6% sulfur) were determined by Chevron. They
estimated that it would cost $47,000,000 to convert a 150,000 bbl/day (24,000
m³/day) refinery.[1] One of the changes that must be made is to replace low-priced
carbon steel in crude and vacuum stills by expensive chrome steel or chrome-clad
steel. If no changes were made the corrosion rate would become excessive, some of
the products would not meet present sulfur specifications (Table 2-1), and there
would be an increase of 500% in the amount of $SO_2$ leaving the stacks per unit of
crude charged. Obviously no petroleum refinery should be constructed until the
composition of the crude oil is known. Also, once a refinery has been constructed to
refine a given oil it should not be expected to process a very different type of crude
efficiently.

Table 2-1

How Feedstock Choice Affects Sulfur Level (Wt. %)

| Product | Typical U.S. Refinery Feeding Gulf Coast Crude | Same Refinery Running Arabian Light | 1973 Spec. | Possible 1975 Spec. |
|---|---|---|---|---|
| Motor gasoline | 0.03 | 0.08 | 0.1 | 0.01-0.03 |
| Jet fuel (kerosene) | 0.04 | 0.15 | 0.12 | 0.05 |
| Diesel No. 2 oil | 0.10 | 0.6 | 0.25 | 0.02 |
| Heavy fuel oil | 0.65 | 3.2 | 0.5-2.0 | 0.3-1.0 |

Source: Prescott, J.H.: "U.S. Refiners Go Sour," *Chemical Engineering*, June 11, 1973, p. 74.

    These examples illustrate how the choice of a site and the design of a plant are
interlinked. In fact, ideally, the site cannot be chosen without designing a plant for
each possible location and then making an economic comparison. Realistically, this
would be too expensive, so the list of potential sites must be reduced before a full
economic evaluation is attempted.
    One of the problems with eliminating plant sites this way is that a location might
be summarily rejected when it would actually be the best. In 1968 the Collier Carbon
& Chemical Company started up a $50,000,000 ammonia-urea plant in Kenai,
Alaska. Consider these facts, which might have easily scared the conventional
engineer. Kenai is located near Anchorage and is about 1,500 miles (2,400 km) from
Seattle. Nearly all supplies and equipment had to be shipped from the 48 contiguous
states or Japan. Each barge trip from Seattle cost $70,000 not including loading and
unloading charges. All other means of transportation also involved boats. Railroad
cars were shipped by sea from Seattle to Seward, Alaska, then went by train to
Moose Pass, where they must be loaded on trucks for the final stage of the journey.

Truck vans were shipped to Whittier, Alaska, via water and then proceeded by land to Kenai.

A dock had to be built where the water temperature is rarely above 40°F (5°C); the tides are 30 ft (9m) high; there is a tidal current of 6 or 7 knots (11-13 km/hr); and the water is so turbid that the divers must be guided mainly by touch. The docks also had to be protected by ice-breakers. These are built into the harbor to prevent large icebergs from hitting the dock before they are reduced into smaller pieces.

The divers who were involved in the construction of these docks and breakers got $540 per day. Since they could only work during slack water, a diver might only be submerged for an hour or two a day, and when he was down there must be another diver above water. The salaries of the other workers were also high but did not compare with the divers'. It was expected that labor charges alone would be 20% greater than if the same work were performed in the contiguous 48 states.

The weather conditions also added headaches. In winter the temperature sometimes gets below −50°F (−45°C) and may remain at −30°F (−35°C) for sustained periods. It is so cold that certain buildings need air locks to prevent excessive heat losses when people enter or leave. Many ordinary construction materials cannot withstand the temperature. The usual mastic sealing compound has to be replaced by an expensive silicone sealant. In addition to this the region has high winds and is earthquake-prone.[2]

Why build there? Large gas and petroleum deposits have been found in the area around Kenai and it is expected that additional oil and gas reserves will be discovered nearby. Natural gas is not only a source for heat and power but also the major raw material in the production of ammonia. Approximately $4 \times 10^7$ BTU ($10^7$ kcal) of energy are required per ton* of ammonia produced. The nearness of the plant to the gas field makes the gas inexpensive.

The other major advantage was that the United States was not to be its only customer. In fact, before the plant was built Japan Gas Chemical contracted to buy half the urea output. Here the advantage of Alaska over any other United States location except Hawaii is a reduction in shipping costs. Kenai is 1,400 miles (2,250 km) closer to Japan than is any California location. This, coupled with the low raw material costs, would make its delivered cost less than most other Japanese fertilizer sources. When all these factors were considered, the Collier Carbon and Chemical Company reasoned it was cheaper to process the natural gas into ammonia at its source even with those difficult climatic and economic conditions than to ship the gas to a more advantageous location and then make ammonia.

## MAJOR SITE LOCATION FACTORS

While many factors can be important in the selection of a plant site, three are usually considered the most important. These are the location of the markets and

---

*Throughout this text the English short ton will be used. Since one metric ton equals 1.102 English short tons, when approximate figures are given no conversion to metric tons will be given.

raw materials and the type of transportation to be used. Any one or all of these factors together may greatly limit the number of sites that are feasible.

### Location of Raw Materials

One possible location is a site near the source of the raw materials. This location should always be one of the sites considered. If a plant is to recover bromine from sea water it will obviously be placed next to the sea. The bromine concentration of sea water is 60 to 70 ppm (parts per million). It is obviously more expensive to transport 1,000,000 pounds of water than 70 pounds of bromine. Whenever the quantity of the product is small compared with the amount of raw materials, the site is placed near the material source.

### Location of Markets

The reverse occurs in the production of foams and consumer items. These plants are usually constructed close to the prospective markets. Insulating materials are often so light that the cost of shipping per ton is very high. Only a small amount of mass can be loaded in a boxcar or truck. The density of polystyrene is 2.4 lb/ft$^3$ (38.5kg/m3). It is made from styrene, which has a density of 56.3 lb/ft$^3$ (902 kg/m$^3$). The foam product occupies 23 times as much space as the styrene, and the polystyrene would cost between 10 and 20 times as much to ship.

Consumer products often are delivered in small shipments to a large number of customers. They also involve packaging in small attractive containers that decrease the amount of product per unit volume and add mass. Since low shipping rates apply only to large bulk shipments going to a single destination, it is often desirable to place production of consumer products near the markets.

Alternately, consumer products could be shipped *en masse* to distribution centers located near the population centers. From here they would be shipped by truck to individual customers. This type of facility has sales advantages. The retailer can be guaranteed that his order will be delivered within 24 hours after its receipt. This allows him to provide excellent customer service with a small warehouse. These centers could receive bulk quantities, but then each one would have to package the product. This would mean a duplication of packaging facilities.

### Transportation

The importance of the cost of transportation has been indicated in the previous paragraphs. The least expensive method of shipping is usually by water; the most expensive is by truck. In between are pipelines and trains. Figures 2-1 through 2-4 illustrate the relative costs of shipping in 1972. The lower costs are available when the transportation company has good opportunities for obtaining a full load on the return trip. The highest occur when the carrier can expect to return empty. The

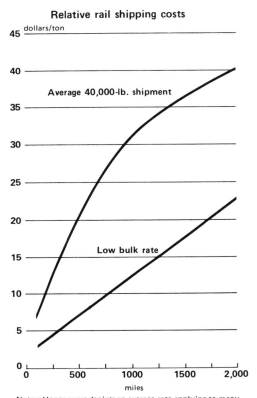

**Relative rail shipping costs**

Notes: Upper curve depicts an average rate applying to many chemicals shipped in 40,000-lb. lots. The "low-bulk" curve represents approximately the cheapest rates available for chemical commodities shipped in bulk, except for the few cases where unit train rates have been established. Actual freight rates are not predicted by the curves because rates for specific chemicals and movements are normally negotiated between the carrier and the shipper.

Figure 2-1  Relative Rail Shipping Costs - Courtesy of Winton, J.M.: "Plant Sites '74," *Chemical Week,* Oct. 17, 1973, p. 45.

difference in costs between various forms of transportation may be expected to increase if the projected energy crisis of the 1970s becomes acute. This is because those forms of transportation that are now the cheapest require the least amounts of energy and manpower. Table 2-2 gives the energy requirements for each mode of transportation. However, logical extrapolations are often faulty because the government often subsidizes various transportation industries. Presently it favors barging and trucking.

While shipping firms must charge everyone having the same circumstances similarly, they can make certain types of deals. The rates given in the tables are not exact because the final rate is negotiated with the transportation company. In these negotiations, concessions may be granted; for example, a railroad may pay for grading and installing a spur. Also, when alternate forms of transportation are

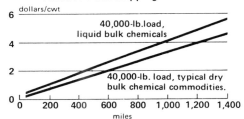

**Relative truck shipping costs**

dollars/cwt

40,000-lb.load,
liquid bulk chemicals

40,000-lb. load, typical dry
bulk chemical commodities.

miles

Notes: Rates apply generally in most states. The costs, however, do not apply to negotiated tariffs, contract trucking and
some commodities for which special tariffs are published.
Rates also do not include a variety of special charges
(i.e., bridge tolls) that are sometimes applicable. The chart
for dry bulk commodities approximates cost of trucking
items such as alum, calcium chloride,coal-tar pitch, phosphate, potash, soda ash, sodium silicate, salt cake and urea.

Figure 2-2    Relative Truck Shipping Costs.  Courtesy of Winton, J.M.: "Plant Sites '74,"
              *Chemical Week,* Oct. 17, 1973, p. 45.

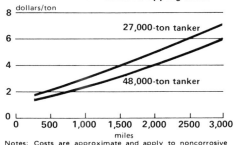

**Chemical ocean tanker shipping costs**

dollars/ton

27,000-ton tanker

48,000-ton tanker

miles

Notes: Costs are approximate and apply to noncorrosive
liquids in nonpressure tanks. The rates shown are averages
from a number of sources and reflect present maritime
conditions. Rates can vary ±15% from curves shown,
depending upon circumstances. For 27,000-ton vessels,
corrosive liquids shipped in nonpressure tanks are 12-40%
higher; shipments of corrosive liquids in pressure tanks are
25-40% higher. For 48,000-ton ships, costs for corrosive
liquids shipped in nonpressure tanks are 12-40% higher;
shipments of corrosive liquids in pressure tanks are 50-70%
higher than charges for noncorrosive liquids in nonpressure
tanks.

Figure 2-3   Chemical ocean Tanker shipping costs.
             Courtesy of Winton, J.M.: "Plant Sites '74," *Chemical Week,* Oct. 17, 1973, p. 45.

### Estimated barge shipping costs

| | (mills per ton-mile) | | | |
| --- | :---: | :---: | :---: | :---: |
| | 250 mi. | 500 mi. | 1,000 mi. | 2,000 mi. |
| Noncorrosive liquids in nonpressure tanks (40,000-60,000-bbl. shipment) | 4.1 | 3.9 | 3.7 | 3.6 |
| Noncorrosive liquids in nonpressure tanks (150,000-bbl. shipment) | 3.2 | 3.1 | 3.0 | 3.0 |
| Corrosive liquids in nonpressure tanks (40,000-bbl. shipment) | 7.5 | 7.3 | 4.9 | 3.9 |
| Corrosive liquids in pressure tanks (40,000-bbl. shipment) | 10.8 | 8.9 | 7.9 | 7.5 |
| Dry bulk solids | 6-14 | 4.5-10.5 | 3.8-7.7 | 3.6-7.5 |

Notes: Above rates are generally classified as "full transportation," including carrier and tow. Lower figures usually reflect contract rates for large movements to ports where return cargoes are usually available. Higher rates reflect "once through" shipments or movements to ports where return cargoes are unlikely to be available. Source: industry estimates.

Figure 2-4  Estimated large Shipping Costs.
Courtesy of Winton, J.M.: "Plant Sites '74," *Chemical Week,* Oct. 17, 1973, p. 45.

grading and installing a spur. Also, when alternate forms of transportation are available the tariffs for everybody may be greatly reduced. Railroad maps are given in references 3 and 4. The rivers that can handle large traffic are illustrated in Figure 2-5.

In general the transportation industry in the United States has been backward. Recently, however, a number of changes have been attempted. The result has been many labor-saving innovations that could affect costs drastically.

A pipeline is the cheapest form of transportation to operate, but it requires a large capital investment and therefore a large throughput. Until 1969 the long-distance pipelines carried almost exclusively natural gas or petroleum products. In that year a pipeline running 850 miles (1,370 km) from Texas to Iowa began carrying anhydrous ammonia to markets in the Midwest. Since then other ammonia pipelines have been completed.

Three factors favored construction of ammonia pipelines. First, over 50% of this country's agricultural nitrogen is used in the Midwest and between 40 and 65% of this total is applied directly to the soil as anhydrous ammonia. Second, the low price of natural gas needed for the production of ammonia favored a Gulf Coast plant site or one near a large gas field. Third, much of the Midwest is inaccessible to cheap barge transportation.

In 1969 Air Products & Chemicals began delivering carbon dioxide and hydrogen to customers in the Houston area via pipeline. There was also talk of shipping methanol by pipeline. A 273-mile pipeline was also opened in 1969 to convey 660 tons/hr of slurried coal from Kayenta, Arizona, to a power plant in southern Nevada. A previous coal pipeline in Ohio closed down in the mid-1960s because it proved to be uneconomical when the railroads reduced their rates.[5]

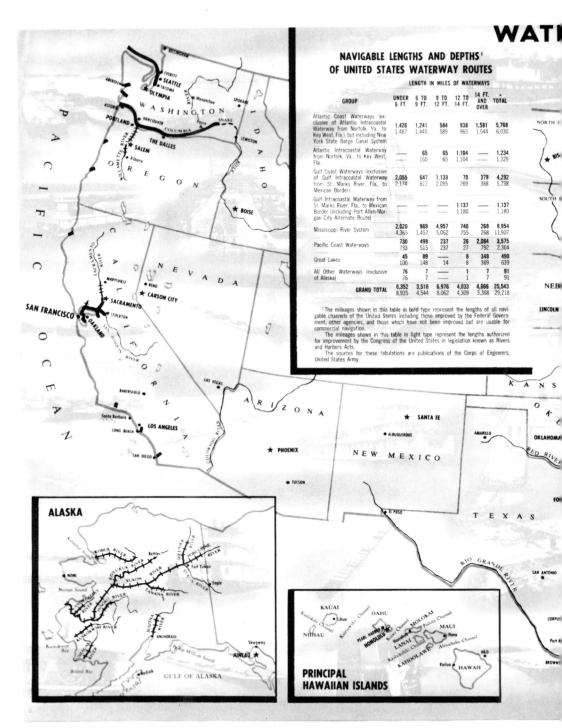

## NAVIGABLE LENGTHS AND DEPTHS¹ OF UNITED STATES WATERWAY ROUTES

LENGTH IN MILES OF WATERWAYS

| GROUP | UNDER 6 FT. | 6 TO 9 FT. | 9 TO 12 FT. | 12 TO 14 FT. | 14 FT. AND OVER | TOTAL |
|---|---|---|---|---|---|---|
| Atlantic Coast Waterways (exclusive of Atlantic Intracoastal Waterway from Norfolk, Va., to Key West, Fla.), but including New York State Barge Canal System | 1,426 / 1,487 | 1,241 / 1,445 | 584 / 589 | 938 / 965 | 1,581 / 1,544 | 5,768 / 6,030 |
| Atlantic Intracoastal Waterway from Norfolk, Va., to Key West, Fla. | — / — | 65 / 160 | 65 / 65 | 1,104 / 1,104 | — / — | 1,234 / 1,329 |
| Gulf Coast Waterways (exclusive of Gulf Intracoastal Waterway from St. Marks River, Fla., to Mexican Border) | 2,055 / 2,174 | 647 / 812 | 1,133 / 2,095 | 79 / 269 | 378 / 388 | 4,292 / 5,738 |
| Gulf Intracoastal Waterway from St. Marks River, Fla., to Mexican Border (including Port Allen-Morgan City Alternate Route) | — / — | — / — | — / — | 1,137 / 1,180 | — / — | 1,137 / 1,180 |
| Mississippi River System | 2,020 / 4,365 | 969 / 1,457 | 4,957 / 5,062 | 740 / 755 | 268 / 268 | 8,954 / 11,907 |
| Pacific Coast Waterways | 730 / 733 | 498 / 515 | 237 / 237 | 26 / 27 | 2,084 / 792 | 3,575 / 2,304 |
| Great Lakes | 45 / 100 | 89 / 148 | — / 14 | 8 / 8 | 348 / 369 | 490 / 639 |
| All Other Waterways (exclusive of Alaska) | 76 / 76 | 7 / 7 | — / — | 1 / 1 | 7 / 7 | 91 / 91 |
| GRAND TOTAL | 6,352 / 8,935 | 3,516 / 4,544 | 6,976 / 8,062 | 4,033 / 4,309 | 4,666 / 3,368 | 25,543 / 29,218 |

¹The mileages shown in this table in bold type represent the lengths of all navigable channels of the United States including those improved by the Federal Government, other agencies, and those which have not been improved but are usable for commercial navigation.

The mileages shown in this table in light type represent the lengths authorized for improvement by the Congress of the United States in legislation known as Rivers and Harbors Acts.

The sources for these tabulations are publications of the Corps of Engineers, United States Army.

**ALASKA**

**PRINCIPAL HAWAIIAN ISLANDS**

Figure 2-5   Inland rivers of the United States which can accommodate barges.
Courtesy of American Waterways Operators, Inc.

# AYS OF THE UNITED STATES

COMMERCIALLY NAVIGABLE
**WATERWAYS**
OF THE
**UNITED STATES**

CONTROLLING DEPTHS

9 FEET OR MORE
UNDER 9 FEET
AUTHORIZED EXTENSIONS

PUBLISHED 1972 BY
THE AMERICAN WATERWAYS OPERATORS, INC.
1250 CONNECTICUT AVENUE, WASHINGTON, D.C. 20036
Compiled from Information Supplied by
CORPS OF ENGINEERS, U.S. ARMY

Table 2-2

Energy Used in Transporting Cargo

| Mode of Transportation | Ton Miles of Cargo per Gallon of Fuel | Metric Ton Kilometers of Cargo per Liter of Fuel |
|---|---|---|
| 100,000-ton supertanker | 900 | 350 |
| Large pipeline | 500 | 200 |
| 200-car freight train | 400 | 150 |
| 100-car freight train | 230 | 90 |
| Inland barge tow | 210 | 80 |
| 15,000-ton containership | 170 | 65 |
| 40-car freight train | 85 | 33 |
| 40-ton truck | 50 | 20 |
| Large cargo jet (747) | 13 | 5 |
| Turboprop air freighter | 5.5 | 2.1 |
| Small cargo jet (707) | 4.5 | 1.7 |
| 165-ton hovercraft | 1.8 | 0.7 |

Source: Rice, R.A.: "System Energy and Future Transportation," *Technology Review*, Jan. 1972, p. 31.

The railroads have introduced the unit train. This is a single train consisting of 75 or more cars periodically going from one source to a specific destination. This is very useful for a power company wanting to ship coal from West Virginia to their plants hundreds of miles away and for chemical companies shipping large amounts of feedstock, for instance ammonia, to processing plants. The savings occur because the train can be scheduled and no switching of individual cars from one freight train to another occurs. The train merely follows the most direct route between two points. It can cut freight costs by as much as 30%.[6]

Unit trains can pose problems to small manufacturers who do not have the volume to use them. This effectively gives the big producer a price advantage; this is frowned upon in Washington. In 1969 the Dow Chemical Company, which was using unit trains to deliver coal from West Virginia to Midland, Michigan, was accused of violating railroad tariff rate. It was claimed Dow paid less than published tariff rates. In an out-of-court settlement Dow paid $350,000 in claims.[7]

The railroads are also encouraging piggyback service. Here flatbed truck trailers are loaded on railroad cars and shipped to distant locations. This allows a combination of truck and rail transportation without having to go through expensive loading and unloading steps. It means the cheaper rail transportation can be used between two widely separated points and yet the cargo can still be delivered to the customer's door even if he does not have a railroad siding.

The railroads have also introduced the jumbo tank car, which can hold 43,000 gallons. Since all tank cars must be individually charged and discharged, increasing the size reduces the number of manual operations per unit volume of material and reduces costs. It also diminishes switching charges.

Cargo ships, which used to require a large number of men, now are highly automated, have less than thirty crew members, and are larger in size. The Gulf Oil Corp. has four oil tankers each of which can handle 326,000 tons of oil. The cargo ships have also begun using containerized cargos. These prepackaged standardized boxes have slashed handling charges by over 50%. They can be loaded directly on trucks or trains, thus eliminating the need for expensive dock storage facilities and greatly reducing loading and unloading charges.

One of the highest costs in ocean shipping is docking. To eliminate this some ships are carrying the cargo on barges that can be removed from the ship without its docking. The barges are then ready for transportation up the various rivers to inland cities.

On the Ohio River and others the number of locks is being decreased and the size of them is being increased. This will reduce costs by allowing faster service and larger tows. The tie-in with ocean ships mentioned above will make plant sites along the major rivers even more attractive.

The trucking industry has convinced some states that a tractor trailer can haul two cargo units on interstate highways. The result is that each truck driver can transport twice as much per man-hour. The interstate highway system itself has sped up trucking service.

These increases in efficiency pose a threat to the workers in the industry. As a result, they may strike, and the final settlement may result in a nullification of many economic benefits. This happened when the railroads converted from steam locomotives to diesels. At that time the union forced the management to keep a fireman on every train whether he was needed or not. Similarly, the settlement of a recent East Coast dock strike provided a setback to the use of containerized units. It stated that all containerized units that entered New York ports could be unpacked and repacked at the union's request and the shipper's expense, if the final destination was less than 50 miles from the dock, or if the container had more than one product within it.[8]

These tactics sometimes backfire on the unions. In the first half of the nineteenth century Chicago was the meatpacking center of the nation. Carl Sandburg had written that his beloved Chicago was "hog butcher for the world." School children all read about the Union Stockyards. It was one of the city's most famous industries and appeared to be a very stable one. After all, there would always be a demand for meat. The Meatpackers Union was, however, very powerful. It negotiated high wages and prevented new methods from being adopted. This, coupled with improvements in refrigeration and the realization that it was cheaper to ship carcasses than live animals, led to all the major packers suspending slaughtering operations in Chicago before 1960.

There are also other clouds on the horizon. In 1971 there were over 10,000 train accidents.[9] Many of these involved hazardous chemicals. There were also similar accidents involving barges, pipelines, and trucks. As a result the federal government and many state governments considered stringent safety regulations. It may take a while, but eventually a much safer, but also more expensive, system will evolve for transporting hazardous chemicals.[10] This extra cost will be passed on to the user.

## OTHER SITE LOCATION FACTORS

Besides the three most important variables, others must be considered. These are given in Table 2-3. For a given plant any one of these may be a reason why a specific location is preferable. Their importance is discussed in the following paragraphs.

Table 2-3

Important Things to Consider When Choosing a Plant Site

1. Transportation
2. Sources and costs of raw materials
3. Prospective markets for products
4. Corporation long range planning
5. Water source—quality and quantity
6. Special incentives
7. Climatic conditions
8. Pollution requirements (Waste disposal)
9. Utilities—cost, quantity and reliability; fuel—costs, reliability and availability
10. Amount of site preparation necessary (site conditions)
11. Construction costs
12. Operating labor
13. Taxes
14. Living conditions
15. Corrosion
16. Expansion possibilities
17. Other factors

### Long-Range Corporate Planning

Most corporations have some long-term goals. Often these goals affect the choosing of a plant site. This means that each plant site is not considered only for itself and that its chosen location might not be the one that would be selected if only the economics of the one plant had been considered. The object of long-range planning is to optimize a whole network of operations instead of each one individually.

The Stanford Research Institute is one of the proponents of long-range planning and has performed comprehensive studies for a number of clients.[11] The planners make proposals to the board of directors, which sets the general philosophy and direction of the company. If there are no strong reasons for making exceptions, their guidelines are followed in selecting a plant site.

One decision they might make is whether to build huge industrial complexes involving many products and processes at one site or to construct many smaller plants. Traditionally the Dow Chemical Company has concentrated most of its plants in Midland, Mich.; Freeport, TX.; and Plaquemine, LA. Dupont, meanwhile, has put plants all around the United States. These are two of the biggest chemical companies in the United States. Concentration of plants at one site allows for greater integration of the processes, permits a concentration of research and development facilities, avoids the duplication of very specialized facilities such as machine and instrument shops, and permits closer scrutiny by top management. It also places all the eggs in one basket. The corporation is much more vulnerable to earthquakes, tidal waves, hurricanes, tornadoes, fire, floods, and strikes. In case of a major disaster, instead of losing at most a few plants, a large percentage of its operations could be destroyed. Placing the plants throughout the country allows each plant to be located optimally. It also makes it easier for a company to recruit employees who may prefer one specific geographical location.

## Water

Water is needed by every processing plant for a number of different purposes. Potable water, which is generally obtained from municipal water systems, is needed for drinking and food preparation. Process water, which does not need to meet the standards set by the Public Health Service, is used in processing operations. It is often obtained from a well or unpolluted lake. Cooling water is the cheapest water available. Its source is usually a nearby river. The only requirement is that it can be easily and inexpensively treated to minimize the fouling of heat exchangers. Cooling water never comes in direct contact with the raw materials or products, whereas process water may.

The plant site must have an adequate amount of each type of water at all times of the year. The peak water demands usually occur during the summer, when rivers and lakes reach their highest temperatures and lowest levels. If nearby towns institute water rationing during parts of the summer, such a location should be dropped from consideration.

Not only the amount and quality but the temperature of the water is important. The size of a heat exchanger is inversely proportional to the temperature difference between the colling water and the material being cooled. Since the plant must run at all times of the year, the heat exchanger must be designed using the maximum cooling water temperature. This will make the temperature difference a minimum and the heat transfer area a maximum. Since the average stream temperatures are highest in the south, the heat exchangers must be larger or the flow rates greater for plants located there. Also, refrigeration systems may need to be installed in the south that are not required in the north. Since refrigeration systems are expensive,

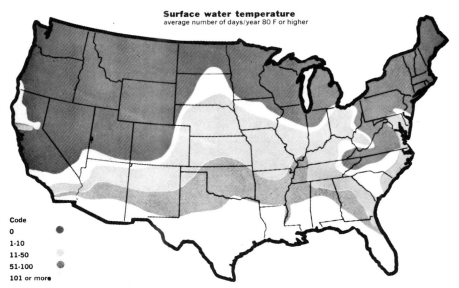

Figure 2-6    Courtesy of Winton, J.M.: "Plant Sites '67," *Chemical Week,* Oct. 28, 1967, p. 88.

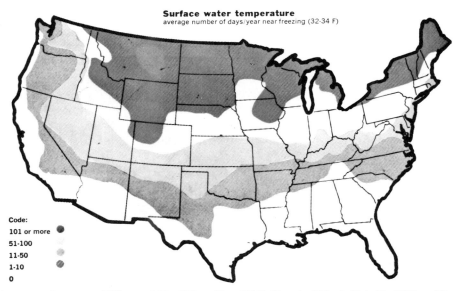

Figure 2-7   Courtesy of Winton, J.M.: "Plant Sites '67," *Chemical Week,* Oct. 28, 1967, p. 88.

this may make a far north location desirable. The average maximum and minimum temperatures of various surface waters are given in Figures 2-6 and 2-7. For more detailed information, see reference 12.

**Special Incentives**

Often the reason a given plant site is chosen is that special incentives have been offered by local authorities. In the mid-1960s, when money for financing was hard to obtain and interest rates were high, tax-free municipal bonds were an important lure. "Tax-free" means the investor does not need to pay taxes on his earnings. This means the bonds can be sold at lower interest rates and the company saves money. In 1967, $1,500,000,000 worth of these industrial bonds were issued. In 1968 the Department of Internal Revenue announced that in the future bonds used to finance private industry would be taxed regardless of who issued them. However, since then various loopholes have developed. Municipal bonds used to finance public projects such as schools, roads, and fire stations are still not taxed, since many communities would be unable to finance these projects at commercial interest rates.

Another special incentive is that of a free port. Usually this is applied for by a company together with a city and state. In a free port the raw materials have no import duty. This means the company can buy the feedstock abroad, process it, and then ship the products abroad without paying duty. However, if the product is sold to someone in the United States the import duty on the product must be paid. This is still frequently a good deal, since the tariff on finished chemicals is often less than that on raw materials. Also, because of inefficiencies, for each pound of raw materials there is less than a pound of salable product.

In the late 1960s the biggest incentives were available in Puerto Rico. As a result, petrochemical investments in Puerto Rico may exceed $1,500,000,000 by 1975. The major baits were tax exemptions and free ports. Companies making products not produced in Puerto Rico previous to 1947 (true for nearly any chemical) could be granted 100% income tax exemptions for up to 17 years. They were also allowed to avoid the import duties on certain raw materials. For instance, in 1970 (naphtha), an important feedstock for producing petrochemicals, could be obtained on the world market at half its selling price in the United States.[3,14]

**Climatic Conditions**

Each part of the United States has different prevalent climatic conditions. When Corn Products Refining Corp. built a plant on the Texas coast they took advantage of the strong prevailing winds off the Gulf of Mexico. The buildings were constructed without walls so that the wind could remove dust and obnoxious odors and prevent the accumulation of pockets of dangerous gases. The floors had a sizable overhang to prevent rain from damaging equipment.

In some parts of the United States special precautions may be necessary: the Gulf and Atlantic Coasts are noted for hurricanes; the plains states have tornadoes; and the highest probability of earthquakes occurs in California and Alaska. Care must also be taken when locating near a river to be certain that flooding, which is always a possibility, will not harm the plant. See references 12 and 15 for detailed climatic information.

**Pollution and Ecological Factors**

Certain areas are unusually susceptible to air pollution disasters. One during 1930, in the Meuse Valley of Belgium, caused the premature death of over 60 people. Another in 1948 at Donora, Pa., caused 20 deaths. A third resulted in the deaths of 22 people at Poza Rica, Mexico. The greatest occurred Dec. 5-9, 1952, in London, England. Over 4,000 people are estimated to have died from respiratory ailments as a result of the smog. Each of these occurred in an industrial valley at the time of a temperature inversion.[16]

A temperature inversion is a climatic condition in which the temperature of the air near the ground is cooler than that above it. This typically occurs on a clear winter evening. The surface of the earth is cooled by energy being radiated into outer space, and the air nearest the ground is cooled by conducting heat to the cold surface. Since hot air rises, there is no tendency for the cold air near the ground to mix with the warmer air above. Should this occur in an enclosed valley, a relatively stagnant mass of air develops. If this air is being polluted by industrial wastes, gasoline fumes, furnace smoke, or other discharges into atmosphere, these will accumulate in the air, since it is not moving.

Figure 2-8 gives the average annual air inversion frequency throughout the United States. Figures 2-9 and 2-10 give a related variable: the average maximum mixing depth. This is the maximum vertical distance through which turbulent mixing occurs. The greater the average mixing depth, the lower is the probability that a dangerous concentration of air pollutants will occur. It is desirable to pick a site having few inversions per year and a high maximum mixing depth. This is especially true if the plant will emit any noxious vapors.

The Japanese, who are very concerned about air pollution, are investigating "floating petroleum refineries." These plants would take on crude oil and produce liquefied petroleum gas (LPG), naphtha, kerosene, and heavy oil while en route to Japan. A conventional tanker would accompany this ship. Both ships would load up with crude and after the refinery ship had processed its own load it would be transferred on the high seas to the tanker, which would transfer its crude to the floating refinery. By the time they reached their destination all the crude would be processed.[17] Any air pollution would occur on the high seas and not directly affect man.

Besides air pollution, stream and thermal pollution must be considered. All coastal and interstate waterways have pollution regulations. These basically set the maximum composition and temperature of plant effluents. As time goes on they will

**Air inversion frequency** (as percent of total hours)

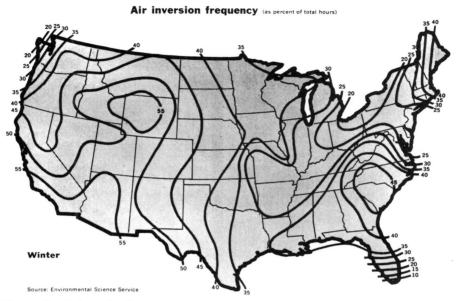

Winter

Source: Environmental Science Service

Figure 2-8    Air inversion frequency.
    Courtesy of Winton, J.M.: "Plant Sites Report '68," *Chemical Week,* Oct. 5, 1968, p. 96.

**Air mixing depths***

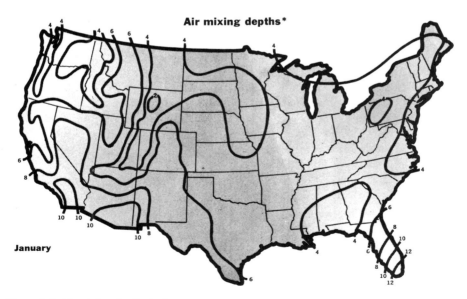

January

Figure 2-9  Air mixing depths in January.
    Courtesy of Winton, J.M.: "Plant Sites Report '68," *Chemical Week,* Oct. 5, 1968, p. 97.

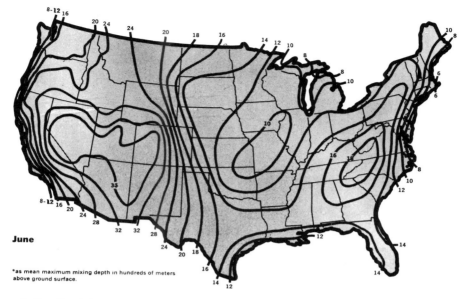

Figure 2-10   Air mixing depths in June.
             Courtesy of Winton, J.M.: "Plant Sites Report '68," *Chemical Week,* Oct. 5, 1968, p. 97.

become more stringent and uniform. As with air pollution, any advantage to be obtained by locating in a state having low standards in these respects will be short-lived. (See Chapter 16.)

Some companies are considering the whole environment in planning large complexes. When planning for a refinery at Point Tupper, Nova Scotia, the British American Oil Company considered how this new industrial center would affect the neighboring towns. This is a very enlightened approach and should be done in all cases. However, with some communities luring industry it will be a while before most companies consider anything but their own needs.

Sometimes paying attention only to the mayor, the governor, and the chamber of commerce can backfire. In the first half of 1967 conservationists caused three different companies to change their plans for building multimillion-dollar plants at specific plant sites. In each case local residents felt that the proposed plant would in some way defile the beauty of the area. In all of them the company had been encouraged to build there by local and state authorities. Such bitterness was generated over a proposed plant in Cascade Locks, Ore., that the citizens recalled all but one of the councilmen and the city manager resigned.[18] Because of the past and present practices of some industries, citizen protests are likely to increase. This will even cause problems for plants that would actually contribute to the community, since industry generally has a bad public image.

**Utilities**

Ever since the major power failure in 1965 that blacked out the northeastern United States, an important site consideration has been the reliability of electrical power systems. For instance, a 5-hour power failure cost a New Jersey firm two days' downtime. In another case the Sun Oil Company's 170,000-barrel-per-day refinery at Marcus Hook, Pa., suffered a $250,000 loss due to a 3 1/2-hour power failure. It was four days before full production could be restored. The problem is that there is no section of the country that has not had a major power failure.

To minimize damage due to power outage, the Celanese Corporation in their plant at Newark, N.J., instituted a policy of always generating half its own power. Merck & Company installed additional auxiliary steam power to insure constant refrigeration for its biochemicals at its West Point, Pa., plant. At Allied Chemical's phenol plant in Frankfort, Pa., electric devices on air compressors and pumps were replaced by steam-operated controls, and diesel generators were installed to maintain cooling water circulation.[19]

Both the quantity and price of utilities are also important. This is especially true for electrochemical plants. Traditionally the Gulf Coast states and those regions covered by the Tennessee Valley Authority and Bonneville Power Administration (Northwest) have had the cheapest power. Now some large nuclear power plants that are being installed offer the promise of cheap power to other regions.

The factors to consider with regard to fuel are the same as for power: quantity, quality, and costs. The costs are given in *Chemical Week*'s annual plant site issue. This subject is covered more completely in Chapter 8.

**Site Conditions**

An ideal chemical plant site is above the flood plain, flat, has good drainage, a high soil-bearing capability (soil that is capable of supporting heavy structures), and consists of 1,000 to 3,000 acres (400 to 1,200 hectares). The number of plant sites that fit this description is rapidly decreasing; those available are increasing in price and often are only 1,000 acres (400 hectares). Costs of $10,000 an acre ($25,000 a hectare) are not uncommon for prime sites. However, usually the prices are under $2,000 an acre ($5,000 a hectare). In 1967 the Olin Corp. paid an estimated $540,000 for 27 acres (11 hectares) of land adjoining its plant site on the Houston Ship Channel.[20] The land was to be used for a plant expansion and modernization. At the other extreme, Dart Industries paid $3,300,000 for 23,000 acres (9,700 hectares) in Kern County, Calif., in 1969.[21]

Because prime plant sites are scarce and expensive, more and more is being spent to correct site deficiencies. One company spent over $1,000,000 to make a usable site out of a 1,000-acre (400-hectare) plot. This high cost of preparing a plant site is one of the major reasons why plants often cost more than expected and why projects are not completed on schedule.

**Construction Costs**

One reason for the high costs of site preparation and for the high costs of building is labor. This usually accounts for 25% of the investment in a new plant. Construction crews are not permanent employees of most companies. They are hired for the length of a construction project. When a company wants an electrician it calls the union, which sends the worker who has spent the longest time waiting for a job. When the job is completed the electrician returns to the union hall and waits for another job. Formerly, even while he was employed he was only paid when he could work. If the weather was inclement this often meant he could not work, since most construction work is done outside. Recently some unions have gotten the employers to pay members a minimum of two hours' salary each day regardless of the weather. Since traditionally the worker had no job security, often had to live away from home, and construction rarely took place in the winter and early spring (recent advances in construction techniques have now made this a year-round operation), the construction workers were paid a very high hourly wage. This is still true.

When there is a rapid increase in industrial and private building, construction labor is usually in short supply. Under these conditions companies must often guarantee two hours' overtime per day and ten hours of overtime per week in order to get workers. All overtime is at least at the time-and-a-half rate, and usually anything over 50 hours per week is at double the base rate. For electricians, all overtime costs the company twice the usual rate. Under these conditions it was not unusual for a man in the building trades to earn over $500 per working week in 1972.

Overtime is especially costly because after laboring eight hours a man's efficiency is less. He is naturally tired. Some employers have claimed that when a worker puts in two hours of overtime they get only about one hour of services. This means a company may pay for twelve hours of work at the base pay rate and effectively get only nine hours of work per day. This could increase the cost of construction labor by one-third or the total cost of constructing the plant by over 8%. The average construction costs index by state is given annually in the "Plant Sites" issue of *Chemical Week* and by the Federal Department of Labor.

**Operating Labor**

The salaries of construction laborers may be an important consideration, but generally the salaries of plant operators are not. In general the chemical industry is highly automated and has relatively few hourly employees. These, however, often need special skills, and the availability of these talents is more important than the hourly wage.

## Taxes

There are as many types of taxes as there are ways to raise money. Each state has different regulations and they change frequently. To keep up with these, reference 22 should be consulted. A brief explanation of a number of different types of taxes follows. The first six are the most common within the United States.

### Income Tax

This is a tax on a person's or company's income. It is often graduated. That is, the rate of the taxation depends on the amount earned. Federal income taxes are based on profits. Before figuring this tax all expenses legtimately incurred may be deducted. Most large corporations in the United States pay around a 48% federal income tax. This means 48% of their profits are taken in taxes by the federal government. Nearly all states and some cities also have income taxes. Some of these are based on total sales, but most are based on profits. Federal income taxes may or may not be considered as an expense.

### Sales Tax

This is a tax levied on a company's sales. It is often a selective tax. For instance, in 1973 only California, New Jersey, and Illinois placed a sales tax on raw materials, while 44 states taxed consumer products. In 1974 there was no federal sales tax.

### Property Tax

This tax is levied on the value of a person's or company's property. The most common form is a tangible property tax levied on the value of the land and the permanent structures erected on it. Traditionally this has supplied the major income for schools and local governments. It is usually greatest in large cities. Tangible property taxes can also be levied on other physical things such as the amount of material being processed or in storage, the quantity of catalysts, equipment, furniture, and the like. Intangible property taxes are taxes on items having no physical value, such as stocks, bonds, cash. money in savings accounts, and good will.

### Franchise Tax

This is essentially the purchase of a permit by a corporation or an individual to do business in a state. The rate of taxation may be a set annual fee based on the size of the company, a certain fee per share of stock, or something else. It is a highly individualistic tax and is usually unimportant in choosing a plant site.

### Severance Tax

This is a fee levied on each unit of material that is extracted from its natural state in the ground. It is applied to oil, gas, coal, metals, and other items.

### Vehicle Fuel Tax

This is added to the price of fuel used in transporting vehicles over public roads. Its original purpose was to provide funds for the construction and maintenance of highways.

### Tariffs (Import Duties)

Tariffs are fees charged on goods entering a country. Tariffs are often used to protect local industries. Some developing countries will try to entice foreign industry by initially waiving all tariffs. The rates vary widely but are usually highest on luxury items. They may be based on the selling price in the country in which the shipment originated, the selling price plus transportation costs, or the average selling price in the country receiving the goods. The United States has used the latter base, and it is referred to as the American Selling Price.

### Value-Added Tax

This tax is similar to a sales tax. However, it is based on the amount by which a company has increased the value of a product, whereas a sales tax is based on the selling price. For instance, if a manufacturer spent $100 on the raw materials used in manufacturing a pump and then sold the pump for $275, the value-added tax would be a certain percentage of $175. For mining concerns the value-added tax would be the same as a sales tax, since minerals in their natural state are assumed to have no value.

### Transmission Tax

A transmission tax is similar to a sales tax. It is levied whenever there is a transfer of funds. If an outside contractor hired an electrical subcontractor who in turn paid an electrician to perform a job, there would be three transmission taxes paid. One would be incurred when the corporation for whom the job was being performed paid the contractor. Another would occur when the contractor paid the subcontractor. The third would occur because the subcontractor paid the electrician. If the transmission tax were 7%, then the total transmission tax would be over 21%. (See the section on compound interest in Chapter 10.)

### Border Tax

A border tax is a tax levied uniformly on all items entering a country. It is essentially a sales tax levied only on foreign goods. It may also be considered a uniform import duty.

### Living Conditions

Low taxes are often felt by the layman and politician to be the most important factor in attracting industry. They often fail to realize that another important factor is living conditions. Generally when taxes are low the services provided by the state are poor. This shows up as below-average educational and health standards and a lack of cultural advantages. One major chemical corporation found it was impossible to hire engineers for a certain location in a state that had low taxes and poor living conditions. To get engineers they offered loyal employees promotions and salary increases if they would transfer to that plant.

The availability of good living conditions is especially important to the wives of salaried employees. Many of these wives are college graduates and are not happy away from the cultural advantages and intellectual stimulation they received in more affluent communities. An unhappy wife often leads to an unhappy husband, which often results in the employee finding another job. As plants become more complex and require more highly trained personnel to run them, this factor will increase in importance. The major reason the Marathon Oil Co. located their corporation research laboratories in Denver, Colo. (1,000 miles or 1,600 km from their headquarters and the center of their petrochemical operations), was because they felt its location would be a major plus in attracting new employees.

The Midwest Research Institute evaluated the quality of life in 1967 and 1973 for each state. These estimates are given in Table 2-4. See reference 23 for more details.

### Corrosion

Once the general area for the plant has been determined, the effect of neighboring industries should be considered when picking the specific site. Their presence may indicate an increased corrosion rate. To illustrate the magnitude of this, Table 2-5 gives the corrosion rates for unpainted carbon steel and zinc for six different locations in Pennsylvania.

Corrosion is also important when a plant is located near the ocean. Table 2-5 also gives the corrosion rates for steel and zinc specimens that were placed 80 and 800 feet from the shore.[24] As a general rule it is best to keep all equipment at least 800 feet from the shore to minimize the effects of corrosion. Similarly, if a plant is to be located on a peninsula, it should be built on the leeward side and not the windward side.

The cost of corrosion damage can be large. One company in the Southwest spent over a million dollars on paint maintenance costs alone. Robert Mears surmised that

Table 2-4

Quality of Life in the United States
(United States = 1.000)

| State | Index 1973 | Rank in 1973 | Rank in 1967 |
|---|---|---|---|
| Alabama | 0.687 | 50 | 48 |
| Alaska | 1.047 | 25 | 34 |
| Arizona | 1.146 | 11 | 23 |
| Arkansas | 0.744 | 44 | 47 |
| California | 1.288 | 1 | 1 |
| Colorado | 1.274 | 2 | 6 |
| Connecticut | 1.226 | 3 | 3 |
| Delaware | 1.100 | 18 | 12 |
| District of Columbia | 1.128 | 14 | NA |
| Florida | 0.904 | 38 | 30 |
| Georgia | 0.752 | 41 | 44 |
| Hawaii | 1.120 | 15 | 14 |
| Idaho | 1.029 | 27 | 28 |
| Illinois | 1.017 | 31 | 11 |
| Indiana | 0.929 | 36 | 25 |
| Iowa | 1.060 | 22 | 10 |
| Kansas | 1.058 | 23 | 26 |
| Kentucky | 0.702 | 48 | 46 |
| Louisiana | 0.736 | 46 | 45 |
| Maine | 0.878 | 39 | 39 |
| Maryland | 1.023 | 30 | 22 |
| Massachusetts | 1.172 | 7 | 4 |
| Michigan | 1.032 | 26 | 16 |
| Minnesota | 1.139 | 13 | 2 |
| Mississippi | 0.698 | 49 | 50 |
| Missouri | 0.864 | 40 | 41 |
| Montana | 1.149 | 9 | 31 |
| Nebraska | 1.109 | 16 | 32 |
| Nevada | 1.094 | 19 | 20 |
| New Hampshire | 0.978 | 34 | 29 |
| New Jersey | 1.087 | 20 | 13 |
| New Mexico | 1.053 | 24 | 38 |
| New York | 1.142 | 12 | 7 |
| North Carolina | 0.710 | 47 | 40 |
| North Dakota | 1.024 | 29 | 19 |
| Ohio | 0.958 | 35 | 17.5 |
| Oklahoma | 0.984 | 33 | 33 |
| Oregon | 1.198 | 5 | 8 |
| Pennsylvania | 1.107 | 17 | 21 |
| Rhode Island | 1.147 | 10 | 15 |
| South Carolina | 0.657 | 51 | 49 |
| South Dakota | 1.008 | 32 | 37 |
| Tennessee | 0.752 | 42 | 42 |
| Texas | 0.916 | 37 | 36 |
| Utah | 1.168 | 8 | 17.5 |
| Vermont | 1.028 | 28 | 27 |
| Virginia | 0.749 | 43 | 35 |
| Washington | 1.217 | 4 | 5 |
| West Virginia | 0.742 | 45 | 43 |
| Wisconsin | 1.064 | 21 | 9 |
| Wyoming | 1.187 | 6 | 24 |

Source: Liu, B.: *The Quality of Life in the United States: 1970 Index, Rating, and Statistics*, Midwest
Research Institute, Kansas City, Mo., 1973.

Table 2-5

Corrosion Rates for Various Locations

| Site | Type of Environment | Corrosion in mills/year* | |
|------|---------------------|-------------|------|
| | | Carbon Steel | Zinc |
| State College, Pa. | Rural | 0.88 | 0.030 |
| Potter County, Pa. | Rural | 0.75 | 0.032 |
| South Bend, Pa. | Semi-rural | 1.21 | 0.046 |
| Monroeville, Pa. | Semi-industrial | 1.785 | 0.049 |
| Pittsburgh, Pa. | Industrial | 1.12 | 0.067 |
| Bethlehem, Pa. | Industrial | 1.37 | 0.033 |
| Kure Beach, N.C. | 800 ft. from shore | 5.34 | 0.052 |
| Kure Beach, N.C. | 80 ft. from shore | 19.5 | 0.164 |

\* multiply by 0.254 to get cm/yr

Source: *Metal Corrosion in the Atmosphere*, American Society for Testing and Materials, Philadelphia, 1968, p. 360+

with an optimum plant location and layout in the same area this maintenance cost could have been cut in half.[25]

## Expansion Possibilities

If the company is considering expanding its operations at the proposed site, it must be determined whether a proposed site can accommodate not only the plant but the contemplated expansions. To do this all factors listed in Table 2-3 must be evaluated to see that for the enlarged plant none are in short supply. All the land necessary should be bought initially. Once the proposed plant site is announced to the public, all the contiguous land will increase in value. This will be especially true if the company ever gives any hint that it would like to expand.

## Other Factors

Under this category are specific items that may be very important at one point in time and totally irrelevant at others. To be aware of these, the process engineer must keep up with engineering periodicals and attend technological meetings. Three specific examples follow.

In the late 1960s Baton Rouge and other areas of Louisiana were noted for labor unrest. Conditions became so bad that in 1967 a dozen companies voluntarily stopped $350,000,000 worth of new construction that was in progress. This specific shutdown, which involved 15,000 workers, was the result of jurisdictional dispute between the teamsters' union and the electrical workers' union. This also caused other companies to change their plans about locating in Louisiana.[26]

Racial violence also was a factor in the late 1960s. One company eliminated both Newark, N.J. and Watts, Calif., from its list of possible locations because of the riots there.

In the early 1970s the major concern was citizens' groups that delayed and prevented companies from building at new sites. This was usually because they feared the companies would do some damage to the ecology or economy of the area. As a result, building adjacent to a site where the company already had a plant became very popular. If the company had proven itself a good neighbor there would usually be a minimal amount of opposition. It was, after all, a member in good standing of that community.

## CASE STUDY: SITE SELECTION FOR A 150,000,000 LB/YR POLYSTYRENE PLANT USING THE SUSPENSION PROCESS

At the end of Chapters 2 through 11 an application of the material presented in the chapter to a specific exemplary task will be presented. This will be the design of a 150,000,000 lb/yr polystyrene plant, which will use the suspension process. The goal of this example will be to provide just enough information so the board of directors can decide whether the plant should be constructed.

Over 3,000,000,000 lb of polystyrene were produced in 1970.[27] The capacity of the plant to be designed is around 5% of this, and it is considered a large plant. The growth rate of polystyrene is predicted to be 11.5% per year[28] between 1969 and 1973. Thus when this plant comes on stream it should not cause any great surplus of polystyrene to occur. For working this example it should be assumed that it is summer, 1971.

### Raw Materials

Polystyrene is made by polymerizing styrene. In the suspension process the styrene is broken up into small droplets which are suspended in water. Various additives aid in controlling this and the reaction rate. These additives amount to about 1% of the styrene added. For high-impact styrene up to 0.15 lb rubber/lb styrene is included. The two major materials needed are water and styrene.

### Raw Material and Product Sources

For simplicity it will be assumed the plant will be located in the United States. The 1969 American sources of styrene are given in Table 2E-1 The 1969 uses for this styrene are given in Table 2E-2. It should be noted that over 50% of the styrene produced is used to make straight and rubber-modified polystyrene. The capacities and locations of the main polystyrene producers are given in Table 2E-3. With the exception of Midland, Mich.; Kobuta, Pa.; Torrence, Calif.; and Penuelas, Puerto Rico, all the styrene is produced in the Gulf Coast states of Louisiana and Texas. At Midland, Dow Chemical uses nearly all the styrene internally. The same is true of Sinclair-Koppers at Kobuta. The capacity of the Shell plant in California is presently not adequate to supply our proposed plant and the other customers on the West Coast. This means styrene must either be imported or come from Texas, Louisiana, or Puerto Rico.

Table 2E-1

Availability of Styrene

| Producer | Location | Capacity (1969 year end) million lb/yr |
|---|---|---|
| Amoco Chemicals | Texas City, Tex. | 800 |
| Cosden Oil & Chemical | Big Spring, Tex. | 100 |
| Cos-Mar | Carville, La. | 500 |
| Dow Chemical | Midland, Mich. | 350 |
| | Freeport, Tex. | 550 |
| El Paso Natural Gas | Odessa, Tex. | 120 |
| Foster Grant | Baton Rouge, La. | 240 |
| Marbon Chemical (division of Borg-Warner) | Baytown, Tex. | 135 |
| Monsanto | Texas City, Tex. | 800 |
| Shell Chemical | Torrance, Calif. | 240 |
| Sinclair-Koppers | Kobuta, Pa. | 430 |
| | Houston, Tex. | 110 |
| Sun Oil's Suntide Refining Co. | Corpus Christi, Tex. | 80 |
| Union Carbide | Seadrift, Tex. | 300 |
| Total | | 4,755 |

Notes: Dow Chemical will have a total capacity at Freeport, Tex., of 1,000,000,000 lb by the end of 1971.

Foster Grant is adding 500,000,000 lb/yr at Baton Rouge to be on stream by mid-1970.

Commonwealth Oil is constructing a 400,000,000 lb/yr plant in Penuelas, Puerto Rico (startup 1971).

Source: "1970 Outlook for Styrene and Its Polymers," *Chemical and Engineering News*, Sept. 22, 1969, p.22.

## Uses and Users of Polystyrene

Polystyrene is used in furniture, packaging, appliances, automobiles, construction, radios, televisions, toys, houseware items, and luggage. Statistics on the states producing the largest number of plastic products are given in Table 2E-4. This indicates that we can expect to have many customers and that most of these will be small companies. The main means for shipping will therefore be by truck and train, and many of these shipments will be small quantities. It therefore would be wise to locate as near as possible to customers. (The styrene can be obtained in large bulk shipments and over 95% is converted into a salable polystyrene.) Note that the

Table 2E-2

Present Uses for Polystyrene

| Market | Total Production (millions of pounds) | Styrene per cent | Styrene Content (millions of pounds) |
|---|---|---|---|
| Straight polystyrene (including foam) | 1170 | 100 | 1170 |
| Rubber-modified polystyrene | 1050 | 90 | 950 |
| Styrene-butadiene rubber | 3100 | 24 | 725 |
| Styrene-butadiene copolymer | 380 | 50 | 190 |
| Acrylonitrile-butadiene-styrene (ABS) | 470 | 60 | 280 |
| Styrene-acrylonitrile (SAN) | 60 | 75 | 45 |
| Unsaturated polyesters | 670 | 15 | 100 |
| Alkyds | 610 | 15 | 90 |
| Exports | | | 750 |
| TOTAL | | | 4300 |

Source: "1970 Outlook for Styrene and Its Polymers," *Chemical and Engineering News*, Sept. 22, 1969, p. 22.

present producers are near the buyer. This would indicate building a plant in the Midwest; the New York-New Jersey-Pennsylvania area; or California. California has already been eliminated.

**Transportation**

To reduce shipping charges for styrene it should be delivered by ship or barge. This further limits sites to along the Mississippi and Illinois Rivers, the Ohio River Basin, the Great Lakes area, or the eastern seaboard. The region bordering the Great Lakes can be eliminated because shipping is curtailed during the winter months. The cost of the extra storage required to store styrene for three months can be shown to make these sites impractical.

It is assumed that any East Coast site would obtain styrene from Puerto Rico or foreign sources. Any site located in the Midwest would obtain its styrene from Louisiana. To compare these sites the cost of styrene from these different sources must be known, as well as import duties and the specifics of the freeport laws which might be involved. A plant of this size would probably contract to receive styrene, and what special price concessions might be obtained would need to be investigated. Since information on this was not available it will be assumed that the only difference in the price of styrene at the site is due to the cost of transportation. The cost of shipping by sea from Puerto Rico to Philadelphia (~1700 miles) is \$4.5/ton=0.23¢/lb. The cost of shipping styrene from Baton Rouge, La. to Cincinnati, Ohio (~900 miles) is \$1.70/ton=0.09¢/lb (see Fig. 2-5).[29]

Table 2E-3

The Market 1969: Our Competitors

| Producer | Location | Capacity (1969 year end) millions of pounds |
|---|---|---|
| Amoco Chemicals | Leominster, Mass. Medina, Ohio Joliet, Ill. Willow Springs, Ill. Torrance, Calif. | 170 |
| BASF | Jamesburg, N.J. | 80* |
| Cosden Oil & Chemical | Big Spring, Tex. | 145 |
| Dart Industries | Holyoke, Mass. Ludlow, Mass. Joliet, Ill. Santa Ana, Calif. | 140 |
| Dow Chemical | Midland, Mich. Allyn's Point, Conn. Hanging Rock, Ohio Torrance, Calif. | 700 |
| Foster Grant | Leominster, Mass. Peru, Ill. | 190 |
| Hammond Plastics | Oxford, Mass. | 25 |
| Howard Industries | Hicksville, N.Y. | 15 |
| Monsanto | Springfield, Mass. Addyston, Ohio Long Beach, Calif. | 375 |
| Richardson Co. | West Haven, Conn. | 50 |
| Shell Chemical | Wallingford, Conn. Marietta, Ohio | 80 |
| Sinclair-Koppers | Kobuta, Pa. | 300 |
| Solar Chemical | Leominster, Mass. | 60 |
| Southern Petrochemicals | Houston, Tex. | 40 |
| Union Carbide | Bound Brook, N.J. Marietta, Ohio | 170 |
| Total | | 2,340 |

* Expandable beads only.

Note: U.S. Steel will build a 200,000,000 lb/yr plant at Haverhill, Ohio, to be on stream early in 1971.

Source: "1970 Outlook for Styrene and Its Polymers," *Chemical and Engineering News*, Sept. 22, 1969, p. 22.

Table 2E-4

Statistics on Plastics Products (1963)

| State | Total Number of Establish- ments | Establish- ments with 20 Employees or More | Value Added by Manufacture Adjusted ($1,000) | Cost of Materials ($1,000) |
|---|---|---|---|---|
| Ohio | 293 | 153 | 189,226 | 174,095 |
| New Jersey | 382 | 172 | 168,035 | 172,008 |
| Illinois | 379 | 175 | 209,395 | 163,367 |
| New York | 664 | 233 | 171,545 | 152,290 |
| California | 591 | 181 | 142,562 | 128,074 |
| Massachusetts | 269 | 136 | 100,910 | 86,946 |
| Michigan | 227 | 86 | 81,501 | 78,981 |
| Indiana | 114 | 48 | 72,486 | 77,918 |
| Pennsylvania | 219 | 87 | 78,217 | 74,723 |
| Texas | 123 | 33 | 31,868 | 35,679 |
| Total Top Ten States | 3,261 | 1,304 | 1,245,745 | 1,144,081 |
| Total U.S. | 4,334 | 1,674 | 1,660,882 | 1,522,899 |

Source: *1963 Census of Manufacturers*, U.S. Dept. of Commerce, Bureau of the Census, p. 30A-10.

Table 2E-5

Site Factors for Selected States

| State | Building Cost Index (New York = 100) | Electricity Cost ¢/kwhr | Labor Cost $/hour | Corporate Income Taxes | Living Cond. (rank)* |
|---|---|---|---|---|---|
| Ohio | 93.5 | 0.20 | 3.60 | 0% | 21 |
| New Jersey | 88.5 | 0.47 | 3.65 | 4.25% | 7 |
| Illinois | 87. | 0.65 | 3.70 | 4% | 6 |
| New York | 90. | 0.47 | 3.32 | 7% | 2 |
| Massachusetts | 86.5 | 0.45 | 3.57 | 7.5% | 4 |
| Indiana | 88. | 0.66 | 3.80 | 2% | 22 |
| Pennsylvania | 88. | 0.40 | 3.26 | 12% | 19 |
| Average | 88.8 | 0.47 | 3.56 | | |

\* These are different than those given in Table 2-4 because a different rating system was used in 1970.

Figures obtained from "Plant Sites," *Chemical Week*, Aug. 19, 1970, and "Plant Sites Report," *Chemical Week*, Oct. 5, 1968.

**Production Expenses**

The cost of the raw materials is probably the most important factor in determining the price of polystyrene. In 1968 styrene sold for 7.75¢/lb[30] while general-purpose polystyrene was selling ten for 12.5¢/lb[31] This means around 60% of the selling price was spent for raw materials. It should be nearer 50%.

A summary of other pertinent costs in the states near customers is given in Table 2E-5. If it were known how each of these items affects the final cost, the various sites could be rated quantitatively. For instance, suppose the average energy costs are 2% of the selling price, before-tax profits are 30%, depreciation costs amount to 5%, and production labor costs are 10%. These items would reduce profits in Illinois by:

$$\left\{ 0.02\frac{0.65}{0.47} + 0.30\frac{0.04}{2} + 0.05\frac{87}{88.8} + 0.10\frac{3.70}{3.56} \right\} \ 14.5¢/lb = 2.70¢/lb$$

This is given for each state as column six in Table 2E-6. The selling price of polystyrene in 1971 was 14.5¢/lb. A factor of one-half is used in the second term of this equation because state income taxes are an expense and reduce federal income taxes by about half the tax paid the state (see Chapter 11 for details). Table 2E-6 gives an idea of the importance of each of the factors. Other factors that should be considered are that land costs near an ocean port would be expected to be greater than those on a river; southeastern Ohio is one of the few regions of the country having a surplus of electrical power; the climatic conditions of those states being considered are similar, although the southern portions of Illinois, Indiana, and Ohio usually have less severe winters; an 8% corporate income tax has been proposed by the governor of Ohio; the pollution regulations of these states are similar.

**Best Site**

When all of the factors are considered the best site seems to be in Ohio. The two sites that seem the most promising are Martins Ferry, Ohio (which is across the Ohio River from Wheeling, W.V.) and Cincinnati, Ohio. The Martins Ferry site would be within 300 miles of Buffalo, N.Y., Toledo, Ohio, and Indianapolis, Ind.; within 200 miles of Cleveland, Ohio and Erie, Pa.; and within 100 miles of Pittsburgh, Pa. There is good rail service and it is near Interstate Highway 70. It is also in the coal mining area where power is cheap. Locating near Cincinnati would increase power costs but it would be around 100 miles from Indianapolis, Ind., and Columbus, Ohio, and within 250 miles of Toledo and Cleveland, Ohio. It would appear that due to reduced power costs the site at Martins Ferry would be best. The Martins Ferry site would be near the heavily industrialized Steubenville, Ohio area,

Table 2E-6

Reduction of Profits in Cents per Pound
Due to Various Costs

| State | Depreciation | Energy Use | Operating Labor | State Income Tax | Total |
|---|---|---|---|---|---|
| Ohio* | 0.76 | 0.12 | 1.47 | 0.0 | 2.35 |
| New Jersey | 0.72 | 0.28 | 1.49 | 0.09 | 2.58 |
| Illinois | 0.71 | 0.39 | 1.51 | 0.09 | 2.70 |
| New York | 0.73 | 0.28 | 1.35 | 0.15 | 2.51 |
| Massachusetts | 0.71 | 0.27 | 1.45 | 0.16 | 2.59 |
| Indiana | 0.72 | 0.40 | 1.55 | 0.04 | 2.71 |
| Pennsylvania | 0.72 | 0.24 | 1.33 | 0.26 | 2.55 |
| Maximum Differential | 0.05 | 0.28 | 0.22 | 0.26 | 0.36 |

\*  If Ohio adopts a proposed 8% income tax its total would be 2.52¢/lb. (In 1974 Ohio had a 4% income tax.)

which on the average has the highest level of air pollution in the country. It is assumed a specific site can be obtained outside the pollution area.

## References

1. "U.S. Refiners Go Sour," *Chemical Engineering*, June 11, 1973, p. 74.
2. "Taming the Alaskan Wilderness," *Chemical Week*, Oct. 28, 1967, p. 121.
3. Low, S.(ed.): *Jane's Freight Containers*, Marston and Co., London, published periodically.
4. Sampson, H. (ed.): *Jane's World Railways*, Marston and Co., London, published periodically.
5. Aude, T.C., Cowper, N.T., Thompson, T.L., Wasp, E.J.: "Slurry Piping Systems Trends, Design Methods, Guidelines, *Chemical Engineering*, June 28, 1971, p. 87.
6. "Unit Trains Trim Acid Costs," *Chemical Week*, Mar. 15, 1969, p. 101.
7. "Business Newsletter," *Chemical Week*, May 25, 1968, p. 24.
8. "CPI Tallies Dock Structure Losses," *Chemical Week*, Mar. 1, 1969, p. 12.
9. Winton, J.: "Trouble Ahead for Chemical Transport," *Chemical Week*, Jan. 31, 1973, p. 25.
10. Woods, W.S.: "Transporting, Loading and Unloading Hazardous Materials," *Chemical Engineering*, June 25, 1973, p. 72.
11. "Plant Sites '67," *Chemical Week*, Oct. 28, 1967, p. 74.
12. Todd, D.K.: *The Water Encyclopedia*, Water Information Center, Port Washington, N.Y., 1970.
13. "Succumbing to 'Lures' of the Islands," *Chemical Week*, Dec. 30, 1967, p. 16.
14. "Puerto Rico, Petrochemical Paradise," *Chemical Engineering Progress*, Apr. 1970, p. 21.
15. *Statistical Abstract of the United States*, U.S. Bureau of the Census, Washington, D.C., published annually.
16. *Encyclopedia Britannica*, vol. 18, p. 185b; vol. 22, p. 358c, 1968.
17. "Chementator," *Chemical Engineering*, Sept. 22, 1969, p. 65.
18. "Chementator," *Chemical Engineering*, June 19, 1967, p. 84.
19. "Ready for Blackouts," *Chemical Week*, June 8, 1968, p. 72.
20. "CPI News Briefs," *Chemical Engineering*, Dec. 18, 1967, p. 145.
21. "Rapid Wrap-up," *Chemical Week*, Oct. 29, 1969, p. 35.

22. *Commerce Clearing House, State Tax Reports*, Commerce Clearing House, Inc., Chicago, published regularly.

23. Liu, B.: *The Quality of Life in the U.S.*, booklet published by Midwest Research Institute, Kansas City, 1973.

24. *Metal Corrosion in the Atmosphere*, American Society for Testing and Materials, Philadelphia, 1968.

25. Mears, R.B.: "Plant Site, Layout Minimize Corrosion," *Chemical Engineering*, Jan. 11, 1960, p. 144.

26. "Chementator," *Chemical Engineering*, July 3, 1967, p. 24.

27. *Modern Plastics Encyclopedia*, McGraw-Hill, New York, 1970-71.

28. "1970 Outlook for Styrene and Its Polymers," *Chemical and Engineering News*, Sept. 22, 1969, p. 22.

29. "Plant Sites," *Chemical Week*, Aug. 19, 1970, p. 66.

30. *Chemical Week*, June 15, 1968, p. 52.

31. *Chemical Week*, Dec. 14, 1968, p. 51.

## Additional References

The annual "Plant Sites" issues of *Chemical Week*.

| | | |
|---|---|---|
| October 28, 1967 | August 19, 1970 | October 17, 1973 |
| October 5, 1968 | October 13, 1971 | October 22, 1974 |
| October 29, 1969 | October 11, 1972 | |

# CHAPTER 3

# The Scope

The first stage in any plant design is the definition of the project. This is referred to as writing the scope. Until the scope is established there can be no control of the project. Without a scope, it is impossible to determine what is germane and what, although interesting and related to the project, is actually extraneous information. Since the major goal of a preliminary process design is to provide a reasonably accurate cost estimate, the scope for such a project must determine what will and will not be included in this estimate. Working without a scope would be like trying to estimate the amount a woman entering a supermarket is going to spend when you don't know if she is doing her weekly shopping or only buying snacks for a party.

The major cause of projects overrunning original estimates and getting cancelled after design engineering is nearly completed is an improperly conceived scope. Unless the engineer is very careful, the initial scope may not include everything that is necessary. Every time something is added, the estimated cost rises. If this projected cost increases too much, the project may be cancelled and the result will be a number of frustrated and disillusioned engineers and scientists.

As an example of an incomplete scope, consider the case of a person who decides to take up snow skiing. He has seen a television program on skiing in the Alps, and thinks he would enjoy it. He goes to his local sporting goods store and inquires about equipment. The salesman tells him that skis can be purchased for $30.00, the poles for $5.00 and a pair of boots for $45.00. After cogitating about it for a while, he decides that for $80.00 it's worth a try. So he goes back to the sporting goods store and buys the skis, boots, and poles for the $80.00 quoted, plus tax. However, before he can ski there are a few added expenses he has not considered. He needs ski bindings to attach the boots to the skis. Good safety bindings cost $25.00 and their installation will cost $7.00. No ski area will allow him on the ski slopes without an Arlberg strap or its equivalent. This is a safety strap that prevents the loss of the ski if the safety binding releases. If a ski gets loose at the top of a hill, it can reach very high speeds on the ski run. Should this runaway ski hit a skier or spectator it could severely injure or kill him. The safety strap will cost $5.00 installed. Since he has spent $45.00 on boots it would seem silly not to buy a $3.50 boot-tree so that the boots will maintain their shape. Next, our friend finds out that the trunk of his car will not hold his 6 ft (2 m) skis. To get them to the ski slopes he will need to buy a $25.00 ski carrier that can be mounted on his car. At this point he has added $65.50

to his initial estimated cost of $80.00. This is an 80% increase over his originally estimated price, and he had firm prices for the initial equipment. But this is not the end. He is going to find that there are more expenses. He will need a variety of waxes ($2.00), straps and a center block for storing the skis ($1.50), and specially designed leather ski gloves for use on rope tows ($8.00). Before long he will also purchase goggles ($4.00), ski pants ($30.00) and a ski parka ($25.00). The total cost is now $216.00 or 170% greater than his original estimate. It does not include tow fees ($4-$9 per day), ski lessons ($8-$13 per hour for a private lesson), transportation to the ski slopes (a distance of maybe 100 to 200 miles each way), meals or lodging.

Although our skiing neophyte was interested in learning how much it would cost him to take up skiing as a hobby, this was not what was written in his initial scope. His initial scope was the purchase of skis, boots, and poles. With this scope he obviously would not be able to achieve his objective.

Another common example of an incomplete scope occurs when a person buys a car. The list price of a car may be around $3,000, but rarely does anyone pay under $3,500, and the price may reach $4,500. Most people do not want just a car. They also want certain extras: a radio, power steering, power brakes, bucket seats, a floor shift, whitewall tires, and so on. When a person asks the price of a car, he wants the dealer to include the extras. The dealer, on the other hand, gives the price of the stripped-down car, since he would probably scare the customer away if he included every possible extra in his quotation.

This same type of reasoning occurs in the chemical industry. When a project is evaluated the researcher would like to see the lowest possible cost attached to it, because then the probability of having it continued is enhanced. Other laboratories have competitive projects in the same way that there are many dealers competing to sell cars, so the projects that look most promising get the most money. On the other hand, management wants a reliable estimate of the total cost so it can decide whether a product can compete in the market place. It is up to the process engineer to see that the scope includes everything that is necessary to produce the required product safely, and nothing that is extraneous.

As an example of what should be included in the scope, suppose the manager of the product department sends you the following message: "We are considering building a plant to produce 60,000 tons per year of chlorine from the brine deposits near Pentwater, Mich. Would you please determine the cost of such a facility." Does this define the scope? No! Definitely not.

In the production of chlorine from brine, caustic is produced. What is to be done with this caustic? Is it to be neutralized and discarded? If so, what acid is to be used and where is the resulting salt to be dumped? Will this cause pollution problems? Maybe there are no federal or state pollution laws being violated, but the fastest way to get restrictive controls is for the chemical companies to ignore the public's feelings. On the other hand, if the caustic is to be purified and sold, how pure should it be, and can the amount produced be sold? When the brine also contains some magnesium salts and bromides, then the scope must indicate if magnesium and bromine are to be recovered, and, if so, what their respective purities should be.

There are different ways of producing chlorine from brine, for example, Dow cells, Hooker cells, and mercury cells. Which process is to be used must be known in order to make an accurate economic evaluation, since the capital costs and operating costs are different for each of these processes. The process engineer may have to investigate the different processes and economically evaluate each before deciding which process is best.

One of the major costs in the electrolytic production of chlorine is electrical power. Should the power be purchased, or is a power generating station to be built? If a power generator is to be built, should it be built large enough so that it can provide power for future expansions and for other existing plants the company may own in the general area? The answers to these questions will greatly affect the amount of capital the company must allocate for the project.

Since the product must be shipped to the customers, there must be transportation and loading facilities. It must be determined if docks, railroad sidings, pipelines, and/or roads need to be constructed, and what types of containers are to be used if all products are not to be shipped in bulk quantities.

While awaiting transportation and customer orders, the materials must be stored. How much storage should be provided for the products and raw materials? Whether a two-week or two-month inventory of products and feed are designed will affect the cost of the facilities, as well as the plant land area.

Another factor that will have a great bearing on the land area required, the plant layout, and the design is whether provision should be made for future expansion. Should enough space be left to add 50 additional cells and to double the storage space? Should the facility be designed so that these additional units can be connected easily into the process system? Should the purification and finishing steps be overdesigned so that when the expansion occurs only new cells need to be added? All these questions must be answered before any attempt can be made to obtain a valid cost estimate.

Let us return to the request of the product department manager to determine the cost of a plant for producing 60,000 tons per year of chlorine from the brine deposits near Pentwater, Mich. Does the process engineer give up because the scope is undefined? If he does, then he can be certain that his hopes for advancement are going to be greatly diminished. Industry is looking for the man who can take a difficult assignment and complete it with a minimum amount of time, money, and effort. Managers are rarely pleased by a man who offers excuses as to why things cannot be done. For the above problem they want an engineer who will recognize that the scope is incomplete, and who will scout out the information needed to complete the scope and the assignment.

By using his engineering experience and common sense, the process engineer can answer many of the above questions himself. The other answers can usually be obtained from one of his co-workers. In every company there are people of varied backgrounds. Usually someone can either provide an answer or suggest a place where the answer may be found.

The scope is a series of assumptions that everyone concerned with the process is expected to question. If one succinctly presents these and makes them generally available, the hope is that any errors will be noted and corrected. For instance, sometimes when "percent" is said the listener may assume it refers to weight percent, whereas the speaker means mole percent. Another example is a formulation that calls for 5 lb of catalyst per 1,000 lb of feed. The engineer may infer this means 5 lb of solution containing 1 lb of active ingredient, where the scientist meant 25 lb of solution or 5 lb of active substance. Anyone who has dealt with people realizes that these communication errors occur all the time.

This is why it is important that the writing of the scope be the first thing done on any project. Then any errors can be corrected before the project has proceeded very far. Even if something is not known, the scope contains the best estimate of it. If this estimate is later found to be wrong, it is then corrected. Thus, the scope becomes a fluid document.

After the scope is approved by all concerned parties, any change is formally made on a paper entitled "Change of Scope." These are distributed to all concerned parties, and are discussed at periodic meetings called to review the status of the project. The information that is necessary to define the scope is given in Table 3-1. The importance of these items will be discussed on the following pages.

## THE PRODUCT

It is presumed that the product(s) to be produced is (are) known. The size of the containers it will be shipped in depends on the size of the expected orders, the facilities the customer has for handling the materials, and the hazardous classification of the material. Material shipped in bulk quantity is cheaper than packaged items, but it requires the customer to have more elaborate unloading and storage facilities. Bulk shipping is only used when large amounts are purchased at one time. Union Carbide will not ship in bulk less than 40,000 lb (18,000 kg) of material.[1] Table 3-2 gives a summary of the maximum bulk shipments possible by various carriers.

Most large standardized containers made of rigid metal are 8 x 8 ft (2.4 x 2.4 m), and are 10, 20, 30, or 40 ft (3, 6, 9, or 12 m) long.[2] These can be transported by truck, boat, or train. Another type of container is a collapsible rubber bag known as a sealdbin. This has the advantage that when it is empty it does not take up much space, hence the cost of returning it to the sender is reduced. It has a 7 ft, 2 in (2.2 m) diameter, and is 8 ft (2.4 m) high. Table 3-3 gives the standard sizes for some small chemical containers. A comparison of costs for shipping by different methods, plus the costs of small containers, is given in reference 3.

## CAPACITY

The capacity of a plant depends, among other things, on how much material can be sold. This is predicted by marketing experts, on the basis of a marketing survey that indicates how much of each product can be sold by the company. This survey must predict what is likely to occur during the next 10-15 years. It must consider end uses, competitors' plans, competing products, market potential, and so on. The

Table 3-1

Items to Be Included in the Scope

1. The product(s) (including package size)
2. Quantity of each product
3. Quality of each product
4. Storage requirements for each product
5. Raw materials for each product
6. Quality of the raw materials
7. Storage requirements for the raw materials
8. By-products
9. Process to be used, including yields and conversions
10. Waste disposal requirements
11. Utilities requirements
12. What provision should be made for future expansions
13. Location of the plant
14. Operating hours per year
15. Completion date
16. Shipping requirements
17. Laboratory requirements
18. Special safety considerations

Table 3-2

Maximum Bulk Shipments by Various Carriers

Petroleum tankers: 4,000,000 bbl of oil*
Cargo ships for chemicals: Up to 290,000 bbl total
Ocean barges: 26,000 tons
River barges: 3,000 tons of liquid
                1,500 tons of dry solids
Railway cars:
      Hopper cars 125 tons (5,800 ft$^3$)
      Tank   cars 100 tons (60,000 gal)
Trucks:** 1,570 ft$^3$ of dry solids
       8,700 gallons of liquid

\* Ships containing over 1 million barrels are too large to enter any U.S. ports.

\*\* Weight limits set by states. See reference 2.

capacity also depends on technical and economic questions that the process engineer must answer.

The final decision on how large the plant will be is made by the board of directors. This, then, is one of the four major decisions made by them. The others are whether, where, and when to build. The factors the engineer must weigh in determining an optimum plant size will be considered next.

Table 3-3

Standard Size Small Containers for Chemicals

| Type | Size | Usable Volume | | Drums/Carload |
|---|---|---|---|---|
| | | ft³ | m³ | |
| Steel drums | 55 gal | 7.35 | 0.208 | 360 |
| | 30 gal | 4.00 | 0.113 | 592 |
| | 16 gal | 2.14 | 0.0606 | 1,225 |
| Fiber drums | 61 gal | 8.15 | 0.231 | 300 |
| | 55 gal | 7.35 | 0.208 | 318 |
| | 47 gal | 6.28 | 0.178 | 424 |
| | 41 gal | 5.48 | 0.155 | 552 |
| | 30 gal | 4.00 | 0.113 | 592 |
| | 15 gal | 2.00 | 0.0566 | 1,272 |
| | 1 gal | 0.1335 | 0.00379 | 17,365 |
| Polyethylene drum | 15 gal | 2.00 | 0.0566 | 1,272 |
| Bags | | 2.00 | 0.0566 | |
| | | 1.33 | 0.0377 | |
| | | 0.84 | 0.0238 | |
| | | 0.12 | 0.0034 | |
| Boxes | 15 x 15 x 22 in | 2.86 | 0.0810 | |
| | 41 x 34 x 36 in | 5.00 | 0.142 | |
| Carboys | 13.5 gal | 1.35 | 0.0382 | |
| | 15 gal | 2.00 | 0.0566 | |
| Pails | 5 gal | 0.67 | 0.0190 | |
| Cans | 1 gal | 0.1335 | 0.00379 | |
| | 1 qt | 0.034 | 0.00096 | |

Source:  Raymus, G.J.: "Evaluating the Options for Packaging Chemical Products," *Chemical Engineering*, Oct. 8, 1973, p. 67.

## Factors Favoring Large Plants

The current trend to build bigger and bigger single-train plants has a number of economic advantages. First, the price of raw materials is inversely related to the quantity bought. Even grocery retailers demonstrate this in principle when they sell a bushel of tomatoes at a lower price per pound than a three-pound bag. Second, generally the cost per unit volume is less when larger equipment is used; a piece of equipment that is twice the size of another does not usually cost twice as much. This is true provided the item is not so large that it must be custom-built, or that it requires an extensive amount of bracing or extra structural support that is not needed for the smaller vessel. Third, the cost of piping, wiring, and instrumenting a larger plant is only slightly more than for a smaller plant.

Fourth, the increased capacity may allow some changes in equipment and/or processing steps that can produce substantial savings. For example, a major reduction in the price of ammonia occurred in the late 1960s because the increase in the size of ammonia plants permitted the replacement of reciprocal compressors with centrifugal compressor. The latter are only used at large flow rates. The centrifugal compressors could also utilize the process heat. This was unused in the smaller plants because the large capital outlay necessary to construct a recovery system could not be justified economically (Fig. 3-1).

**Ammonia from methane: biggest size, lowest cost**

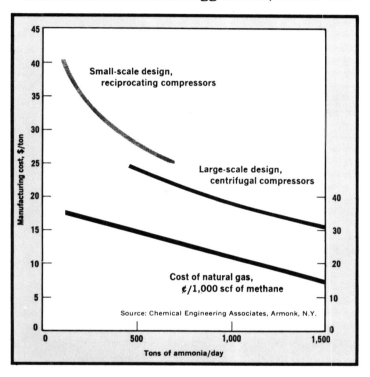

Figure 3-1   Cost vs. capacity curve for ammonia in 1967.
Source: "Bell Tolls for Little Plants," *Chemical Week,* Oct. 28, 1967, p. 127.

Fifth, if the system is merely enlarged it does not require any more operating personnel than a smaller unit. For this reason, a larger so-called single-train plant would be preferable to two duplicate plants side by side, even if there were no other savings.

The importance of size on the economics of ammonia production can be seen from Figure 3-1 and Table 3-4, which was developed in 1967 by G. Russell James, general manager of Chemical Engineering Associates (Armonk, N.Y.)[4] Before 1969, a 400-tons-per-day plant was large. Now it can barely compete even if it is updated technologically.

Table 3-4

Cost for a Single Train Ammonia Plant

| Plant Size, tons/day | Manufacturing Cost, $/ton | Storage, $/ton | Transfer, $/ton | Terminal, $/ton | Total Plant Cost, $/ton |
|---|---|---|---|---|---|
| 100 | 40.50 | 3.00 | 1.50 | 0.50 | 45.50 |
| 400 | 30.20 | 1.85 | 0.38 | 0.12 | 32.55 |
| 600 | 22.80 | 1.23 | 0.26 | 0.09 | 24.38 |
| 1,000 | 19.50 | 1.15 | 0.24 | 0.08 | 20.97 |
| 1,500 | 15.25 | 1.00 | 0.23 | 0.07 | 16.55 |

Source: "Bell Tolls for Little Plants," *Chemical Week*, Oct. 28, 1967, p. 127.

Sometimes a company will build a larger plant because it expects the reduced prices will open new markets. This, of course, would not be done without some strong indications that the new markets would develop. In 1969, Hercules announced that nitrosyl chloride, which was then sold in cylinders at $5-$8 per pound, would be sold in bulk at $2-$3 per pound. It was also stated that when the sales exceeded 1,000,000 lb/yr the price was expected to drop below $0.50 per pound. They had built a plant that could produce 1,000,000 lb/yr but only had a market of 200,000 lb/yr. Their marketing experts estimated that the reduction in price could cause a five-fold increase in the demand in only three years.[5]

## Factors Limiting Plant Size

Sometimes technological problems forbid the size of an operation to exceed a certain value. For instance, in the batch polymerization of polyethylene and polystyrene, it is important to maintain the temperature below a critical value, because otherwise the material will be damaged. Since this is an exothermic reaction, it means the energy must be removed as fast as it is formed. If it is not, the temperature will begin to rise, which will increase the rate of polymerization. This will result in an acceleration of the temperature rise and the result will be a discolored batch. This requirement establishes a limit on the size of the reactor. The practical significance is demonstrated in the polystyrene case-study example following Chapter 5.

In other cases, it is the maximum possible size of the equipment itself that places limits upon plant capacity. For example, the size of glass-lined reactors (where the whole vessel must be annealed in an oven) is limited by the size of the largest oven available to the equipment manufacturers.

## Factors Favoring Small Plants

Despite all the advantages of large plants described above, the largest possible plant is not always best. One reason for not building a large plant is that the prospective sales volume would not warrant it. A plant is designed to run at full capacity. If it does not, the costs per pound increase.

In 1969, the chemical industry was operating at an average 81% of capacity. For ammonia the rate was even less than this. This meant that the large new plants were probably just breaking even, but the smaller, less efficient ones, which had to run at full capacity to show a profit, were losing money. The result was that about 20 of the smaller and less efficient ammonia plants were shut down.

The major disadvantage of large plants is their vulnerability to large losses. In 1967 an explosion and fire in a Cities Service oil refinery at Lake Charles killed 7 employees and injured 14.[6] The damage and business interruption costs exceeded $30,000,000. Usually the losses are not this large. However, in 1966 there were 20 fires in the chemical and petroleum industry, which caused damages in excess of $250,000.[7] Even if there is no fire, the failure of a bearing on an ammonia compressor can cause the plant to shut down for a number of days: two days for cool down, one day for repairs, two days for startup. The loss in sales from this interruption alone could exceed $50,000 per day, or a total of $250,000.[8]

This downtime can be avoided by installing alternate pieces of equipment that can be immediately put into service when an item fails. If this is done for all pieces of equipment, it is almost equivalent to building a spare plant. The cost of this would of course be prohibitive. Usually for large plants a spare piece of equipment will be installed only when the equipment is expected to break down frequently.

## Multipurpose Plants

Not all plants produce one major product. Some plants, called multipurpose plants, may use the same equipment to produce a number of different products. This might be true for insecticides or herbicides. The major market for these chemicals is in the spring and early summer. If another product could be produced for fall and winter consumption, such as a de-icer additive for gasolines, then the plant could be operated all year with a minimum amount of storage. Such a plant would probably not be the optimum plant for either process, but the company's over-all profits would be greater than if two separate plants were built.

A different but related problem arose in producing drugs to combat coccidiosis, a disease found in chickens. It was found that the protozoan parasites that cause the disease develop an immunity to the drugs given the chicken. However, if the drug is stopped the immunity disappears. The solution is to alternate two different drugs. The designer must now decide whether to build one bigger plant and produce each drug in alternate years, or build two smaller plants, one for each drug, and alternate marketing regions.

A more common decision concerns the production of a material made to different specifications in one plant. In the production of polyvinyl chloride (PVC) there are many different possible products. The average molecular weight may differ, as well as the range of molecular weights. It may be sold in pellets or powder, may or may not be colored, and it may or may not have certain impurities present. The permutations of these variables are infinite. However, practically, a company cannot cheaply make a product that is custom-tailored for each customer. To make an inexpensive product, it must be mass-produced. This means only a limited number of products can be offered for sale. Most manufacturers will make 10-30 different clear PVC resins. Each has different properties and each is put to a different end use. The manufacture of each of these resins requires different processing conditions.

When a multiproduct plant is constructed, the amount of each product to be made each year must be included in the scope. This is called the product mix. It is important because the product mix determines the size of much of the equipment. One resin may require a reaction cycle of three hours, while another takes six hours. If the majority of the product is the former resin, a smaller reactor is required than if the majority were the latter.

## QUALITY

The number and types of processing steps are directly related to the purity of the product and raw materials. This means the capital and operating costs are very dependent on the quality that is specified. The design engineer must determine whether it is cheaper to buy a relatively cheap, impure raw material and then purify it, or buy an expensive, clean feed stock that can be used without further modification.

When only small quantities of a chemical are needed, it is usually best to avoid any purification steps and buy a more expensive raw material. The manufacturer can generally purify the material at a lower cost per pound because he is processing large quantities.

When the buyer is planning to purchase large quantities of a raw material, it may be more economical if he specifies the cheapest raw material available. He can then selectively remove only those substances that may be harmful in his particular process. The raw material manufacturer, in this case, might be spending more per pound to purify the material, because he wants to produce a product that meets the requirements of a large number of customers. Generally, as the desired purity rises the cost for the purification operations increases exponentially.

The product specifications should be set at the time the capacity is decided. The product should be pure enough that it will have a wide customer appeal, and yet not so pure that the cost of the purification steps unnecessarily increases the price. The ideal specifications from the producer's point of view are those that permit the largest amount of product to be sold at a reasonable profit.

For some drugs, especially experimental ones, not only is each batch analyzed separately but it can never be combined with other batches. When it is packaged, it is labeled with the batch number and the analysis. For these pharmaceuticals a record must be retained of the complete production history, as well as of the results of all tests. All samples taken from the batch must also be kept. Then, should some odd reaction occur in a patient, all the information about the batch is available to assist research personnel in determining its cause. This required isolation, of course, increases costs.

## RAW MATERIAL STORAGE

The purpose of the raw material storage facilities is to make certain the plant never shuts down, under normal conditions, because of a lack of raw materials. The most important considerations are the nearness and reliability of the supplier and the size of the shipment. The designer must estimate for the proposed plant what will be the minimum and maximum elapsed time between the placing of an order for a raw material and its delivery to the plant site. The difference between these two times determines how much feed must be on hand when the order is placed. The maximum amount that can be present when the order arrives, plus the amount ordered, sets the minimum size of the storage facility. This is usually given in terms of hours or days, instead of pounds or tons (kg). The average transit time using different modes of transportation is given in Table 3-5.

Table 3-5

Average Transit Times for Chemicals in 1962

| Method | Miles/Day |
|--------|-----------|
| Pipeline | 65 |
| Barge | 100 |
| Rail | 200 |
| Ship tanker | 300 |
| Truck | 700 |

Source: Scheyer, R.H.: "Costs of Transporting Chemicals," *Chemical Engineering*, Sept. 3, 1962, p.158.

### Example 3-1

A plant is being designed that will require 20,000 lb of feedstock per day. A supplier has said he could guarantee that any order would be filled within 15 days of its receipt. It will be shipped by rail in 36,000-gallon jumbo tank cars. The time en route could be anywhere between 2 and 5 days. The specific gravity of the feedstock is 0.85.

If the supplier shipped the feedstock immediately, it could arrive in 2 days. He also could delay loading the tank car for 15 days, and the rail shipment could take 5 days. If the order were placed on the Saturday preceding a major holiday, there might be a 3-day delay before the supplier receives it. The maximum time between ordering and receipt of the feedstock would then be 23 days. The minimum time is 2 days. The number of days of feed a jumbo tank car will supply is:

$$36{,}000 \text{ gal} \times \frac{1 \text{ ft}^3}{7.48 \text{ gal}} \times (0.85 \times 62.4) \text{lb/ft}^3 \times \frac{1 \text{ day}}{20{,}000 \text{ lb}} = 13 \text{ days}$$

The total storage needed is $23 + 13 - 2 = 34$ days.

Instead of performing these calculations on smaller items like inhibitors, dyes, or catalysts, the storage requirements are usually set at either 60 or 90 days. One company, however, purchased an 11-year supply of a necessary additive. The chemical was used in very small amounts and the only manufacturer had announced it would cease producing it because of the small volume of sales. The company that needed the chemical then decided to buy a "lifetime" supply.

In large-volume plants the storage capacity might be a few hours or less. This is especially true of satellite plants supplied by pipelines. If the supplier has adequate storage, the plant does not need a large storage system any more than the average city resident needs a cold-water storage tank. One of the Proctor & Gamble detergent plants operates on this basis. It is supplied by rail and has only a few hours' storage for feed and products. The material is pumped directly into the reactors from the tank cars. The product, in consumer packages, is loaded directly onto boxcars for shipment to distribution centers nearer the consumer on a prearranged schedule. Should the railroad have a strike, the plant would be shut down.

In determining the size of storage systems the possibility of strikes or major disasters, such as fires, earthquakes, or riots, that may cut off feed stocks for a long period of time is not considered. The probability of such events occurring is not great enough to warrant the added expense that would be involved. If the disaster strikes only one plant, the raw material can usually be obtained from another source.

If it is expected that a supplier may have a strike, the company may stockpile items. Some needed materials like coal are dumped on a cleared piece of ground. Others may be stored in box cars or tank cars if the present storage facilities are full. This is expensive storage, but it is only a temporary situation. Actually, most chemical plants do not close if the hourly workers strike. The salaried employees run the plant, and the supply of chemicals is not stopped, although it may be reduced.

## PRODUCT STORAGE

Like raw material storage, the product storage must be large enough that it does not impede production. It is not very economical to shut a plant down because the warehouses are full and then four weeks later refuse orders because the plant cannot produce enough.

The amount of product storage depends on whether the product has a steady market or one that varies with the season of the year. Ammonia can only be sold for fertilizer in the spring and the early summer. Many fertilizer companies have tried to convince farmers that they should fertilize in the fall, and offer them a price reduction if they will buy in the fall. Soil scientists have, however, shown that even considering the price reductions, the farmer should not spread ammonia on his field at that time, as most of it is lost before the spring arrives. Therefore, any ammonia plant selling directly to farmers must have a storage volume greater than half the yearly plant capacity. The opposite extreme is the detergent plant mentioned previously, which only required a few hours' product storage.

For plants that produce a variety of different products using the same equipment, the storage requirements for each product can be different. If they are different, the product with the smallest volume should have the maximum days of storage. For instance, for a plant producing a variety of PVC resins, a 60-day supply of the smaller-volume resins may be desirable, whereas 20 days' supply of the larger-volume resins may be adequate. This is a compromise. If the goal of the designer were to minimize the storage space, the products should be produced as the orders are received. This would involve switching production from one product to another, very often. However, if the designer wished to maximize the throughput of good product, the number of these changes should be minimized. This is because each time a change of products occurs the equipment and lines must be cleaned out, which takes time, and there is a danger of the previous product contaminating the new one. Thus, when the number of changes is increased, more downtime occurs and the probability of a product not meeting specifications is greater. Increasing the number of days' storage for the lower-volume products decreases the number of times per year these products must be made. This decreases the downtime and the probability of off-grade product. It also does not substantially increase the storage facilities, because the production rate per year is small.

The amount of product storage is also dependent on the company policy. One sulfuric acid supplier promises a customer's order will be filled within 24 hours of its receipt. This requires a larger storage capacity than if the producer had promised the acid would be shipped within two weeks.

If the product is sold in different size containers, it may be stored mainly in bulk quantities and packaged after the receipt of a customer's order, or it can be packaged immediately after it is made. The type and amount of storage will be affected by the decision.

## THE PROCESS

The *unit ratio material balance* is often included in the scope. For discussion of this, see Chapter 4. Any information that might be useful for designing equipment of optimizing operating conditons should also be included.

## WASTE DISPOSAL, UTILITIES, SHIPPING
## AND LABORATORY REQUIREMENTS

The reason why waste disposal, utilities, shipping and laboratory requirements are included in the scope is to establish firmly which of these facilities must be constructed, which services will be bought, and which are already present. One company might contract to have a nearby city process its wastes through its sewage system. Another might build its own treatment system, while a third might be able to handle these wastes in its present treatment plant. The mistaken omission of these peripheral items can cause the capital costs to greatly exceed the estimates.

## PLANS FOR FUTURE EXPANSION

If a company's long-range plans call for the expansion of the proposed plant in a few years, the company may substantially reduce the cost of that expansion by making provisions for it in the current plans.

Anyone who has remodeled a house is very aware of this. Consider the example of a man who builds a house and doesn't have the capital to include a second bath. He realizes, however, that when his children become teen-agers he will definitely need another one. Therefore, if he is wise he will design the house so that there is space for another bath. He will also make sure that the hot and cold water lines, a drain, and a stack that can service the second bath are installed. These provisions for the future will only increase the cost of the house slightly. Then five years later when he installs the second bath he merely needs to buy the fixtures and attach them to the installed plumbing.

If no plans had been made in advance the owner would find the task of adding a bath much more inconvenient and expensive. Parts of the walls and floor would need to be torn up to install the necessary piping. These walls and floors would of course have to be replaced and stained or painted. In five years the walls would have faded so this would necessitate painting and/or staining a number of rooms.

When a plant is designed, a unit may be designed with excess capacity to facilitate a future expansion. For instance, consider a batch process that has a continuous purification step, such as distillation or extraction. If it is planned to expand this plant a few years after it starts up, it might be wise to put the extra capacity in the continuous unit and to plan a layout that will easily allow the addition of other batch units.

Building some units in a plant with excess capacity in anticipation of an expansion is not always wise, since it gives a strong economic advantage to plans for expanding production at the present plant. Should an expansion never occur, the money spent in planning for it would be wasted. It basically eliminates the possible marketing advantages that might result from building two widely separated plants. It also may economically favor the enlarging of an old plant when hindsight would show the best course of action to have been to make no plans for expansion, and eventually build a new plant using new technology.

## HOURS OF OPERATION

The operating hours per year are usually 7,900 hours (90% of 8,760, the number of hours in a year) for small plants involving new technology, and 8,300 hours (95% of 8,760) or more for large plants with well-documented processes. By operating 24 hours a day rather than 8 hours, the same throughput can be maintained with processing equipment one-third the size. Thus, the initial capital investment is greatly reduced. Also, for continuous processes and highly integrated processes, costly and time-consuming startups and shutdowns are minimized.

With small plants using batch processes this may not be the best policy. Just as there is a maximum size for standard equipment, there is a size below which the cost is almost constant. In this case, it may be more economical to buy larger equipment and save on manpower by having only one shift of workers. Often, in this low-volume region, by using larger equipment the same number of operators can process three times as much material.

## COMPLETION DATE

The completion date has an effect on the cost, if for no other reason than the change of prices and salaries with time. In the 1960s there was approximately a 2.5% increase in plant construction costs per year and a similar increase in equipment costs. Between 1970 and 1974, the rates were 3 and 5% respectively.

If the board of directors insists that the plant be rushed to completion at the earliest possible date for the reasons given in Chapter 2, it can greatly increase the cost of the plant. It will mean paying premium prices to speed up equipment delivery and doing things in a more expensive way to save time. Some examples are given in Chapter 13. It usually takes at least two, and often three, years after the scope is written before the plant is ready for startup.

## SAFETY

Safety considerations may require burying a tank, fire walls separating parts of the plant, a sprinkler system throughout, storage facilities 500 yd (450 m) from the processing equipment, and so on. This is made a part of the scope, to make certain from the beginning of the project that safety is a major design consideration. Placing it here also means that it is the concern of everyone involved in the project, not just the process engineer. By recording all the dangerous aspects of the chemicals and processes, it also reduces the probability that some safety feature will be inadvertently omitted. Such an omission could cause serious injuries to men and equipment. This topic is covered more extensively in the next chapter.

## CASE STUDY: SCOPE FOR A 150,000,000 LB/YR
## POLYSTYRENE PLANT USING THE SUSPENSION PROCESS

There are three large-volume types of polystyrene. These are general purpose (GPPS), sometimes called straight polystyrene; medium impact MIPS), containing about 5% rubber; and high impact (HIPS), containing up to 0.15 lb rubber/lb styrene. From Table 2E-2 it can be seen that around 50% of polystyrene produced is rubber-modified. However, at present the GPPS market is growing at a higher rate.[9] Therefore 60% of the product will be GPPS, 20% will be MIPS, and 20% will be HIPS.

### Polystyrene Storage and Shipping

The product will be shipped in bulk by truck and hopper cars as well as in 50 lb bags, 200 lb fiber cartons, and 1,000 lb boxes.[10] Since there are many small customers, it will be assumed that 30% is shipped in 50 lb bags, and 15% each in 200 lb cartons and 1,000 lb boxes. Most of these small customers will not have a large storage facility. To provide good service, a 60-day supply of each product stored in bags, cartons, and drums will be specified. A 25-day storage facility for bulk materials will be constructed.

### Physical Properties

The physical properties of polystyrene depend upon the specific reaction components, the mass ratios of the components, and the conditions at which the reaction occurs. These will be discussed later. The impurities remaining in the polystyrene also affect the properties. For instance, the heat distortion temperature may be as low as 70°C if there is unreacted styrene present. It is normally between 90 and 95°C. Therefore the maximum percentage of styrene that will be allowed in the product is 0.01%. Careful drying is also necessary if the polystyrene is to be extruded. For this application the polystyrene must contain a maximum of 0.03-0.05% water.[11] We will set 0.03% as the maximum amount of water allowed. The specifications for the polystyrene are given in Table 3E-1. Different types of rubbers may be used for making impact polystyrenes.[12] We shall use polybutadiene.

Table 3E-1

Chemical Composition of Product

|                  | GPPS    | MIPS    | HIPS    |
|------------------|---------|---------|---------|
| Polystyrene      | 99.96%  | ∿95%    | ∿88%    |
| Rubber           | 0.0%    | 5%      | 12%     |
| Water            | <0.03%  | <0.03%  | <0.03%  |
| Styrene          | <0.01%  | <0.01%  | <0.01%  |
| Other impurities | <0.01%  | <0.03%  | <0.03%  |

## Operating Hours

The number of operating hours per year will be assumed to be 8,300. This plant is large and the technology is well developed.

## Styrene Storage

The styrene will be obtained by barge from Louisiana (approx. 1,250 miles away). For GPPS, assuming a 3% loss in processing, the amount of styrene needed per hour is:

$$\frac{150,000,000\,(\text{lb/yr}) \times 1.03}{8,300\,(\text{hr/yr})} = 18,600 \text{ lb/hr}$$

Barges carry between 1,000 and 3,000 tons each.[13] On large rivers a single tow may consist of up to 12 barges. A barge containing 1,000 tons of styrene will last:

$$\frac{1,000 \text{ tons} \times 2,000\,(\text{lb/ton})\,(1 \text{ day/24 hr})}{18,600\,(\text{lb/hr})} = 4.5 \text{ days}$$

A 3,000-ton barge load will last 13.5 days. It is assumed that an agreement can be made with a supplier to ship the styrene within 5 days after the order is received and that after leaving Louisiana a shipment will take 10-15 days to reach the plant. Under ideal conditions it will take 1 day after the order is received to obtain a barge and load it. This means it will take a minimum of 11 days between sending the order and delivery of the styrene. If the order arrived at the styrene plant on the Saturday of a 3-day weekend and the maximum order delays occurred (5 days before a barge was loaded and the trip took 15 days) then the shipment would arrive 23 days after the order was sent. The difference between these times is 12 days. This means that when a large barge is used a storage capacity of 12 + 13.5 or 26 days is required.

For a smaller load of 1,000 tons the storage capacity should be 17 days. Since a barge shipment would be needed every 4.5 days, undoubtedly automatic ordering procedures would be instituted and some of the possible time delays could be eliminated. See Chapters 10 and 11 for methods to determine which size barge shipment is best. A 17-day styrene storage capacity will be assumed.

## The Suspension Process

There are many different ways of making polystyrene using the suspension process. Most producers use a batch process, although there are no technical reasons why a continuous process could not work.[10] For this study a batch processing scheme will be used. In the suspension process a number of small styrene drops 0.15-0.50 mm in diameter are suspended in water. The reaction occurs within these drops. To aid in the formation of the proper size drops a suspending agent is used, and to keep them at that size a stabilizing agent is added. A catalyst is used to control the reaction rate.

James Church[14] gives the following general information about the suspension process. Some typical suspension agents are methyl cellulose, ethyl cellulose, and polyacrylic acids. (Smith[10] in addition lists polyvinyl alcohol, sulfonated polystyrene, and polyvinylpyrollidone). Their concentration in the suspension is between 0.01 and 0.5% of the monomer charged. The stabilizing agents are often insoluble inorganics such as calcium carbonate, calcium phosphates, or bentonite clay. (Smith[10] in addition lists barium sulfate, calcium oxalate, and aluminum hydroxide). These are present in smaller amounts than the suspending agents. The catalysts are usually peroxides. The most common ones are benzoyl, diacetyl, lauroyl, caproyl, and tert-butyl. Their concentration varies from 0.1 to 0.5% of the monomer charged. The ratio of monomer to dispersing medium is between 10 and 40%. The parts per 100 parts of monomer that are typical for a polystyrene system are given in Table 3E-2 along with the temperature and cycle time. Anderson[15]

Table 3E-2

Typical Formulations Used for the
Batch Suspension Process for Polystyrene

|  | Parts | |
| --- | --- | --- |
|  | Smith* | Church** |
| Styrene | 100 | 100 |
| Water | 400 | 68 |
| Calcium phosphate | 0.2 |  |
| Tricalcium phosphate |  | 0.77 |
| Methyl cellulose | 0.5 |  |
| Dodecylbenzene sulfonate |  | 0.00256 |
| Diacetyl peroxide | 0.3 |  |
| Benzoyl peroxide |  | 0.204 |
| Reaction temperature (°F) | 194° | 190-200° |
| Cycle time (hours) | 6.5 | 3-4 |

\*   Smith, W.M.: *Manufacture of Plastics*, Reinhold, New York, 1964, p. 410 — as obtained from Grim patent issued to Koppers (U.S. Patent 2,715,118), Aug. 1955.

\*\*  Church, J.M.: "Suspension Polymerization," *Chemical Engineering*, Aug. 1, 1966, p. 79.

states the reaction time varies from 6 to 20 hours, depending on the desired product.

Obviously more than one recipe is successful. Some companies produce over 10 different kinds of polystyrene. None will reveal their reaction mix or conditions.

Ordinarily a company building such a plant would either have a research laboratory or production facility that can provide the correct formula and operating conditions, or it would purchase this information. For this case study a compromise between the two mixes of Table 3E-2 was made. The actual numbers are given in Figures 4E-1, 4E-2, and 4E-3.

## SCOPE SUMMARY

*Product*
> 90,000 lb/yr of general-purpose polystyrene
> 30,000 lb/yr of medium-impact polystyrene
> 30,000 lb/yr of high-impact polystyrene

*Product Purity*
> Given in Table 3E-1

*Product Package Size*
> 50 lb bags (30% of each product)
> 200 lb fiber drums (15% of each product)
> 1,000 lb cardboard cartons, plastic-lined (15% of each product)
> Bulk shipments by truck or hopper car (40% of each product)

*Product Color*
> 100% colorless product

*Product Storage*
> 60 days for bags, drums, and cartons
> 25 days for bulk storage

*Raw Materials & Purity*
> Styrene (see Table 3E-3)
> Hydroxy apatite (tricalcium phosphate) (tech grade)
> Dodecylbenzene sulfonate (tech grade)
> Benzoyl peroxide containing 50% water (see safety section)
> Polybutadiene (polymer grade)
> Hydrochloric acid

*Raw Material Storage*
> Styrene — 17 days
> Benzoyl peroxide — 15 days
> Hydrochloric acid — 30 days
> All other materials — 60 days

*Raw Material Obtained in the Following Size Containers*
    Styrene by barge in 2,200,000 lb shipments
    Hydroxyl apatite in 50 lb bags
    Dodecylbenzene sulfonate in 5 gal cans
    Benzoyl peroxide in 300 lb drums
    Polybutadiene in 50 lb bales
    Hydrochloric acid in tank truck shipments

Table 3E-3

Polymerization-Grade Styrene

| | |
|---|---|
| Styrene        99.6% | |
| Polymer      (none) | |
| Aldehydes as CHO | 10 ppm |
| Peroxides as $H_2O_2$ | 5 ppm |
| Chlorides as Cl | 10 ppm |
| Sulfur as S | 10 ppm |
| *p-tert* butylcatechol (TBC) | 12 ppm |

Source: Bikales, N.M. (ed.): *Kirk-Othmer Encyclopedia of Chemical Technology*, Wiley, New York, 1970,
        vol. 13, p. 144.

*Byproduct*

The 3% of the material which will be offgrade will be sold to toy manufacturers.

*Waste Disposal Requirements*

A primary and secondary treatment plant to handle all process water. Any solid waste that cannot be sold will be used for landfill. All air laden with polystyrene dust will be sent through bag filters before it is discharged to the atmosphere.

*Utilities Requirements*

Power will be purchased from a nearby company. (One of the major reasons for locating here was the presence of low-cost, plentiful power.) A gas-fired plant for 125 psig steam will be built. This must be able to supply enough power to operate agitators and cooling-water pumps associated with the reactors when there is a power failure. Gas will be purchased from a local company. Drinking water will be purchased from the community of Martins Ferry, Ohio. Process and cooling water will be obtained from the Ohio River. Both will require treatment before they can be used in the plant.

*Future Expansions*

A 50% expansion 5 years after startup is expected because of rapid growth in the use of polystyrene and the excellent choice of a site.

*Plant Location*

Martins Ferry, Ohio, on the Ohio River

*Operating Hours Per Year*

8,300

*Completion Date*

October 1974

*Shipping Requirements*

1. Railroad spur into the plant
2. Road into the plant (plant within a block of a paved road)
3. Barge dock capable of handling one barge
4. Ground level warehouse

*Laboratory Requirements*

1. Instron® for testing tensile strength, stress, strain, and creep characteristics
2. Injection molding machine for testing products
3. Impact testing machine (notched IZOD)
4. Ultracentrifuge for obtaining average molecular weight from viscosity measurement in 90% toluene
5. Extruder for testing products

*Safety Considerations*

The human threshold limit value for styrene vapor in air is 100 ppm. It was set by the American Conference of Governmental Industrial Hygienists. Repeated contact with the monomer can produce skin irritations.[16]

Styrene liquid and vapor are flammable. At room temperature styrene does not have a large enough vapor pressure to form an explosive mixture. However, above 86° F it can be explosive. Since polymerization can occur and create heat in a storage vessel, a refrigeration system should be used for bulk storage vessels when temperatures regularly exceed 80° F. The reactor area where styrene will be at elevated temperatures should be properly isolated from the rest of the plant to prevent possible fires or explosions. All equipment in that area should be grounded and explosion-proof motors should be used. All open flames should be banned in the styrene reactor and storage areas.[16]

Organic peroxides have a low toxicity. The diacetyl peroxides are sensitive to heat, friction, and shock and may detonate upon the slightest mechanical disturbance. Benzoyl peroxide dust may explode easily by friction. Anyone designing this plant should obtain and use the publication *Properties and Essential Information for the Safe Handling and Use of Benzoyl Peroxide.*[17]

Benzoyl peroxide as a pure solid is classified as a deflagration hazard. When it is a solid containing about 30% water it is an intermediate fire hazard. As a paste (50% peroxide) it is a low fire or negligible hazard. See reference 18 for a definition of hazard classifications. Benzoyl peroxide containing 50% water will be purchased. It should be stored in a separate cool area, since all peroxides have short half-lives.

## References

1. "Containers Cut Shipping Costs for Small Loads," *Chemical Week*, Dec. 23, 1970, p. 43.
2. *Jane's Freight Containers*, Sampson Low Marston, London, published periodically.
3. Raymus, G.J.: "Evaluating the Options for Packaging Chemical Products," *Chemical Engineering*, Oct. 8, 1973, p. 67.
4. "Bell Tolls for Little Plants," *Chemical Week*, Oct. 28, 1967, p. 127.
5. "Market News Letter," *Chemical Week*, May 17, 1969, p. 99.
6. "Chementator," *Chemical Engineering*, Aug. 28, 1967, p. 56.
7. "Huge Plants Add to Insurers' Anxiety," *Chemical and Engineering News*, Oct. 23, 1967, p. 30.
8. "Chementator," *Chemical Engineering*, Mar. 13, 1967, p. 86.
9. *Oil, Paint and Drug Reporter*, Apr. 18, 1966, p. 9.
10. Smith, W.M.: *Manufacture of Plastics*, Reinhold, New York, 1964, pp. 21, 424, 435.
11. *Kirk-Othmer Encyclopedia of Chemical Technology*, Ed. 2, Wiley, New York, vol. 19, p. 110.
12. Deland, D.L., Purdom, J.R., Schoneman, D.P.: "Elastomers for High Impact Polystyrene," *Chemical Engineering Progress*, July 1967, p. 118.
13. Hay, W.H.: *An Introduction to Transport Engineering*, Wiley, New York, 1961.
14. Church, J.M.: *"Suspension Polymerization,"* *Chemical Engineering*, Aug. 1, 1966, p. 79.
15. Anderson, E.V., Brown, R., Belton, C.E.: "Styrene — Crude Oil to Polymer," *Industrial and Engineering Chemistry*, July 1960, p. 550.
16. *Kirk-Othmer Encyclopedia of Chemical Technology*, op. cit., p. 72.
17. Chemical Safety Data Sheet SD81, Manufacturing Chemists Association, Washington, D.C., 1960.
18. Noller, D.C., et al., "A Relative Hazard Classification of Organic Peroxides," *Industrial and Engineering Chemistry*, Dec. 1964, p. 18.

# CHAPTER 4

# Process Design and Safety

Once the scope has been written (or often while it is being written), the process engineer begins the process design. Process design is the selection and ordering of the processing steps and the setting of process conditions. This is a highly innovative activity and is the portion of plant design where, potentially, the largest savings can be realized. By eliminating a processing step one saves all the equipment, maintenance, and processing costs that would have been incurred had that step been included. The proper placement of the various steps involved can result in smaller, less expensive equipment and fewer losses. The wise choice of operating conditions can eliminate the need for much expensive peripheral equipment such as refrigeration or vacuum-producing units. Through considering all the possible alternatives and selecting the best, savings of over $1,000,000 can often be realized during the process design stage.

## CHEMISTRY

Whenever chemical reactions occur these are the key to the process design. The engineer must be aware of what kinds of reactions are possible. He must also keep in mind that there are no such things as pure reactants, nor does the stream emerging from his reaction vessel ever contain just the desired product. Nearly always, a number of reactions occur and other products than those desired are produced. The engineer's purpose in investigating the reaction step is to increase the yields of desired products while reducing the quantity of unwanted substances.

To do this, not only must he know the chemistry of the reactions but he must know the rates at which the reactions occur and what affects those rates. The study of this is called chemical kinetics. By the proper choice of raw materials and operating conditions for the reaction stage the process designer can manipulate the ratio of products formed. One major variable is the temperature. An increase in temperature usually causes the reaction rates to increase, but some increase faster than others. Thus, the product mix in the reactor is dependent on the temperature. The pressure and the time the material spends in the reactor also affects the results. In the gaseous phase a high pressure will impede those steps in which the number of moles is increased and assist those in which the number of moles is decreased. A third way of modifying the reaction is to use a catalyst that selectively favors a given reaction. Other ways are to add inert materials or to change the feed ratios.

The reacting conditions, along with the raw material and desired product purity, determine the type and size of the separation steps that are required. Hence, their importance cannot be underestimated. Consider the selection of the operating conditions for a polymerization reactor. Since reactions generally occur more rapidly at higher temperatures, one way to reduce the total volume of the reactors and hence their cost is to increase the temperature. However, if a certain temperature is exceeded, charring will occur. Should this happen, an expensive processing step would need to be installed for removing the charred material, and then some scheme would need to be devised for disposing of this material. If these are the only considerations, then the reaction temperature should be set and closely controlled just below the point where charring first occurs.

In the production of the herbicide 2,4,5-T (2,4,5 tetrachlorophenoxyacetic acid), a dioxin (2,3,7,8 tetrachlorodibenzodioxine) is often formed. Not only has it been called the second most lethal chemical ever discovered but it can also produce birth defects. During the early 1970s there were court battles over whether 2,4,5-T should be banned because of the possible presence of the dioxin. The Dow Chemical Company maintained that no detectable dioxin was produced in its process and that therefore its product was safe and should not be taken off the market. In this instance, the removal and hence concentration of the dioxin would pose problems of such a magnitude that the only feasible reaction conditions are the ones that produce no dioxin.

## SEPARATIONS

Once the reaction conditions have been decided upon, the feed preparation steps and product purification steps must be determined. The designer must decide how much of which compounds must be removed from the feed and product streams. The latter has already been set by the product composition specified in the scope. The former is often determined by how the impurities affect the reaction. For instance, when platinum catalysts are used all sulfur and heavy metals must be removed or this very expensive catalyst will be poisoned.

### Types of Separations

All separations are based on a difference in some property. The separation of the compounds given in Table 4-1 is done by distillation. It is based on the fact that compounds with different vapor pressures will have different compositions in the vapor and liquid phases. The magnitude of this difference, and hence the ease of separation, is directly related to the difference in the vapor pressures. This can be determined from the boiling-point differences. Among the six groups of compounds

Table 4-1

Commercial Yields from a Cracking Furnace
Used in Production of Ethylene

| Compounds | % of Output | Boiling Point |
|---|---|---|
| $CH_4$ & $H_2$ | 24 | $< -161°C$ |
| $C_2H_4$ | 33.5 | $-104°C$ |
| $C_2H_6$ | 6.0 | $-89°C$ |
| $C_3H_6$ | 12.5 | $-48°C$ |
| $C_3H_8$ | 10.0 | $-42°C$ |
| $C_4$'s and heavier | 14.0 | $> -5°C$ |

Source: *Kirk-Othmer Encyclopedia of Chemical Technology*, Ed. 2, Wiley, New York, vol. 8, p. 509.

given in Table 4-1 the most difficult separation will be that of propylene from propane.

A list of separation processes and the properties that are exploited by them is given in Table 4-2. Just as for distillation, for all the processes the greater the difference in the magnitude of the separative property, the easier it is to perform the separation. Of course, various complications can negate this generality. For instance, the formation of an azeotrope or the tendency to foam may eliminate the use of distillation even though there is a reasonable difference in boiling points.

The engineer is charged with deciding which of the separation processes should be used and what the process conditions should be. The most popular separation process in chemical engineering is distillation. The most data are available for it, and it has been the most extensively studied. Also, when there are no complications, the temperature difference is at least 10°F (5°C), and it can be performed at reasonable temperatures and pressures, it is usually the least expensive.

As a rule it is best to avoid separation steps, such as extraction, that involve the addition of a compound not already present. The reason for this rule is that extra processing steps will be required to recover the foreign compound so it can be reused. This means it is a costly operation. Whenever possible the number of processing steps should be reduced, not increased.

Other separations to avoid are those requiring either very high or very low temperatures and pressures, because these operations frequently involve expensive equipment. Where this cannot be avoided and there is a choice, high pressures and temperatures are preferable to vacuum and cyrogenic operations.

The conditions at which the separations are performed depend on the properties of the materials. Let us suppose we wish to separate n butane from n pentane. Table 4-3 gives the boiling points of these compounds. When possible the pressure in a distillation column is usually kept close to atmospheric. Since all multistage distillation columns require reflux, which is obtained by condensing the exiting vapor stream, if the top of the column were producing nearly pure butane the condensing temperature would be around 31°F($-1°C$). To obtain condensation at this temperature a coolant is needed at a temperature at least 10°F (5°C) cooler. This means that

Table 4-2

Physical Properties (in Addition to Diffusivity)
on Which Separation Processes Are Based

| Property | Separations Based on the Property |
|---|---|
| Vapor pressure | Distillation, sublimation, evaporation |
| Solubility | Crystallization, gas absorption, leaching |
| Solubility and density | Liquid extraction |
| Chemical affinity (Van der Waal bonding) | Adsorption, hypersorption, chromatography, foam separation |
| Adsorption and electrical charge | Ion exchange |
| Electric charge | Electrodialysis, electrolytic ion exchange |
| Molecular size and shape | Molecular sieves, membrane permeation |
| Vapor pressure and velocity | Molecular distillation |
| Velocity | Gaseous diffusion, thermal diffusion |
| Particle size | Filtration, sieves |

Source: Henley, E.J., Stauffin, H.K.: *Stagewise Process Design*, Wiley, New York, 1963, p. 5.

Table 4-3

The Boiling Point in Degrees Centigrade
of 1-3 butadiene, n butane, and n pentane

| Compound | Vapor Pressure psia (mm Hg) | | | | | | | |
|---|---|---|---|---|---|---|---|---|
| | 3.87 (200) | 7.74 (400) | 14.7 (760) | 29.4 (1,520) | 73.5 (3,800) | 147 (7,600) | 294 (15,200) | 441 (22,800) |
| 1-3 butadiene | —33.9 | —19.3 | —4.5 | +15.3 | 47.0 | 76.0 | 114.0 | 139.8 |
| n butane | —31.2 | —16.3 | —0.5 | +18.8 | 50.0 | 79.5 | 116.0 | 140.0 |
| n pentane | 1.9 | 18.5 | 36.1 | 58.0 | 92.4 | 124.7 | 164.3 | 191.5 |

Source: Perry, J.H. (ed.): *Chemical Engineer's Handbook*, Ed. 4, McGraw-Hill, N.Y., 1963, Section 3.

an expensive refrigeration system is required. To avoid this the pressure at which the distillation column is operating should be raised. Suppose cooling water is available even during the hottest summer months at 85°F (29.4°C). Then if a 20°F difference between the coolant and the condensing pentane is allowed, the condensing temperature should be set at a minimum of 105°F (40.5°C). This means that pressure of around 4 atmospheres should be specified at the top of the tower.

If instead we are separating n butane from 1-3 isobutane, a similar problem arises (see Table 4-3) However, this situation is complicated by the fact that there is a small difference in boiling points — a maximum of 7.2°F (4.0°C) — and this difference is also a function of the pressure within the distillation column. The easiest separation occurs at a pressure of 1 atm. For this situation an economic evaluation would need to be run to determine whether to operate the column at 1 atm, so that a smaller column could be used but refrigeration equipment is needed, or to run at a pressure of 4 atm in a large column without refrigeration equipment.

In some cases a material decomposes at a given temperature. If this decomposition is to be avoided, the temperature at which processing occurs must be kept below that point. When distillation is used, the hottest point is in the reboiler. The only way to be absolutely sure that no point in the reboiler exceeds the decomposition temperature is to make certain the temperature of the heating medium does not exceed it. The temperature of the heating medium sets the temperature at the bottom of the column, and from this the pressures and temperatures within the column may be estimated (see Chapter 8). Under these circumstances, the column may have to operate at below-atmospheric pressure.

For some substances safety considerations play a large role in deciding operating conditions. For instance, the pressure within equipment containing a toxic substance may be kept below atmospheric pressure to prevent it from coming in contact with employees. On the other hand, some highly explosive materials may be kept a pressure above atmospheric so that air cannot enter the equipment and cause an explosion.

**Order for Separations**

The order in which separations are performed can vary, but some general rules can be set forth.[1] First the corrosive or hazardous materials should be separated out. Next, the separation steps that remove large quantities of materials or divide a stream into two or more large-volume streams should be considered. These steps, by reducing the amount of material in a stream, reduce the size of the following separation equipment.

In the processing of petroleum the first step is the removal of salt water. The presence of salt water in any processing steps would mean that expensive corrosion-resistant materials are required for those steps. This would greatly increase the price of the equipment (see Chapter 9). After removing the salt water, the next major separation is the crude still where the feed is split into six or more large-volume streams to reduce the size of future processing equipment.

No unnecessary separations should be made. In a refinery no attempt is made to separate the streams into their individual compounds. Instead, several streams each containing a number of compounds are produced. These are blended together to produce a desirable product. If instead a complete separation of compounds had been made followed by a recombination of them to produce gasoline, fuel oil, aviation fuel, lubricating oils, and so forth, the cost of the end products would be much greater. Partial separations followed by blending are more economical than total separations.

As a general rule difficult or expensive separations should be performed last, since by that time less total material will be involved. Consider Table 4-1, which gives the product mix obtained in a cracking furnace of an ethylene plant and the normal boiling points of the compounds. Suppose it is desired to separate the six groups listed in the table using distillation. The separation of ethylene from ethane and propylene from propane will be the most difficult because they have the smallest boiling-point differences. Therefore, these steps should be performed last.

As a different type of situation consider the following: suppose a feed has an impurity that has nearly the same boiling point as a reactant. Further, suppose the impurity does not affect the reaction and is itself unchanged by the processing steps. Here it may be wise to perform the distillation after the reaction has occurred. The advantage of doing this is that after the reaction step the total amount of reactant plus the inert substance will be less. Therefore less total energy is needed and smaller equipment can be purchased.

## UNIT RATIO MATERIAL BALANCE

After the processing steps have been selected and ordered, the amount and composition of each process stream entering and leaving each unit must be specified. This is an accounting procedure. It assumes a steady-state operation. That is, at any point in space there is no change occurring with respect to time.

If we neglect the case of nuclear reactions, this means a balance must be run over every chemical element that is present. However, when no chemical reactions are occurring in a given unit, a component rather than an element balance is run. This is then checked by running an over-all balance to determine if the total amount of material that enters each unit also leaves each unit.

The unit ratio material balance is based on the production of one pound of salable product. This basis is used because it is independent of the plant size and because the use of numbers near one minimizes the possibility of future calculation errors.

The material balance is presented on a block flow sheet so that the reader can graphically visualize what is happening. An example is given in Figure 4E-1. Each major operation appears as a block. No attempt is made to identify the specific pieces of equipment or to size them. The blocks are interconnected with flow lines, which indicate for each substance where it enters the process, what path it follows, and where it is eventually discharged. These flow lines are keyed to a chart that gives the composition and amount of each stream in the form of a unit ratio material balance. A material balance should be given for each product made by a multipurpose plant.

Whenever there are not enough data available to complete the material balance, the engineer should determine whether his company has any currently operating plants that have similar processing steps. If so, by assuming his plant will be similar he can at least obtain a ballpark estimate. This should be checked with the engineers who operate the existing plant.

When the object of the process engineer's study is to estimate the processing costs of a competitor's process, he may find it especially difficult to find quantitative data. Under these circumstances he should assume high yields and low losses. Then if the results show his company's proposed process superior to its competitor's, he knows his company is in a good economic position.

Even for older operating processes data are often lacking. It may be well known that the over-all material losses are 5%, but how these are distributed between the various operations may still involve a large amount of guesswork. This is usually further complicated by proposed innovations which do not appear in the older plant.

## DETAILED FLOW SHEET

After completion of the unit ratio material balance, a detailed flow sheet is constructed. This is a sketch of the system that shows all the equipment that is necessary to operate the plant, all process lines, and indications of where utilities are needed. It is not drawn to scale nor does it show spatial relationships. It includes all pumps, agitators, air filters, heat exchangers, hoists, elevators, lift trucks, blowers, and mixers as well as distillation columns, reactors, storage tanks, unloading docks, and steam boilers. Generally anything as large or as expensive as a pump is included. Piping and electrical details are excluded.

Figure 4-1 gives the symbols that should be used. If the equipment is not given in the symbol list, it is drawn to look like itself.

The items on the flow sheet are coded by letter and number, so that the equipment on the flow sheet can be identified with a specific item in the equipment list. Table 4-4 gives the code letter associated with a specific type of equipment. The numbers following the letter may be just a sequential listing in no specific order or the first number can refer to a given area of the plant. For instance, the feed storage areas might be designated 0, the feed preparation area 1, the reactor area 2, and so on. The advantage of the latter method is that it permits the engineer to determine more quickly the specific location on the flow sheet of a given item in the equipment list. This can be important, since plants contain thousands of items. The coding in Figure 4E-4 follows this method.

The flow sheet allows the engineer to visualize what is occurring — to follow the incoming material from the time it enters the plant area through storage, purification, reaction, separation, and packaging until finally it leaves as a finished product. At each step the process engineer must mentally place himself in the plant to be certain that nothing is omitted.

For instance, if the material arrives via tank car he must visualize how the material is going to get to the storage tank. First he must realize that most fluids are removed through openings in the top of a tank car. The bottom openings are used mainly for washing out the cars prior to their being refilled. Acid tank cars by law do

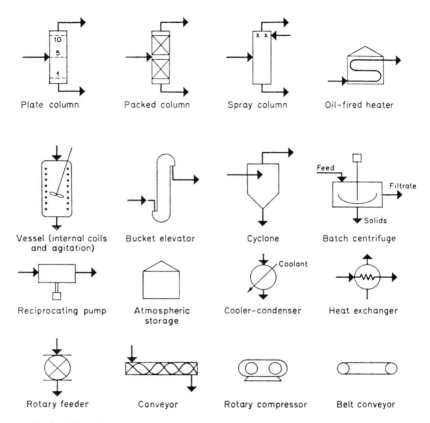

Figure 4-1    Typical flowsheet symbols.
Courtesy Backhurst, J.R., Barker, J.H.: *Process Plant Design,* Heinemann Educational Books, Ltd., London, 1973.

not even have a bottom opening.[2] When a pump is used to transfer the material, the engineer must determine how the pump can be primed. If air pressure is to be used to push the material out, he must decide if the air is to be treated before it can be released to the atmosphere. It probably contains some vapors that would contaminate the atmosphere. Does this require a scrubbing system? This close scrutinization is necessary for each step in the process.

A system for unloading tank cars of a low-boiling compound such as methyl chloride is given in Figure 4-2. The forwarding pump shown in the figure is not necessary, but its presence is desirable, since it speeds up the unloading process. A feasible alternate system would be to use an inert gas instead of compressed methyl chloride. Air cannot be used, because it forms an explosive mixture with methyl chloride if the percentage of air is between 8.25 and 18.7%.[3] Methyl chloride is also highly flammable (the flash point is below 32°F (0°C)). After unloading, the tank car

Table 4-4

Some Symbols Used in Equipment Lists

| | |
|---|---|
| Ag | Agitator |
| B | Blower |
| C | Compressor |
| CT | Cooling Tower |
| CV | Conveyor |
| CY | Cyclone |
| D | Drum or Tank |
| DR | Dryer |
| E | Heat Exchanger |
| F | Fan |
| Fi | Filter |
| M | Motor |
| P | Pump |
| R | Reactor |
| RV | Rotary Valve |
| S | Separator |

must be left under positive methyl chloride or inert gas pressure. This is to prevent air from leaking into the tank car and forming an explosive mixture. For handling such flammable compressed gases the Interstate Commerce Commission (ICC) has set up specific regulations (ICC Sec. 174-560 to 174-563). For other flammable, explosive, or toxic compounds the ICC has regulations also.

When the engineer is designing storage facilities he must consider the climatic conditions. For storage systems, the designer must determine whether a heater is necessary for high-boiling compounds to prevent freezing in winter. For low-boiling compounds he must decide whether a condenser should be installed on the tank to lower the pressure on hot summer days.

In some cases the decision whether storage vessels will be equipped with a vapor recovery system has been determined by the United States Environmental Protection Agency (EPA). In 1973 it set the standards[4,5] for all petroleum liquids that are stored in vessels of more than 65,000 gal (245 m$^3$). It states that if the vapor pressure is greater than 11.1 psia (570 mm Hg) a vapor recovery system or its equivalent must be installed on any new tanks. If the vapor pressure is between 1.52 psia (78 mm Hg) and 11.1 psia (570 mm Hg), a floating head tank may be used or a vapor recovery system may be installed. Since the former is cheaper it will usually be selected. Below 1.52 psia (78 mm Hg) only a conservation vent or its equivalent is required.

The EPA is developing a whole series of standards and will be updating the present ones. To keep up with these changes the engineer must yearly obtain a copy of the *Code of Federal Regulations*, Title 40 (Environmental Protection Agency), Chapter 1, part 60.[6] He can keep abreast of interim changes by checking a supplement to the *Federal Register*[7] that lists all the changes in the *Code of Federal*

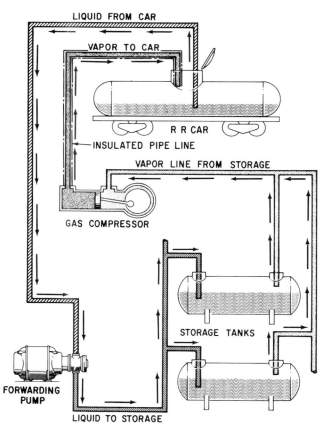

Figure 4-2   A typical tank car unloading system for unloading compounds with low boiling points. Courtesy Dow Chemical Company, U.S.A.

*Regulations* that have occurred since it was last issued and tells where in the *Federal Register*[8] these changes can be found.

One of the most effective means of transferring powder from one location to another is a pneumatic conveying system. In these systems the powder is introduced into a moving air stream. The air carries it to its destination, where the solids are separated out. The designer must determine how to introduce the powder to the air stream, which is under pressure. He must design it so that the air cannot escape into the feed tank and blow the feed out the top. Not all the powder will be removed at its desired destination, so a filter must be installed before the air can be discharged from the system. Care must also be taken so that the powder does not get into the blower, where a spark could ignite it. A series of flow diagrams for a pneumatic conveying system are shown in Figures 4-3 through 4-7.[9]

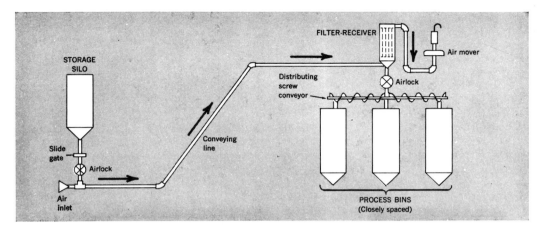

Figure 4-3 A vacuum system for transferring solids from storage silos to one receiver.
Courtesy Kraus, M.N.: "Pneumatic Conveyors," *Chemical Engineering,* Oct. 13, 1969, p. 60.

## SAFETY

It is during the construction of the detailed flow sheet that safety begins to affect the design. In the scope some concerns about safety were expressed. These and other general principles are now put to use.

Whenever powders are transported, they should be tested to determine whether there is a strong possibility of a fire or explosion. If the potential exists and there is a high probability of extensive damage, then a preventive device must be able to detect and snuff out the explosion in less than 0.1 sec after the initial blast occurs.

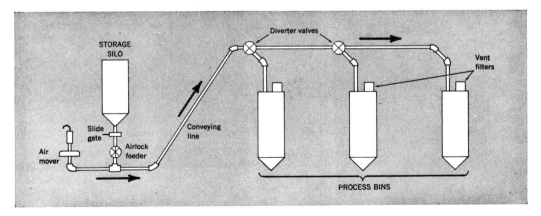

Figure 4-4 A positive pressure system which can deliver product to many receivers.
Courtesy Kraus, M.N.: "Pneumatic Conveyors," *Chemical Engineering,* Oct. 13, 1969, p. 61.

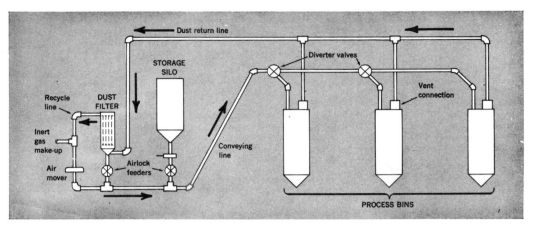

Figure 4-5    A closed loop system which can operate under either vacuum or pressure.
Courtesy Kraus, M.N.: "Pneumatic Conveyors," *Chemical Engineering,* Oct. 13, 1969,
p. 61.

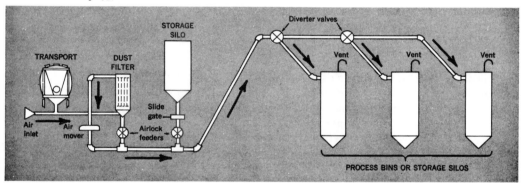

Figure 4-6    A combination vacuum-pressure system. A vacuum is used to withdraw material from a
hopper car and positive pressure is used to transport it to storage silos.
Courtesy Kraus, M.N.: "Pneumatic Conveyors," *Chemical Engineering,* Oct. 13, 1969,
p. 61.

Fenwal Incorporated makes such a system. It consists of a series of pressure or
radiation detectors and suppressors. A detector, after receiving the signal that an
explosion has begun, fires a suppressor (often water and bromo-chloromethane) at
speeds up to 600 ft/sec (200 m/sec), which snuffs out the explosion or fire. Thus a
damaging explosion is prevented by an explosion. The equipment usually must be
cleaned out if a suppressor discharges, and often the batch of material being
processed must be discarded because of the resulting contamination. It does,
however, prevent damage to the equipment and nearby personnel. The disadvantage is that the full system may cost over $100,000.

As another example, consider vessels that are under a positive pressure. For all
pressure vessels, including storage tanks, a vent system must be installed to protect
the vessel from rupturing. The vent goes to either a flare or a blowdown tank. These

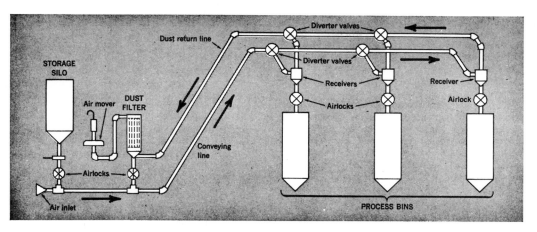

Figure 4-7   A parallel dust return system which permits a vacuum system to distribute material to more than one receiver.
Courtesy Kraus, M.N.: "Pneumatic Conveyors," *Chemical Engineering,* Oct. 13, 1969, p. 60.

are located a distance from the other pieces of equipment and in an unfrequented area. The vent system is designed so that at a given pressure a pressure-relief device will release. This opens the vessel to a specially sized vent line. To design the vent line it first must be determined from chemical reaction rates and heat transfer rates how much gas must be removed per unit time in order to prevent the pressure in the vessel from exceeding the design pressure .In doing this one should assume that the gas is leaving at a velocity of half the speed of sound. This gives a safety factor of two to the design. The pressure drop in the line at these conditions is assumed to be the design pressure minus atmospheric pressure.

Unless the precautions given in the previous two examples are taken, some employees could be injured or killed. Even if this did not occur the financial loss could be large, since if critical pieces of equipment were damaged it might take months before they could be repaired or replaced and before production could be resumed. For each day the plant does not run a number of expenses continue (salaries, depreciation, taxes, insurance). There is also the problem of supplying customers. If a customer goes to another supplier it may be difficult to lure him back. As a result, the company may furnish him with product made at a distant plant and not charge him the extra transportation expenses, or may buy a competitor's product and sell it to the customer at less than the purchase price.

Insurance costs also depend on the safety precautions taken. When a homeowner buys fire insurance the cost depends on the type of construction and the nearness and effectiveness of fire protection equipment, as well as the value of house and its furnishings. To determine the effect of these factors a statistical analysis is made of the factors contributing to fire losses. Insurance companies are noted for not losing money, and so rates also depend on the past record of those insured. In determining automobile insurance rates the age and the number of previous accidents and traffic violations are considered. In fact, people who have had a number of accidents find it very difficult to obtain insurance at any price. The same is true with industries.

But still, the most compelling reason for having a safe plant should be to protect the employees. Industrial corporations have not always felt this way, but they have now been forced to accept the Williams Steiger Occupational Safety and Health Act of 1970 (OSHA).[10] The Congress of the United States declared the purpose of the act was "to assure as far as possible every workingman and woman in the nation safe and healthful working conditions and to preserve our human resources." It requires that "each employer shall furnish to each of his employees employment and a place of employment which are free from recognized hazards that are causing or are likely to cause death or serious physical harm to his employees."

Under this act the government is charged with setting up standards and checking to see they are followed. Anyone can request that a plant be inspected to see if it is in violation of the rules. If violations are found the company must make whatever changes are requested and may also be fined. Failure to correct possible injurious conditions can result in the plant being closed.

This act places all responsibility for safety on the employer. If the law requires that a hard hat or safety goggles be worn when a given task is performed, and a worker who has been issued these devices refuses to wear them, the company can be fined and given a citation.

Under this act the designer is charged with building an inherently safe plant. He is charged with building it in accordance with the best safety standards available. No plant should be designed that requires employees to wear earmuffs or ear plugs or requires that temporary barriers should be erected for safety purposes when safety can be achieved by some other means.

To assist the engineer in this effort a large number of organizations have developed safety standards and suggestions. In a series of three articles Burklin[11] lists these organizations and discusses their functions. He lists the subject areas in which they may be of assistance and lists a number of their publications that might be useful to the process engineer. Many of the codes developed by these organizations have been adopted as federal standards.

On May 29, 1971, the standards of the OSHA act were published in the *Federal Register*. A revised version of these was published on October 18, 1972.[12] This includes regulations on the exposure of employees to hazardous chemicals. For some chemicals an absolute maximum concentration that may be present in the atmosphere is given. For others the standard merely says the 8-hr weighted average may not exceed a given level. For some substances both limits are given, plus a peak limit for a given period of time. For instance, benzene has an 8-hr weighted average maximum of 10 ppm and an accepted ceiling concentration of 25 ppm. However, for 10 min of an 8-hr shift it is permissible for the concentration to reach 50 ppm.

The code also sets maximum noise and radiation exposure levels. It requires that all stored liquified petroleum gases have an agent added that will give a distinct odor as a warning against leaks. It gives codes for the storage of flammable or combustible liquids. For these substances it prescribes the minimum distances between storage vessels, between the vessels and the property line, and between the vessels

and buildings. These distances depend on whether the material has boilover characteristics, whether the liquid is unstable, the type and size of tank it is to be stored in,
the type of protection provided for the tank, and whether the tank is above ground
or buried. Standards are also given for compressed gases such as nitrous oxide,
hydrogen, oxygen, and acetylene. Specific regulations for such industries as tex-
tiles and pulp and paper are promulgated.

These regulations must be compared with those of the Environmental Protection
Agency. Since the goals of the two federal agencies are different it cannot be
predicted which standards will be more stringent. All that can be said is that both
must be met.

The degree of detail present throughout the OSHA regulations can be illustrated
by the following items from the section on sanitary standards. There is a requirement that toilet facilities should be located within 200 ft (70 m) of an employee's
normal work area and that he should not be required to climb more than one flight of
stairs to reach them. The toilet facilities must provide hot and cold water, hand
soap, and towels. Not only does it state the ratio of water closets to persons but it
states that the walls must be at least 6 ft high and must be a maximum of 1 ft off the
ground. Further, each cubicle must have a door latch and a clothes hanger.

When toxic materials or injurious dusts may be present, it specifies that a
separate lunchroom must be provided. If the maximum number of people using the
lunchroom is less than 25, 13 ft$^2$ (1.4 m$^2$) per person must be provided. The required
square footage gradually decreases to 10 ft$^2$ (1 m$^2$) when the number of people
exceeds 150.

Just as with the Environmental Protection Agency (EPA) standards, these will
also be added to and revised. They are given in the *Code of Federal Regulations,*
under Title 29 (Labor), Chapter 17, Part 1910.[13] To keep up to date one should
follow the same procedure given previously for EPA standards.

To enable the government to determine where hazardous areas may be, each
company is required to keep track of all injuries and fully document their causes and
what is being done to prevent a recurrence. For comparative purposes the company
usually also determines the disabling injury frequency rate. This is the number of
lost-time injuries per million man-hours worked. A lost-time injury occurs
whenever an employee is unable to report to work at his next scheduled time as a
result of an accident that occurred while he was working. A lost-time injury could
occur if an employee carrying hot coffee tripped and burned himself. Another
lost-time injury would result if a person were killed in a boiler explosion. This
method of accounting does not take into account the severity of the injury. One
graphic way of looking at this statistic is to consider that the average person works
40 hours a week or approximately 2,000 hours a year. Then a lost-time injury rate of
5 would set the probability of any employee having a lost-time injury at 1% per year.
In an attempt to measure the gravity of the injuries, a severity rate is calculated in
terms of days an employee is unable to report for work due to injuries per million
man-hours worked. To compute this, fatalities and permanent total disabilities are
arbitrarily assessed at 6,000 days per case. Also, when permanent impairment of
some employees' facility occurs, in addition to the days missed a small allowance is

added to account for any loss in the employees' efficiency that may be due to the impairment. The record for a selected group of industries is given in Table 4-5.

According to the Bureau of Labor Statistics, the injury frequency rate for all manufacturing companies rose from 11.8 in 1960 to 15.3 in 1970. The National Safety Council estimates that there are around 15,000 job-related deaths each year and another 2,300,000 workers suffer disabling injuries. The total cost associated with these accidents is nearly $9,500,000,000/yr.[14] These figures are conservative, since they include only those companies that belong to the National Safety Council, and these companies are considered the most safety-conscious ones.

Table 4-5

1970 Safety Record for Various Industries

| Industry | Frequency Rate Disabling Injuries per 1,000,000 Man-Hours | Severity Rate Time Charges per 1,000,000 Man-Hours |
|---|---|---|
| Federal civilian employees | 6.6 | 554 |
| Electric, gas, and sanitary services | 6.6 | 813 |
| Electrical machinery equipment and supplies | 8.1 | 333 |
| Chemicals and allied products | 8.5 | 562 |
| Textile-mill products | 10.4 | 579 |
| Petroleum refining and related industries | 11.3 | 1116 |
| Paper and allied products | 13.9 | 937 |
| Machinery (except electrical) | 14.0 | 583 |
| Manufacturing (U.S. average) | 15.2 | 759 |
| Primary metal industries | 16.9 | 1128 |
| Rubber and miscellaneous plastics products | 18.6 | 795 |
| Fabricated metal products | 22.4 | 1003 |
| Metal mining and milling | 23.7 | 3238 |
| Stone, clay, and glass products | 23.8 | 1540 |
| Nonmetal mining and milling | 24.1 | 2624 |
| Contract construction | 28.0 | 2100 |
| Food and kindred products | 28.8 | 1156 |
| Coal mining and preparation | 41.6 | 7792 |

Source: *Statistical Abstract of the United States — 1973*, U.S. Department of Commerce, U.S. Government Printing Office, Washington, D.C., 1973.

For their own benefit, companies record not only injuries but near misses. These are accidents that could have, but did not, result in a lost-time injury. Upon analyzing these, problem areas can be discovered and improvements made before a major disaster occurs. It can also be determined which men are accident-prone.

These are people whose psychological makeup causes them to attract trouble. When such a man is detected, he must be placed in a position where the probability of his being injured or causing an injury is very low.

In order to promote safety, companies have contests, conduct periodic safety meetings, give prizes, award plaques, and conduct big advertising campaigns. Managers feel that making the employee aware of safety will help improve the company's over-all performance. Experience has shown them to be correct.

Not only does the government have standards but, as noted previously, before a plant is insured the insuring company requires certain safeguards. Insurers will also suggest many others that, if adopted, may result in a lower rate. Since it is impossible to design a plant in which no accidents can occur, the engineer must always weigh the cost of the unrequired safeguards against the probability of an accident. If the cost is high and the probability is low, he may accept the risk. The insurance companies will naturally try to get as many of these measures installed as possible, since this reduces their risks.

Even with all this help the engineer must still scrutinize his plant and try to anticipate what type of losses can occur. This study usually begins by noting that nearly all losses are due to explosion, fire, and/or mechanical failure. Then the possibility of each type of loss is evaluated.

In determining the probability of fire and explosions it is important to be able to classify chemicals by risk. According to these classifications, various appropriate safeguards can be installed. The Dow Chemical Company has developed a process safety guide[15] in which a Fire and Explosion Index is calculated for each unit of the plant. This unit may be a processing area such as the reaction area or a physical region such as a storage area or a finishing building. The index is based on the *material factor* of the most hazardous material present in significant quantities. "Significant" means that the substance is present in a high enough concentration to represent a true hazard. For instance, when a hazardous material is a reactant it may be present in such small concentrations after the reaction-vessel stage that it poses no threat to man or equipment. It then would be the most hazardous substance only in the storage, feed, and reactor areas, and another material would be the most hazardous in the other areas.

The *material factor* is a number between 1 and 20 that indicates the susceptibility of a compound or mixture to fire or explosion. A list of these factors for specific compounds is given in reference 15 along with their flash points, autoignition temperatures, and explosive limits. This factor is then adjusted for special material hazards such as the presence of oxidizing materials or spontaneous heating, general process hazards such as reactions or physical changes, and special processing hazards such as high or low pressures or temperatures. The result is the Fire and Explosion Index.

The protective features recommended depend on this index and are given in reference 15. For instance, for an area subject only to fire an explosion or blast wall is not required unless the Fire and Explosion Index exceeds 40. For an index below 20 a ventilation rate that would result in total change of the air in the building every

30 min is adequate. If the index is between 20 and 40 the air should be completely changed every 6 min. When the index exceeds 40 this should be done every 4 min.

The National Fire Protection Association classifies liquids by their explosion and flame-propagation abilities.[16],[17] These ratings are then used to specify the type of electrical equipment required. These standards have been adopted by OSHA. Woinsky[18] gives a procedure for obtaining the material classifications of individual compounds and mixtures.

The Manufacturing Chemists Association uses a different method.[19] It determines an explosion and a fire-hazard classification for each unit. Browning[20-23] presents a method for calculating the possible losses to a plant based on the MCA code and the determined probability of a mechanical failure. He calls this Systems Safety Analysis. It will determine whether more or less protection against possible losses is desirable. The first step is to determine all the possible events that could contribute to a given loss and assign a relative possibility to each. A loss analysis diagram is then constructed that indicates the relationship between these events. Sometimes two or three events must all occur before a loss occurs. Sometimes only one. The probability of the loss is then calculated and compared with a previously chosen maximum tolerable probability. If it is greater, more safeguards must be included. If it is less, the system is acceptable. The problem is determining the maximum and individual probability. This method does, however, attempt to quantify what has previously been done by intuition.

The father of this system was the so-called *fault tree* that was developed for the U.S. missile program. The developers ran into the problem of testing the electric circuits of the Minute Man missiles. No one wants a nuclear warhead accidentally fired into space. Yet all the electric circuits had to be tested so that in case of an attack the missiles could be relied on. The fault tree was a method of predicting the probability of an unplanned launch as a result of testing. If the probability were high then either another way would have to be found to test the circuits or more safety devices would have to be installed.

With chemical plants becoming larger, the potential for loss becomes greater. *Chemical Engineering Progress*[24] reported that from 1960 through 1966 the chemical and allied industries had 205 "large losses." This resulted in damages of $165,000,000, 116 fatalities, and 1,133 nonfatal injuries. Table 4-6 gives a breakdown of the causes. The National Fire Protection Association reported that in 1968 there were 4,100 fires and explosions in the chemical process industries and an estimated $27,000,000 in damage was done. By 1971 the number of incidents had dropped over 25% to 3,100 but the damage had nearly tripled to $74,000,000.[25] It is no wonder chemical and insurance companies are concerned.

Ammonia plants provide an example of how bigness can pose safety problems. In Chapter 3 it was noted that advances in technology had made larger plants very economical and had resulted in the closing of many smaller plants. In 1963 the big plants began producing ammonia. The next few years were disastrous, with ammonia plants accounting for the greatest share of chemical plant losses. As a result the deductible amounts in the insurance contracts were greatly increased. In 1963 a $50,000-deductible clause was high for ammonia plants. This meant that for any

Table 4-6

A Study of 317 Case Histories of Major Losses
in the Chemical and Allied Industries

| Hazard Factor | Times Assigned | % |
|---|---|---|
| Equipment failures | 143 | 31.1 |
| Inadequate material evaluation | 93 | 20.2 |
| Operational failures | 79 | 17.2 |
| Chemical process problems | 49 | 10.6 |
| Ineffective loss prev. program | 37 | 8.0 |
| Material movement problems | 20 | 4.4 |
| Plant site problems | 16 | 3.5 |
| Structures not in conformity with use requirements | 14 | 3.0 |
| Inadequate plant layout and spacing | 9 | 2.0 |
| | 460 | 100.0 |

Source: "Equipment Failures Prime Culprit in Plant Losses," *Chemical Engineering Progress*, Feb. 1969, p. 47.

given catastrophe the insurance company would pay for all damages in excess of $50,000. The insured must pay for the first $50,000. By 1967 a $1,000,000-deductible clause was not uncommon and the Hartford Steam Boiler Inspection and Insurance Company, the largest insurer of ammonia plants, said, "The financial loss to our company has been so devastating over the past three years as to cause our management to consider withdrawal from insuring all ammonia plants."[26]

Safety is important. It must always be a prime concern of the process engineer.

## CASE STUDY: PROCESS DESIGN FOR A 150,000,000 LB/YR POLYSTYRENE PLANT USING THE SUSPENSION PROCESS

The unit ratio material balances and the flow diagram are shown in Figures 4E-1, 4E-2, 4E-3 and 4E-4. Much of what appears is a direct result of assumptions presented in the scope. Some of the reasoning that was used follows.

### Reactor and Washing Areas

The key to the successful production of polystyrene is the reaction. The ratio of materials to be used is a compromise of the literature values, as was discussed in the section on the scope. All the authors discussing suspension polymerization say the reaction should be allowed to go to completion. (Removing and recycling the unreacted styrene would be more expensive.) It will be assumed that this means 99.8% of the styrene is reacted, and that this can be accomplished by using an average of the temperatures and cycle times given in Table 2E-2.

Figure 4E-1

Unit Ratio Material Balance for General Purpose Polystyrene (GPPS)

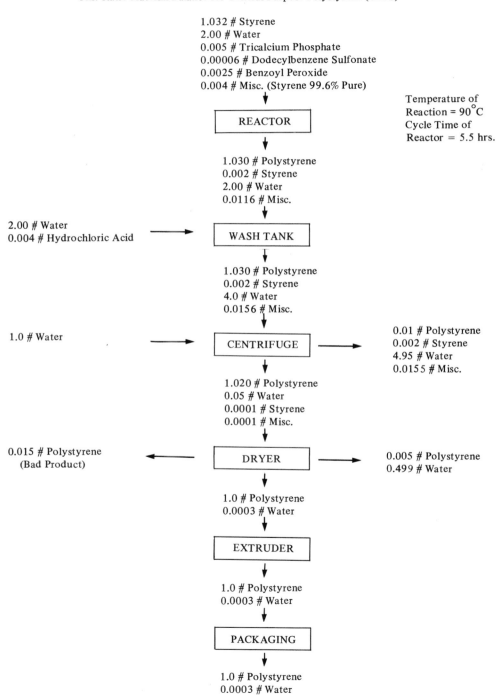

Figure 4E-2
Unit Ratio Material Balance for Medium Impact Polystyrene (MIPS)

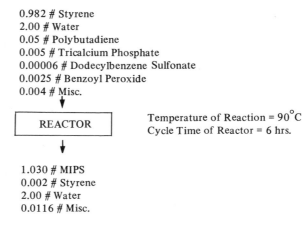

0.982 # Styrene
2.00 # Water
0.05 # Polybutadiene
0.005 # Tricalcium Phosphate
0.00006 # Dodecylbenzene Sulfonate
0.0025 # Benzoyl Peroxide
0.004 # Misc.

REACTOR

Temperature of Reaction = 90°C
Cycle Time of Reactor = 6 hrs.

1.030 # MIPS
0.002 # Styrene
2.00 # Water
0.0116 # Misc.

From here on the flow sheet is the same as Figure 4E-1 if polystyrene is replaced by MIPS.

Figure 4E-3
Unit Ratio Material Balance for High Impact Polystyrene (HIPS)

0.912 # Styrene
2.00 # Water
0.12 # Polybutadiene
0.005 # Tricalcium Phosphate
0.00006 # Dodecylbenzene Sulfonate
0.0025 # Benzoyl Peroxide
0.004 # Misc.

REACTOR

Temperature of Reaction = 90°C
Cycle Time of Reactor = 6.5 hr.

1.030 # HIPS
0.002 # Styrene
2.00 # Water
0.0116 # Misc.

From here on the flow sheet is the same as Figure 4E-1 if polystyrene is replaced by HIPS.

Washing will be used to remove this residual styrene, since we have a water suspension. This makes the assumed amount of styrene very important, because it will determine the amount of wash water that must be added. Enough water must be present to dissolve all the styrene that remains. At 25°C the solubility of styrene in water is 0.032%. It should be greater at higher temperatures. Assume the reactor products will cool to 60°C in the wash tank. At this temperature the solubility of styrene in water is assumed to be 0.050%. This means that for each pound of styrene originally charged, 4 lb of water must be present in the wash solution to make it

Figure 4E-4 Flow Diagram for a 150,000,000 lb. Per Year Polystyrene Plant Using the Suspension Process

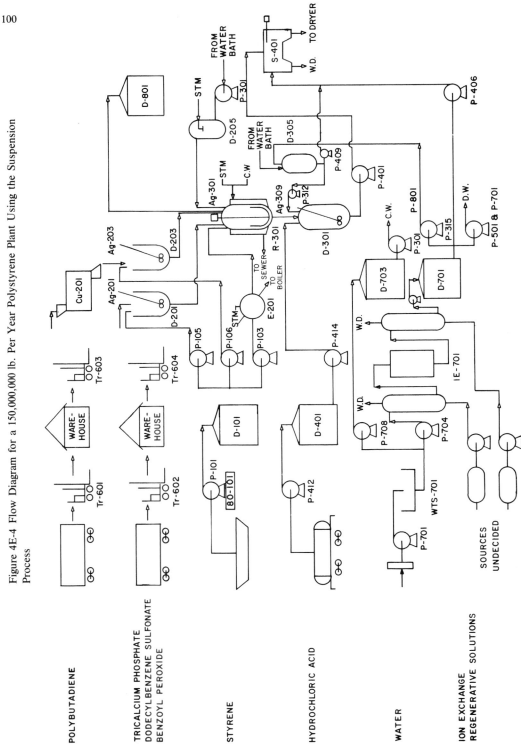

101

*(Figure 4E-4 continued)*

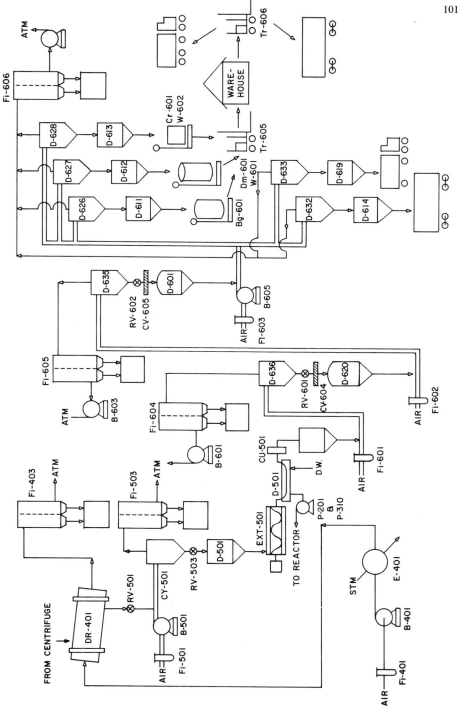

possible for all the unreacted styrene to dissolve in the water (0.002 lb styrene unreacted per lb styrene charged).

Before the polystyrene is separated from the water and impurities, the insoluble inorganics must be dissolved. This is done by adding enough dilute hydrochloric acid to react with the tricalcium phosphate and make a soluble product. The acid will also react with any remaining peroxide. However, nearly all the peroxide will have already reacted or decomposed, since its half-life is 2.1 hr at 85°C.[27]

The washing could be done in the reactor or in separate tanks. In the proposed scheme, it is done in separate tanks. These wash tanks also provide the feed to the continuous purification system that follows. A continuous purification system is used because usually continuous processing is cheaper than batch processing. Thus, the wash tanks serve also as holdup tanks. Since the wash tanks are much less expensive than the reactors (see Chapter 9), it is expected that money will be saved by reducing the time the material spends in the reactors, and hence the number of reactors needed.

### Feed Preparation

To reduce the cycle time of the reactors, the entering water and styrene will be preheated. The temperatures of the input streams will be set so as to obtain the desired reaction temperature. The water entering the reactor will be heated to 95°C. A sparger is used, since it is the most effective means of transferring heat. The water entering the reactor and that entering the steam boilers comes from the same source, so their cost and purity are similar. The meter that controls the amount of water entering the reactor must follow the sparger. Otherwise, the steam that is condensed would not be included in the amount of water added, and the wrong ratio of styrene to water would be obtained.

The bulk of the styrene is to be heated to 85°C before being charged. This is done in a vertical double-pipe heat exchanger, which is directly above the reactor. To prevent polymerization from occurring in the heat exchanger or piping system, there are to be no obstructions between this heat exchanger and the reactors.

The catalyst, rubber stabilizer, and suspending agent are premixed in styrene and discharged by gravity into the reactor. Since different mixtures are required for each product, these will be added to the styrene manually. This mixture will not be preheated, since it might polymerize.

### Purification Steps and Extrusion

If the water can be removed using physical separation processes, then the styrene and other impurities dissolved in it will also be discharged. A centrifuge with a washing step will be used to do this. According to Anderson,[28] the material leaving the centrifuge has 1-5% water. The higher value will be assumed.

The final purification step is drying. The polystyrene leaving this unit must meet the specifications set in the scope (0.03% water). The blower for the hot air entering the dryer may be placed either before the air heater or after the bag filter. If it is placed after the bag filter, whenever a hole develops in the filter polystyrene dust can get onto the blades, or even into the motor of the blower. Also, the air at this point will be hotter and contain more water. For these reasons, it was decided to place the blower before the air heater.

There is one more processing step to be added. This is the extrusion of the product into 1/8-in pellets. It is done to make it easier for the customer to handle the product.

**Waste**

It was assumed that 3% of polystyrene would be removed from the process in airveying, drying, centrifuging, transferring, or as bad product. At least 95% of that which is lost in processing must be intercepted before it leaves the plant. Most of it can be recovered and sold as off-grade material. This waste is split among the various streams leaving the processing area.

**Use of Energy**

In Chapter 8 it will be determined how best to use all the energy available. For instance, what should be done with the water leaving the polystyrene cooling baths and/or the steam condensate? Both these streams are hotter than the incoming process water. Should the water from the cooling baths be used as feed to the reactors? Should the steam condensate be returned to the boilers or sent to the reactors?

**Refrigeration of Styrene**

There is disagreement in the literature over whether the styrene should be refrigerated. The *Encyclopedia of Chemical Technology* says, "In climates where temperatures in excess of 80°F are common, bulk storage of monomer should be refrigerated."[29] For our plant site the temperature during the summer months is above 86°F 5% of the time.[30] However, many plants downstream of this site on the Ohio River do not refrigerate their styrene storage tanks. Therefore, since refrigeration systems are expensive, none will be installed.

## CHANGE OF SCOPE

The refrigeration system for styrene is to be deleted.

### References

1. Rudd, D.F.: "Process Synthesis," *Chemical Engineering Education*, vol. 7, no. 1, Winter 1973, p. 44.
2. Wood, W.S.: "Transporting, Loading and Unloading Hazardous Materials," *Chemical Engineering*, June 25, 1973, p. 72.
3. Weast, R.C. (ed.): *Handbook of Physics and Chemistry*, Ed. 53, CRC Press, Cleveland, 1972.
4. "Standards of Performance for New Stationary Sources," *Federal Register*, vol. 38, no. 111, part II, Washington, D.C., June 11, 1973, p. 15406.
5. *Background Information for Proposed New Source Performance Standards*, vol. 1, U.S. Environmental Protection Agency, Research Triangle Park, N.C., June 1973.

6. *Code of Federal Regulations*, Title 40, Chap. 1, part 60, United States Government Printing Office, Washington, D.C. (published annually).

7. *Federal Register* Supplement: Code of Federal Regulations; List of Code of Federal Regulations (CFR) Sections Affected, United States Government Printing Office, Washington, D.C. (published throughout the year).

8. *Federal Register*, United States Government Printing Office, Washington, D.C. (published almost daily).

9. Kraus, M.N.: "Pneumatic Conveyors," *Chemical Engineering*, Oct. 13, 1969, p. 59.

10. *United States Statutes at Large*, U.S. Government Printing Office, Washington, D.C., vol. 84, part 2, 1970, p. 1590.

11. Burklin, C.R.: "Safety Standards Codes and Practices for Plant Design," *Chemical Engineering*, Oct. 2, 1972, p. 56; Oct. 16, 1972, p. 113; Nov. 13, 1972, p. 143.

12. "Occupational Safety and Health Standards," *Federal Register*, vol. 37, no. 202, Oct. 18, 1972, p. 22101.

13. *Code of Federal Regulations*, Title 29, Chap. 17, part 1910, United States Government Printing Office, Washington, D.C. (published annually).

14. Piombino, A.J.: "When the Safety Sleuth Comes to Call," *Chemical Week*, June 6, 1973, p. 41.

15. "Dow's Process Safety Guide," in *Chemical Engineering Progress Reprint Manual*, American Institute of Chemical Engineers, New York, 1966.

16. Le Vine, R.Y.: "Electrical Safety in Process Plants," *Chemical Engineering*, May 1, 1972, p. 50.

17. *National Fire Codes*, vol. 1, National Fire Protection Association, Boston (published annually).

18. Woinsky, S.G.: "Predicting Flammable Material Characteristics," *Chemical Engineering*, Nov. 27, 1972, p. 81.

19. *Guidelines for Risk Evaluation and Loss Prevention in Chemical Plants*, Manufacturing Chemist's Association, Washington, D.C., 1970.

20. Browning, R.L.: "Analyzing Industrial Risks," *Chemical Engineering*, Oct. 20, 1969, p. 109.

21. Browning, R.L.: "Calculating Loss Exposures," *Chemical Engineering*, Nov. 17, 1969, p. 239.

22. Browning, R.L.: "Estimating Loss Probabilities," *Chemical Engineering*, Dec. 15, 1969, p. 135.

23. Browning, R.L.: "Finding the Critical Path to Loss," *Chemical Engineering*, Jan. 26, 1970, p. 119.

24. "Equipment Failures Prime Culprits in Plant Losses," *Chemical Engineering Progress*, Feb. 1969, p. 47.

25. Spiegelman, A.: "OSHA and the Chemical Industry," *Chemical Engineering*, June 18, 1973, p. 158.

26. "Frank Look at Big Plant Problems," *Chemical Week*, Oct. 7, 1967, p. 67.

27. Mageli, O.L., Kolcznski, J.R.: "Organic Peroxides," *Industrial and Engineering Chemistry*, Mar. 1966, p. 25.

28. Anderson, E.V., Brown, R., Belton, C.E.: "Styrene — Crude Oil to Polymer," *Industrial and Engineering Chemistry*, July 1960, p. 550.

29. *Kirk-Othmer Encyclopedia of Chemical Technology*, Ed. 2, Wiley, New York, 1969, vol. 19, p. 73.

30. *Evaluated Weather Data for Cooling Equipment Design. Addendum No. 1, Winter and Summer Data*, Fluor Products Company, Inc., 1964.

## Additional References

*Toxic Substances List*, National Institute of Occupational Safety and Health, Government Printing Office, Washington, D.C., 1973.

Sax, N.I.: *Dangerous Properties of Industrial Materials*, Ed. 2, Reinhold, New York, 1963.

Nichols, R.A.: "Hydrocarbon-Vapor Recovery," *Chemical Engineering*, Mar. 5, 1973, p. 85.

Strassberger, F.: "Polymer-Plant Engineering: Materials Handling and Compounding of Plastics," *Chemical Engineering*, Apr. 3, 1972, p. 81.

Oringer, K.: "Current Practice in Polymer Recovery Operations," *Chemical Engineering*, Mar. 20, 1972, p. 96.

Harmer, D.E., Ballantine, D.S.: "Applying Radiation to Chemical Processing," *Chemical Engineering*, May 3, 1971, p. 91 (costs given).

Kirven, J.B., Handke, D.P.: "Plan Safety for New Refining Facilities," *Hydrocarbon Processing*, June 1973, p. 121.

Vervalin, C.H.: "Who's Publishing on Fire and Safety," *Hydrocarbon Processing*, Jan. 1973, p. 128.

# CHAPTER 5

# Equipment List

The equipment list is a compilation of all the equipment that costs more than a pump, together with enough information about each piece for its cost to be estimated. The list is developed directly from the detailed flow sheet. Each piece of equipment shown on the flow sheet should appear on the equipment list. The symbols assigned to each piece of equipment are included on the equipment list to provide a cross-reference between the list and the detailed flow sheet.

To determine the cost of purchasing and installing each piece of equipment, the following major factors must be specified:

1. Specific type of equipment
2. Size and / or capacity
3. Material of construction
4. Operating pressure
5. Maximum temperature if the equipment is to run above the ambient temperature
6. Minimum temperature if the equipment is to be refrigerated
7. Whether insulation is required
8. Corrosion allowances if they are large
9. Special features such as jackets on heat exchangers or special conditions such as the burying of a storage tank

The effect of these items on the cost is discussed in Chapter 9.

A typical entry in the equipment list appears below:

V-03 Horizontal Methanol Storage Tank. 20,000 gal, 11 ft diameter, 31 ft long, carbon steel, 10 psig, no insulation

In metric units this is:

V-03 Horizontal Methanol Storage Tank. 76 m³, 3.35 m diameter, 9.5 m long, carbon steel, 1.8 kg/cm², no insulation

The V-03 is the equipment number. The type of equipment is a horizontal storage tank. It is used to store methanol. The capacity is 20,000 gal (76 m³) and the size is 11 ft (3.35 m) in diameter by 31 ft (9.5 m) long. The tank is to be made of carbon steel and will not be insulated. It should be built to withstand a pressure of 10 psig (1.8 kg / cm²). Any vessel with a rating of under 15 psig (2 kg / cm²) is not considered a pressure vessel. No information is given on temperature or corrosion allowances because neither of these factors presents any special problem. The rate of corrosion is small enough that no special corrosion allowance is needed, and the temperature will be near that of the surroundings.

## SIZING OF EQUIPMENT

The chemical engineering undergraduate spends most of his time sizing equipment. Usually in the problems assigned the type of equipment to be used is specified. For a distillation column the student would be told whether it is a bubble cap, a sieve plate, a valve tray, a packed column, or something else, and then asked to size it for a desired separation. In other cases he would be given the size of the specific equipment and asked to determine what the output would be for a given input.

In design, neither the type of equipment nor the size is given, and, as the reader has already discovered if he read the section in Chapter 4 on the unit ratio material balance, he also must specify the output conditions. In fact, the approach to sizing is very different, since it is usually not done from a rigorous theoretical viewpoint. To follow this approach for a preliminary process design in all cases would be very impractical. First, the design is not finalized, and any changes that occur may make the previous calculations meaningless. Second, the purpose of the preliminary design is merely to provide enough information to determine whether it is economically feasible to build such a plant. Since over 50% of the projects will not meet this criterion, any extra money spent in obtaining precise information is wasted money.

To aid the engineer in sizing equipment, various rules of thumb (see end of chapter), nomographs, and simplified formulae are used. Manufacturers' catalogs can also be very helpful in providing methods for sizing. The best source for these is the *Visual Search Microfilm Catalog* (VSMF).[1] This microfilm file of manufacturers' information is totally updated yearly. It permits a large amount of well-indexed current information to be stored in a minimum of space.

When the engineer has a difficult or time-consuming problem, he may ask company experts to assist him. These people are charged with becoming familiar with the latest developments in a given area. They often are in charge of the detailed design of specific equipment. Because of their experience, they often can provide a quick answer that will be adequate for a preliminary design.

If an answer is still not forthcoming, an equipment manufacturer may be asked to provide the information. The vendor should be told that this is a preliminary budget estimate so he will not provide a detailed quotation. That may take a couple of man-weeks (a man-week is the time spent by a man working one week), while the budget estimate may only take a few man-days. The added expense for the detailed quote will have to be absorbed by the vendor and will eventually result in increased prices for the buyer.

Sometimes not enough is known about the processing rates for even a vendor to size the equipment. In this case it may be possible for the engineer to have the equipment manufacturer obtain the necessary data.

For instance, the Swenson Evaporator Company maintains a pilot plant for sizing crystallization equipment. They will rent it to prospective customers, who can then obtain rate information without building a pilot plant themselves. Swenson employees run the plant and collect the data and samples. The customer provides an adequate supply of raw materials and sends observers. Usually any analysis of samples is done by the customer at his own plant. This is a cheap way to obtain data. It is especially useful for studies on processes whose design is mainly an art, such as crystallization and filtration.

One problem that frequently arises when processes for new products are being designed is that the physical properties of these compounds are unknown. Often these unknown properties may need to be estimated before the equipment can be sized. A list of references on estimating physical properties is given at the end of this chapter.

All calculations done by the design engineer are retained. They are often appended to the original copy of the preliminary design report, which is filed in the process engineering department. This allows engineers working on the project in the future to determine what assumptions have been made, and permits them to check for possible errors. These calculations are also valuable after the plant is built, because then they can be used to determine the accuracy of the equations that were used, and to assess whether the assumptions that were made are correct. This is the only way to correct calculation procedures and to decide if, and how much, oversizing of equipment should be specified in the future.

## One or Two Trains

At this point, the designer must make a critical decision for a continuous process. He must decide whether the plant will be a single-train plant or a dual-train plant. This means he must decide whether he will design a process in which all the material goes through each unit, or whether two identical plants will be built side by side, and half the flow will go through each. Actually, this decision can be made on a unit-by-unit basis. The advantage of the single-train plant is that it costs less to build and operate. Its disadvantage is that a failure of any item in the plant may cause the shutdown of the entire plant. For a dual-train plant, only half the plant would be inoperative. To reduce the probability that a single-train plant will shut down by an equipment failure, duplicates of many inexpensive items are purchased as spares, and are often installed in parallel with the operating ones. Then if one fails, merely opening and closing a few valves and switches allows the spare to take its place, and the whole process need not be shut down. This procedure may even be followed for expensive items if it is known that they have frequent failures.

To determine the reliability of a manufacturer's equipment, the engineer may request that the manufacturer supply him with the information, or he may ask for the names of companies that have used the equipment for a long time. Then by contacting these users he can get some idea of the percentage of the time the equipment is likely to be inoperative. On the basis of this response, he must decide whether to buy only one or two items. If he chooses the latter, he must then decide if each is to be large enough to handle the total flow rate, or if they are to be run concurrently with each handling half the flow rate. Obviously, economics (see Chapter 10) is a very important factor.

### Bulk Storage of Materials

After the approximate storage requirements have been calculated, the size of the storage vessels must be obtained. Table 5-1 gives the types of vessels commonly used for gas storage. In general, the storage of large quantities of gases is hazardous and costly, and should be avoided.[2]

Table 5-1

Vessels for Gas Storage
Operating Pressures Psig

| Volume, ft³: | 0-1 | 1-30 | 30-100 | 100-150 | 150-300 | Over 300 |
|---|---|---|---|---|---|---|
| 0-1,000 | Tank | Tank or drum | Drum[4] | Drum | Drum | Cylinder[1] |
| 1,000-5,000 | Tank | Drum | Drum | Drum | Cylinder | Cylinder |
| 5,000-17,000 | Tank or gas holder[3] | Spheroid[2] | Sphere[2] | Sphere | Not economical | Not economical |
| 17,000-500,000 | Gas holder | Spheroid | Sphere | Not economical | Not economical | Not economical |

[1]Gas cylinder. A welded or forged vessel usually with a ratio of length to diameter from 5:1 to 50:1, usually with hemispherical or other specially designed heads.
[2]Spheroids and spheres are built in standard sizes from 32 to 120 ft dia. See manufacturer's data books, viz: Chicago Bridge and Iron Works.
[3]Gas holders are built in standard sizes from about 10,000 to 10,000,000 ft³ capacity.
[4]Drums usually have a length that is three times the diameter.

Source: House, F.F.: "An Engineer's Guide to Process-Plant Layout," *Chemical Engineering*, July 28, 1969, p. 126.

Liquids are most economically stored in bulk containers. When large quantities above 25,000 gal must be stored, the tanks should be constructed to the dimensions given by the American Petroleum Institute Standards (see Table 5-2). These tanks must be field-erected. For smaller quantities more economical shop-constructed tanks should be specified. Field fabrication is always more expensive than shop fabrication. A shop-constructed item, in general, must be less than 11.5 ft (3.5 m) in diameter, so that it can be shipped by truck or train. For any piece of equipment that must be shop-fabricated, or for which the cost of field construction is prohibitive, this limitation should be noted. When barges may be used for transportation, this limitation does not hold.

An example showing how the size of the methanol tank listed previously was obtained follows.

Table 5-2

A Selection of Typical Sizes
of API Field Constructed Tanks

| Tank Diameter | | Approx. Capacity | | Height | | Volume | |
|---|---|---|---|---|---|---|---|
| ft | m | gal/ft | $m^3/m$ | ft | m | gallons | $m^3$ |
| 15 | 4.6 | 1320 | 16.4 | 18 | 5.5 | 23,700 | 90 |
| 20 | 6.1 | 2350 | 28.0 | 18 | 5.5 | 42,500 | 161 |
| 25 | 7.6 | 3670 | 45.6 | 18 | 5.5 | 66,000 | 250 |
| 25 | 7.6 | 3670 | 45.6 | 24 | 7.3 | 88,000 | 334 |
| 30 | 9.1 | 5290 | 65.6 | 24 | 7.3 | 127,000 | 481 |
| 35 | 10.7 | 7190 | 89.3 | 30 | 9.1 | 216,000 | 819 |
| 45 | 13.7 | 11900 | 148 | 36 | 11.0 | 429,000 | 1625 |
| 70 | 21.3 | 28800 | 358 | 36 | 11.0 | 1040,000 | 3940 |
| 100 | 30.5 | 58700 | 728 | 36 | 11.0 | 2110,000 | 8000 |
| 120 | 36.6 | 84500 | 1050 | 48 | 14.6 | 4060,000 | 15400 |
| 180 | 54.9 | 190000 | 2380 | 48 | 14.6 | 9150,000 | 34700 |

Source: "Welded Steel Tanks for Oil Storage," American Petroleum Institute, Washington, D.C., 1973.

**Example 5-1**

Size a methanol storage tank for a plant producing 10,000,000 lb of product. From the scope and unit ratio material balance, the following information is obtained: 0.3 lb of methanol is required for each pound of product; a 15-day storage capacity is specified for methanol; and the plant will operate 8,300 hours per year.

Pounds of methanol used per day:

$$\frac{0.3 \text{ lb methanol}}{\text{lb product}} \times \frac{10,000,000 \text{ lb product}}{\text{yr}} \times \frac{1 \text{ yr}}{8,300 \text{ hr}} \times \frac{24 \text{ hr}}{\text{day}} = 8,675$$

Pounds of methanol that must be stored:

$$8,675 \text{ lb / day} \times 15 \text{ days } 130,000 \text{ lb}$$

The specific gravity of methanol (ref. 3) = $0.792^{20/5}$
Gallons of methanol that must be stored:

$$130,000 \text{ lb} \times \frac{1 \text{ ft}^3}{(0.792 \times 62.4) \text{ lb}} \times \frac{7.48 \text{ gal}}{\text{ft}^3} = 19,700 \text{ gal}$$

Size of storage tank, assuming the length is 3 times the diameter and it will only be filled to a maximum of 90% of the capacity:

$$\frac{\Pi D^2 L}{4} = \frac{3\Pi D^3}{4} = 19{,}700 \text{ gal} \times \frac{1 \text{ ft}^3}{7.48 \text{ gal}} \times \frac{1}{0.90} = 2{,}930 \text{ ft}^3$$

$$D = 10.72 \text{ or } 11 \text{ ft}$$
$$L = 31 \text{ ft (to give the correct volume)}$$

The width of the tank in the example is less than 11.5 ft (3.5 m), so it could be shipped by railroad to the site. It should also be noted that when figuring the methanol flow rate the number of operating hours per year was used. This is equivalent to assuming that the plant will operate continuously whenever possible, which is exactly what everyone hopes will be true.

Solids usually should not be stored in bulk containers. If a large quantity is to be stored, it is usually best to pile it on the ground. Hercules Inc., when they wanted to store ammonium nitrate, which is hygroscopic, built a large aluminum building which was kept under positive pressure. The solid was dumped onto the floor. The building was large enough to bring trucks and railroad cars inside. These were loaded by an end loader of the same type that is seen on most highway construction jobs.

Frederick F. House[2] has suggested the following guidelines for designing solid storage bins:

1. Make the storage bin a plain cylindrical tank. Rectangular bins require elaborate bracing, are very costly to fabricate, and tend to "pack" in corners.

2. Provide one large bin wherever possible, rather than multiple small bins. This saves supports, materials, fabrication cost, and conveyors.

3. Make the cylindrical bin of a diameter that can be completely fabricated in a shop. The usual limit of diameter is about 11 ft 6 in[3.5 m] and of practical length about 30 ft [9 m] (or 3,000 cu ft capacity [85 m³]).

4. Remember that coarse, uniform-particle-size materials flow easily, whereas fine, relatively uniform materials are almost fluid. The greater the distribution of particle size in a mixture, the greater the tendency to compact and to resist flow.

5. Bin-discharging bottoms are subject so stoppage due to "bridging" and packing of granular materials. To minimize problems, make the bottom an eccentric cone with one straight vertical side. The total included angle of the cone should be not over 45 deg. for easy-flowing materials, 30 deg. for materials that do not flow readily.

## Intermediate Storage

There are several purposes for intermediate storage tanks. One is to average out the short-range fluctuations in the feed to a unit. This is important where the following units are hard to control and it is desired to have the feed fluctuations minimized. Another is to make certain the feed is uninterrupted when for some reason a unit upstream is temporarily disabled, for example by a pump failure. This is important for units such as distillation columns that take a while to reach steady-state conditions. This storage also gives more time for safely and optimally shutting down a unit, either when an emergency occurs or for routine maintenance. It also allows the plant to be started sequentially rather than trying to start it all at once. This means shutdowns and startups can be accomplished with fewer people.

Intermediate storage can, however, be very dangerous if the material being stored is flammable.[4] It is in a region where operating equipment is located and where men are working. Hence this is where accidents are more likely to occur. For these reasons, when dealing with flammable substances, all towers and tanks should be designed to hold a minimum of flammables.

For nonflammable materials, a very rough rule of thumb would be to design these tanks to hold the equivalent of one hour's storage.

## PLANNING FOR FUTURE EXPANSIONS

In Chapter 3 there was a discussion of the advantages and disadvantages of planning for future expansions. Here it will be assumed that a decision has been made to expand the plant at some future date. The process designer must now decide where extra capacity should be initially built into the plant. The advantage of doing this is that it is cheaper to construct and to operate single rather than duplicate facilities. Single facilities also occupy less space. One disadvantage is that the initial cost of the plant is higher, and if the expansion never occurs the extra money spent has been wasted. Another disadvantage is that it may be more difficult and expensive to operate than one that was designed to operate at full capacity.

In general, when the equipment is either difficult to operate or very expensive, it is desirable to specify equipment large enough to handle the expansion. Whenever this is done, calculations must be made to determine whether the equipment can produce the desired result for both the initial and the expanded conditions. Whenever this is not the case, the equipment should be sized for the initial rate. Then, when the expansion occurs, another unit may be added in parallel or in series. Plans for this should be made at the initial stages of design, so that adequate space can be reserved for this equipment. Much of the material that follows on specific types of equipment was obtained from an article by James M. Robertson.[5]

### Distillation Towers

Distillation columns are expensive items in any plant, and are tricky to control. They should initially be built large enough to accommodate a proposed expansion. The reboilers, condensers, and pumps, however, do not need to be designed to handle any more than the initial throughput. Figure 5-1 shows how the auxiliary system may be expanded by placing similar equipment in parallel when the plant capacity is increased.

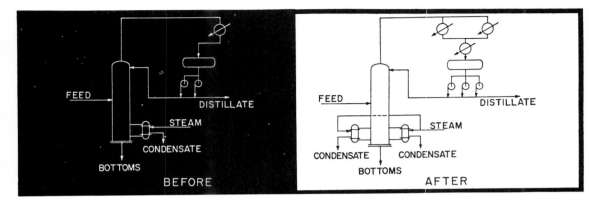

Figure 5-1  Auxiliary units for a distillation column before and after expansion.
  Source: Robertson, J.M.: "Plan Small for Expansion," *Chemical Engineering Progress,*
  Sept., 1967, p. 87.

To operate the larger column at a reduced rate probably will not be too difficult if a bubble cap is specified. These columns have wide stable operating ranges. The only difficulties are that the tray efficiency may be low and the liquid holdup relatively large. This latter is especially undesirable if, as has been noted, the liquid is flammable, or if some undesired reaction takes place at the elevated temperatures within the column.

If a sieve, dual-flow, or grid-tray column is used, the only way to operate the column in a stable manner at the low initial flow rates is to blank off part of the trays. This increases the vapor velocity through the mixing section, and assures good contact and an efficient separation. These blanks can be removed at the time of the expansion.

**Packed Towers**

Packed towers are used mainly for absorption, but may also be used for distillation. They are sensitive to vapor flow rates, but this problem can be successfully circumvented by changing packing size. Robertson[6] notes that for one column an increase in capacity of 60% was obtained when 1 in (2.5 cm) packing was replaced with 2 in (5 cm) packing of the same type. However, more of the large packing was required since it has a higher H.C.T.P. (height of packing equivalent to a theoretical tray). There were also differences in the pressure drop.

**Chemical Reactors**

The output of a chemical reactor depends, among other things, on the residence time (the average length of time the material is in the reactor), the temperature, and the fluid dynamics. These conditions may make designing a single reactor to handle both the initial and final flow rates undesirable. However, sometimes this can be done by installing a vertical reactor and initially maintaining it only partially full. In this way the residence time can be maintained constant. However, this affects the heat transfer area if a jacketed vessel is used, and it affects the flow characteristics when agitators are used. Also, the reduced flow rates will affect the heat transferred in any peripheral equipment and piping, as well as in the reactor. Here, both flow conditions and heat transfer rates must be checked to see if the reactor can be easily operated at the initial and expanded rates.

## MATERIALS OF CONSTRUCTION

The materials of construction are usually dictated by the chemicals present, the operating conditions, and the end use of the products. The standard material of construction is carbon steel. It is usually the cheapest metal that can be used in fabricating equipment. However, it generally is not used below -50°F (-45.6°C) because of a loss of ductility and impact strength, nor is it generally used above 950°F (510°C), because of excessive scaling rates. It is not good for mildly acidic conditions, because under these circumstances corrosion is greatly accelerated.

Carbon steel cannot be used in the production of most polymers because even trace amounts of iron discolor the product. In polymer manufacturing, usually glass-lined or stainless steel equipment must be used. Glass-lined or stainless steel equipment is required for any product that may be ingested by man. This includes all foods, pharmaceuticals, and food additives.

Section 23 of the fourth edition of the *Chemical Engineer's Handbook* has an excellent presentation of data concerning materials of construction and corrosion.

## TEMPERATURE AND PRESSURE

The temperatures and pressures in the processing equipment are a function of the processing conditions and the properties of the substances used. For instance, the highest temperature in a distillation column can be estimated by knowing the boiling points of all the key components at the operating pressure. The highest boiling point will be close to the highest temperature that will occur within the column.

The temperatures and pressures in the storage system depend on the physical properties of what is being stored and the weather conditions. The local weather bureau can supply information on the average high and low temperatures for each month, as well as the extremes. A synopsis of this information is given in references 7 and 8.

For an uninsulated tank located in the middle of an empty field, the actual temperatures in the tank can exceed these limits by 20°F (11°C). On a hot summer day when such a tank is directly receiving solar radiation, the temperature could reach 125°F (52°C). It is for this reason that tank cars are usually designed to withstand temperatures of 130°F (54°C). This temperature could be reached if the car were placed on an unshaded siding in Texas during the middle of the summer. During a clear, cold winter night, the reverse can happen. The uninsulated tank, instead of receiving radiant energy, is itself radiating energy to the universe. It can thus become colder than the surrounding air.

# LABORATORY EQUIPMENT

In general, the laboratory in a plant is mainly a quality control laboratory. It will consist of all the off-line equipment necessary to determine whether the product and raw materials meet the desired specifications, and whether all the waste streams meet the criteria set by local, federal, and state authorities.

The equipment list should include all laboratory equipment that is going to cost more than $1,000. This would include special testing devices such as an Instron®, an infrared spectrometer, or a chromatograph.

# COMPLETION OF EQUIPMENT LIST

The equipment list cannot be completed at this point in the design. The heat exchangers and pumps cannot be sized until an energy balance is completed. The energy balance, in turn, cannot be performed until the layout is established, and this depends on the approximate size of the equipment. Thus, all equipment except for heat exchangers and pumps will be sized at this point in the design. The rest will be sized after the energy balance is complete.

# RULES OF THUMB

It would be impractical to include in this book the methods for sizing each different type of equipment that might be used in a chemical plant. Instead, a number of rules of thumb have been included that may be helpful in quickly sizing equipment. These rules are, of course, generalizations, and will not apply in all circumstances. A list of general and specific references is also included. For more information, the reader should consult the abstracts listed at the end of Chapter 1.

## Vessels

1. Vessels below 500 gal (1.9 m³) are never more than 85% filled.
2. Vessels above 500 gal (1.9 m³) are never more than 90% filled.
3. Liquid in quantities less than 1,000 gal (3.8 m³) is stored in vertical tanks mounted on legs.[9]
4. Liquid in quantities between 1,000 and 10,000 gal (3.8 and 38 m³) is stored in horizontal tanks mounted on a concrete foundation.[10]
5. Liquid in quantities exceeding 10,000 gal (3.8 m³) is stored in vertical tanks mounted on a concrete foundation.[9]

6. The optimum diameter for a shop-constructed tank[10] is given by

$$D = 0.74 \ (V)^{1/3}$$
$$D = \text{diameter in ft (m)}$$
$$V = \text{volume in ft}^3 \ (m^3)$$

**Agitators**

1. Typical horsepower requirements for agitators.[11,12]

| | |
|---|---|
| Blending vegetable oils | 1 hp / 100,000 lb (0.5 hp / 100,000 kg) |
| Blending gasoline | 0.3 hp / 100 bbl (0.019 hp / m³) |
| Clay dispersion | 10-12 hp / 1,000 gal (2.6-2.9 hp / m³) |
| Fermentation (pharmaceutical) | 3-10 hp / 1,000 gal (0.8-2.6 hp / m³) |
| Suspension polymerization | 6-7 hp / 1,000 gal (1.6-1.8 hp / m³) |
| Emulsion polymerization | 3-10 hp / 1,000 gal (0.8-2.6 hp / m³) |
| Solution polymerization | 15-40 hp / 1,000 gal (4.0-10.5 hp / m³) |

2. Types of agitators to use for fluids of various viscosities:[13]
   Propeller agitators — fluids below 3,000 cp
   Turbine agitators — fluids at 3,000-50,000 cp
   Paddle agitators — fluids at 50,000-90,000 cp
   Modified paddle agitators — fluids at 90,000-1,000,000 cp

**Columns**

1. The height of columns should not exceed 175 ft, because of foundation and wind-loading problems.
2. For distillation columns, as a safety factor, the number of trays should be increased 5-10%.[13]
3. Distillation column trays are usually 18 or 24 in (0.45 or 0.6 m) apart.
4. For columns less than 3 ft (1 m) in diameter, packed towers are usually used because of high cost involved in fabricating the small trays.[14]
5. Figure 5-2 gives a preliminary estimate of the tower diameter for a packed column.[5]
6. For packed column separations using Pall rings, the height of a transfer unit (HTU) may be estimated as follows:
   For non-halogenated organics having specific gravities less than 1, the HTU is 1-1.5 ft (0.3-0.45 m) for 1 in rings (2.5 cm) and 1.5-2.5 ft (0.45-0.75 m) for 2 in rings (5.0 cm). If one of the organic compounds does not meet these specifications, then the HTU is 2.0-3.0 ft (0.6-0.9 m) for 1 in rings and 2.5-3.5 ft (0.75-1.0 m) for 2 in rings.[15]

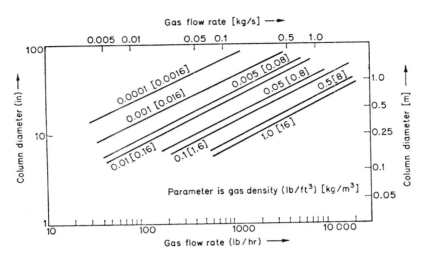

Figure 5-2    Estimation of packed column diameters.
Courtesy of Backhurst, J.R., Harker, J.H.: *Process Plant Design,"* Heinemann Educational Books LTD, London, 1973.

## Vapor-Liquid Separators

1. Horizontal separators should be designed for liquid holding times of 2-10 min. The height of the vapor space should be at least 1 ft, or 20% of the diameter if that is larger.[16]
2. Vertical separators should be designed for liquid holding times of 2-5 min. The height of the vapor space should be at least 1 ft, or 75% of the height if that is greater.[17]
3. Vertical separators should have a maximum superficial vapor velocity[17] given by the following equations:

$$v_{max} = 0.295 \sqrt{(\rho_L - \rho_v)/\rho_V}\ \text{(with mesh)}$$

$$v_{max} = 0.11 \sqrt{(\rho_L - \rho_V)/\rho V}\ \text{(without mesh)}$$

$v_{max}$ = maximum superficial vapor velocity in ft/sec
(multiply above by 0.3048 to get m/sec)

$\rho_L$ = density of liquid

$\rho_V$ = density of vapor

**Dryers**[18]

The evaporative capacity of dryers is given below in lb water / hr ft² of heat transfer surface (kg water/hr m² of heat transfer surface):

| | | |
|---|---|---|
| Drum dryer | 1-10 | (5-50) |
| Rotary direct heat dryer | 0.2-7 | (1-35) |
| Rotary indirect heat dryer | 1.0-12 | (5-60) |
| Rotary steam tube dryer | 1.0-12 | (5-60) |
| Rotary direct and indirect heat dryer | 4-9 | (20-45) |
| Rotary louver dryer | 0.3-15 | (1.5-75) |
| Tunnel (conveyer) dryer | 0.3-7 | (1.5-35) |
| Rotary shelf (conveyer) dryer | 0.1-2 | (0.5-10) |
| Trough (conveyer) dryer | 0.1-3 | (0.5-15) |
| Vibratory (conveyer) dryer | 0.1-20 | (0.5-100) |
| Turbo (conveyer) dryer | 0.2-2 | (1.0-10) |
| Spray dryer | 0.1-3 | (0.5-15) |

## CASE STUDY: MAJOR EQUIPMENT REQUIRED FOR A 150,000,000 LB/YR POLYSTYRENE PLANT USING THE SUSPENSION PROCESS

Table 5E-1 gives a partial equipment list. The key to the numbers is given in Table 5E-2. The reasoning behind this equipment list follows.

To aid us in the calculations the following figures will be helpful:

Pounds of polystyrene produced per year = 150,000,000 lb

Pounds of polystyrene produced per hour = $150,000,000 \, \dfrac{\text{lb}}{\text{yr}} \times \dfrac{1 \text{ yr}}{8,300 \text{ hr}} = 18,050 \text{ lb/hr}$

The properties of styrene:
  density[19] = 56.3 lb/ft³ @ 20°C
  viscosity[20] = 0.763 cp @ 20°C
  heat capacity[21] = 0.4039 BTU/lb/°F @ 20°C

The properties of polystyrene:
  density[21] = 65.5 lb/ft³
  heat capacity = 0.32 BTU/lb°F
  thermal conductivity = 0.058-0.080 BTU/hr ft°F

The summer temperature that is exceeded only 5% of the time in Martins Ferry = 86°F

The amounts of each stream can be related to the rate of polystyrene produced by multiplying by the numbers in the unit ratio material balance (Figs. 4E-1, 4E-2, and 4E-3).

Table 5E-1

Partial Equipment List
(to be completed when energy balance is run)

| Item No. | No. Reqd. | Description |
|---|---|---|
| BD-101 | 1 | Barge Dock, 215 ft long. |
| D-101 to D-103 | 3 | Styrene Storage. 429,000 gal. D = 45 ft, L = 36 ft, steel lined with epoxy except for bottom and lowest 2 ft of sides, which is coated with Dimetcote®. Outside painted with aluminum, insulated. |
| D-201 & D-202 | 2 | Additive Mixing Tank, stainless steel clad, 70 gal. |
| Ag-201 & Ag-202 | 2 | Agitator for T-201, T-202, stainless steel, < 1 hp. |
| D-203 to D-206 | 4 | Rubber Dissolving Tank, jacketed, stainless steel clad, 1,000 gallons. |
| Ag-203 to Ag-206 | 4 | Agitator for D-203 etc., stainless steel, < 1 hp. |
| Cu-201 | 1 | Cutter for bales of rubber, 4,000 lb/hr. |
| R-301 to R-308 | 8 | Reactor — 6,750 gal, I.D. = 8.33 ft, L = 16.67 ft. Jacketed, stainless steel, insulated. |
| Ag-301 to Ag-308 | 8 | Agitators for reactors, 120 hp, anchor type, stainless steel, 40 rpm. |
| D-301 to D-304 | 4 | Hold Tank — 22,000 gal, D = 11.5 ft, L = 27 ft. Stainless steel, insulated. |
| Ag-309 to Ag-312 | 4 | Agitators for hold tank, propeller type, 50 hp. Stainless steel. |
| S-401 & S-402 | 2 | Centrifuge. Scroll discharge centrifugal sedimentor/filter, bowl diam. 32 in, stainless steel, 60 hp. |
| Dr-401 & Dr-402 | 2 | Dryer, D = 6.5 ft, L = 38 ft, carbon steel, insulated. |
| Fi-401 & Fi-402 | 2 | Air Filter, 3,600 in², dry, throwaway type. |
| Fi-403 & Fi-404 | 2 | Bag Filter, 6,000 ft² to process 8,100 ft³/min. |
| D-401 | 1 | Hydrochloric Acid Storage Tank, D = 11.5 ft, L = 23 ft, stainless steel. |
| Ext-501 to Ext-509 | 9 | Extruder, vented single screw, D = 8 in, L = 16 ft, stainless steel, 180 hp, insulated. |
| Cu-501 to Cu-509 | 9 | Precision Knife Cutter, knife edge 2 ft, 60 hp. |
| D-501 to D-509 | 9 | Water Bath, L = 5 ft, H = 2 ft, W = 2 ft, aluminum. |
| D-510 to D-518 | 9 | Feed Storage for Extruder, D = 4.5 ft, L = 9 ft, aluminum. |
| D-620 to D-624 | 5 | Storage for Testing, D = 11.5 ft, L = 26 ft, aluminum. |

(*Table 5E-1 continued on following page*)

Table 5E-1 (*continued*)

Partial Equipment List
(to be completed when energy balance is run)

| Item No. | No. Reqd. | Description |
|---|---|---|
| D-601 to D-610 | 10 | Bulk Storage for Product — conical hopper, D = 27 ft, L = 41 ft, aluminum. |
| Bg-601 | 1 | Automatic Bagging and Palletizing Unit, 9 bags/min. |
| D-611 | 1 | Storage for Automatic Bagger, D = 8.5 ft, L = 17 ft, aluminum. |
| D-612 | 1 | Storage Hopper for Drums, D = 8.5 ft, L = 8.5 ft, aluminum. |
| Dm-601 | 1 | Drum Filling Station. |
| D-613 | 1 | Storage Hopper for Cartons, D = 8.5 ft, L = 8.5 ft, aluminum. |
| Cr-601 | 1 | Carton Filling Station. |
| Tr-601 to Tr-612 | 12 | Fork-Lift Trucks, 5,000 lb. |
| W-601 | 1 | Scale for Drums. |
| W-602 | 1 | Scale for Cartons. |
| Cv-601 | 1 | Conveyor for Drums, 30 ft long, 20 in wide. |
| Cv-602 | 1 | Conveyor for Cartons, 30 ft long, 40 in wide. |
| D-614 to D-619 | 6 | Storage for Bulk Shipments, D = 11.5 ft, L = 65 ft, aluminum. |
| Vf-601 to Vf-633 | 33 | Vibrating Feeders for Solids Storage Tanks. |
| El-801 | 1 | Elevator — 10,000 lb capacity to rise 75 ft. |
| D-801 | 1 | Blowdown Tank — 24,800 gal, carbon steel. |

Table 5E-2

Area Equipment Numbers

100 — Receiving and Storage of Raw Materials
200 — Feed Preparation
300 — Reactors and Hold-up Tanks
400 — Product Purification
500 — Product Finishing
600 — Product Storage and Shipping
700 — Utilities
800 — Miscellaneous

## Barge Dock BD-101

The dock must be large enough to tie up one barge. It should be 10 ft longer on each end than the length of a standard barge. A 1,000-ton barge is 195 ft long,[22] so the dock should be 215 ft long. A width of 20 ft seems reasonable.

## Styrene Storage D-101

The styrene storage capacity is given by the scope as 17 days. To convert to a weight basis, the pounds of styrene used for each separate product must be multiplied by the average amount of the product that would be made during the time period. The weight of styrene the storage facilities must be able to contain is:

$$17 \text{ days} \times \frac{18,050 \text{ lb}}{\text{hr}} \times \frac{24 \text{ hr}}{\text{day}} \times$$

$$\left[ \left( \frac{1.032 \text{ lb styrene}}{\text{lb GPPS}} \times \frac{0.60 \text{ lb GPPS}}{\text{lb P.S.}} \right) \right.$$

$$+ \left( \frac{0.982 \text{ lb styrene}}{\text{lb MIPS}} \times \frac{0.20 \text{ lb MIPS}}{\text{lb P.S.}} \right)$$

$$\left. + \left( \frac{0.912 \text{ lb styrene}}{\text{lb HIPS}} \times \frac{0.20 \text{ lb HIPS}}{\text{lb P.S.}} \right) \right]$$

$$= 7,360,000 \text{ lb styrene}$$

The volume of styrene storage units is:

$$\frac{7,360,000}{(56.3 \text{ lb/ft}^3)0.90} = 145,500 \text{ ft}^3 = 1,088,000 \text{ gal}$$

This assumes the tanks can be filled to 90% of their capacity.

It is desirable to have at least 3 tanks so they can be periodically cleaned without affecting production. If the standard size tanks given in Table 5-2 have been specified by the engineering staff as the optimal sizes, then 3 tanks of 424,000 gal or 5 tanks holding 216,000 gal could be constructed (see Problems following Chapter 10). An economic analysis show that the former is the best.

To prevent contamination, the tanks will be lined with epoxy, except for the bottom and the lower 2 ft of the sides. The lower portions will be lined with an inorganic zinc silicate such as Dimetcote® to prevent rust formation and to allow static charges that may develop in the liquid to drain off through the tank.[23]

The tanks will be well insulated and will be pained with a reflective paint to keep temperatures low in the summer. Styrene freezes at —31°C, so no provision will be made to heat the tanks in winter.

## Reactors R-301

As a vessel of a given shape increases in size, both the surface area and the volume increase, but they do not increase at the same rate. For a sphere the surface area is a function of the diameter squared and the volume is a function of the diameter cubed. This is also true for a cylinder whose height is a multiple of its diameter. The polymerization of styrene is an exothermic reaction. The amount of energy released at any time is dependent on the volume of the reactor, and the rate of removal of that heat is dependent on the surface area. Unless the heat is removed, the temperature will rise and the reaction rate will increase. The result will be an uncontrolled reaction that not only may ruin the batch but could also damage the reactor and might cause a fire or explosion to occur.

Therefore, there is a maximum size reactor for each set of reaction conditions. This size will now be calculated. The maximum rate of heat production will be determined first.

The heat of polymerization is 290 BTU/lb

The weight of styrene in the reactor $= \rho V \dfrac{1.032}{3.044}$

$\rho$ = density of mixture = 58 lb/ft$^3$ (1/3 of way between the density of water and styrene)

$V$ = volume of reactor $= \dfrac{\pi D^2 L}{4}$

It will be assumed that the reactor is generally 90% full and the height is twice the diameter.

The energy released by polymerization $= 58 \times \dfrac{\pi D^3}{2} \times \dfrac{1.032}{3.044} \times 290(0.90)$

$\qquad = 8{,}060 \ D^3 \ \text{BTU}$

All this energy must be removed as it is formed. If it takes 1 hr to charge and discharge the reactor and it takes 0.5 hr to initiate the reaction, then all the energy is released and must be removed in 4 hr (the cycle time for GPPS is 5.5 hr).

$\qquad \text{Average BTU produced per hour} = \dfrac{8{,}060 \ D^3}{4} = 2{,}015 \ D^3 \ \dfrac{\text{BTU}}{\text{hr}}$

However, the reaction rate is not uniform. The maximum reaction rate must be known to calculate the area needed for heat exchange. This can and should be determined in a laboratory. For the suspension polymerization of polystyrene at 80°C using 0.5% benzoyl peroxide in an inert atmosphere, the reaction takes 4.5 hr to reach completion and the maximum conversion rate is 20% in 0.5 hr.[24] Although our conditions are different, it will be assumed that our maximum conversion rate is also around twice the average rate.

The maximum BTU produced per hour = 2,015 $D^3$ x 2 = 4,030 $D^3$

One way to remove heat is reflux cooling. A low-boiling constituent (water in our case) is boiled off, condensed, and then returned to the reactor. Another technique is recirculation; a portion of the suspension is withdrawn and passed through an external heat exchanger before being returned to the reactor. Both of these methods can pose problems, since polymerization often occurs inside the heat exchangers. Neither will be used. Instead, it will be assumed that all energy must be transferred through the wall of the reactor. A reasonable over-all heat transfer coefficient between the reaction mixture and the cooling water in the jacket is 50 BTU / hr ft² °F.[25]

The rate of heat removal is $Q = UA\Delta t_o$

where:  U = over-all heat transfer coefficient = 50 BTU/hr ft²°F
A = area of heat transfer = $2\pi DL$
$\Delta t_o$ = average temperature driving force between coolant and suspension

Since 95% of the time the air temperature is below 86° F, it will be assumed that the inlet cooling water temperature never exceeds 85°F. The reaction temperature has been set at 194°F (see Scope). If it is assumed the maximum cooling water temperature rise is 10°F, then

$$\Delta t_o = 194 - 90 = 104°F$$

The area of heat transfer is the area covered by the suspension. This can be estimated to be the bottom plus 90% of the sides. (The reactor is only 90% full.) The following calculation is not exact, since the reactor will have a dished bottom.

$$A \sim 0.9\pi DL + \frac{\pi D^2}{4} = (1.8 + 0.25)\pi D^2 = 2.05\pi D^2$$

$$A \sim 2.05\pi D^2 = \frac{Q}{U\Delta t_o} = 4,030 \ D^3/(50 \times 104)$$

$$D = 8.33$$

Volume of material in tank = $0.90 \frac{\pi D^3}{2}$ = 814 ft³

Volume of tank = $\frac{\pi D^3}{2}$ = 904 ft³ = 6,750 gal

Note: if L = D, the volume of the tank would be 4,760 gallons, and if L = 3D, the volume of the tank would be 8,920 gallons.

In the *Encyclopedia of Polymer Technology and Science* the following statement appears: "In a suspension polymerization of styrene in a 5000 gallon reactor, the lowest coolant temperature required is 120°F."[26] the reaction conditions were the same as those specified above, but the average coolant temperature was 120°F instead of 90°F the maximum reactor size would be 2,420 gals. Since no conditions are given, it may be assumed that our assumptions are conservative. Possibly a U = 60 BTU/hr ft²°F or a maximum reaction rate of 1.8 times the average would be better estimates. If both of these are substituted for our original conditions, a 16,110 gal reactor is possible. If the coolant temperature were 120°, these new assumptions would indicate a maximum reactor size of 5,770 gal. This shows the critical importance of obtaining accurate rate information.

A 6,750 gal reactor will be specified. Its dimensions will be 8.33 ft in diameter and 16.67 ft high. These dimensions were chosen because it is generally cheaper to install and operate a smaller number of large units than a large number of small units (see Chapter 9). This philosophy will be followed throughout the design of the plant.

The pounds of styrene processed per reactor per hour are:

$$\frac{814 \text{ ft}^3 \times 58 \text{ lb/ft}^3 \times 1.032 \text{ lb styrene}}{5.5 \text{ hr/cycle} \times 3.044 \text{ lb suspension}} = 2,900 \text{ lb/hr}$$

Number of GPPS reactors required is:

$$\frac{18,050 \text{ lb P.S./hr} \times 1.032 \text{ lb styrene/lb P.S.} \times 0.60 \text{ lb GPPS/lb P.S.}}{2,900 \text{ lb styrene/hr}} =$$

3.86 or 4 reactors.

All the previous calculations have been done using GPPS. It will be assumed that the same conditions apply MIPS and HIPS except that the reaction times are different. For economic purposes the same size reactor will be used for each product. For MIPS the reaction takes 0.5 hr longer; however, only one-third as much product is planned.

Number of MIPS reactors required=
3.86 x (0.20 / 0.60) x (6 hr / 5.5 hr)=1.41
Number of HIPS reactors required=
3.86 x (0.20 / 0.60) x (6.5 hr / 5.5 hr)=1.52

Together we need 3 reactors for MIPS and HIPS, making a total of 7 reactors needed. An eighth will be installed as a spare. This will allow full production to continue if cleaning out the reactors becomes more of a problem than expected. Obviously a standard size reactor will be chosen; it may be somewhat below the size calculated. (A similar statement can be made for each piece of equipment listed.)

There are conflicting statements in the literature on whether stainless steel or glass-lined reactors should be used. Stainless steel will be used because some producers use it and it is cheaper. Since the material charged to the reactors will be hot, the reactors will be minimally insulated to protect employees.

### Agitator for Reactor Ag-301

Church[27] suggest using a paddle or anchor type stirrer of medium speed at 20-60 rpm. Since styrene is only slightly soluble in water and all the other substances are present in very small quantities, it is assumed that the properties of water, the continuous medium, can be used in the design of the agitator. The size of the motor will then be doubled to correct for any errors and then doubled again to obtain the necessary power for startup.

$$\text{Reynolds number} = \frac{D_A{}^2 N \rho}{\mu} = 18,000,000$$

$D_A$ = diameter of agitator = 8 ft
$N$ = revolutions per sec = 0.67
$\rho$ = density = 60.3 $lb_m/ft^3$
$\mu$ = viscosity = 0.318 x 6.72 x $10^{-4}$ $lb_m$/ft sec
Power[28] $\sim$ 1.70 $N^3 D_A{}^5/550$ = 30 hp

(for a flat paddle—no figures are given for anchor type.) Specify a 120 hp agitator.

### Additive Mixing Tank and Agitator D-201 Ag-201

The additive feed tank must be large enough to handle all additives plus a carrier solution of styrene. The pounds of dodecylbenzene sulfonate, tricalcium phosphate, and benzoyl peroxide (50% water) used per batch are

$$[0.005 + 0.00006 + 0.0025(2.0)] \ \frac{18,050 \times 5.5 \ \text{hr/batch}}{7 \ \text{reactors}} = 142 \ \text{lb}$$

Assume 2 lb of styrene are to be used to carry each 1 lb of additive into the reactor. If the average density is that of styrene (it will be greater than this), then:

$$\text{Volume of tank} = \frac{426}{56.3 \times (0.85)} = 8.90 \ \text{ft}^3 = 66.6 \ \text{gal}$$

Use a nominal 70 gal tank. A small agitator of less than 1 hp will be used to mix the materials.

### Rubber Dissolving Tank and Agitator D-203 Ag-203

The rubber dissolving tank will be designed for HIPS, since this product requires more rubber.

$$\text{Pounds of rubber per batch for HIPS} = 0.12 \times \frac{18,050}{7} \times 6.5 = 2,050 \ \text{lb}$$

Assume 2 lb of stryene are to be used for each 1 lb of rubber; then:

$$\text{Volume of tank} = \frac{6,150}{56.3 \times (0.85)} = 128.5 \text{ ft}^3 = 960 \text{ gal}$$

A 1,000 gal tank will be used.

It sometimes takes up to 2 hr for the rubber to dissolve.[29] This means 4 tanks will be needed.

The tank will be jacketed with steam to heat the solution to 120°F before it is discharged. A small agitator of less than 1 hp will be used as a mixer.

### Hold (Wash) Tanks D-301

Before the hold tanks can be specified it must be determined whether one, two, or three products are to be made simultaneously. If only one product is made at a time, a large single-train centrifuging and drying system can be constructed. However, when two or three products are produced in separate trains at different rates, it would appear that the desire to have both similar equipment and optimally sized equipment for each train cannot be met. Therefore, for this plant only one product will be made at a time. Further, two full trains will be constructed. This is to prevent the plant from fully shutting down because of the failure of a piece of equipment. When the proposed expansion occurs another full train will be added.

One purpose of the hold tanks is to a buffer between the batch reactors and the continuous processing units that follow. Initially it woull appear that 2 hold tanks should be specified—one for each train. Each of these would be fed by 4 reactors. In case of a power failure it would be necessary to discharge those reactors that could have runaway reactions into the hold tanks immediately. At most, 4 reactors should be at this state. If properly sequenced this is 2 for each hold tank. Since a hold tank may be expected to contain at least 1 and possibly 2 reactor loads plus wash water, the system should be designed to hold 4 reactor loads plus an equivalent amount of wash water.

$$\text{Volume} = \frac{4}{7} \frac{1}{0.9} \frac{(18,050 \text{ lb P.S./hr} \times 5.5 \text{ hr/batch} \times 5.05 \text{ lb mix/lb P.S.})}{58 \text{ lb/ft}^3}$$

Volume = 15,500 ft$^3$ = 41,000 gal

Assuming that the maximum width for a shop-constructed tank is 11.5 ft, this gives a height of 53 ft. This seems absurd. Instead, specify 4 hold tanks each 27 ft high and 11.5 ft in diameter.

The wash tanks will be made of the same materials as the reactors, stainless steel. They will be insulated to protect employees.

### Hydrochloric Acid Storage D-401

Hydrochloric acid is to be obtained in the cheapest form possible. This is as 18 Baume acid, which is 27.9% hydrochloric acid (specific gravity $=1.142$). The material will be shipped by 8,000 gal tank truck from a site less than 200 miles away.

> Pounds of HCl required per day =
> 18,050 lb P.S./hr x 24 hr/day x 0.004 lb HCl/lb P.S. = 1,730 lb/day
> Pounds of HCl on a tank truck =
> 8,000 gal (1 ft³/7.48 gal) 1.14 x 62.4 lb/ft³ x 0.279 = 21,200 lb

One tank truck will supply enough HCl to last for 12.25 days. Hydrochloric acid is readily available. It should take less than 7 days after an order is received before it is loaded into a truck. The truck will take less than a day to travel between plants. Therefore a storage capacity of 16,000 gal will be adequate. If the tank diameter is 11.5 ft the height will be 23 ft. The tank will be constructed of stainless steel and will not be insulated.

### Centrifuge C-401

The choice of a centrifuge was made by reviewing the criteria set forth by Charles M. Ambler.[30] In a suspension polymerization the average size particle produced is between 50-1,000 microns in diameter.[31] A scroll discharge centrifugal sedimentor / filter was chosen because it can handle fines with only small losses, the product is relatively dry, and the wash can be separated from the mother liquor. The size must again be obtained experimentally. Perry[32] suggests that a Helical conveyor centrifuge having a bowl diameter of 32 in will handle 3-6 tons of solids and 250 gal of liquid per minute, which meets our specifications. It has 60 hp motor.

### Dryer DR-401

Usually some type of air dryer is used to remove the remaining water. The most popular are rotary and flash dryers. A rotary dryer will be specified. Care must be taken that the polymer does not exceed 185°F, or its heat distortion properties will be affected.[33] Therefore this will be chosen as the exit temperature of the air, and the air flow will be parallel with the polymer flow. The air will enter at 300°F.[34] It is assumed that the solids will enter at around 70°F (room temperature) and leave at around 175°F To estimate the size the following equations will be used.[34]

$$V = \frac{Q_t}{U_a \Delta t_m} \qquad\qquad U_a = \frac{20G^{0.16}}{D}$$

where $Q_t$ = total energy transferred, BTU/hr
$U_a$ = volumetric heat transfer coefficient, BTU/hr ft³°F

$\Delta t_m$ = log mean temperature difference between hot gases and material, °F

G = air mass velocity, lb/(hr) (ft² of dryer cross section)

D = dryer diameter, ft

V = volume of dryer, ft³

$$Q_t \sim 1.020 \, \frac{\text{lb P.S.}}{\text{lb Pdt}} \left( \frac{18,050}{2} \, \frac{\text{lb}}{\text{hr}} \right) \left( 0.32 \, \frac{\text{BTU}}{\text{lb°F}} \right) \left( 175°F - 70°F \right)$$

$$+ \, 0.05 \, \frac{\text{lb H}_2\text{O}}{\text{lb Pdt}} \left( \frac{18,050}{2} \, \frac{\text{lb}}{\text{hr}} \right) \left( 1,140 \, \frac{\text{BTU}}{\text{lb}} - 38 \, \frac{\text{BTU}}{\text{lb}} \right)$$

$$= 810,000 \text{ BTU/hr}$$

$$\Delta t_m \sim \frac{(300 - 70) - (185 - 175)}{\ln \dfrac{230}{10}} = 70°F$$

The minimum air velocity is set by the particle size. A flow rate of 1,000 lb / hr ft² is adequate for 420-micron particles. This will be used.[34] The minimum velocity is used since it gives the smallest dryer.

The amount of air required is determined by the amount of energy the 300°F air must supply to remove the moisture from the polystyrene.

$$m = Q_t/Cp\Delta t$$

where Cp = heat capacity of air = 0.237 BTU/lb°F

$\Delta t$ = difference in air temperature entering and leaving dryer

$Q_t$ = heat transferred in dryer = 810,000 BTU/hr

m = mass flow rate of air

m = 810,000/(0.237) (300 − 185) = 29,700 lb/hr

A check must be made to determine if the air can contain the water without becoming supersaturated.

$$\text{Maximum amount of water that can be in air} = 29,700 \, \frac{\text{lb}}{\text{hr}} \times 0.730 \, \frac{\text{lb H}_2\text{O}}{\text{lb air}}$$

$$= 21,700 \text{ lb H}_2\text{O/hr}$$

$$\text{The pounds of water to be removed} = \frac{18,050}{2} \times 0.0497 = 450 \text{ lb H}_2\text{O/hr}$$

This amount of air is adequate. Add 10% to account for possible heat losses.

Mass flow rate = 1.1 x 29,700 = 32,700 lb/hr

If the mass velocity is 1,000 lb/hr ft² then

$$\text{Area of dryer} = \frac{32,700}{1,000} = 32.7 \text{ ft}^2$$

Diameter of dryer = 6.45 ft

$$V = \frac{810,000 \times 6.45}{20 \ (1,000)^{0.16} \times 70} = 1,242 = \pi D^2 L/4$$

L = 37.8 ft

Commercial dryers have diameters of 3-10 ft and an L/D ratio of 4 to 10.[32] So this is reasonable.

**Air Filter before Heat Exchanger Fi-401**

This is to remove the dust in the incoming air before it reaches the dryer. It differs from the filter following the dryer in that the air entering has less foreign material. A dry throwaway type of filter will be used because it is the most efficient and cheapest.[35] For effectiveness the velocity of the air cannot be very great. The size is determined by using a standard face velocity of 350 ft/min.[35]

$$\text{Volumetric flow rate} = \frac{32,700 \text{ lb/hr}}{0.0750 \text{ lb/ft}^3} = 435,000 \text{ ft}^3/\text{hr}$$

$$= 7,250 \text{ ft}^3/\text{min}$$

$$\text{Surface area of filter} = \frac{7,250 \text{ ft}^3/\text{min}}{350 \text{ ft/min}} = 20.7 \text{ ft}^2$$

Eight standard 20 x 20 in unit filters are needed for each train. Specify 9 and place in a square arrangement, 60 x 60 in.

**Bag Filter Fi-403**

See reference 36 for discussion of various types of dust collection equipment. Fabric filters will be used in order to recover all the fines in a dry state. These should recover 99% of the material larger than 0.2 microns.[36] A continuous-envelope fabric filter will be used, since this seems to have the lowest annual costs. For this filter ft³ of surface area is required for about each 2 ft³/min of air.

$$\text{Surface area of filter} = 7,250 \text{ ft}^3/\text{min} \times \frac{1 \text{ ft}^2}{2 \text{ ft}^3/\text{min}} \ \frac{647°F}{530°F} = 4,430 \text{ ft}^2$$

It is always best to overdesign this facility. Therefore a filter having a surface area of 6,000 ft² will be designated for each train.

**Extruder Ext-501**

The extruder capacity versus diameter is given by the following equations.[37]

For an extruder having a length-to-diameter ratio of 20:1

$$r = 277d - 494$$

For an extruder having a length-to-diameter ratio of 28:1

$$r = 325d - 500$$

where r = rate of throughput, lb/hr
d = diameter of extruder, in

The most popular length-to-diameter (L/d) ratios are 20:1 and 24:1. The standard inside diameters are 1.5 in, 2.5 in, 3.5 in, 4.5 in, 6 in, and 8 in.[37] For an 8 in diameter extruder using and L/d of 20:1, r=1,722 lb/hr. If an L/d of 28:1 is used, r=2,300 lb/hr. By interpolation for an L/d of 24:1, r ~ 2,000 lb /hr.

The largest of the popular standard sizes will be specified. Nine 8 in extruders with an L/d=24:1 will be needed for processing the polystyrene leaving the drier.

The horsepower required for the extruder[39] is:

$$P = 5.3 \times 10^{-4} \, r \, C(T_e - 80)$$

where P = power, horsepower
C = mean specific heat including heat of melting over the range 80-$T_e$, BTU/lb°F
r = throughput, lb/hr
$T_e$ = temperature of the melt entering die, °F

For high-impact polystyrene[38], C=0.42 BTU/lb °F. The temperature at the die entrance[39] is given below.

| | |
|---|---|
| regular polystyrene | 390-410° F |
| medium impact | 375-390° F |
| high impact | 365-380° F |

The horsepower is the greatest for regular polystyrene because it has the greatest melt temperature. Using the mean specific heat for polystyrene:

$$P = 5.3 \times 10^{-4} \times 2{,}000 \text{ lb/hr} \times 0.40 \times (410 - 80) = 140 \text{ hp}$$

Extruders should be equipped with 25% over the adequate horsepower.[40] Therefore, specify the horsepower as 180. For polystyrene a vented extrusion process is suggested.[40]

### Feed Storage for Extruders D-510

A 2 hr storage capacity before each extruder will be specified. The bulk density of polystyrene is around 35.5 lb/ft³.

$$\text{Volume} = 2{,}000 \; \frac{\text{lb}}{\text{hr}} \times \frac{1}{0.85} \times \frac{2 \text{ hr}}{35.5 \text{ lb/ft}^3} = 132.8 \text{ ft}^3$$

If the length equals twice the diameter, then the diameter will be 4.5 ft and the length will be 9 ft.

These units will be constructed of aluminum, since it is cheaper than stainless steel.

### Temporary Storage While Waiting for Testing D-620

Assume one day of storage is sufficient for testing.

$$\text{Volume of storage} = \frac{18{,}050 \text{ lb/hr}}{0.9} \times \frac{24 \text{ hr}}{\text{day}} \times \frac{1}{35.5 \text{ lb/ft}^3} = 13{,}550 \text{ ft}^3$$

Assume each bin is 11.5 ft in diameter.

Total height = 130 ft

Specify 5 silos 26 ft high made of aluminum.

### Water Bath D-501

These are usually 3-5 ft long.[41] They will be specified as 5 ft long, 2 ft wide, and 2 ft deep. They will be made of aluminum.

### Cutter Cu-501

A precision knife cutter whose cutting surface is made of stainless steel will be designated. The knife edge should be 24 in long. The product is to be ⅛ in cylinders. The product rate is 2,000 lb/hr through each extruder. This should require about a 60 hp motor.[41]

### Bulk Storage D-601

The scope specifies 25 days' storage. This will be in silos. The amount stored this way is 40% of the total product. The bulk density of polystyrene[41] is around 35.5 lb/ft³.

$$\text{Volume of GPPS storage} = \frac{18{,}050}{0.90} \; \frac{\text{lb}}{\text{hr}} \times \frac{24 \text{ hr}}{\text{day}} \times 25 \text{ days} \times \frac{0.4}{35.5 \text{ lb/ft}^3}$$

$$= 136{,}000 \text{ ft}^3$$

$$\text{Volume of storage for HIPS \& MIPS} = \frac{(0.2)(136{,}000)}{0.60} = 45{,}300 \text{ ft}^3$$

Assuming there will be 2 bulk storage containers for both HIPS and MIPS, and that all the storage units will be the same size:

Volume of each of 10 storage units=22,700 ft³

If each were 11.5 ft in diameter, they would be 218 ft high. This is impractical, so field-erected silos are necessary. Assuming L=1.5 D, the diameter is 26.8 ft.

Each of these units will have a vibrating bin feeder at the bottom. To prevent contamination of the product, each will be made of aluminum.

## Packaging Bg-601

To avoid paying a shift premium and to cut down on the amount of shift supervision, the packaging operations will be designed so they can be performed on the day shift. A 35 hr operating schedule per week will be assumed.

$$\text{Capacity of bagger} = \frac{168}{35} \times 18{,}050 \times 0.30 = 26{,}000 \text{ lb/hr}$$

This is about 9 bags per minute, or one pallet of 40 bags every 4.5 minutes. An automatic bagging and palletizing unit will be specified to reduce operating costs. The drum and carton filling operations will be done manually (see Fig. 5E-1).

Figure 5E-1   A manual container filling operation. The operator manually controls the amount of caustic soda beads being changed into the special container from the storage container which is mounted above it. The container is sitting on a scale.
Courtesy of the Dow Chemical Company U.S.A.

**Fork-Lift Trucks Tr-601**

A fork-lift truck will be used to take the pallets from the packaging area to the warehouse. Two trucks will be needed to keep pace with the automatic bagging equipment. If 2 each are used for storing cartons and drums, 1 for handling raw materials, and 5 for loading railroad cars and trucks, a total of 12 will be required.

**Storage before Bagger D-611**

Assume a 1 hr storage capacity.

$$\text{Storage volume} = \frac{26,000 \text{ lb/hr}}{0.85 \times 35.5 \text{ lb/ft}^3} = 860 \text{ ft}^3$$

If the length is twice the diameter, the length will be 17 ft. Similar calculations were performed for the storage preceding the packaging facilities for cartons and drums.

**Storage for Bulk Shipments D-614**

The largest bulk shipment will be in a railroad hopper car that has a capacity of 5,700 ft$^3$ (202,000 lb of P.S.).[42] One tank 20% larger than this will be constructed for each product and will be positioned above the railroad track.

Three more will be constructed to supply trucks. These will be 11.5 ft in diameter and 65 ft high.

**Blowdown Tank D-801**

The same reasoning applies here as for the wash tanks. It must be able to hold 4 full reactor loads.

$$V = \frac{4}{7} \frac{1}{0.90} \frac{(18,050 \text{ lb P.S./hr} \times 5.5 \text{ hr/batch} \times 3.044 \text{ lb mix/lb P.S.})}{58 \text{ lb/ft}^3}$$

$$V = 3,310 \text{ ft}^3 = 24,800 \text{ gal}$$

It is to be constructed of carbon steel.

## CHANGE OF SCOPE

1. A spare reactor will be installed.
2. Only one product will be made at a time.
3. Two trains will be constructed instead of one large train.
4. The bulk storage required is assumed not to include that required for charging hopper cars and trucks.
5. The storage before the extruders will hold a 2 hr supply of polystyrene.

6. Temporary storage of one day will be provided so tests on polystyrene can be run.
7. 18° Baume hydrochloric acid will be purchased in 8,000 gal tank cars.
8. 25 days' storage for hydrochloric acid will be provided.

## References

1. *Visual Search Microfilm File,* Information Handling Services, Inc., Englewood, Colo. (updated yearly).
2. House, F.F.: "An Engineer's Guide to Process-Plant Layout," *Chemical Engineering,* July 28, 1969, p. 120.
3. Perry, J.H. (ed.): *Chemical Engineers' Handbook,* Ed. 4, McGraw-Hill, New York, 1963, section 3 p. 36.
4. Preddy, D.L.: "Guidelines for Safety and Loss Prevention," *Chemical Engineering,* Apr. 21, 1969, p. 95.
5. Robertson, J.M.: "Plan Small for Expansion," *Chemical Engineering Progress,* Sept. 1967, p. 87.
6. Robertson, J.M.: "Design for Expansion," *Chemical Engineering,* Oct. 13, 1969, p. 87.
7. *Evaluated Weather Data for Cooling Equipment Design. Addendum No. 1, Winter and Summer Data,* Fluor Products Inc., Santa Rosa, 1964.
8. *Statistical Abstract of the United States,* U.S. Government Printing Office, Washington, D.C., published yearly.
9. Schmidt, R.G.: *Practical Manual of Chemical Plant Equipment,* Chemical Publishing Co., New York, 1967, p. 11.
10. MacCary, R.R.: "How to Select Pressure-Vessel Size," *Chemical Engineering,* Oct. 17, 1960, p. 187.
11. Parker, N.H.: "Mixing," *Chemical Engineering,* June 8, 1964, p. 165.
12. Schlegel, W.F.: "Design and Scaleup of Polymerization Reactors," *Chemical Engineering,* Mar. 20, 1972, p. 88.
13. McLaren, D.B., Upchurch, J.C.: "Guide to Trouble-Free Distillation," *Chemical Engineering,* June 1, 1970, p. 139.
14. Wall, K.J.: *Chemical Process Engineering,* July 1967, p. 56. Cited by Backhurst, J.R., Harker, J.H.: *Process Plant Design,* American Elsevier, New York, 1973.
15. Eckert, J.S.: "Design Techniques for Sizing Packed Columns," *Chemical Engineering Progress,* Sept. 1961, p. 54.
16. Scheiman, A.D.: "Horizontal Vapor Liquid Separators," *Hydrocarbon Processing and Petroleum Refiner,* May 1964, p. 155.
17. Scheiman, A.D.: "Size Vapor-Liquid Separators Quicker by Nomograph," *Hydrocarbon Processing and Petroleum Refiner,* Oct. 1963, p. 165.
18. Parker, N.H.: "Aids to Dryer Selection," *Chemical Engineering,* June 24, 1963, p. 115.
19. Perry, J.H.: *op. cit.,* Section 3, p. 41.
20. *Kirk-Othmer Encyclopedia of Chemical Technology,* Ed. 2, vol. 19, Wiley, New York, 1969, p. 56.
21. Bikales, N.M. (ed.): *Encyclopedia of Polymer Science and Technology,* Wiley, New York, 1970, vol. 13, p. 244.
22. Perry, J.H.: *op. cit.,* Section 6, p. 71.
23. Shelly, P.B., Sills, E.J.: "Monomer Storage and Protection," *Chemical Engineering Progress,* Apr. 1969, p. 29.
24. Schildkneckt, C.E. (ed.): *Polymer Processes,* Interscience, New York, 1956, pp. 97-98.
25. Bikales, N.M.: *op. cit.,* vol. 11, p. 296.
26. Bikales, N.M.: *op. cit.,* vol. 11, p. 291.
27. Church, J.M.: "Suspension Polymerization," *Chemical Engineering,* Aug. 1, 1966, p. 79.
28. Perry, J.H.: *op. cit.,* Section 19, pp. 14-15.

29. Deland, D.L., Purdon, J.R., Schoneman, D.P.: "Elastomers for High Impact Polystyrene," *Chemical Engineering Progress,* July 1967, p. 118.
30. Ambler, C.M.: "Centrifuge Separation," *Chemical Engineering,* Feb. 15, 1971, pp. 55-62.
31. Smith, W.M.: *Manufacture of Plastics,* Reinhold, New York, 1964, p. 407.
32. Perry, J.H.: *op. cit.,* Section 19, p. 92.
33. Anderson, E.V., Brown, R., Belton, C.E.: "Styrene — Crude Oil to Polymer," *Industrial and Engineering Chemistry,* July 1960, p. 550.
34. Perry, J.H.: *op. cit.,* Section 20, pp. 19-20.
35. Perry, J.H.: *op. cit,* Section 20, p. 95.
36. Sargent, G.D.: "Dust Collection Equipment," *Chemical Engineering,* Jan. 27, 1969, pp. 131-150.
37. *Modern Plastics Encyclopedia,* McGraw-Hill, New York, 1970-71, p. 480.
38. Carley, J.M.: "Introduction to Plastics Extrusion," *Chemical Engineering* Symposium Series, vol. 60, no. 49, 1964, p. 38.
39. Teach, W.C., Kiessling, G.C.: *Polystyrene,* Reinhold, New York, 1960, p. 85.
40. Sweetapple, L.: "How to Buy Extrusion Processing Equipment," *Plastics Technology,* mid-Sept. 1970, p. 71.
41. Schenkel, G.: *Plastics Extrusion Technology and Theory,* American Elsevier, New York, 1968, p. 380.
42. Uncles, R.F.: "Containers and Packaging," *Chemical Engineering,* Oct. 13, 1969, p. 87.

# Physical Property References

**Books**

Dean, J.A.: *Lange's Handbook of Chemistry,* Ed. 11, McGraw-Hill, New York, 1973.
Weast, R.C., Selby, S.M.: *CRC Handbook of Chemistry and Physics,* Ed. 54, CRC Press, Cleveland, 1973.
Perry, J.H. (ed.): *Chemical Engineers' Handbook,* Ed. 5, McGraw-Hill, New York, 1973.
Touloukian, Y.S. (ed.): *Thermophysical Properties of Matter,* IFE / Plenum Data Corp., New York, 1970, onward (all volumes not yet published).
    Vol. 1, 2, 3, Thermal Conductivity
    Vol. 4, 5, 6, Specific Heat
    Vol. 7, 8, 9, Thermal Radioactive Properties
    Vol. 10, Thermal Diffusivity
    Vol. 11, Viscosity
    Vol. 12, 13, Thermal Expansion
Canjar, L.W., Manning, F.S.: *Thermodynamic Properties and Reduced Correlations for Gases,* Gulf, Houston, 1967.
*Technical Data Book – Petroleum Refining,* American Petroleum Institute, New York, 1966.
Washburn, E.W.: *International Critical Tables of Numerical Data, Physics, Chemistry and Technology,* vol. 1-8 and index, McGraw-Hill, New York, 1926-1933.
Reid, R.C., Sherwood, T.K.: *The Properties of Gases and Liquids,* Ed. 2, McGraw-Hill, New York, 1966.
Timmemans, J.: *Physico-Chemical Constants of Pure Organic Compounds,* vol. 1 & 2, Elsevier, New York, 1960-1965.
Gallant, R.W.: *Physical Properties of Hydrocarbons,* vol. 1 & 2, Gulf, Houston, 1968.
*Engineering Data Book,* Ed. 8, Natural Gas Processors Suppliers Association, Tulsa, 1966.
Hala, E.P., Fried, V.: *Vapor-Liquid Equilibrium,* Pergamon, New York, 1967.

## Articles

*Hydrocarbon Processing* series on "Thermodynamic Properties" — Published frequently.
Gold, P.I., Ogle, G.J.: in *Chemical Engineering:* "Estimating Thermophysical Properties of Liquids,"
  Oct. 7, 1968, p. 152.
  "Critical Properties," Nov. 4, 1968, p. 185.
  "Density Molar Volume, Thermal Expansion," Nov. 18, 1968, p. 170.
  "Boiling, Freezing and Triple Point Temperatures," Jan. 13, 1969, p. 119.
  "Latent Heat of Vaporization," Feb. 24, 1969, p. 109.
  "Enthalpy, Gibbs Free Energy of Formation," Mar. 10, 1969, p. 122.
  "Heat Capacity," Apr. 7, 1969, p. 130.
  "Surface Tension," May 19, 1969, p. 192.
  "Compressibility, Velocity of Sound," June 30, 1969, p. 129.
  "Viscosity," July 14, 1969, p. 121.
  "Parachor — Others," Aug. 11, 1969, p. 97.
  "Vapor Pressure," Sept. 8, 1969, p. 141.
"Estimating Physical Properties," *Chemical Engineering Progress,* July 1967, p. 37.
Frith, K.M.: "Your Computer Can Help You Estimate Physical Property Data," *Chemical Engineering,* Feb. 21, 1972, p. 72.
Holmes, J.T., Baerns, M.A.: "Predicting Physical Properties of Gases and Gas Mixtures," *Chemical Engineering,* May 24, 1965, p. 103.
Hall, K.R., Yarborough, L.: "A New Equation of State for Z-factor Calculations," *Oil and Gas Journal,* June 18, 1973, pp. 82-92.
Hall, K.R., Yarborough, L.: "New Simple Correlation for Predicting Critical Volume," *Chemical Engineering,* Nov. 1, 1971, p. 76.
Smoot, L.D.: "Estimate Transport Coefficients," *Chemical Engineering,* Oct. 16, 1961, p. 187.
Alves, G.E., Brugmann, E.W.: "Estimate Viscosities by Comparison with Known Materials," *Chemical Engineering,* Sept. 18, 1961, p. 181.
Weintraub, M., Corey, P.E.: "High Temperature Viscosity of Gases Estimated Quickly," *Chemical Engineering,* Oct. 23, 1967, p. 204.
Pachaiyappan, V., Ibrahim, S.H., Kuloor, N.R.: "Simple Correlation for Determining Viscosity of Organic Liquids," *Chemical Engineering,* May 22, 1967, p. 193.
Pachaiyappan, V., Ibrahim, S.H., Kuloor, N.R.: "A New Correlation for Thermal Conductivity," *Chemical Engineering,* Feb. 13, 1967, p. 140.
Pachaiyappan, V. Ibrahim, S.H., Kuloor, N.R.: "A New Correlation for Liquid Surface Tension," *Chemical Engineering,* Oct. 23, 1967, p. 172.
Think, T.P., Duran, J.L., Ramalho, R.S., Kaliaguine, S.: "Equations Improve Cp* Predictions," *Hydrocarbon Processing,* Jan. 1971, p. 98.
Swanson, A.C., Cheuh, C.F.: "Estimating Heat Capacity," *Chemical Engineering Progress,* July 1973, p. 83.
Dimplon, W.: "Estimating Specific Heat of Liquid Mixtures," *Chemical Engineering,* Oct. 2, 1972, p. 64.
Procopio, J.M., Jr., Su, G.J.: "Calculating Latent Heat of Vaporization," *Chemical Engineering,* June 3, 1968, p. 101.
Mathur, B.C., Kuloor, N.R.: "Get Latent Heat from Mole Weight," *Hydrocarbon Processing,* Feb. 1971, p. 106.
King, F.G., Naylor, J.: "Nomograph for Fast Estimation of Heat of Vaporization," *Chemical Engineering,* July 13, 1970, p. 118.
Hilado, C.J., Clark, S.W.: "Autoignition Temperatures of Organic Chemicals," *Chemical Engineering,* Sept. 4, 1972, p. 75.

Jelinek, J., Hlavacek, V.: "Compute Boiling Points Faster," *Hydrocarbon Processing*, Aug. 1971, p. 135.

Osburn, J.O., Markovic, P.L.: "Calculating Henry's Law Constant for Gases in Organic Liquids," *Chemical Engineering*, Aug. 25, 1969, p. 105.

Horvath, P.J.: "Graphical Predictions of Ternary Azeotropes." *Chemical Engineering*, Mar. 20, 1961, p. 159.

Klein, I: "How to Predict Ternary Azeotropes," *Chemical Engineering*, Nov. 14, 1960, p. 233.

Shulman, W.: "Vapor Pressure of Organics," *Chemical Engineering*, Dec. 12, 1960, p. 180.

Staples, B.G., Procopio, J.M., Jr.: "Vapor Pressure Data for Common Acids at High Temperature," *Chemical Engineering*, Nov. 16, 1970, p. 113.

Chidambarams, S., Narsimham, G.: "Generalized Chart Gives Activity Coefficient Correction Factors," *Chemical Engineering*, Nov. 23, 1964, p. 135.

Van Vorst, W.D.: "Make Your Own Diagram to Estimate Enthalpy Changes of Real Gases," *Chemical Engineering*, June 19, 1967, p. 229.

Fries, H., Buthod, P.: "Equilibrium Computations with Speed and Precision," *Oil and Gas Journal*, Aug. 20, 1973, p. 71.

Throne, J.L., Griskey, R.G.: "Heating Values and Thermochemical Properties of Plastics," *Modern Plastics*, Nov. 1972, p. 96.

## Additional References

### General

Ludwig, E.E.: *Applied Process Design for Chemical and Petrochemical Plants*, vol. 1, 2, 3, Gulf, Houston, 1964, 1964, 1965.

Perry, J.H. (ed.): *Chemical Engineers' Handbook*, Ed. 5, McGraw-Hill, New York, 1974.

McCabe, W.L., Smith, J.C.: *Unit Operations of Chemical Engineering*, McGraw-Hill, New York, 1967.

Kuong, J.F.: *Applied Nomography*, vol. 1, 2, 3, Gulf, Houston, 1965, 1968, 1969.

Bourton, K.: *Chemical and Process Engineering Unit Operations*, Plenum, New York, 1967 (a bibliographic guide to references, media, and specific unit operations).

Backhurst, J.R., Harker, J.H.: *Process Plant Design*, American Elsevier, New York, 1973.

Desk book issues of *Chemical Engineering*:

| "Engineering Materials" | Oct. 12, 1970 |
| "Solids Separations" | Feb. 15, 1971 |
| "Engineering Materials | Dec. 4, 1971 |

(lists equipment and/ or materials manufacturers)

Hengstebeck, R.J., Jr.: *Distillation – Principles and Design Practice*, Reinhold, New York, 1961.

Van Winkle, M.: *Distillation*, McGraw-Hill, New York, N.Y., 1967.

Jordan, D.G.: *Chemical Process Development*, vol. 1 & 2, Interscience, New York, 1968.

### Materials

Nelson, G.A. (ed.): *Corrosion Data Survey – 1967 Edition*, National Association of Corrosion Engineers, Houston, 1968.

Tesmen, A.B.: "Materials of Construction for Process Plants — 1," *Chemical Engineering*, Feb. 19, 1973, p. 140.

Henthorne, M.: "Material Selection for Corrosion Control," *Chemical Engineering*, Mar. 6, 1972, p. 113.

Geerligns, H.G., Neiuwenhuizen, D.H.v.: "Selecting Construction Materials for High Pressure Processes,". . *Oil and Gas Journal*, Nov. 6, 1972. p. 58.

Swandby, R.K.: "Corrosion Charts: Guides to Materials Selection," *Chemical Engineering*, Nov. 12, 1962, p. 186.

Fenner, O.: "Plastics for Process Equipment," *Chemical Engineering*, Nov. 9, 1962, p. 170.

Hoffman, C.H.: "Wood Tank Engineering," *Chemical Engineering*, Apr. 17, 1972, p. 120.

## Mixers

Penny, W.R.: "Recent Trends in Mixing Equipment," *Chemical Engineering,* Mar. 22, 1971, p. 86.

Parker, N.: "Mixing," *Chemical Engineering,* June 8, 1964, p. 165.

Ho, C.F., Kwanga, A: "A Guide to Designing Special Agitators," *Chemical Engineering,* July 23, 1973, p. 95.

Harris, L.S.: "Jet Eductor Mixers," *Chemical Engineering,* Oct. 10, 1966, p. 216.

Holland, F.A.: "Scale-Up of Liquid Mixing Systems," *Chemical Engineering,* Sept. 17, 1962, p. 179.

Chen, S.J., MacDonald, A.R.: "Motionless Mixers for Viscous Polymers," *Chemical Engineering,* Mar. 19, 1973, p. 105.

Oldshue, J.Y.: "How to Specify Mixers," *Hydrocarbon Processing,* Oct. 1969, pp. 73-80.

Gretton, A.T.: "A Critical Evaluation of Power Requirements in Agitated Systems," *Chemical Engineering,* Jan. 20, 1964, pp. 145-148.

Weber, A.P.: "Selecting Propeller Mixers," *Chemical Engineering,* Sept. 2, 1963, pp. 91-98.

Fischer, J.J.: "Solid-Solid Blending," *Chemical Engineering,* Aug. 8, 1960, p. 107.

## Reactors

Barona, N., Prengle, H.W., Jr.: "Design Reactors This Way for Liquid-Phase Processes," *Hydrocarbon Processing,* Mar. 1973, p. 63.

Schlegel, W.F.: "Design and Scaleup of Polymerization Reactors," *Chemical Engineering,* Mar. 20, 1972, p. 88.

Barona, N., Prengle, H.W., Jr.: "Design Reactors This Way for Liquid-Phase Processes," *Hydrocarbon Processing,* Dec. 1973, p. 73.

## Distillation

Hengstebeck, R.J.: "An Improved Shortcut for Calculating Difficult Multicomponent Distillations," *Chemical Engineering,* Jan. 13, 1969, p. 115.

Lowenstein, J.G.: "Sizing Distillation Columns," *Industrial and Engineering Chemistry,* Oct. 1961, p. 44A.

Gallagher, J.L.: "Estimate Column Size by Nomogram," *Hydrocarbon Processing and Petroleum Refiner,* June 1963, p. 151.

Holland, F.A., Brinkerhoff, R., Carlston, R.C.: "Designing Many-Plate Distillation Columns," *Chemical Engineering,* Feb. 18, 1963, p. 153.

Van Winkle, M., Todd, W.G.: "Optimum Fractionation Design by Simple Graphical Methods," *Chemical Engineering,* Sept. 20, 1971, p. 136.

MacFarland, S.A., Sigmund, P.M., Van Winkle, M.: "Predict Distillation Efficiency," *Hydrocarbon Processing,* July 1972, p. 111.

Guerreri, G., Peri, B., Seneci, F.: "Comparing Distillation Designs," *Hydrocarbon Processing,* Dec. 1972, p. 77.

Van Winkle, M., Todd, W.G.: "Minimizing Distillation Costs Via Graphical Techniques," *Chemical Engineering,* Mar. 6, 1972, p. 105.

Ellerbe, R.W.: "Batch Distillation Basics," *Chemical Engineering,* Mar. 28, 1973, p. 110.

Treybal, R.E.: "A Simple Method for Batch Distillation," *Chemical Engineering,* Mar. 4, 1974, p. 105.

Ellerbe, R.W.: "Steam Distillation Basics," *Chemical Engineering,* Mar. 4, 1974, p. 105.

Koppel, P.M.: "Fast Way to Solve Problems for Batch Distillations," *Chemical Engineering,* Oct. 16, 1972, p. 109.

Luyben, W.L.: "Azeotropic Tower Design by Graph," *Hydrocarbon Processing,* Jan. 1973. p. 109.

## Gas absorption

Zenz, F.A.: "Designing Gas-Absorption Towers," *Chemical Engineering,* Nov. 13, 1972, p. 120.

Fair, J.R.: "Sorption Processes for Gas Separation," *Chemical Engineering,* July 14, 1969, p. 90.

Rowland, C.H., Grens, E.A., II: "Design Absorbers: Use Real Stages," *Hydrocarbon Processing,* Sept. 1971, p. 201.

Chen, N.H.: "New Equation Gives Tower Diameter," *Chemical Engineering,* Feb. 5, 1962, p. 109.

"Chart for Height of Transfer Unit in Wetted Wall Columns," *Chemical Engineering,* May 10, 1965, p. 193.

Guerreri, G., King, C.J.: "Design Falling Film Absorbers," *Hydrocarbon Processing,* Jan. 1974, p. 131.

## Extraction

Hanson, C.: "Solvent Extraction," *Chemical Engineering,* Aug. 26, 1968, p. 76.

Rickles, R.N.: "Liquid-Solid Extraction," *Chemical Engineering,* Mar. 15, 1965, p. 157.

Oberg, A.G., Jones, S.C.: "Liquid-Liquid Extraction," *Chemical Engineering,* July 22, 1963, p. 119.

Nemunaitis, R.R., Eckert, J.S., Foote, E.H., Rollison, L.R.: "Packed Liquid-Liquid Extractors," *Chemical Engineering Progress,* Nov. 1971, p. 60.

## Dryers

Sloan, C.E., Wheelock, T.D., Tsao, G.T.: "Drying," *Chemical Engineering,* June 19, 1967, p. 167.

Parker, N.H.: "Aids to Dryer Selection," *Chemical Engineering,* June 24, 1963, p. 115.

Belcher, D.W., Smith, D.A., Cook, E.M.: "Design and Use of Spray Dryers," *Chemical Engineering,* Sept. 30, 1963, p. 83; Oct. 14, 1963, p. 201.

## Crystallization

Garrett, D.E., Rosenbaum, G.P.: "Crystallization," *Chemical Engineering,* Aug. 11, 1958, pp. 125-140.

Powers, J.E.: "Recent Advances in Crystallization," *Hydrocarbon Processing,* Dec. 1966, p. 97.

Randolph, A.D.: "Crystallization," *Chemical Engineering,* May 4, 1960, p. 80.

Wilson, D.B.: "Crystallization," *Chemical Engineering,* Dec. 6, 1965, p. 119.

## Solid Liquid Separations

Ambler, C.M.: "How to Select the Optimum Centrifuge," *Chemical Engineering,* Oct. 20, 1969, p. 96.

Porter, H.F., Flood, J.E., Rennie, F.W.: "Improving Solid-Liquid Separations," *Chemical Engineering,* June 20, 1966, p. 141.

Thrush, R.E., Honeychurch, R.W.: "How to Specify Centrifuges," *Hydrocarbon Processing,* Oct. 1969, p. 81.

Flood, J.E., Porter, J.G., Rennie, F.W.: "Centrifugation Equipment," *Chemical Engineering,* June 20, 1966, p. 190.

Purchas, D.B.: "Guide to Trouble-Free Plant Operation — Filtration," *Chemical Engineering,* June 26, 1972, p. 88.

Hauslein, R.H.: "Ultra Fine Filtration of Bulk Solid," *Chemical Engineering Progress,* May 1971, p. 82.

Roberts, E.J., Stavenger, P., Bowersox, J.P., Walton, A.K., Mehta, M.: "Solids Concentration," *Chemical Engineering,* June 29, 1970, p. 52.

Dollinger, L.L., Jr.: "How to Specify Filters," *Hydrocarbon Processing,* Oct. 1969, pp. 88-92.

Flood, J.E., Porter, H.F., Rennie, F.W.: "Filtration Practice Today," *Chemical Engineering,* June 20, 1966, p. 163.

Anderson, A.A., Sparkman, J.F.: "Review Sedimentation Theory," *Chemical Engineering,* Nov. 2, 1959, pp. 75-80.

Browning, J.E.: "Agglomeration," *Chemical Engineering,* Dec. 4, 1967, p. 147.

Special Issue — many articles: *Chemical Engineering,* June 20, 1966, pp. 139-170.

## Solids Handling

Gluck, G.E.: "Design Tips for Pneumatic Conveyors," *Hydrocarbon Processing,* Oct. 1968, p. 88.

Condolios, E., Chapus, E.: "Transporting Solid Materials in Pipelines," *Chemical Engineering,* June 24, 1963, p. 93, p. 131; July 22, 1963, p. 145.

Fischer, John: "Practical Pneumatic Conveyor Design," *Chemical Engineering,* June 2, 1958, p. 114.

**Crushing, Grinding, and Screening**

Ratcliffe, A.; "Crushing and Grinding," *Chemical Engineering,* July 10, 1972, p. 62.

Stern, A.L.: "A Guide to Crushing and Grinding Practice," *Chemical Engineering,* Dec. 10, 1962, p. 120.

Gluck, S.E.: "Gyratory, Circular Motion and Special Action Screens," *Chemical Engineering,* Oct. 25, 1965, p. 131.

Matthews, C.W.: "Screening," *Chemical Engineering,* July 10, 1972, p. 77.

**Discharge Times**

Saunders, M.J.: "Evacuation Time of Gaseous Systems Calculated Quickly," *Chemical Engineering,* Nov. 20, 1967, p. 166.

Elshout, R.: "Graphs Determine Time to Drain Vessels," *Chemical Engineering,* Sept. 23, 1968, p. 246.

Tate, R.W.: "Estimating Liquid Discharge from Pressurized Vessels," *Chemical Engineering,* Nov. 2, 1970, p. 126.

**Miscellaneous**

Sargent, G.D.: "Dust Collection Equipment," *Chemical Engineering,* Jan, 27, 1969, p. 130.

Burke, A.J.: "Weighing Bulk Materials in the Process Industries," *Chemical Engineering,* Mar. 5, 1973, p. 66.

Johnston, W.A.: "Designing Fixed-Bed Adsorption Columns," *Chemical Engineering,* Nov. 27, 1972, p. 87.

Lemlich, R.: "Foam Fractionation," *Chemical Engineering,* Dec. 16, 1968, pp. 95-102.

Frantz, J.F.: "Design for Fluidization," *Chemical Engineering,* Sept. 17, 1962, p. 161; Oct. 1, 1962, p. 89; Oct 29, 1962, p. 103.

Considine, D.M.: "Process Weighing," *Chemical Engineering,* Aug. 17, 1964, pp. 113-132.

# CHAPTER 6

# Layout

The laying out of a plant is still an art rather than a science. It involves the placing of equipment so that the following are minimized: (1) damage to persons and property in case of a fire or explosion; (2) maintenance costs; (3) the number of people required to operate the plant; (4) other operating costs; (5) construction costs; (6) the cost of the planned future revision or expansion.

All of these goals cannot be met. For example, to reduce potential losses in case of fire, the plant should be spread out, but this would also result in higher pumping costs, and might increase manpower needs. The engineer must decide within the guidelines set by his company which of the aforementioned items are most important.

The discussion of the layout of a totally new plant will be followed by a discussion of the expansion of older facilities.

## NEW PLANT LAYOUT

The first thing that should be done is to determine the direction of the prevailing wind. This can be done by consulting Weather Bureau records. In the United States the prevailing winds are often from the west. Wind direction will determine the general location of many things. All equipment that may spill flammable materials should be located on the downwind side. Then if a spill occurs the prevailing winds are not apt to carry any vapors over the plant, where they could be ignited by an open flame or a hot surface.

For a similar reason the powerhouse, boilers, water pumping, and air supply facilities should be located 250 ft (75 m) from the rest of the plant, and on the upwind side. This is to minimize the possibility that these facilities will be damaged in case of a major spill. This is especially important for the first two items, where there are usually open flames.

Every precaution should be taken to prevent the disruption of utilities, since this could mean the failure of pumps, agitators, and instrumentation. For this reason, it may also be wise to separate the boilers and furnaces from the other utilities. Then, should the fired equipment explode, the other utilities will not be damaged.

In this respect the engineer must also consider all neighboring facilities. More than one plant has been badly damaged because of spills at another company. In

1970 Du Pont filed a $454,008 damage suit against the Matador Chemical Company. The lawsuit charged that as a result of an explosion at the Matador plant in Orange, Texas, Du Pont's power supply was interrupted. This resulted in a crash shutdown, which damaged and reduced the life of some cracking heaters. Matador Chemical was asked to pay for damages as well as for production losses due to the shutdown.[1]

Other facilities that are generally placed upwind of operating units are plant offices, mechanical shops, and central laboratories. All of these involve a number of people who need to be protected. Also shops and laboratories frequently produce sparks and flames that would ignite flammable gases. Laboratories that are used primarily for quality control are sometimes located in the production area. A list of items that should be placed downwind of the processing facilities is given in Table 6-1 and Table 6-2.

Table 6-1

Items That Should Be Located Upwind of the Plant

| | |
|---|---|
| Plant offices | Electrical Substation |
| Central laboratories | Water treatment plant |
| Mechanical and other shops | Cooling tower |
| Office building | Air compressors |
| Cafeteria | Parking lot |
| Storehouse | Main water pumps |
| Medical building | Warehouses that contain non- |
| Change house | hazardous, nonexplosive, and |
| Fire station | nonflammable materials |
| Boiler house | Fired heaters |
| Electrical powerhouse | All ignition sources |

Table 6-2

Items That Should Be Located Downwind of the Plant

Equipment that may spill inflammable materials
Blowdown tanks
Burning flares
Settling ponds

## Storage Facilities

Tank farms and warehouses that contain nonhazardous, nonflammable, and nonexplosive materials should be located upwind of the plant. Those that do not fit this category should not be located downwind of the plant, where they could be damaged and possibly destroyed by a major spill in the processing area. Nor should they be located upwind of the plant where, if they spilled some of their contents, the processing area might be damaged. They should be located at least 250 ft (75m) to the side of any processing area.[2] Some authorities suggest this should be 500 ft. The same reasoning applies to hazardous shipping and receiving areas.

Sometimes storage tanks are located on a hill, in order to allow the gravity feeding of tank cars. Care must be taken uder these circumstances to see that any slopover cannot flow into the processing, utilities, or service areas in case of a tank fire.[3,4]

**Spacing of Items**

The OSHA has standards for hazardous materials that give the minimum distances between containers and the distance between these items and the property line, public roads, and buildings. These depend on the characteristics of the material, the type and size of the container, whether the tank is above ground or buried, and what type of protection is provided. Specific details are provided for compressed gas equipment containing acetylene-air, hydrogen-oxygen, and nitrous oxide, as well as liquefied petroleum gases. They also prohibit the storage and location of vessels containing flammable and combustible materials inside buildings, unless special precautions are taken. The *Code of Federal Regulations* should be consulted for details (see Chapter 4 for how to do this).

Some general guidelines for minimum distances between various items are given in Tables 6-3 and 6-4. From these guidelines and those presented previously, the approximate sites of the processing, utilities, waste pollution, storage, and service areas of the plant may be located.

Again, the major reason for including the layout in the preliminary plant design is so the transporting equipment and buildings may be sized, to make certain that no needed equipment is omitted and that the chosen plant site will be large enough. At this point, since most of the energy transfer equipment has not been sized, only its approximate location can be given.

**Processing Area**

There are two ways of laying out a processing area. The *grouped layout* places all similar pieces of equipment adjacent. This provides for ease of operation and switching from one unit to another. For instance, if there are 10 batch reactors, these would all be placed in the same general area, and could be watched by a minimum of operators; if they were spread out over a wide area, more operators might be needed. This type of scheme is best for large plants.

The *flow line layout* uses the train or line system, which locates all the equipment in the order in which it occurs on the flow sheet. This minimizes the length of transfer lines and, therefore, reduces the energy needed to transport materials. This system is used extensively in the pharmaceutical industry, where each batch of a drug that is produced must be kept separate from all other batches. In other industries it is used mainly for small-volume products.[3]

Often, instead of using the grouped or flow line layout exclusively, a combination that best suits the specific situation is used.

Table 6-3

# OIL INSURANCE ASSOCIATION

## General Recommendations for Spacing in Petrochemical Plants

MINIMUM DISTANCE IN FEET

| | Process Unit—HH | Process Unit—LH | Tank Farms—HH | Tank Farms—LH | Product Whses—LH | Ship'g. & Rec'g.—HH | Ship'g. & Rec'g.—LH | Service Buildings | Boiler Area | Fire Pumps | Emergency Controls | Water Spray Controls | Turret Nozzles | Emergency Flares | Pilot Plants | Large Cooling Towers | Fire Hydrants | Fired Process Heaters | |
|---|---|---|---|---|---|---|---|---|---|---|---|---|---|---|---|---|---|---|---|
| Process Unit— HIGH HAZARD [B] | 200 | | | | | | | | | 250 | 100 | 50 [5] | 50 - 100 to Center of Target | For 100' Flare that is 25' above Surrounding Equipment, Use 300' | 200 | 150 | 50 to 250 | 50 to 100 | |
| Process Unit— LOW HAZARD | 100 | 50 | | | | | | | | 150 | 50 | | | | 200 | 100 | | 50 | |
| Tank Farms— HIGH HAZARD [C] | 250 [1] | 250 [1] | 1½ Dia. Larger | | | | | | | 250 | | 100 [5] | | | 250 | 250 | | 200 | |
| Tank Farms— LOW HAZARD | 200 [2] | 100 [3] | One Dia. Larger | ½ Dia. Larger | | | | | | 200 | | | | | 200 | 200 | | 200 | |
| Product Warehouse LOW HAZARD [D] | 150 | 50 [4] | 250 [1] | 100 [3] | 50 [4] | | | | | 200 | | | | | 200 | 150 | | 100 | |
| Shipping & Receiving— HIGH HAZARD [E] | 200 | 200 | 150 [2] | 100 [3] | 150 | 50 | | | | 150 | 100 | 50 [5] | | | 200 | 200 | | 200 | |
| Shipping & Receiving— LOW HAZARD | 150 | 100 | 100 | 50 | 20 | 50 | — | | | 100 | 50 | | | | 150 | 150 | | 100 | |
| Service Buildings • | 200 | 100 | 200 | 100 | 100 | 150 | 100 | See Bldg. Chart | | 100 | | | | | 200 | 100 | | 100 | |
| Boiler Area | 200 | 150 | 200 | 150 | 100 | 200 | 100 | 100 | — | — | | | | | 200 | 100 | | 100 | |

## Recommended Spacing Within Process Units

| | React. | Comp. | Tanks | Fract. Equip. | Cont. Rooms |
|---|---|---|---|---|---|
| Reactor | 25 [6] | | | | |
| Small Compressor House or Pump House | 40• [6] | | | | |
| Intermediate Stge. Tanks High Hazard Rundown-Feed | 100 to 200 | 100 to 200 | One Dia. [7] | | |
| Fractionation Equipment | 50 | 30 | 100 | | |
| Control Rooms • | 50 to 100 [6] | 50 to 100 | 100 | 50 to 100 | 10 |

•Control houses serving unusually large or hazardous units and central control houses for multiple units or housing computer equipment, require greater spacing and may require blast-resistant construction.

A. Distance between process units is measured from battery limits.

B. A high hazard process unit has explosion classification under petrochemical schedule of E-4 or E-5.

C. High hazard tanks are class "D" under the above schedule. Class "E" requires special consideration.

D. High hazard product warehouses contain unstable materials, low flash flammable liquids or highly combustible solids. These require special consideration.

E. High hazard shipping and receiving denotes stable materials with flash point below 110° F.

F. High hazard shipping and receiving of unstable materials requires special consideration.

G. Service buildings include offices, gate houses, change houses, laboratories, shops, garages, maintenance warehouses, cafeterias, hospitals, etc.
Experimentals laboratories classify as process units.

• H. Keep open flames 100' from vapor hazard area.

I. Deviation from these distances requires special protective installations such as fixed foam systems, water spray, automatic sprinklers, fire-system grading of 4 or better, or superior construction.

J. In borderline cases, high value requires high hazard classification.

K. Vertical storage tanks should be individually diked. If not, capacity in single dike should not exceed 25,000 bbls. For horizontal storage tanks, maximum is 400,000 gallons per group, with 100' between groups, or other suitable arrangement.

1. For specific vertical tank, use 5 diameters.
2. For specific vertical tank, use 4 diameters.
3. For specific vertical tank, use 3 diameters.
4. Standard firewall and sprinklered warehouse acceptable. Limit warehouse to maximum 25,000 sq. ft. floor area.
5. Two stations desirable.
6. Barricades desirable for hazardous reactors.
7. Over 100,000 gallons requires special consideration.

August 1972

Courtesy of the Oil Insurance Association.

# Table 6-4

## OIL INSURANCE ASSOCIATION
### General Recommendations for Spacing in Refineries

**MINIMUM DISTANCE IN FEET**

| | Service Buildings | Process Units | Boilers, Utilities, Elect. Gen. Equip., etc. | Fired Process Heaters | Process Vessels, Fract. Equip., etc. | Gas Compressor Houses | Large Oil Pump Houses | Control Houses | Cooling Towers | Controls for Dropout, Steam Snuff. & Spray | Blowdown Drums, Flare Stacks | Product Stge. Tanks | Rundown Tanks | Blending Tanks | Hazardous Log. & Unlog. Facilities, Incl. Docks | Fire Pumps | Turret Nozzles | Fire Hydrants | Fire Equip. Houses |
|---|---|---|---|---|---|---|---|---|---|---|---|---|---|---|---|---|---|---|---|
| Service Buildings | See Bldg. Chart[10] | | | | | | | | | | | | | | | | | | |
| Process Units | 100 | 50 to 100[3] | | | | | | | | | | | | | | 50 to 250 | 50 to 100[1] | 100 | |
| Boilers, Utility & Elect. Generating Equipment, etc. | 100 | 100 | --- | | | | | | | | | | | | | 50 to 250 | 100 | | |
| Fired Process Heaters | 100[2] | 50 | 100 | 25[2] | | | | | | | | | | | | 50 to 250 | 50 to 100 | 100 | |
| Process Vessels, Fractionating Equipment, etc. | 100 | — | — | 100 | 50[2] | | | | | | | | | | | 50 to 250 | 50 to 100 | | |
| Gas Compressor Houses | 100 | 100 | — | 30 | See Bldg. Chart | 30 | | | | | | | | | | 50 | 100 | | |
| Large Oil Pump Houses | 100 | 100[2] | 100 | 20 | 30 | 30 | See Bldg. Chart | | | | | | | | | 50 to 250 | 50 to 100 | | |
| Control Houses* | | 100 | 100[2] | 50 | 50 | 50 | See Bldg. Chart | | | | | | | | | 50 to 250 | 50 to 100 | | |
| Cooling Towers | 50 to 100 | 100 | 100 | 100 | 100 | 50 to 100 | 50 to 100 | 25 to 50[6] | | | | | | | | 50 to 250 | 100 | | |
| Dropout Controls, Steam Snuffing, & Water Spray Controls | | 50 | 50 | 50 | 20 | 30 | 50[7] | 50 | 50 | | | | | | | 50 to 250 | 100 to 200 | | |
| Blowdown Drums & Flare Stacks | 200 to 300[8] | 200 to 300 | 200 to 300 | 200 to 300 | 200 to 300 | 200 to 300 | 200 to 300 | 200 to 300 | 200 to 300 | — | | | | | | 100 | 250 | | |
| Product Storage Tanks | 250 | 250 | 250 | 250 | 250 | 250 | 250 | 250 | 250 | See Note[9] | See Note[9] | | | | | — | 300 | | |
| Rundown Tanks | 200 | 200 | 200 | 200 | 200 | 200 | 200 | 200 | 200 | See Note[9] | See Note[9] | See Note[9] | | | | — | 300 | | |
| Blending Tanks | 200 | 200 | 200 | 200 | 200 | 200 | 200 | 200 | 200 | See Note[9] | See Note[9] | 250 | 250 | | | — | 250 | | |
| Hazardous Loading & Unloading Facilities, Including Docks | 200 | 200 | 200 | 200 | 200 | 200 | 200 | 200 | 200 | 250 | 250 | 300 | 300 | 300 | | — | 250 | | |
| Fire Pumps | 50 to 100 | 250 | 250 | 250 | 250 | 100 | — | — | — | — | — | 300 | 300 | — | — | — | — | | |

*Control houses serving unusually large or hazardous units and central control rooms for multiple units or housing computer equipment, require greater spacing and may require blast-resistant construction.

**Notes:**

1. Special consideration should be given to the installation of fire hydrants and turret nozzles.
2. Small open flame devices should be located not less than 100' from any vapor-hazardous area.
3. Between battery limits.
4. Tanks over 10,000 bbls. capacity—250'; tanks less than 10,000 bbls. capacity—150'.
5. Tanks with capacities in excess of 5000 bbls —200'; tanks less than 5000 bbls.—100'.
6. 25' to 50', considering area.
7. Controls may be installed adjacent to or inside, to serve as a shield.
8. Flare stacks less than 75' in height should be 300' distance; with stacks over 75' in height 200' distance.
9. Tanks with capacities up to 10,000 bbls. should be spaced 1/2 dia. apart; tanks from 10,000 to 50,000 bbls. capacity, space 1 dia. apart; and tanks over 50,000 bbls. should be spaced 1 1/2 dia. apart. Tanks over 250,000 bbls. require special consideration.
10. Service buildings include: offices, change houses, maint. whses., cafeterias, labs, hospitals, garages, except as specifically provided for as indicated.
11. Propane tank batteries, preferably, should be isolated to more remote sections of plant, and "aimed" away from major plant values or occupancies. Spheres also should be remotely located whenever possible.

August, 1972

Courtesy of the Oil Insurance Association.

145

## Placing of Equipment

Once a general scheme is decided upon, the processing area is divided into unit areas. These are usually the general processing areas given in Chapter 1, or suitable subdivisions of those areas. Then, taking into account what has previously been presented in this chapter, these areas are placed in their approximate locations on a plot plan of the site. The units should be grouped so that the number of operating personnel is minimized. All items on the equipment list should be placed to scale on the plot plan. A small box should indicate each pump and heat exchanger, since they have not been sized. After they have been sized, the layout will be adjusted to accommodate them, if adequate space has not been allocated. Each piece of equipment should be designated by the same symbol used in the equipment list.

The maximum loss concept must also be considered. Some companies place a limit on the maximum loss that can be expected if a fire or explosion occurs. This means that only a certain amount of equipment can be placed in any given area, and that it must be physically separated from other areas. This can be done by providing fire walls, wide spaces between areas, and other fire-localizing designs.

At this point the location of the control panel(s) should be decided. Usually this should be a central location. This permits those watching the control panels to quickly investigate and determine the cause of any problems that might arise. As plants become more automated, it may be desirable to have two or more processes controlled from one location; this could reduce the number of operators required. In this case the control room should be located in a relatively unexposed area near the edge of the processing area, but away from fired heaters.[4] This is to protect both the employees and the equipment.

## Elevation

If there is no special reason for elevating equipment, it should be placed on the ground level. The superstructure to support an elevated piece of equipment is expensive. It can also be a hazard should there be an earthquake, fire, or explosion. Then it might collapse and destroy the equipment it is supporting as well as that nearby.

Some pieces of equipment will be elevated to simplify the plant operations. An example of this is the gravity feed of reactors from elevated tanks. This eliminates the need for some materials-handling equipment.

Other pieces may have to be elevated to enable the system to operate. A steam jet ejector with an intercondenser that is used to produce a vacuum must be located above a 34 ft (10 m) barometric leg. Condensate receivers and holding tanks frequently must be located high enough to provide an adequate net positive suction head (NPSH) for the pump below. For many pumps an NPSH of at least 14 ft (4.2 m) $H_2O$ is desirable. Others can operate when the NPSH is only 6 ft (2 m) $H_2O$. See Chapter 8 for a method of calculating NPSH.

The third reason for elevating equipment is safety. In making explosive materials, such as TNT, the reactor is located above a large tank of water. Then if the mixture in the reactor gets too hot and is in danger of exploding, a quick-opening valve below the reactor is opened and the whole batch is dumped into the water. An emergency water tank may need to be elevated so that, in case of a power failure, cooling water to the plant will continue to flow, and there will be water available should a fire occur. Sometimes this tank is located on a nearby hill.

An elevation plan should be drawn to scale showing the vertical relationships of all elevated equipment. These drawings, as well as the plot plan, are usually sketched by the engineer and then redrawn to scale by a draftsman.

## Maintenance

Maintenance costs are very large in the chemical industry. In some cases the cost of maintenance exceeds the company's profit. The magnitude of maintenance costs is given in Chapter 8. The engineer must design to reduce these costs.

He should determine what types of equipment need to be serviced by mobile cranes. These pieces of equipment will need to be located on the perimeters of the plant or on a roadway. The roadways along which the crane will travel must have adequate overhead and horizontal clearances. A 12 ft (3.6 m) vertical clearance is necessary.[4]

Adequate space must be left around all equipment so that it can be easily serviced and operated. For instance, a floating head heat exchanger must have enough clear space so that the tube bundle can be removed from the shell and taken elsewhere for repairs. One company had problems with a heat exchanger during startup. They tried to remove the tube bundle, but found that they had not allowed adequate space and had to knock an opening in a brick wall. They replaced the wall with a door so that they would not need to remove any more walls when they needed to service it again.

For tanks containing coils or agitators, enough headroom must be provided so that these can be removed. Table 6-5 gives some general clearances for preliminary layouts.

## Construction and Building

Proper placing of equipment can result in large savings during the construction of the plant. For instance, large columns that are field-erected should be located at one end of the site so that they can be built, welded, and tested without interfering with the construction of the rest of the plant.

## Railroads, Roadways, and Pipe Racks

The main purpose of railroads is to provide an inexpensive means for obtaining raw materials and for shipping products. This means that they should be close to raw material and / or product storage. Buildings and loading docks should be set back 8 ft (2.4 m) from the center of the railroad track. Spurs and switches should be laid out with a 100 ft. (30 m) radius.[3] Roads are used not only for these purposes, but

Table 6-5

Clearances for Preliminary Layout

| Clearance, Ft.* | H | O |
|---|---|---|
| Main roads to battery limits (BL) | 30 | 18 |
| Secondary roads, accessways to BL | 25 | 16 |
| Railroads to BL | 50 | 23 |
| Main pipe rack (accessway under) | 15 | 16 |
| Secondary pipe rack | 10 | 12 |
| All other overhead piping | — | 7 |
| Clearance between: | | |
|   Small pump bases, < 25 hp | 2½ | 12 |
|   Large pump bases, > 25 hp | 3 | 14 |
|   Compressors and nearest equipment | 10 | † |
|   Adjacent vertical vessels | 10 | — |
|   Adjacent horiz. vessels, < 10-ft. dia | 4 | 4 |
|   Adjacent horiz. vessels, >10-ft. dia | 8 | — |
|   Adjacent horiz. heat exchangers | 4 | 3 |
| Shell of fired heater and nearest equipment | 50 | — |
| Control houses and reactor or main equipment structures | 30 | — |

* H = Horizontal, O = Overhead

† As needed for maintenance

Source:   House, F.F.: An Engineer's Guide to Process-Plant Layouts," *Chemical Engineering,* July 28, 1969, p. 120.

to provide access for fire-fighting equipment and other emergency vehicles, and for maintenance equipment. This means that there should be a road around the perimeter of the site. No roads should deadend. For safety's sake there should be two ways to reach every location. All major traffic should be kept away from the processing areas. It is wise to locate all loading and unloading facilities, as well as plant offices, warehouses, and personnel facilities, near the main road, to minimize traffic congestion within the plant.

All roadways that are used frequently should be blacktopped. They should be 20 ft (6 m) wide to allow two-way traffic. All turns should have a minimum inner radius of curvature of 20 ft (6 m) and a minimum outer radius of 40 ft (12 m) to provide adequate turning room for large trucks.[3]

Pipe racks (Fig. 6-4) are an elevated collection of pipes that transport utilities as well as raw material, product, and waste streams from one part of the plant to another. They may also be used to transfer information to and from control centers. Placing all the pipes together simplifies their construction and, later, the location of problems. Nothing should be located under pipe racks, since if leaks occur they may damage equipment.

## Planning for Future Expansion and Improvements

In the last chapter the design of equipment for proposed future expansions was discussed. Obviously, if the equipment has been overdesigned to meet the anticipated future expansion, no extra space needs to be provided. If, however, additional equipment will be required, space should be allocated for it. The net result will be an increase in the initial cost of construction and some increase in material transfer costs, because the transfer lines will be longer.

Robertson[5] cited a cost increase of 3% in the initial cost of building a plant if the linear distance between all parts was increased 25%. With such a small increase in costs, even when an expansion is not planned it is usually wise to allow plenty of space between units. This will permit the plant engineers to install improvements in the future to increase yields, eliminate bottlenecks, and improve the stability of the process. These improvements cannot be anticipated where extra space will be needed.

## Buildings

Included with the layout of the plant is the decision as to what types of buildings are to be constructed, and the size of each. When laying out buildings, a standard size bay (area in which there are no structural supports) is 20 ft x 20 ft (6m x 6m). Under normal conditions a 20 ft (6 m) span does not need any center supports. The extension of the bay in one direction can be done inexpensively. This only increases the amount of steel in the long girders, and requires stronger supports.[6]

Lavatories, change rooms, cafeterias, and medical facilities are all located inside buildings. The minimum size of these facilities is dictated by OHSA. It depends on the number of men employed.

Research laboratories and office buildings are usually not included in the preliminary cost estimate. However, if they are contemplated their location should be indicated on the plot plan.

## Processing Buildings

Quality control laboratories are a necessary part of any plant, and must be included in all cost estimates. Adequate space must be provided in them for performing all tests, and for cleaning and storing laboratory sampling and testing containers.

The processing units of most large chemical plants today are not located inside buildings. This is true as far north as Michigan. The only equipment enclosed in buildings is that which must be protected from the weather, or batch equipment that requires constant attention from operators. Much of the batch equipment used today does not fit this category. It is highly automated and does not need to be enclosed.

When buildings are used, the ceilings generally vary from 14 to 20 ft (4 to 6 m). Space must be allowed above process vessels for piping and for access to valves.

One rule of thumb is to make the floor-to-floor heights 8-10 ft (approximately 3m) higher than the sides of a dished-head vertical tank.[6]

Packaging equipment generally must be in an enclosed building, and is often located at one end of the warehouse. If the material being packaged is hazardous, either this operation will be performed in a separate building, or a firewall will separate it from any processing or storage areas.

### Warehouses

The engineer must decide whether warehouses should be at ground level or at dock level. The latter facilitates loading trains and trucks, but costs 15-20% more than one placed on the ground. It is usually difficult to justify the added expense of a dock-high warehouse.

To size the amount of space needed for a warehouse, it must be determined how much is to be stored in what size containers. The container sizes that will be used are obtained from the scope. Liquids are generally stored in bulk containers. No more than a week's supply of liquid stored in drums should be planned. Solids, on the other hand, are frequently stored in smaller containers or in a pile on the ground.

Having decided what is to be stored in a warehouse, the engineer can now approximately size it. For instance, suppose he has decided to store the product in 50lb (23 kg) bags. These are usually stored on pallets that contain 5 bags on each of 8 rows. This means each pallet contains 2,000 pounds (900 kg) of material. If a fork-lift truck is used, these can be stacked 3 pallets high, in rows about 3 deep. If they are stacked higher they tend to be unstable. The standard pallet sizes are given in Table 6-6. The most common size is 40 x 48 in (107 x 122 cm). The rows must be 1 ft (0.3 m) apart in order to permit fork lifts to operate properly. This information is used to size a plant in the example below.

Automatic storage and retrieving equipment can substantially cut down on storage space and the number of operators needed. It is especially useful where there is a large variety of products.

### Example 6-1

A warehouse is to store 500,000 lb of polyvinyl chloride (PVC) in 50 lb bags. How large a warehouse is needed?
Determine bag size.
"Bags vary from 24 in. to 36 in. long, 14 in. to 17 in. wide and 6 in. to 10 in. thick."[7] Suppose there will be 5 bags on every row arranged as below.

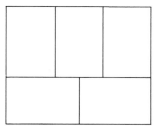

This means that: 2L = 3W
where L = length of a bag
       W = width of a bag
Also 3W = side of a pallet
W + L = side of a pallet
W can vary from 14 to 17 in, or, 3W = 42-51 in. The only standard pallet sizes in this range are 42 in and 48 in. This means W = 14 in or 16 in.
2L = 3 = 42 or 48 in
L = 21 in or 24 in
21 in is small for length.
So make the bags 16 in wide and 24 in long.
Pallet size would be 40 in x 48 in.
The thickness of the bag would be

WLt$\rho$ = 50 lb

where t = thickness
       $\rho$ = bulk density of PVC[8] = 0.45 x 62.4 lb/ft$^3$
       $$t = \frac{50 \times 144}{0.45 \times 62.4 \times 16 \times 24} = 0.667 \text{ ft} = 8 \text{ in}$$

Determine height of 3 pallets (maximum stable height) containing 2,000 lb of PVC.

$$\text{number of bags} = \frac{2,000}{50} = 40$$

$$\text{rows of bags on pallets} = \frac{40}{5} = 8$$

$$\text{height of bags on pallet} = 8 \times \frac{8}{12} = 5.33 \text{ ft}$$

This means the height of pallet and bag $\sim$ 6 ft.
Or the total height of two pallets $\sim$ 18 ft.

This means that to allow space for operating a lift, for voids between bags, and for heating pipes in the warehouse the ceiling should be approximately 24 ft high.

$$\text{The number of pallets needed} = \frac{500,000}{2,000} = 250$$

$$\text{The number of stacks of pallets} = \frac{250}{3} = 84$$

If the pallets are stacked 3 rows deep on either side of an aisle, the length of the aisles would be (84/6) x 4 56 ft.
Assume the width of an aisle is 12 ft.
The area occupied 56 (12 + 6 x 3.33) = 1,800 ft².
This allows for no room between pallets, no cross aisles, no office space, no dock space, no space for servicing lift trucks, no toilets, and so on. From 1,000 to 1,500 ft²

will be needed for these, so assume a 3,000 ft² warehouse is needed. If a packaging area is to be included these space requirements will need to be increased.

Table 6-6

Pallet Sizes

| Inches | Inches |
| --- | --- |
| 24 x 32 | 48 x 60 |
| 32 x 40 | 48 x 72 |
| 36 x 42 | 36 x 36 |
| 32 x 48 | 42 x 42 |
| 36 x 48 | 48 x 48 |
| 40 x 48* | |

* most common size

Source: American Standards Association, from Perry, J.H. (ed.): *Chemical Engineers' Handbook*, Ed. 4, McGraw-Hill, New York, 1963, Section 7, p. 37.

**Control Rooms**

The control center(s) and the electrical switching room are always located in an enclosed building. It is important that both of these services be maintained so that the plant can be shut down in an orderly manner in the case of an emergency. Therefore these buildings must be built so that should an external explosion occur the room will not collapse and destroy the control center and switching center. To avoid this, either the structure must have 3-4 ft (1-1.2 m) thick walls, or the roof must be supported independently of the walls. The Humble Oil and Refining Co. has specified that the building withstand a 400 psf (2,000 kg / m²) external explosive force.[2]

To keep any flammable or explosive vapors from entering the building, it is frequently slightly pressurized. This prevents the possibility of an internal explosion.

## EXPANSION AND IMPROVEMENTS OF EXISTING FACILITIES

When the engineer is laying out an expansion or improvement of an existing facility, he should first get copies of the engineering flow sheets and plot plans of the existing site. Then he must check these to see that they are correct. Often many changes have never been recorded on the drawings. He must then determine how best the proposed changes can be implemented within the restrictions imposed by the present layout. The principles used are the same ones that have been discussed for new plants in this chapter. For a crowded plant, this is like a jigsaw puzzle.

Instead of using old drawings, C.F. Brawn & Co. takes a photograph of the area from an airplane. 1,200 ft (400 m) above the plant. The prints are enlarged to a scale of 1 in 10 ft (3 m). This can easily be done, because certain distances can be measured exactly on the ground. An advantage of this method is that it avoids any drafting errors or omissions.[9]

Old and new plants alike must meet OSHA standards. For many older plants this may mean a major revamping of the facilities. Sometimes, when engineering design cannot meet the standards, the employees may be required to wear protective clothing, ear plugs, or other items.

## CASE STUDY: LAYOUT AND WAREHOUSE REQUIREMENTS FOR A 150,000,000 LB/YR POLYSTYRENE PLANT USING THE SUSPENSION PROCESS

The layout is given in Figurer 6E-1 through 6E-6. Some general considerations that influenced the plans follow:

1. Space was set aside for a whole new train. This appears as dotted lines in the figures.
2. The prevailing wind in the summer comes from the northwest and in the winter comes from the west.[10]
3. The blowdown tank is located on the south side of the plant where winds will not generally carry any spills over the plant.
4. The utilities and waste treatment areas are located on the north side of the plant where they will be upwind of the plant.
5. The styrene storage will be located on the south side of the plant 300 ft from the river and the dock. It will be 300 ft from the processing area.
6. The warehouse and bulk storage will be located on the west side, upwind from the plant and styrene storage. They will be at least 250 ft from the reactor area.
7. The reactor and feed preparation area will be on the east side of the plant 200 ft from the river.
8. The other processing areas will be between the reactor area and the warehouse. They will be over 200 ft from the reactor area.

Some specific considerations follow:

1. There must be enough headroom above the reactor to remove the agitator.
2. There must be enough room to remove the screw from the extruder.
3. Gravity feed is to be used for charging additives to the reactor, for discharging the reactor to the hold tanks, and for feeding the dryer.
4. Each of the styrene storage tanks will have a dike around it that is capable of containing the tank's contents when it is full.

### Additive Storage

The storage capacity by component is:
  Benzoyl peroxide — 32,500 lb (108 drums)
  Tricalcium phosphate — 130,00 lb (65 pallets containing 2,000 lb each)
  Dodecylbenzene sulfonate — 1,560 lb (approximately 40 5-gal containers)
  Polybutadiene — 884,000 lb (17,680 bales)

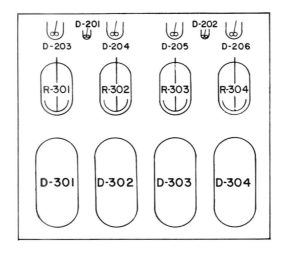

Figure 6E-1   Reactor building — front view.

The space requirement for each material is:
  Benzoyl peroxide — 125 ft², approx. 9 ft x 27 ft (assuming 4 drums on a pallet 3 ft x 3 ft stacked 3 high)
  Tricalcium phosphate — 440 ft², approx. 10 ft x 31 ft (assuming a 40 in x 48 in pallet stacked 3 high)
  Dodecylbenzene sulfonate — 20 ft²
  Polybutadiene — 1,800 ft², approx. 9 ft x 200 ft (assume 8 ft high stacking)

If two 12 ft roadways are used and all pallets are stacked 2 deep, then the warehouse would be approximately 60 ft x 80 ft. This leaves enough space at one end for a roadway. Some extra space will be needed for ion exchange resins and regenerants.
Specify it to be 60 ft x 100 ft. (This permits the use of 20 ft spans.)

**Product Warehouse**

The warehouse must be able to accommodate:
  7,800,000 lb of P.S. in 50 lb bags
  3,900,000 lb of P.S. in 200 drums
  3,900,000 lb of P.S. in 1,000 lb cartons
If calculations similar to those in Example 6-1 are performed:
  Area needed for bags 833 ft x 36 ft (stacked 3 high, 3 deep)
  Area need for drums 542 ft x 36 ft (assuming a 42 in x 42 in pallet stacked 4 high)
  Area need for cartons 421 ft x 36 ft (height of packed area 13.5 ft)
The total area is 65,000 ft². This area should be increased by 5% for extra aisles, an office, and storage for fork lifts. Specify a building 180 ft x 380 ft.

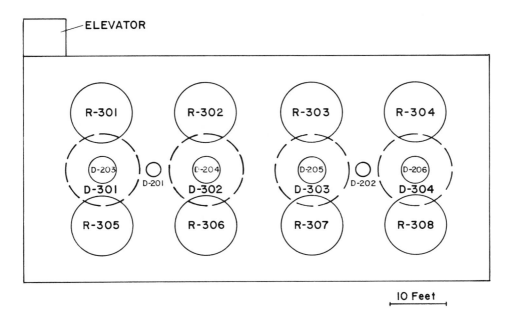

Figure 6E-2   Reactor building - top view.

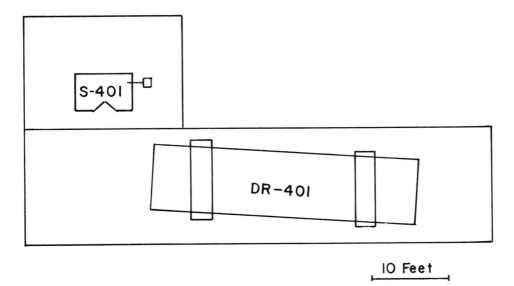

Figure 6E-3   Finishing building: drying area. Front view.

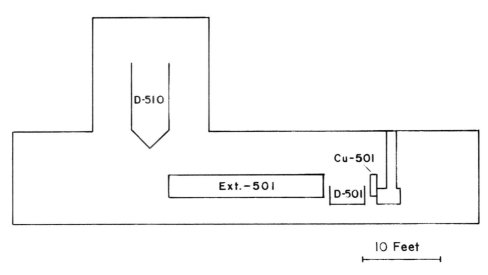

Figure 6E-4   Finishing building: extruder area front view.

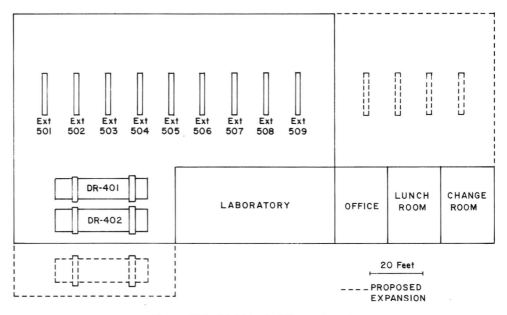

Figure 6E-5   Finishing building — top view.

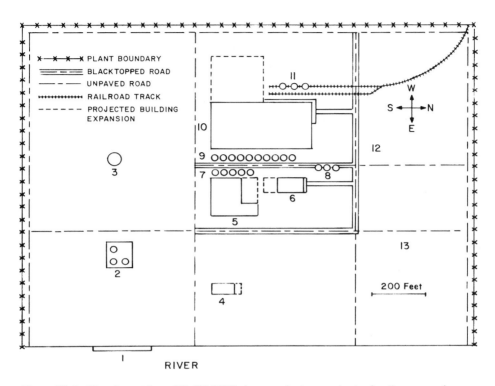

Figure 6E-6    Plant layout for a 150,000,000 lb / year polystyrene plant using the suspension process.

Key to Figure 6E-6
Plant Layout

1. Barge Dock    BD-101
2. Styrene Storage    D-101 - D-103
3. Blowdown Tank
4. Reactor Building
5. Finishing Building
6. Raw Materials Warehouse
7. Temporary Styrene Storage    D-519 - D-523
8. Truck Bulk Loading Storage    D-614 - D-616
9. Bulk Product Storage    D-601 - D-610
10. Product Warehouse
11. Railroad Bulk Loading Storage    D-617 - D-619
12. Parking Lot
13. Utilities and Waste Treatment Area

    The outer dimensions of the plant are 1,700 ft x 1,250 ft.

# References

1. "Chementator," *Chemical Engineering,* July 27, 1970, p. 80.
2. "Safety vs. Automation," *Chemical Engineering,* Jan. 3, 1966, p. 64.
3. House, F.F.: "An Engineer's Guide to Process-Plant Layout," *Chemical Engineering,* July 28, 1969, p. 120.
4. *Safety Recommendations for New Refining, Chemical Gas and Gas Liquids Plants,* Gulf Oil Corp., Pittsburgh, 1971.
5. Robertson, J.M.: "Design for Expansion," *Chemical Engineering,* May 6, 1968, p. 187.
6. Thompson, D.: "Rational Approach to Plant Design," *Chemical Engineering,* Dec. 28, 1959, p. 73.
7. Perry, J.H. (ed): *Chemical Engineers, Handbook,* Ed. 4, McGraw-Hill, New York, 1963, Section 7, p. 36.
8. Dow Chemical Company Product Information Sheet.
9. "Aerial Photos Help Engineers to Plot a Plant Modernization," *Chemical Engineering,* Nov. 22, 1965, p. 89.
10. *Evaluated Weather Data for Cooling Equipment Design. Addendum No. 1, Winter and Summer Data,* Fluor Products Company, Inc., Santa Rosa, Calif., 1964

# CHAPTER 7

# Process Control and Instrumentation

One of the basic concepts in chemical engineering is the existence of a steady state for a flow system. This implies that if all inputs plus all operating and environmental factors are held constant over a long enough period of time, all the variables will attain constant values at a given point in space. It is tacitly assumed in most undergraduate chemical engineering courses. If steady-state conditions could be easily attained and maintained, there would be no reason to mention the subject of process control in this book. However, in any process there are changes that will affect the uniformity of products, by-products, and/or costs unless some corrections are made.

Some factors change continually, such as the temperature of the ambient air and the temperature of cooling water. The temperature of cooling water may vary from near freezing in winter to 80°F (27°C) in the summer. Even on a given day this temperature may vary between 5AM and 3PM by 10°F (5°C). The temperature of the feed streams coming from storage vessels may differ by twice this amount between the same hours. These changes affect the amount of energy transferred, since the rate of heat transfer in a heat exchanger depends on the difference between the temperatures of the two streams. If the temperature of the cooling water increases, then its flow rate must be increased if it is to remove the same amount of energy per unit time from the other stream in the exchanger.

Besides the environmental changes, there are usually changes in the feed composition. It has already been noted that when raw materials are obtained from different sources they will vary considerably. Even those obtained from the same source will differ from batch to batch. There is such a wide variation in the composition of crude petroleum that if a knowledgeable person analyzes a sample he can tell not only the region of the world it came from, but from which oil field the sample was obtained. While most environmental changes are gradual, the changes in feed composition can be very abrupt when the operator switches from one storage tank to another.

Another type of abrupt upset occurs when there is a failure of some utility or machine. Power failures have been discussed in Chapter 2. The cutting off of cooling water can be just as disastrous.

The purpose of process control is to assure whenever possible that the plant can continue to operate safely, efficiently, and profitably regardless of what upsets occur. When this is not possible, its goal is to shut the plant down safely and/or warn the operators so that they may take appropriate action.

Control is considered at this point because there is an interaction between process control and design. A plant designed solely on the basis of a steady-state analysis may be very difficult to control. Therefore, the process engineer must consider what problems are likely to arise and how best to cope with them. Before he can do this, however, the engineer must decide which variables must be controlled.

## PRODUCT QUALITY

The scope has specified the quality of the product. To obtain this quality certain items must be accurately controlled. The process engineer must look at the process and determine what steps control what qualities.

In the production of polyvinyl chloride by the emulsion process, the percentages of catalyst, wetting agent, initiator, and solvent all affect the properties of the resultant polymer.[1] They must be carefully metered into the reaction vessel. The vinyl chloride used must also be very pure. Either the scope must specify that the purchased raw material shall meet certain specifications, or some purification equipment must be installed so that the required quality can be obtained.

For many reactions the temperature and pressure determine the mix of the products. Under these circumstances it is very necessary to control these variables accurately.

The stream leaving a reactor not only contains the product but may also include a series of by-products, unreacted raw material, solvents and various catalysts, surfactants, initiators, and so on. The unwanted material is removed from the product in a series of separations. Since each separation step is intended to remove certain compounds, the composition of the leaving stream must be monitored and controlled if the product quality is to be maintained constant. In the production of ethylene, $C_3$ compounds, $C_4$ compounds, and $C_5$ compounds are all removed separately and are sold as by-products. Each of these by-product streams must also meet certain specifications and therefore must also be monitored and controlled (see Table 4-1).

## PRODUCT QUANTITY

Besides the quality of various streams, their quantity must also be controlled. If the product bins are nearly full the production rate must be slowed down. Later, after a number of shipments to customers have been made, the rate may be increased. This is called *material balance control*.

Often the throughputs of the various process steps in a plant are different, even though on paper they were designed to be the same. This could result in an inadequate amount of feed to one unit while for another unit the feed rate is too great to be handled properly. Again, some type of material balance control is necessary.

In a petroleum refinery a large number of different products are produced, and the demand for some of these products is seasonal. For instance, there is not much need for residential fuel oil in the summer. The price of products also varies from day to day. To optimize the company's profit, it is therefore necessary periodically to vary the amount of each product produced. This can be done by changing the amounts of material sent to cracking units and reformers and by changing the conditions in these and other process steps. Some petroleum companies provide a computer with the data on market prices, current inventories, and crude oil compositions. The computer output then specifies the operating conditions that will yield the greatest profit for the company. The computer could then make the changes in these conditions directly, or this could be done manually.

## PLANT SAFETY

The importance of safety has been discussed in Chapter 4. At this point, it is necessary to look at each stage of the process to see what might go wrong and, if this happened, what might be done to prevent any mishap from occurring.

For instance, a level gauge in a tank with an alarm on it could warn an operator that unless something is done quickly, the tank may overflow. A spill not only wastes material, but could present a fire hazard as well as a hazard to the environment. A similar alarm system might also alert the operator that the tank is nearly empty. This may be important because a pump downstream may burn up if it continues to operate with no feed.

In some cases more drastic action may be necessary. For instance, when the temperature of a reactor exceeds a given value a reaction inhibitor may be added to the mixture. Or, as noted before, when small solid particles are being handled in the presence of air, a fire-suppression system may be installed that will quickly snuff out any incipient explosion.

## MANUAL OR AUTOMATIC CONTROL

Most systems can be controlled manually or automatically. The modern trend is to automate the process as much as possible. One reason is that automatic controllers always respond the same way to changes, whereas men are erratic. Controllers may work for years with only minor maintenance, whereas a man fatigues easily. This means that while controllers may not produce a better product than an alert man, they can, in the long run, produce a more uniform product, with less waste and fewer accidents.

The use of controllers may also reduce over-all expenses. The average operator in a chemical plant, when fringe benefits are included, costs the company over $5.00 per hour. This is equivalent to $43,800 per year for an operating position. By the methods given in Chapters 10 and 11, this can be shown to be equivalent to a net present value of −$320,000 (assumes money is valued at 8%). The average controller, installed, costs between $3,000 and $6,000. This means that a large number of controllers can be installed, maintained, and operated for the price of one operating position.

Some plants can be operated essentially without any people. However, for safety purposes there usually are two employees per shift. Then if some mishap should occur to one man, the other can obtain help. This implies that a plant can be over-automated. The operators can become bored if they do not have some tasks to perform. If these are make-work tasks, the operators will rapidly determine this, and either the tasks will be ignored or the reports will be falsified. To keep these men alert, and sufficiently knowledgeable and involved so that they can respond quickly and properly when an emergency arises, it may be best not to automate the plant totally.

## CONTROL SYSTEM

A control system consists of four stages. First, the item to be controlled must be measured. This reading must then be compared with some desired value, called the *set point*. Depending on the result of this comparison, a decision must be made whether some variable(s) in the process should be changed. Then if a change is indicated, the amount of change required must be determined and it must be instituted. The comparison, decision making and size change determination are considered part of the controller.

At this stage of design, the details of the whole control system need not be specified. It is only necessary to determine what variables are to be measured, which are to be controlled, and how this is to be accomplished.

The controller and its quantitative interaction with the system will not be covered in this text. Numerous books have been written about this, and most chemical engineering curricula have a course that is devoted solely to the topic.

## VARIABLES TO BE MEASURED

The ideal variable to measure is one that can be monitored easily, inexpensively, quickly, and accurately. The variables that usually meet these qualifications are pressure, temperature, level, voltage, speed, and weight. When possible the values of other variables are obtained from measurements of these variables. For example, the flow rate of a stream is often determined by measuring the pressure difference across a constriction in a pipeline. However, the correlation between pressure drop and flow is also affected by changes in fluid density, pressure, and composition. If a more accurate measurement is desired the temperature, pressure, and composition may also be measured and a correction applied to the value obtained solely from the pressure difference. To do this would require the addition of an analog or digital computer to control scheme, as well as additional sensing devices. This would mean a considerable increase in cost and complexity, which is unwarranted unless the increase in accuracy is demanded.

Composition is another variable that is often measured indirectly. A temperature-sensing device is often used at the top of an atmospheric distillation column to indicate the composition. If the material being measured is a binary mixture, then from the phase rule it can be shown that this is a very accurate procedure, provided equilibrium exists between the gas and liquid phases and the pressure is constant. However, if more than two compounds are present this procedure will be inadequate unless the composition of all but two of the components are held constant. The process designer must decide whether this is a reasonable assumption.

There are other, more direct means of measuring compositions. One is the use of a semicontinuous device such as a gas chromatograph. These instruments analyze a sample obtained from the process. Until the analysis of that sample is complete, another sample cannot be processed. For some compounds it may take 15 minutes or longer to process the sample. In some processes with short response times such delays may not be permissible. Another disadvantage is that chromatographs have an installed cost at least eight times greater than that of a temperature-measuring control system. They also require more maintenance and have a greater operating cost.

A list of onstream process analyzers, operating principles, manufacturers, costs, advantages, and disadvantages is given in reference 2. A list of process instrument elements with their accuracy and principles of operation is given in reference 3.

Sometimes it is not possible to use even a semicontinuous onstream analyzer, and a sample must be analyzed in a laboratory. For example, the blow-molding characteristics of a plastic must be tested off-line. This often requires that the product be temporarily stored until the laboratory results are obtained.

## FINAL CONTROL ELEMENT

To be able to control implies that there is some means of manipulating a variable. The element that makes the change in the variable is called the final control element. This is a valve, switch, or other item that is activated by the controller in order to maintain the measured variable at some desired value or within some set limits. For instance, the level in Figure 7-1 is controlled by opening and closing a valve that changes the flow rate out of the tank. The valve is the final control element.

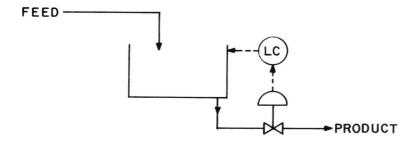

Figure 7-1  Feedback level control system.

## CONTROL AND INSTRUMENTATION SYMBOLS

A list of symbols to assist the designer in representing control and instrumentation schemes has been standardized by the Instrument Society of America.[4] A partial selection and some examples are given in Tables 7-1 and 7-2. The symbols are placed in circles on a flow diagram (Fig. 7-1). A line from the circle to the flow line or piece of equipment indicates its location. The first symbol indicates what is being measured (such as temperature, pressure, level) and the following letters indicate what is to be done with the measurement (such as record, control, transmit). The number given within the circle together with letters identifies the specific instrument. The numbers are usually assigned in the same way equipment numbers are assigned. A horizontal line through the center of the circle indicates that the instrument is mounted on a panel board. Absence of such a line indicates that the instrument is mounted on or near the thing it is monitoring.

Table 7-1

Instrument and Control Symbols

| Symbol | Meaning |
|--------|---------|
| | Measuring Devices |
| E | Voltage |
| F | Flow Rate |
| FQ | Flow Integrator or Totalizer |
| L | Level |
| P | Pressure |
| Pd | Pressure Differential |
| pH | $-\log [H_3O^+]$ |
| S | Speed |
| T | Temperature |
| W | Weight |
| | Functions Performed |
| C | Control |
| HA | High Alarm |
| LA | Low Alarm |
| H/LA | High and Low Alarm |
| I | Indicate |
| R | Record |
| HS | High Switch |
| LS | Low Switch |
| H/LS | High and Low Switch |
| V | Valve |

These symbols, with the exception of pH and Pd, are standardized symbols set by the Instrument Society of America in monograph no. ISA-S5.1.[5]

Table 7-2

Examples of Instrument Symbols

| Symbol | Meaning |
|--------|---------|
| FI | Flow indicator |
| LRC | Level-recording controller |
| LICV | Valve for a level-indicating controller |
| LHS | High-level switch |
| PLA | Low-pressure alarm |
| PIC | Pressure-indicating controller |
| TH/LS | High/low temperature switch |
| TRC | Temperature-recording controller |
| TRCV | Valve for a temperature-recording controller |
| WR | Weight recorder |

Table 7-3

Instrument Line Symbols

All lines shall be fine in relation to process piping lines

(1) Connection to process, or mechanical link, or instrument supply*

(2) Pneumatic signal †, or undefined signal for process flow diagrams

(3) Electric signal

(4) Capillary tubing (filled system)

(5) Hydraulic signal

(6) Electromagnetic § or sonic signal (without wiring or tubing)

When a control system is being used, the designer must indicate the transmission lines that will connect the measuring element to the final control element. The notation for these transmission lines is given in Table 7-3. The final control element has the same code letters and numbers as the controller that regulates it, plus an additional letter to indicate whether it is a valve, switch, or other device.

## AVERAGING VERSUS SET POINT CONTROL

The type of problem studied extensively in chemical engineering process control classes is one in which it is desired to keep a measured quantity as close to a specific predetermined value, the set point, as is physically possible. This is the type of control used to maintain product quality. It is referred to as *set point control*. The usual example is a tank in which it is desired to keep the liquid level constant. This is accomplished by measuring the level and, if it is too high, opening a valve that will increase the efflux of fluid from the tank. The amount the valve is open is decreased when the level begins to diminish (Fig. 7-1). This means that the flow exiting from the system is varying with the level in the tank. Any process downstream of the tank will then be confronted with a widely varying flow rate.

For many plants this is a very poor system, since usually the engineer does not care about the level in the tank—he is more interested in maintaining a constant flow to the processing step. However, should he ignore the level in the tank and use a constant-volume pump to maintain a constant flow rate, the tank may either overflow or become emptied. Neither of these possibilities is acceptable. For this case the use of *averaging flow* is desirable. For averaging flow, the designer first determines the maximum and minimum permissible heights of the fluid in the tank. He then sets the flow rate out of the tank when the level is at its maximum to be somewhat larger than the greatest input flow rate. The flow rate out when the level is at its minimum is set somewhat below the lowest input flow rate. The output flow rate at any intermediate level is then given by the following equation:

$$F = F_{min} + \frac{(F_{max} - F_{min})}{(L_{max} - L_{min})} (L - L_{min})$$

where $F$ = flow rate out of tank

$F_{max}$ = a flow rate somewhat greater than the maximum input flow rate

$F_{min}$ = a flow rate somewhat less than the minimum input flow rate, or zero

$L$ = level in the tank

$L_{max}$ = maximum desired level in the tank

$L_{min}$ = minimum desired level in the tank

Averaging control damps out fluctuations. It responds slowly to changes. The damping is enhanced by increasing the size of the tank. This type of control differs from product quality control, which should respond quickly. Therefore, in product quality control loop holdup should be minimized, while for averaging control it should be large. Averaging control is indicated on diagrams by two lines from the system to the measuring device, as shown in Figure 7-2.

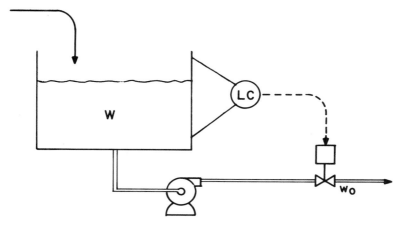

Figure 7-2   Material balance control in the direction of flow (Averaging Control).
Source: Buckley, P.S.: "Input of Process Control Advances on Process Design," paper presented at A.I.Ch.E. Process Control Workshop in Memphis, Tenn., Feb. 5, 1964.

Page Buckley[5] suggests that for averaging control the surge tank should be designed so that the volume divided by the maximum flow rate is at least 10 times the average frequency variation of the input. This implies a knowledge of the system dynamics. There is no steady-state method available for sizing these tanks. Buckley[5] further notes that when an averaging technique was adopted to reduce the feed-rate and composition fluctuations prior to a distillation column, the utility costs were reduced, there were fewer losses, and the column capacity was increased 5-10%.

## MATERIAL BALANCE CONTROL

The amount of product sold may vary seasonally or randomly. The usual way to adjust for this is to increase or decrease the feed rate to the system, according to expected demand for the product or the size of the product inventory. The altering of the feed rate, in turn, increases or decreases the feed to each of the succeeding units, and eventually the output is changed. This is diagrammed in Figure 7-3 and is called *mass balance control in the feed direction.*

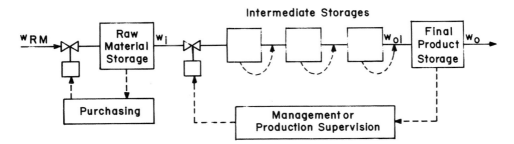

Figure 7-3    Overall material balance control with intermediate material balance controls in the direction of flow.
Source: Buckley, P.S.: "Input of Process Control Advances on Process Design," paper presented at A.I.Ch.E. Process Control Workshop in Memphis, Tenn., Feb. 5, 1964.

Buckley[6,7] has pointed out that this type of control results in a large product inventory, because of the long lag in the control system. He suggests that the system shown in Figure 7-4 be installed. This is known as *mass balance control opposite to the stream direction*. Here each unit controls the output of the previous unit. Buckley claims that a process with six intermediate stages that uses mass balance control in the feed direction may require a 25- to 30-fold greater product inventory than the other scheme. This, of course, assumes that the plant can operate efficiently over a very wide range of production rates and that it has been designed for the peak instead of the average rate. The control becomes very complicated if there are two or more product streams leaving one unit. This is discussed in reference 8. These two types of control are further illustrated in Figures 7-2 and 7-5.[9]

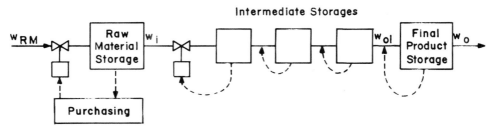

Figure 7-4    Overall material balance control in direction opposite to flow.
Source: Buckley, P.S.: "Input of Process Control Advances on Process Design," paper presented at A.I.Ch.E. Process Control Workshop in Memphis, Tenn., Feb. 5, 1964.

### TEMPERED HEAT TRANSFER

The driving force in any heat exchanger is the temperature difference. The rate of heat transfer can be quickly changed by changing this difference. A tempered system is designed to quickly and accurately control the temperature of an input stream. This system requires two sources of feed. One must be above the desired temperature and the other below it. These are mixed together to obtain the desired temperature. This can be done by using a ratio controller, as is illustrated in Figure 7-6.

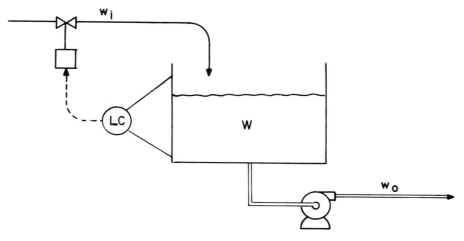

Figure 7-5   Material balance control in the direction opposite to flow.
Source: Buckley, P.S.: "Input of Process Control Advances on Process Design," paper presented at A.I.Ch.E. Process Control Workshop in Memphis, Tenn., Feb. 5, 1964.

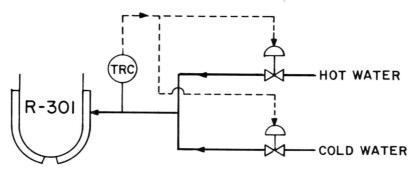

Figure 7-6   Tempered water system.

A modification of this scheme that can be used when only one stream is available is given in Figure 7-7. It is useful for controlling the input temperature to a reactor. For this system the greater part of a process stream is heated to a temperature above that desired. This is then mixed with a portion (usually around 15%) of the same stream that has not been heated. This system requires a larger heat exchanger than would be required if the whole stream went through the exchanger and the output temperature were controlled by the flow rate of the utility stream.[10]

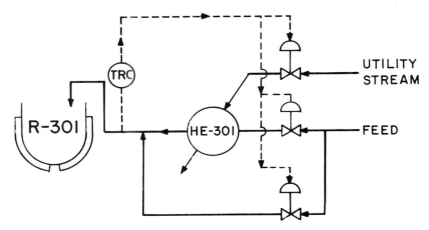

Figure 7-7    Precise temperature control of the feed to a reactor.

## CASCADE CONTROL

In one typical situation the temperature of the product stream is controlled by manipulating a valve that regulates the amount of steam entering an exchanger (Fig. 7-8). Should the upstream steam pressure increase, this will increase the flow rate of steam through the control valve. The steam pressure in the exchanger will increase, which in turn will increase the rate of heat transfer to the process stream. The result will be a change in the product temperature due to a change of the steam pressure. The system will eventually return to the desired temperature.

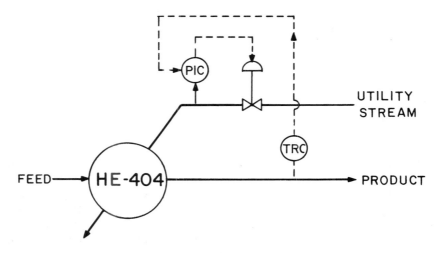

Figure 7-8    Feedback temperature control.

If very close control is desired, then any disturbance due to steam pressure changes should be minimized. Figure 7-9 shows how this can be done using a cascade control system. In this case, the temperature of the process stream is measured and compared to its desired value, as before. The output of the controller, however, instead of affecting the control valve, regulates the set point of a second controller, the steam-pressure controller. This controller compares the set point determined by the first controller with the pressure downstream of the steam valve. If there are any differences it then adjusts the steam valve. If the downstream pressure changes, a correction in the control valve is made immediately, instead of waiting for a product temperature change. Should the output temperature of the process stream rise, this would cause a set point change of the steam-pressure controller, which would cause a decrease in the steam pressure in the heat exchanger. Cascade control is very useful when the variation in the quality of a utility or other manipulable stream can cause deviations from the desired output.

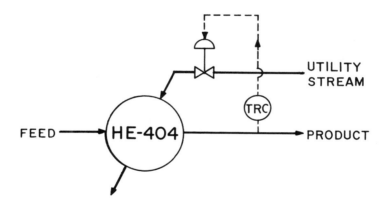

Figure 7-9   Cascade temperature control.

## FEEDFORWARD CONTROL

When close control is desired, usually the variable that is to be closely controlled is monitored and no changes are made until the measurement differs from what is desired. This is feedback control. It obviously is not an ideal system, since the controller can only react to changes. A better system would be one that anticipates a change and takes corrective action that ensures an unvarying output. This is a feedforward control system. This type of control is very advantageous when the input variables have a wide range of variation.

Since it is impractical to measure everything that may affect the output variables, even when feedforward control is used feedback control is also included. Figure 7-10 shows how a feedforward system might be used on a waste neutralizer.[11] The purpose of the waste neutralizer is to make certain that the streams leaving the plant are neutralized. First, all the streams are combined together and the feed rate and

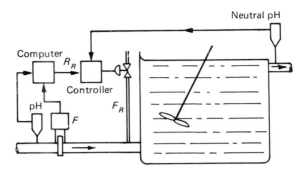

Figure 7-10  A control system for the neutralization of waste by reagent R.
Source: Friedman, P.G., Moore, J.A.: "For Process Control Select the Key Variables,"
*Chemical Engineering,* June 12, 1972, p. 90.

acid content (pH) are measured before they enter the neutralizer. The amount of reagent necessary to neutralize this feed is determined and is added. A feedback system is also included; it either increases or decreases the calculated amount of base added, depending on the exit value of the acid content (pH).

## BLENDING

The neutralizer in the previous example might be controlled differently if the main fluctuation in the load occurs in one or two of the streams. Instead of combining all the streams together before they enter the neutralizer, those streams that vary widely might enter an additional holding tank, where they would be neutralized using traditional feedback control. They would then be added to the main neutralizer, which also has a feedback controller. Which system is best can be determined by running an economic analysis (see Chapters 10 and 11).

Both of the control schemes for the neutralizer took measurements on the major varying streams before they were diluted in the large blending tank. This is usually desirable because once the streams are mixed the measurable differences are smaller, and the possibility of noise (the equivalent of static in radio signals) affecting the measurement accuracy is greater.

## DIGITAL CONTROL

All of the control schemes mentioned previously can be accomplished using digital control or the traditional analog control methods. For many processes there is at present no economic justification for a digital control system if only physical disturbances are to be considered. One exception to this is a multiproduct batch processing plant, in which a number of operations are required.

Digital computers, however, can do some things the analog equipment cannot. They offer the opportunity of using more advanced control concepts. These are not necessary for 90% of the processes, but they can increase production and reduce waste for some hard-to-manage processes. For multiproduct plants, direct digital control can be used to optimize the product mix with regard to current selling prices, inventories, and projected demand. Computers can also perform other useful functions while controlling. They can replace clerks by maintaining inventory and product-quality records, running economic balances, and simplifying billing and ordering techniques and record-keeping. The computer can also provide rapid and safe automatic startup and shutdown procedures.

The major stumbling blocks in the way of direct digital control are that hardware costs are high and engineers and operators are unfamiliar with the computers. Price is a deterrent even though between the late 1950s and 1968 the cost of process control computers dropped[12] by a factor of 10. However, this is not the only cost. For a computer in the hands of knowledgeable engineers, software expenses will amount to as much as hardware costs. This may increase to 4 times as much for experimental projects.[13] Of the total computer control project costs, 20-50% are connected with measuring, programming, and controlling the signals coming to and from the process.[14] The instrument maintenance costs will also increase about 20% if a computer is used to control the process. Another factor not included in the costs is that some plants have had startup times exceeding one year.[14] In 1969 Baker and Weber estimated that a computer system for a batch plant would cost between $400,000 and $900,000.[15] This includes the costs for the computer and its associated hardware instruments, engineering, and programming.

Another area of concern, which has been alluded to before, is that when the computer acts like Big Brother the operator will become less familiar with the process. Then should something go wrong, he will be slower and less able to cope with an emergency. Huters[16] suggests that this can be corrected by proper signals and displays within the control room. This of course means an increase in costs.

No definitive answers can be given as to which system will be best in the future, since this is a rapidly progressing area. Only the questions can be raised.


## PNEUMATIC VERSUS ELECTRONIC EQUIPMENT

The death knell for pneumatic control equipment has been predicted for at least the past 15 years. So far this has not happened, but it is still predicted. The major reason why pneumatic equipment is so popular is that the pneumatic control valve is cheap and requires little maintenance. The pneumatic system also has the advantage of posing no problems in the presence of flammable substances. (Extreme care must be exercised if electrical signals are used in such environments.) One major problem with pneumatic systems is the delay encountered in sending a pneumatic signal over 300 ft (90 m). However, this can usually be avoided by mounting the controller next to the unit instead of in the control room. This does not affect the monitoring of the process, which can still be done in a remote location.

The electronic system is obviously preferable if an online computer is to be used. Pneumatic equipment could be used, but the cost of interfacing equipment would make it more expensive.

## CASE STUDY: INSTRUMENTATION AND CONTROL FOR A 150,000,000 LB/YR POLYSTYRENE PLANT USING THE SUSPENSION PROCESS

Figure 7E-1 shows how the plant will be instrumented and controlled. An explanation of why these specific schemes were used follows.

### Analog vs. Digital Control

Since this is a batch system, it might be advisable to use direct digital control. Undoubtedly the throughput could be increased over that with the more traditional analog control system. However, the initial costs and maintenance expenses would also increase. To fully instrument the system would also greatly complicate the equipment required, especially for feeding the reactors (this is discussed later). An economic balance should be run to determine whether this is feasible. I feel it would not be warranted, and have chosen to instrument the plant in the traditional way.

### Reactor

The quality of the product is dependent upon the amount and composition of all inputs to the reactor and the temperature within the reactor. The controls connected with these items must keep the variables as close as possible to the desired value. This could be done entirely with instrumentation. However, this would be rather complicated, and, as noted before, would greatly increase the capital and maintenance costs. The catalyst, rubber stabilizer, and suspending agent will be weighed manually, and then charged directly to the mixing tanks. The styrene will be automatically metered into the additive mixing tanks, the rubber dissolving tanks, and the reactors. The water will be automatically metered into the reactors. This will be done with a positive displacement meter.[17] When the desired amount has been charged to a vessel, the meter will close a valve in the inlet line and shut off the pump supplying the material. The pump will be activated by an operator in the control room when it is time to prepare another batch. The material in the additive mix tank and the rubber-dissolving tank will be discharged into the reactor by a solenoid valve, which will be operated from the control room.

The reactants entering the system need not be purified before they are used. However, they will need to be periodically checked in the laboratory to determine that they meet the specifications set by the scope. Any material not meeting specifications will be returned to the supplier.

The temperature of the reactor could theoretically be controlled by changing the flow rate or the temperature of the water in the jacket. It will now be shown that the former is impractical. The over-all heat transfer coefficient is given in the major equipment section as around 50 BTU/hr ft²°F or greater. This means that the major resistance to heat transfer is the film on the inside of the reaction vessel.

If the thermal resistance of the stainless-steel wall is ignored, then

$$U = \frac{1}{\dfrac{1}{h_i} + \dfrac{1}{h_o}}$$

where    $h_i$ =  inside heat transfer coefficient—reactant fluid to stainless
          steel wall

  $h_o$ =  outside heat transfer coefficient—heat transfer medium to
         stainless steel wall

  $U$ =  over-all heat transfer coefficient for the jacketed vessel

The over-all heat transfer coefficient will next be determined for values of the outside heat transfer coefficient that differ by a factor of 2.

When $h_o$ = 200 and $h_i$ = 60, then U = 46. When $h_o$ = 400 and $h_i$ = 60, then U = 52. The result of this change is a 13% increase in the amount of energy transferred. Further note that to double the outside heat transfer coefficient requires more than a doubling of the flow rate through the jacket. This indicates that the temperature in the reactor cannot be adequately controlled by changing the flow rate of water to the jacket. A change in the flow rate barely changes the rate of heat transfer. The only practical means of control is to regulate the incoming temperature. A tempered water system will be used. This requires two sources of water at different temperatures. One should be at a temperature below the lowest ever desired. The other should be at a temperature greater than will be needed. These two streams are then mixed together to give the desired temperature.

Cascade control, along with ratio control, is used to control the temperature. The cold-water line is to have an air-to-close control valve. In case of failure in the air supply, the valve would open fully and a runaway reaction would be prevented. The hot-water line will have an air-to-open valve for similar reasons. After the two streams are mixed, the temperature will be measured. If it is above the desired temperature, the amount of air supplied to the valves will be reduced. This will increase the cold-water flow rate, and decrease the hot-water throughput. The result will be a reduction in the inlet water temperature. The desired temperature will be determined from a measurement of the reactor temperature. A deviation from the desired temperature will cause the set point of the second controller to be changed. This will result in a change of the inlet water temperature.

The cold-water supply for the tempered water system will be ordinary cooling water. No attempt will be made to keep its temperature constant. The hot-water temperature will be maintained constant by opening and closing the steam input to the hot-water storage tank. Close control is not necessary.

The temperature of the water entering the reactor will be controlled by sparging steam directly into the feed tank. The entering styrene temperature is to be controlled by manipulating the steam pressure on the shell side of the styrene heat exchanger.

**Styrene Storage Tanks**

The styrene storage tanks will be equipped with level indicators and a high-level alarm and switch. If the level in the tanks becomes too high, the feed will automatically be switched to one of the other tanks. The temperature of the styrene will be monitored. If the temperature should exceed 86°F, the operator will be alerted by an alarm bell. He can then take any action deemed appropriate.

176

Figure 7E-1 Piping and Instrument Diagram for a 150,000,000 lb/year Polystyrene Plant Using the Suspension Process. [a]

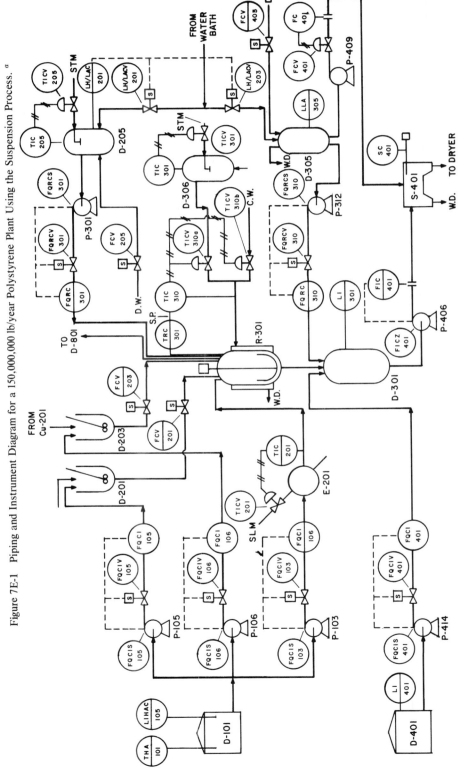

Figure 7E-1 continued on following page.

(Figure 7E-1 continued)

177

$^a$ Special Symbols and Unlisted Equipment appearing in Figure 7E-1: D-306 Tank supplying hot water to the tempered water system: FICZ-401 Voltage regulator on a variable-speed motor, which is manipulated to maintain a constant flow rate to the centrifuge; TRCZ-405 Voltage regulator on the blower, which is manipulated to maintain a constant exiting wet-bulb temperature for the dryer; SIZ-501 Synchronized motor on the cutter, which is controlled by the extruder screw speed.

## Wash Tank

The water and hydrochloric acid charged to the wash tanks will be regulated in the same way as the amounts of styrene and water are to be metered into the reactors.

## Centrifuge

Centrifuges are designed for a given feed rate, and the rate will be maintained close to that value by varying the speed of the centrifugal pumps. The rate of rotation of the centrifuge must also be controlled. Some type of warning and shutdown system should be included. It should indicate when there are excessive vibrations and when there has been a failure of some needed utility.[18] It should alert an operator by ringing a bell or causing a siren to blow, and safely shut the system down. Often these devices are supplied by the centrifuge manufacturer. In our case, it will be specified that these controls are to be included with the centrifuge when it is purchased.

## Dryer

To prevent bubble formation during extrusion, the polystyrene leaving the dryer must contain less than 0.05% water (see the scope). It would be desirable if the moisture content of the polystyrene could be continuously measured, either directly or indirectly. Unfortunately, this cannot be done reliably, so instead the mass and heat-transfer driving forces will be controlled. The dry and wet bulb temperatures of the gases leaving the dryer will be measured and used to control the inlet temperature and throughput of the air. The temperature of the polymer cannot exceed 185°F, or the heat distortion properties will be affected. Therefore, the exit air temperature will be controlled at 185°F. If it gets too high, the inlet air temperature will be reduced using a cascade control system. The temperature of air leaving the air heat exchanger will be controlled by regulating the pressure (and, hence, the temperature) of the steam in the jacket of the air heat exchanger. The desired pressure will be determined by the exit temperature of the airstream from the dryer. The flow rate of the air through the system will be controlled by the wet-bulb temperature of the airstream leaving the dryer. The set point will be determined experimentally during startup.

An averaging means of control will be used to regulate the feed rate to the dryer.

## Extruder

An extruder is a complicated device to control. Often the barrel is divided into three sections, and the temperature at the exit of each section determines the additional amount of electrical energy to be supplied. Most of the energy for heating is provided by the screw. The throughput is usually set by the rate at which the screw rotates, and is maintained constant. Work is currently being done on the effect of extruder operating conditions on product quality. Preliminary conclusions indicate that conditions should be kept as constant as possible if reproducible results are desired.

Usually when an extruder is purchased the controls are included. Therefore, they will not be indicated on the control diagram.

## Cutter and Water Bath

When the polystyrene leaves the dryer, it is in strands of ⅛ in. diameter. These strands are cooled in a water bath, and then sliced into ⅛ in. lengths. The cutter must be synchronized with the extruder output so that the correct length is obtained.

The water bath must not be allowed to exceed a given temperature. Usually it is controlled somewhat below this temperature by regulating the rate at which the cooling water enters. The level of the fluid and the material balance is determined by an overflow pipe on the tank.

## Product and Testing Storage

The product and testing storage silos will have high-level alarms that automatically switch the feed to another vessel when a given level is exceeded.

## Packaging

The feed tanks for all the packaging systems will be equipped with a high-level alarm and a high-level feed shutoff.

The automatic bagging and palletizing unit will be purchased with all the controls attached. The drum- and carton-filling stations will require accurate weighing devices that automatically meter a prescribed amount into each container. The other operations will be essentially manual.

The bulk loading station will be capable of automatically weighing any set amount of material into a truck or hopper car.

## Conveying Systems

The conveying systems will be operated manually.

## Ion Exchange

The flow rate through the ion exchanger will be determined using averaging control based on the amount of deionized water in the storage tanks. The controls will be purchased with the unit, not designated separately.

## Steam Generator

The complete generating system, including controls, will be purchased as a package.

## Water Purification

The rate of water pumped from the river will be determined by the water level in the storage facilities following the sand filter.

## Water Distribution

If the level of the water in the hot-water storage tanks (D-205) or the wash-water tank (D-305) gets below a prescribed level, an alarm will sound. The operator will then start pumping water from the deionized water storage tanks into them.

All the water from the steam condensate lines and the water baths (D-501) will be pumped into the hot-water storage tanks, unless they are full. In that case, a control valve will divert the water from the water baths into the wash-water tanks. An overflow pipe will send any excess water from these tanks to the waste treatment facilities.

## References

1. Smith, W.M.: *Manufacture of Plastics*, Reinhold, New York, 1964, p. 308.
2. Brown, J.E.: "Onstream Process Analyzers," *Chemical Engineering*, May 6, 1968, p. 164.
3. "Process Instrument Elements," *Chemical Engineering*, June 2, 1969, pp. 137-164.
4. *Instrument Symbols and Identification*, ISA-S5.1, Instrument Society of America, Pittsburgh, 1973.
5. Buckley, P.S.: "A System Approach to Process Design," presented to the Richmond, Va., AIChE, March 21, 1967.
6. Buckley, P.S.: "Impact of Process Control Advances on Process Design," *Chemical Engineering Progress*, Aug. 1964, pp. 62-67.
7. Buckley, P.S.: *Techniques of Process Control*, Wiley, New York, 1964.
8. Rijnsdorp, J.E.: "Chemical Process Systems and Automatic Control," *Chemical Engineering Progress*, July, 1967, pp. 99-100.
9. Buckley, P.S.: "Impact of Process Control Advances on Process Design," talk given at the AIChE Process Control Workshop in Memphis, Tenn., Feb. 5, 1964.
10. Gould, L.A.: *Chemical Process Control: Theory and Applications*, Addison Wesley, Reading, Mass., 1969, p. 183.
11. Friedman, P.G., Moore, J.A.: "For Process Control Select the Key Variable," *Chemical Engineering*, June 12, 1972, p. 85.
12. "How to Cash in on Process Control Computers," *Chemical Week*, Apr. 20, 1968, p. 71.
13. Moore, J.F., Gardner, N.F.: "Process Control in the 1970's," *Chemical Engineering*, June 2, 1969, p. 94.
14. Lawrence, J.A., Buster, A.A.: "Computer Process Interface," *Chemical Engineering*, June 26, 1972, p. 102.
15. Baker, W., Weber, J.C.: "Direct Digital Control of Batch Processes Pays Off," *Chemical Engineering*, Dec. 15, 1969, p. 121.
16. Huters, W.A.: "Process Control System Planning and Analysis," *Chemical Engineering Progress*, Apr. 1968, p. 47.
17. Spolidoro, E.F.: "Comparing Positive Displacement Meters," *Chemical Engineering*, June 3, 1968, p. 91.
18. Landis, D.M.: "Process Control of Centrifuge Operations," *Chemical Engineering Progress*, Jan. 1970, p. 51.

## Additional References

Coughanowr, D.R., Koppel, L.B.: *Process Systems Analysis and Control*, McGraw-Hill, New York, 1965.
Buckley, P.S.: *Techniques of Process Control*, Wiley, New York, 1964.
Gould, L.A.: "Chemical Process Control: Theory and Applications," Addison Wesley, Reading, Mass., 1969.
Deskbook Issue of *Chemical Engineering*: "Instrumentation and Process Control," Sept. 11, 1972 (lists equipment manufacturers).

# CHAPTER 8

# Energy and Utility Balances and Manpower Needs

The chemical industries use a large portion of the total energy generated in the United States. In 1970 the chemical process industries used 27% of all the electrical energy available, and this percentage was expected to increase.[1] Dow Chemical estimated the cost of energy to be about 10% of its operating costs. For the industry as a whole in 1965, about 3% of sales costs were attributed to the cost of purchased energy.[2] On the same basis, 14% was spent for labor, 6% for capital investments, 6% for administration, and 3% for research and development. For a modern ethylene plant, the equipment inside battery limits (see Chapter 9) involved directly with energy and its transfer amounted to over 75% of the cost of the purchased equipment. A summary is given in Table 8-1. For an air separation plant the energy costs are 35%-50% of the total cost.[1]

Table 8-1

Purchased Equipment Costs for an Ethylene Plant

| Item | % of Total Equipment Costs |
|------|----------------------------|
| Heat exchangers | 24.0 |
| Furnaces | 32.0 |
| Drums | 7.4 |
| Distillation towers | 8.6 |
| Pumps and compressors | 28.0 |
| Total purchased equipment | 100.0 |

Source:  Miller, R., Jr.: "Process Energy Systems," *Chemical Engineering*, May 20, 1968, p. 130.

All of this means that the use of energy is a very important consideration in designing a plant. This has become especially true since 1970, when a general awareness began to develop that the United States was entering a decade of energy crises. In 1972 the natural gas shortage became so acute that in many areas of the country the gas companies were refusing to accept new customers, and even reduced their supply of gas to customers with valid contractual agreements by as much as 10%.[3] The United Gas Pipe Line Company admitted that it had to curtail deliveries by 10%-25%. On some days the delivery was only 50% of that usually delivered to industrial customers.[4] The havoc this can cause to customers who rely on gas for energy as well as a raw material could cause some companies to stop operating whole plants. In fact, Shell Chemical announced that the decreasing availability of natural gas was a contributing factor in its decision to close its two ammonia plants on the West Coast. The Dow Chemical Company, in a similar move, delayed the completion of a $20,000,000 magnesium plant at Dallasport, WA from 1971 to 1975, citing among other reasons its doubt that there was an adequate supply of electric power and natural gas.[3]

Not only is there a shortage of gas, but the cost of energy is rising. In the Houston area energy cost approximately 20¢/1,000,000 BTU (80¢/1,000,000 kcal) in the 1960s. Its estimated cost for future plants is now 65¢/1,000,000 BTU (260¢/1,000,000 kcal).[1,3] This is not only an increase in energy costs; for a large segment of the petrochemical industry it means an increase in raw material costs.

## CONSERVATION OF ENERGY

The present shortage of gas and the projected shortage of power has caused the chemical industry to become aware of how it can conserve the power it now has. For any new plants the process will be carefully scrutinized to see that power usage is minimized. To accomplish this, the design engineer must be aware of all the places within the plant where energy is needed and of all the sources of energy available.

The most efficient plant in terms of energy is one that uses the energy over and over again. For instance, if the exiting bottoms from a distillation column are hot, this energy might be used to heat up the feed, reducing the energy that must be supplied in the reboiler. Another use of the same principle is placing a heat exchanger in the stack of a furnace. Here the hot exiting gases can transfer their energy to another process stream or to the incoming air.

The goal of these systems is to use as much as possible of the thermal energy that is available above the ambient temperature. They are called *direct energy-recovery systems*. They do two things: they reduce the amount of energy that must be supplied and also reduce the amount of cooling water that is necessary. This, in turn, can also decrease the amount of thermal pollution.

In pursuing this goal the engineer may end up with four or five streams transferring energy to one other stream. When this is done, as a general rule the coolest stream should contact the cold feed first and the warmest should be in the last heat exchanger. This allows the greatest amount of energy to be transferred. Exceptions to this rule may occur if the flow rates of the streams are widely different.

How large the exchanger should be, which in turn determines how much heat is transferred, is dependent on an economic analysis. If the total amount of energy that *could* be transferred *were* transferred, an infinitely large exchanger would be required. Two rules of thumb are that the greatest temperature difference in an exchanger between two streams, one gaining and one losing energy, should be at least 36°F (20°C) and the minimum temperature difference should be at least 10°F (5°C). When the energy is being transferred between two process streams, the minimum temperature difference should be increased to 36°F (20°C);[5] the hot stream should always be at least 36°F (20°C) warmer than the stream being heated. If the temperatures are too near each other the exchanger will have to be so large as to be uneconomical.

While temperature is a measure of the level of thermal energy, pressure is a measure of the level of mechanical or kinetic energy. Whenever there are streams at high pressure, they may be used to power devices directly or to produce electricity. It is more efficient to directly power a pump or compressor, but this requires the correct matching of the power available to the power required.

Sometimes a motor is coupled with the directly powered source for startup purposes as well as to obtain more stability in the operation. The motor uses electrical power only when enough energy is not provided by the high-pressure sources.

The amount of energy that may be recovered from liquid streams by using a hydraulic turbine is given in Figure 8-1. Similarly, Figure 8-2 gives the results when expansion turbines are used to recover energy from gas streams.

If after fully exploiting these possibilities there is still energy unused, the process engineer should consider making steam. This steam could be used in a steam turbine to create mechanical energy or directly to obtain thermal energy. Generating steam is usually less efficient than direct energy conversion because it involves an intermediate processing step. This means more equipment and an extra transfer of energy. Since this transfer is never 100% efficient, it is best to use direct transfer processes, where they are feasible, first. Steam, however, does have one big advantage—it can easily and safely be transported from one part of a plant to another. Ryle Miller, Jr. has written a good article that covers these subjects in more detail (see reference 2). He also gives exchanger and turbine costs and gas turbine sizes.

In many cases the optimization of the energy exchange systems may require a modification of the proposed operating conditions as well as the material balance. In fact, some companies have suggested that the energy flow diagram be developed before the material flowsheet.[1]

**Energy Balances**

After all the possibilities of conserving and transferring energy within the process have been considered, the engineer must determine what other energy must be supplied to and removed from the process. This requires a total energy balance for the whole process as well as for the individual units.

The flow rate, pressure, and temperature of each stream must be specified. This has already been done in part by constructing the unit ratio material balance. It must be extended to all energy transfer systems by using material and energy balances.

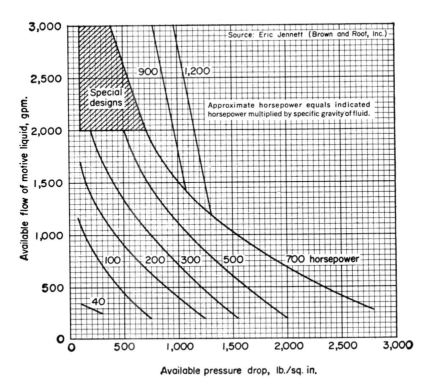

Figure 8-1   The energy that may be recovered from fluids by using pump-turbines.
Source: Miller, R.J., Jr.: "Process Energy Systems," *Chemical Engineering,* May 20, 1968, p. 130.

Since ambient air and water inlet temperatures differ depending on the time of day and month of the year, an energy balance over all air coolers and all equipment using cooling water or air must be made using summer and winter conditions before utility requirements or equipment sizes can be determined. For instance, in sizing a cooling-water pump the warmest probable conditions will be used, as this will guarantee that even under the worst conditions the plant can be operated. On the other hand, when sizing heaters it is noted that the greatest energy transfer requirements will occur during the winter months, and the equipment must be designed to operate under these conditions. Reference 6 gives the temperature that is exceeded only 1% of the time during the summer and the temperature that is exceeded 99% of the time in winter for various locations, as well as humidity data that can be used for sizing cooling towers. The average number of days per year the surface water temperature is near 32°F (0°C) is given in reference 7. This source also gives the number of days per year that the surface water temperature exceeds 80°F (27°C). These data can be used to estimate the temperature of cooling water obtained from lakes, rivers, or shallow wells. If a deep underground supply of water is used, it may not vary much in temperature throughout the year.

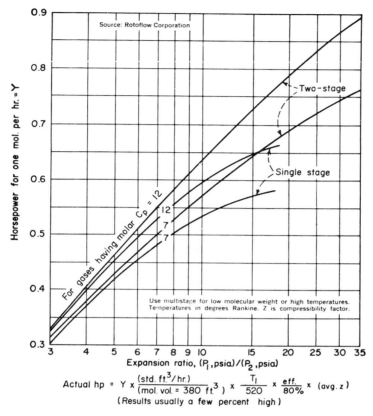

Figure 8-2  The energy that may be recovered from gases by using turbo-expanders.
Source: Miller, R.J., Jr.: "Process Energy Systems," *Chemical Engineering*, May 20, 1968, p. 130.

## Sources of Energy

The major sources of energy are electricity, coal, gas, and oil. There are predicted and actual shortages for all of these except coal. If coal is selected, either some means of removing sulfur must be included or a low-sulfur coal must be specified. In 1973 low-sulfur coal was in short supply and commanded a high premium price. Much of it is used in the steel industry, where it is a necessity. Electricity is usually the most expensive of the sources, but if it is purchased the capital costs are low. It is the most convenient source. If either oil or gas is selected as an energy source, it might be wise to specify equipment that could use either feed, since oil could be in short supply in the near future and/or new supplies of synthetic natural gas could alleviate the gas shortage of the mid-1970s.

**Energy Transfer Media**

If the fuel cannot be used to supply the energy required directly, some medium must be selected for transferring it to the process vessels. The most common medium is steam. Most plants have steam available at several different pressures, for instance, 450, 125, and 25 psig (32, 9, and 1.8 kg/cm²). Since the feed water for the boilers is usually deionized water, when the steam is not directly charged to the process the condensate is returned to the boiler to make more steam. There is always some water loss, so some makeup is always necessary. When the steam temperature is above 500°F (260°C) the boiling point increases by less than 20°F (10°C) for each 100psi (7 kg/cm²) increase in pressure. Since an increase in pressure means an increase in cost, other media have become popular at high temperatures. Dowtherm A® is the most commonly used indirect heat transfer medium in the temperature range of 400-750°F (200-400°C). It is a mixture of diphenyl and diphenyloxide that has a boiling point of 496°F (257°C) and is chemically stable below 750°F (400°C).

Other indirect heat-transfer media and the temperature ranges for which they are used are hot water (35-400°F; 2-200°C), mercury (600-1000°F; 315-540°C), molten inorganic salts (300-1100°F; 150-600°C), and mineral oils (30-600°F; −1-315°C). The properties of these materials are given in the *Chemical Engineer's Handbook.*[8] References 9 and 10 give the properties of some other substances.

**Cooling Water**

The most common cooling medium is water obtained from a nearby river, lake, or well. To protect against fouling and corrosion, this water is usually not heated above 158°F (70°C).[11] In some systems where water is plentiful, it is used once and then discharged into a stream or lake. If it picks up too much energy, some of this energy may need to be removed before it can leave the plant site. State and federal regulations regarding thermal pollution should be checked. Chapter 16 gives information on how this can be done.

When water is scarce or when it must be extensively treated, the water, after being heated, is cooled and recirculated using either an open or a closed system. In the open recirculation system (see Fig. 8-3), the water, after being heated, is sent to a cooling tower or spray pond. In the cooling tower the water is sprayed into an airstream that is drawn through the tower by large fans. When a spray pond is used water is just sprayed into the air, which moves because of winds and natural convection. In either case, the water comes in direct contact with air. Since the water is warm, some of it vaporizes into the air. The energy required for vaporization comes mainly from the water, which thus is cooled.

The temperature of the cooled water can at best approach the wet-bulb temperature of the air. Practically, the engineer usually designs on the basis that the water will be 8-13°F (4.5-7°C) above the wet-bulb temperature.[12] Since this is an open system, and oxygen is picked up with each pass through the system, corrosion and the growth of micro-organisms within the system are facilitated. For this reason, pretreatment of the water is necessary.

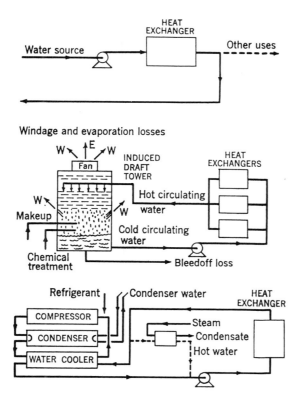

Figure 8-3    Three basic types of cooling water systems. Top: the once-through system where the
cooling water is used once and then discharged. Middle: the open recirculation system
where the water is cooled and recycled through a system in which it comes in direct contact
with air. Bottom: a closed recirculation system where the water is cooled and recycled
without coming in direct contact with the atmosphere.
Source: Silverstein, R.M., Curtis, S.D.: "Cooling Water," *Chemical Engineering*, Aug. 9,
1971, p. 88.

With each pass through the system, some water is vaporized, so that some
makeup water is necessary. The amount of water evaporated can be determined by
running an energy balance. As water is added the concentration of minerals and
other substances in the water increases, since they are not removed by evaporation
and every pound of makeup water adds some more. To counteract this build-up,
some water must be continuously removed from the system. This is known as
blowdown. As a rule of thumb, the blowdown is about 0.3% of the water being
recirculated for each 10°F (5°C) of cooling that occurs within the tower. This
assumes a solids concentration in the water of 4-5 times that in the makeup water. In
places where water is scarce and hard, a deionization system may need to be
installed.[13]

The closed recirculation system (see Fig. 8-3) is used when it is necessary or economical to produce cooling water at a temperature below that feasible with an open system, or it is important to prevent foreign matter from getting into the water.[14] Here secondary heat exchangers are used to remove the energy. The only water loss is due to leaks, and since it is an isolated system, any water pretreating that is required is minimal.

### Cooling-Water Treatment

Before the water enters the heat exchanger, it often must be treated to prevent corrosion from occurring, deposits from forming, and/or the growth of micro-organisms on the heat transfer surfaces. Chlorine and/or a nonoxidizing biocide may be used to solve the last problem. Lists of biocides are given in references 14 and 15. A dispersant such as modified tannins, lignosulfonates, or polyelectrolytes may be added to prevent particle agglomeration and to make the particles less adherent to metal surfaces. Hardness stabilizers, often polyphosphates, may be necessary to prevent hardness salts from precipitating out.[16] A large variety of corrosion inhibitors is available, and they are discussed in reference 14. Some costs for these chemicals are given in Table 8-2. In all these instances, the engineer must be certain that what he is adding will not harm the environment when the water is eventually discharged.

If the water is very muddy, it may be necessary to clarify and/or filter it. The simplest form of clarification is passage through a large tank that reduces the velocity nearly to zero and has a large enough residence time to allow most of the undissolved materials to settle out. Often this process is speeded up by adding flocculents such as alum, sodium aluminate, or iron salts.[16] The material settling out is removed and must then be disposed of in some manner. Since, with the exception of the flocculent, this is not material added by processing, it may often be used for landfill.

### Air Cooling

Another medium that may be used for cooling is air. The transfer of energy occurs in air coolers, which are nothing more than a direct means of transferring energy from a fluid to the surrounding air. They do not require any cooling water or any of the pumps and cooling towers connected with a water system, and hence require less energy to operate. This factor will become more important as energy costs rise. They are an ideal choice when high-quality cooling water is in short supply.

Their major disadvantages are that they require more space, they have a higher initial capital cost, and the coolest temperature that can be attained is 20 to 30°F (10 to 15°C) above the ambient air temperature.[17] For the same amount of heat transfer, an installed carbon-steel shell-and-tube exchanger will cost about one-third as much as an air cooler. This difference diminishes as more expensive materials are used.[18] A modification of the air cooler, called the wet-surface air cooler, over-comes some of the above-mentioned disadvantages. It can reduce the temperature that can be attained to nearly the ambient temperature, and there are some claims

Table 8-2

Cooling Water Treatment Costs

| Treating Costs for Once-Through System | |
| --- | --- |
| Flow, GPM | Daily Cost for Antifoulant, $ |
| 1,000 | 0.50–1.50 |
| 10,000 | 5.00–15.00 |
| 50,000 | 25.00–75.00 |

| Treating Costs for Open-Circulating System | | | |
| --- | --- | --- | --- |
| Circulation Rate, GPM | Inhibitor | Daily costs, $<br>Antifoulant | Biocide |
| 1,500 | 0.50–2.00 | 0.15–1.00 | 0.10–0.20 |
| 3,000 | 1.00–4.00 | 0.30–2.00 | 0.20–0.40 |
| 6,000 | 2.00–8.00 | 0.60–4.00 | 0.40–0.80 |
| 30,000 | 10.00–40.00 | 3.00–20.00 | 0.50–3.00 |

| Treating Costs for Closed Systems | |
| --- | --- |
| System Capacity, gal | Inhibitor Cost, $ |
| 1,000 | 10–50 |
| 10,000 | 100–500 |

Source:   Silverstein, R.M., Curtis, S.D.: "Cooling Water," *Chemical Engineering*, Aug. 9, 1971, p. 84.

that it has lower capital and operating costs than a complete system involving a cooling tower and a shell-and-tube heat exchanger.[19] A wet-surface air cooler (see Fig. 8-4) differs from the more common dry-surface air cooler in the way energy is transferred. For both, the air flows on the outside of the tubes. In the dry-surface exchanger energy is transferred to the air by conduction and convection from the metal surface of the tubes. For a wet-surface heat exchanger the energy is first transferred to water, which continuously flows over the surface of the heat-exchanger tubes. The water is in direct contact with the air flowing through the unit, and as the water is heated its vapor pressure increases and some of it is vaporized into the surrounding air. The cooling caused by the vaporization of the water keeps the water temperature close to the wet-bulb temperature.

Wet-surface air coolers should not be used where the surface temperature of the exchanger exceeds 150°F (65°C), because scaling will occur.[19] The pretreatment of the water is similar to that used in the open-loop cooling systems discussed previously.[19] The amount of blowdown necessary depends on the water properties and the operating conditions, but as an approximation the rule of thumb given before can be used.

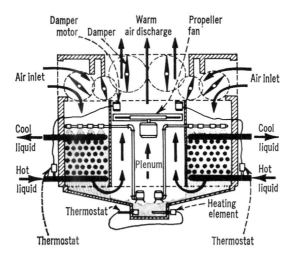

Figure 8-4    A wet surface air cooler.
Water is sprayed over the surface of the tubes. This evaporates into the passing air and cools the liquid inside the tubes.
Source: Kals, W.: "Wet Surface Aircoolers," *Chemical Engineering,* July 26, 1971, p. 91.

### Refrigeration Systems

Refrigeration systems are used when a low enough temperature cannot be obtained using air or water cooling systems. They are not designed from scratch for a preliminary plant design. It is usually assumed that a complete system will be purchased. One notable exception is when either a feed or a product is itself a refrigerant. When a system is to be purchased, the temperatures involved, the energy that must be removed, and the type of system that will be used must be determined. The first two items are dictated by the process chosen. The energy removed has usually been given as tons of refrigeration needed. A ton of refrigeration is the amount of energy necessary to produce one ton of ice per day from water at 32°F (0°C). This is equivalent to the production of 1,200 BTU/hr or 200 BTU/min (0.844 kcal/sec). The British unit of refrigeration is 1 kcal/sec or 237.6 BTU/min.

There are two common refrigeration systems, mechanical and thermal compression. In the mechanical system work is done in compressing a gas, the refrigerant. The energy thus added plus the amount of refrigeration required must be removed in a condenser, usually by cooling water. The calculations necessary and some typical values are given in references 20 and 21.

The thermal compression or absorption refrigeration systems are less common. They do not require a compressor. Their energy source is steam, natural gas, or waste heat. This system requires much more cooling water than the previous one, but may be economical if a large amount of waste heat is available.[21] See references 20 and 21 to determine heat and cooling water requirements.

## SIZING ENERGY EQUIPMENT

When the energy balance is complete, or even while it is being calculated, the engineer should complete the details of the equipment list by sizing all the remaining equipment that is larger than a pump. Again, only the information needed to specify the cost of the item should be obtained for a preliminary design.

### Furnaces and Direct-Fired Process Heaters

The most logical place to begin is to size the furnaces and direct-fired heaters. Often the chemistry of the process has dictated the conditions, but frequently they can be modified in order to conserve energy usage. The burning of waste materials should also be considered as a means of both disposing of unwanted by-products and reducing fuel requirements. For each of these units, the amount of fuel needed per pound of product should be determined.

The flux[9] for fired steam boilers is often between 40,000 and 50,000 BTU/hr ft² (110,000-135,000 kcal/hr m²), while that for organic fluids is 5,000-12,000 BTU/hr ft² (13,000-33,000 kcal/hr m²).

### Heat Exchangers

The major things to specify for heat exchangers are the materials of construction and the heat-transfer area required. Generally, streams containing materials that can precipitate out or form a scale are placed on the tube side. If this is not a factor, it is generally best to place the stream flowing at the highest velocity on the tube side. Usually a 20% improvement in the corrected mean temperature difference can be realized if this is done.[2]

For a preliminary design, except for very large or expensive heat exchangers, it is usually adequate to use approximate heat transfer coefficients. These can be found in references 22, 23 and 24. When calculating individual heat transfer coefficients, it may simplify calculations to note that for streams that have a viscosity greater than 5 cp the tube-size coefficient is two or three times what the shell-side coefficient would be for the same material.[2] This is often the deciding factor in determining which fluid should flow within the tubes.

Another factor that may determine whether a material should flow inside or outside the tube is the material of construction. If expensive alloys are required for only one of the fluids, it is cheaper to plate the inside of the tubes than the shell side.

If the engineer decides to be more precise in calculating heat transfer coefficients, he may assume that the heat exchanger is constructed using 3/4 in or 1 in O. D. (outside diameter) tubing. Also, the velocity should be maintained above 3 ft/sec (1 m/sec) if there are any particles in the fluid that might settle out.[25] The heat transfer coefficients can be calculated using the equations given in reference 5.

Fouling factors that may be assumed are 1,000 BTU/hr ft² °F (5,000 kcal/hr m² °C for condensers, coolers and heaters and 500 BTU/hr ft² °F (2500 kcal/hr m² °C) for reboilers and vaporizers.

### Pumps

Pumps supply the energy to transport liquids, sometimes containing solids, from one point in a plant to another. The centrifugal pump is the most commonly used. It is generally chosen when the viscosity of the fluid is less than 2,000 cp, the flow rate is greater than 5 GPM (5 gal/min; 1.0 m³/hr), the fluid contains less than 5% entrained gases, and the attrition of any solids is not a problem.[26,27] The other major category of pumps is positive displacement pumps. These pumps, besides supplying energy, can be used for metering or proportioning. Their major characteristic is that the amount of fluid moved is nearly independent of the pressure increase across the pump. For the centrifugal pump the flow rate is dependent on the pressure increase. Positive displacement pumps are not generally designed to operate above a flow rate of 4,000 GPM (900 m³/hr).[28]

For a pump, the type, capacity (usually as gal/min or m³/hr) brake horsepower, and material of construction need to be specified. The liquid horsepower (LHP) of a pump is the energy that must be supplied to the fluid. The brake horsepower (BHP) is the energy that must be supplied to the pump. The ratio of the former to the latter expressed as a percentage is the pump efficiency (EFF).

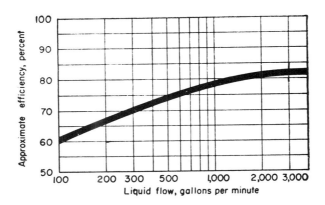

Figure 8-5    The approximate efficiencies of process pumps.
Source: Miller, R.J., Jr.: "Process Energy Systems," *Chemical Engineering*, May 20, 1968, p. 130.

$$EFF = \frac{LHP}{BHP} \times 100 \qquad (1)$$

Figure 8-5 gives the efficiency of centrifugal pumps and Figure 8-6 gives the efficiency of the motors that often drive them.

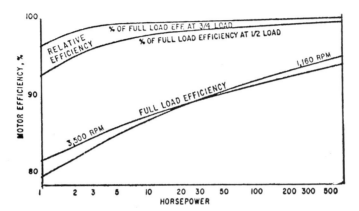

Figure 8-6  Efficiencies of commercial induction motors.
Source: Jacobs, J.K.: "How to Select and Specify Process Pumps," *Hydrocarbon Processing,* June 1965, p. 122.

The energy that must be supplied to the fluid depends on the distance the substance must be transported, the height to which it must be raised, the difference in pressure between the point where the material enters the system and where it leaves, the velocity at which the material is transported, and the obstructions in its path. The first two items can be obtained from the plant layout and the third from the energy balance. A rule of thumb says that the optimal velocity for liquids is 5-10 ft/sec (1.5-3 m/sec). The remaining item to be calculated is the pressure loss due to the friction that occurs as the material flows through pipes and pieces of equipment. The reader may have learned how to calculate precisely the pressure drop through pipes, valves, fittings, and the like. Since most of these items are not specified at the preliminary stages of design, that procedure cannot be followed. Therefore the pressure drop due to friction is assumed to be 1-2 lb/in$^2$ per 100 ft of pipe (0.25-0.5 kg/cm$^2$ per 100 m of pipe) in a liquid system where the viscosity is near that of water.

If a control valve is included, the pressure drop in the whole system should be increased by 50%. If a valve is to control a flow rate adequately, a major portion of the pressure drop must be across that valve. If it is not, a change in the percentage it is open will not have much effect on the flow rate, which is what the valve is supposed to regulate.

The pressure drop across some pieces of equipment is given in Tables 8-3 and 8-4. For others the engineer may query the manufacturer, estimate it from similarity to other pieces of equipment whose approximate pressure drop is known, calculate it from detailed formulas obtained from the literature, or actually run tests on a model and scale up the results.

Table 8-3

Approximate Pressure Drop for Liquids

| | |
|---|---|
| Control Valves | 10 psi (0.7 kg/cm$^2$) or one-third of the total pressure drop in the system, whichever is larger |
| Heat Exchangers | 1/10 the absolute pressure for vacuum systems<br>1/2 the average gauge pressure for pressure systems where the pressure is less than 10 psig (25 psia or 1.75 kg/cm$^2$)<br>5-10 psi if the pressure exceeds 10 psig (1.75 kg/cm$^2$)<br>20 psi (1.5 kg/cm$^2$) if there are 3 shells in series<br>25 psi (1.75 kg/cm$^2$) for 4 or more shells in series |
| Furnaces | 25-40 psi (1.75-2.8 kg/cm$^2$) |
| Miscellaneous | at least 5 psi (0.35 kg/cm$^2$) for each different unit |

Sources: Younger, A.H., Ruiter, J.L.: "Selecting Pumps and Compressors," *Chemical Engineering*, June 26, 1961, p. 117.
Lord, R.C., et al.: "Design of Heat Exchangers," *Chemical Engineering*, Jan. 26, 1970, p. 96.

Table 8-4

Approximate Pressure Drop in Gases

| | |
|---|---|
| Control Valves | 5 psi (0.35 kg/cm$^2$) or one-third of the total pressure drop in the system, whichever is larger |
| Heat Exchangers | 5 psi (0.35 kg/cm$^2$) for each shell |
| Intercoolers (between compressor stages) | 5 psi (0.35 kg/cm$^2$) for each intercooler |
| Miscellaneous | 5 psi (0.35 kg/cm$^2$) for each unit |

Source : Younger, A.H., Ruiter, J.L.: "Selecting Pumps and Compressors," *Chemical Engineering*, June 26, 1961, p. 117.

The pressure drop across the pump is calculated using the following equation:

$$\Delta P_p = P_{out} - P_{in} + \rho \frac{\Delta v^2}{2g_c} + \Delta Z \rho \frac{g}{g_c} + \Delta P_L + \Delta P_E \qquad (2)$$

where
$\Delta P_p$ = pressure drop across the pump
$P_{in}$ = pressure at the entrance to the flow system
$P_{out}$ = pressure at the exit from the flow system
$\Delta v^2$ = $v_{out}^2 - v_{in}^2$
$v_{in}$ = velocity of fluid just before entering the flow system
$v_{out}$ = velocity of fluid immediately after leaving the flow system
$\Delta Z$ = $Z_{out} - Z_{in}$
$Z_{in}$ = height of the entrance to the flow system
$Z_{out}$ = height of the exit to the flow system
$\Delta P_L$ = pressure drop due to friction in the piping
$\Delta P_E$ = pressure drop due to equipment in the flow systems
$\rho$ = density of the fluid
$g_c$ = Newton's law conversion factor, $\dfrac{32.16 \text{ ft lb}_m}{\text{lb}_f \text{ sec}^2}$ , $\dfrac{9.81 \text{ kg}_m \text{ m}}{\text{kg}_f \text{ sec}^2}$

$g$ = acceleration of gravity

For liquids having a velocity of less than 7 ft/sec (2 m/sec) the kinetic energy term $\rho \Delta v^2/g_c^2$ can usually be ignored. The liquid horsepower is obtained from the pressure drop:

$$\text{LHP} = \frac{v \Delta P_p}{1,713} = \frac{v' \Delta P_p'}{27.4} \qquad (3)$$

where
$v$ = velocity in gal/min (GPM)
LHP = liquid horsepower*
$\Delta P_p$ = pressure drop in psi
$v'$ = velocity in m³/hr
$\Delta P_p'$ = pressure drop in kg/cm²

Some designers suggest adding 10% to the calculated liquid horsepower to allow for changes in the process. For reflux pumps the LHP should be increased 35% beyond that calculated.[26] This is because the amount of reflux is a determinant in the amount of separation possible and the pump should not limit the options available to the plant operator.

---

*   The horsepower used throughout this text is the English horsepower. A metric horsepower = 0.986 horsepower (English).

The next important thing the designer must calculate is the net positive suction head (NPSH). This is the difference between pressure of the fluid at the entrance to the pump (often referred to as the pump suction) and the vapor pressure of the fluid:

$$\text{NPSH} = P_s - P_v \tag{4}$$

where   NPSH = net positive suction head
  $P_s$   =  pressure at pump entrance
  $P_v$   =  vapor pressure of liquid being pumped

Obviously the NPSH must be positive, or the liquid would be vaporized and the pump would be filled with gas. Since a pump is designed to transport liquids, if this happened it would just spin in its housing and no transfer would be accomplished. There is an increase in velocity as the liquid enters most pumps. This conversion of pressure energy to kinetic energy may reduce the pressure enough to cause fluids that have a positive NPSH to vaporize. Therefore, each pump has some minimum NPSH below which it will not operate properly. For most pumps an NPSH of 14 ft (4.2 m) of fluid is adequate. Some positive displacement pumps can operate at an NPSH of 6 ft (2 m). Use equation 5 to calculate the NPSH for each pump to be specified.

$$\text{NPSH} = P_{in} - \rho \, \frac{\Delta v_1^2}{2g_c} - \Delta Z_1 \rho \, \frac{g}{g_c} - \Delta P_{L_1} - \Delta P_{E_1} - P_V = P_s - P_V \tag{5}$$

where   $\Delta v_1^2$  =  $v_p^2 - v_{in}^2$
  $v_p$   =  velocity at the pump entrance
  $\Delta Z_1$  =  $Z_p - Z_{in}$
  $Z_p$   =  height of the pump entrance
  $\Delta P_{L_1}$ =  pressure drop due to friction in the piping between the entrance and the pump
  $\Delta P_{E_1}$ =  pressure drop due to equipment between the entrance and the pump

If the NPSH is not adequate, a change in the elevation of equipment may be necessary. To meet this requirement the receivers that collect condensate are placed about 15 ft above the ground. The pump is then placed directly below at ground level.

The process engineer must decide which operations require spare pumps. Obviously, if a system contains only one pump and that pump fails, there is no way to transfer the fluid and the process must eventually be shut down. If a spare pump is in place the process can be continued with almost no interruption. However, the initial cost of the plant is also increased by the cost of the pump and the price of its installation. For large-capacity systems an alternative is to place two pumps that can operate at about 55%-60% of the desired capacity in parallel with each other. Then when one fails the process merely needs to be cut back until repairs can be made.

**Fans, Blowers, Compressors**

Fans, blowers, and compressors are essentially gas pumps. Fans are centrifugal machines that produce a pressure drop of less than 60 in $H_2O$ ($0.15$ kg/cm²). Blowers operate to about 30 psig ($3.0$ kg/cm²). Compressors can produce pressure up to 45,000 psig ($3,000$ kg/cm²). The purpose of all these items is to increase the pressure of a gas.

This can be done by reducing its volume or by increasing its velocity so that the kinetic energy created can be converted into potential energy (pressure). Devices with the former action are known as positive-displacement machines, and those with the latter are either centrifugal or axial devices. Positive-displacement compressors are not used when the flow exceeds 10,000 scfm (standard cubic feet per minute or 17,000 m³/hr). Centrifugal compressors are not used for flow rates below 500 scfm (850 m³/hr).[29] An axial compressor is used when a high flow rate is desired[27] (10,000-1,000,000 scfm or 17,000-1,700,000 m³/hr; $\Delta P <$ 100 psi or 7 kg/cm²).

Since the density of a gas is strongly dependent on the temperature and pressure, knowing only the volumetric flow rate is not adequate. For this reason the flow is given above as so many standard cubic feet per minute (scfm). This is the volume of fluid that would be transferred at a temperature of 60°F (15.6°C) and a pressure of 14.7 psia (1.033 kg/cm²). For design purposes, this rate is usually increased by 5%.[27]

For a fan the changes in pressure and temperature are small enough that the incompressible fluid flow equations given for pumps may be used. In fact, for most gas systems, if the pressure drop is less than 40% of absolute upstream pressure the fluids can be treated as incompressible.[30]

To determine the frictional pressure drop in standard steel pipe for this case, Figure 8-7 may be used. For nonstandard pipes see reference 31. This source also gives tables and figures for determining the pressure drop for water, air, and steam. Since the number of pipe fittings is unknown, this figure should be increased by at least 25%. The ideal pipe size may be approximated by assuming that the gas travels at the optimum velocity given in Table 8-5. When the incompressible equations cannot be used, the reader should consult references 23, 30, and 32.

FRICTION LOSS PER 100 FEET IN INCHES OF MERCURY - PIPE/AIR DENSITY 0.075 POUNDS PER CUBIC FEET

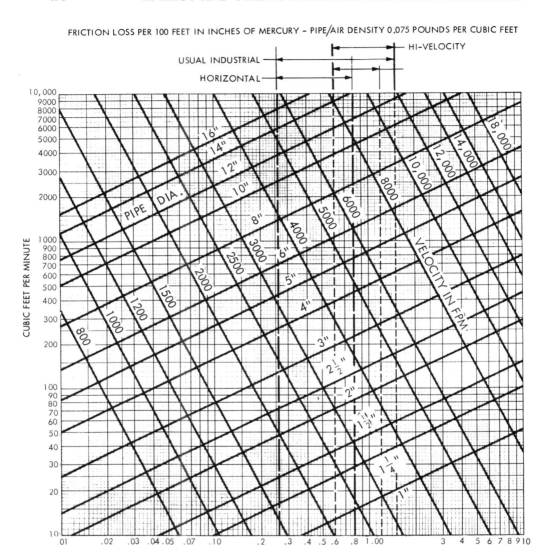

FRICTION LOSS PER 100 FT. OF PIPE IN INCHES OF MERCURY

| 1 INCH HG = 0.4912 PSI | V = VELOCITY IN FEET | VK = 14,786 WHEN P = INCHES H G |
|---|---|---|
| = 13.63 INCHES WATER | PER MINUTE | = 21,094 WHEN P = PSI |
| = 1.136 FEET WATER | VK = CONSTANT BASED ON AIR DENSITY | = 4,005 WHEN P = INCHES WATER |
| V = VK √P | OF .07495 POUNDS PER CUBIC FEET | P = PRESSURE |

Figure 8-7    Pressure drop for air having a density of 0.075 pounds per cubic foot flowing in 100 feet of straight pipe. For other conditions and gases the pressure drop may be assumed proportional to the gas density.

Courtesy of Hoffman Air and Filtration Division of Clarkson Industries, Inc.

Table 8-5

Optimum Velocity for Turbulent Flow  in Steel Pipe

| Fluid Density Lb/Ft³ | Optimum Velocity Ft/Sec | Optimum Velocity Ft/Sec |
|---|---|---|
| 100 | 5.1 | 8 |
| 50 | 6.2 | 10 |
| 10 | 10.1 | 16 |
| 1 | 19.5 | 31 |
| 0.1 | 39 | 59 |
| 0.01 | 78 | 112 |

Source:   Simpson, L.L.: "Sizing Piping for Process Plants," *Chemical Engineering*, June 17, 1968, p. 193.

The theoretical horsepower requirements for an adiabatic compressor (equivalent to LHP) can be obtained from the following equation.[33]

$$\text{HP} = \frac{k}{(k-1)} \frac{\text{NRT}_1 Z}{550} \left[ \left(\frac{P_2}{P_1}\right)^{(k-1)/k} - 1 \right] \qquad (6)$$

$$\text{HP}' = \frac{k}{(k-1)} \frac{N'R'T'_1 Z}{75} \left[ \left(\frac{P_2}{P_1}\right)^{(k-1)/k} - 1 \right] \qquad (6a)$$

where   $\text{HP}$   =   theoretical adiabatic horsepower
$\text{HP}'$   =   theoretical adiabatic metric horsepower
$k$   =   $C_p/C_v$
$C_p$   =   heat capacity at constant pressure
$C_v$   =   heat capacity at constant volume
$P_2$   =   entering pressure
$P_1$   =   downstream exiting pressure
$T_1$   =   inlet temperature in $°R$
$T'_1$   =   inlet temperature in $°K$
$R$   =   gas constant = 1,542 ft $\text{lb}_f/°R$ lb mole
$R'$   =   gas constant = 846 kg m/ $°K$ kg mole
$Z$   =   gas compressibility factor at inlet conditions (near 1.0 except at high pressures)
$N$   =   gas flow rate, lb moles/sec
$N'$   =   gas flow rate, kg moles/sec

The maximum efficiency of fans is 82-90%.[34] Figure 8-8 gives the efficiency of reciprocating compressors. Centrifugal compressors are generally from 5-20% less efficient than reciprocating ones.[30]

Large centrifugal compressors are usually so expensive that it is generally not economically feasible to specify spares.[29] To put two of these in parallel can result in surging, due to their flat operating characteristics, unless a control scheme such as that given in reference 35 is specified.

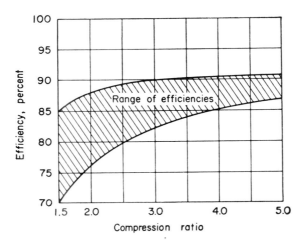

Figure 8-8    Reciprocating compressor efficiencies.
             Source: Miller, R.J., Jr.: "Process Energy Systems," *Chemical Engineering*, May 20,
             1968, p. 130.

### Pneumatic Conveying System

In a pneumatic conveying system, air or some other gas is used to transfer solids from one place to another. These systems are entirely enclosed, hence the product loss is small, contamination is minimized, and the problem of dust emission to the atmosphere is greatly reduced. There are many different systems; only a few will be presented here.

All the systems must have a means of bringing the solids and the gas stream together. Since the gas is under either vacuum or pressure, some sort of airlock is needed in order to prevent air either from being sucked into the system or from being blown into and through the feeding system. The most common solution is a rotary valve (see Fig. 8-9) driven by a motor.

After the material reaches its destination, the gas and solids must be separated. When this is done in a cyclone separator (see Fig. 8-10) the gas stream enters a large container from the side tangentially. The solids are thrown against the side by centrifugal force, fall under the force of gravity, and exit through the bottom. The gas exits at the top. Again some type of airlock is needed to remove the solids. One

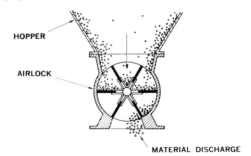

Figure 8-9 Cutaway drawing of a rotary valve. The valve is turned at a given rate by a
motor.
Courtesy of MicroPul Division of the United States Filter Corporation.

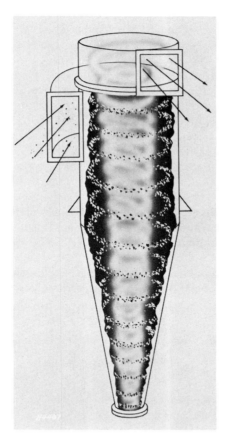

Figure 8-10 A cyclone separator. The air enters tangentially and the solids are thrown out to the sides
and leave through the bottom. The air exits at the top either horizontally as shown or,
frequently, vertically.
Courtesy of American Air Filter Company, Inc.

answer is a gate that opens when the weight above it is great enough and then, after most of the material is discharged, is closed by a spring. Not all the solids will be removed from the gas stream, so a fabric filter must be installed in the exiting gas line.

If air is used as the transporting medium, before it enters the system from the surroundings it is usually cleaned by a filter, and it may also be dried. The energy needed to move the air is provided by a blower. A positive-displacement blower is usually specified, since it can provide enough head to move the material in case the line plugs up.[36]

Air can either be pulled through the system as in Figure 4-3, or it can be pushed through the system as in Figure 4-4. The former is a vacuum system and is normally used to deliver material to a single receiver. Note that in the figure a screw conveyor is used to distribute the solids from the receiver to the three bins depicted. One advantage of the vacuum system is that because it is at a pressure below atmospheric pressure no material can escape from the ductwork. A system whereby a vacuum system can be used with a number of receivers and only one dust filter is given in Figure 4-7.

Figure 4-4 shows a typical system under positive pressure. It differs from the vacuum system in that the material enters from one source and is distributed directly to several tanks. In this case no cyclone separator is used; the air laden with solids enters the process bins directly. The decrease in velocity of the stream and its change in direction will cause most of the solids to drop out. For this system each receiver must have a filter to remove the remaining solids. Note that the blower is placed at the air entrance, instead of after the filter as in the vacuum system. Should a bag in the fiber filter break, no dust will get into the blower or its motor. Another advantage is that no contaminants from the atmosphere can enter the system when it is under positive pressure, except through the air inlet system.

When a fluid other than air is used, the carrier gas usually must be recycled. Such a closed-loop system is shown in Figure 4-5. This can be operated with a vacuum or a positive pressure.

For all the systems the air velocity must be great enough so that the solids do not drop out onto the side of the pipe. The minimum velocity[36] necessary is

$$V_{min} = 910 \left( \frac{\rho_s}{\rho_s + 62.3} \right) (D_s)^{0.60} \tag{7}$$

where   $V_{min}$   =   minimum velocity—velocity below which some particles will drop out, ft/sec
         $\rho_s$   =   density of solids being transported, lb/ft$^2$
         $D_s$   =   diameter of largest particle being transported, ft

$$V'_{min} = 570 \left( \frac{\rho'_s}{\rho'_s + 1} \right) (D'_s)^{0.60}$$

(7a)

where    $V'_{min}$ = minimum velocity, m/sec

$\rho'_s$ = density of solids being transported, g/cm³

$D'_s$ = diameter of largest particle being transported, m

The optimal velocity is probably close to that for air alone or approximately 70 ft/sec (20 m/sec). Of course, a higher value must be used if the minimal velocity is greater than this figure.

The pressure drop in the ductwork is first figured on the basis that no solids are present. This result is then corrected using Figure 8-11 to obtain the true pressure drop. The weight ratios of solids to air vary from 1:1 to 20:1. The flow rate of the solids in the positive-pressure system is much greater than for the typical negative system.[37] The pressure drop in the fiber filters is generally about 4 in $H_2O$ (100 kg/m²), and that in cyclone separator varies from 1 to 3 in $H_2O$ (25-75 kg/m²).[37]

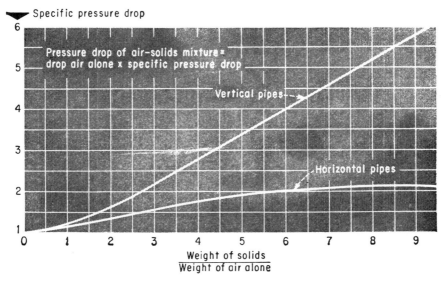

Figure 8-11   Ratio of the pressure drop for fluids containing solids to that for the fluids without solids. Source: Constance, J.D.: "Calculating Pressure Drops in Pnematic Conveying Lines," *Chemical Engineering*, Mar. 15, 1965, p. 200.

### Slurry Piping Systems

For some solids water may be a better medium for transportation than air. This is because the material may already be in a water suspension, such as sludges for sewage treatment plants or the product following filtration of crystallization operations. A water suspension may also be desired for some future processing step, for instance flotation. This has already been mentioned in Chapter 2 as a means of transporting coal, ores, pulps, and other substances from mines to processing or user facilities. Their design procedures are given in reference 38.

### Solids-Handling Equipment

There are many other different types of solids-handling devices, ranging from dumpsters to screw and belt conveyors and bucket elevators. The range and variety of equipment make this subject too large to be adequately covered by this book. The reader is referred to the voluminous literature on the subject. For sizing equipment the most helpful source is often the manufacturers' manuals.

## PLANNING FOR EXPANSION

The same general philosophy espoused in Chapter 5 regarding expansion applies here. Some specific cases follow.

### Heat Exchangers

If the optimum size heat exchanger for the initial plant is installed, an additional exchanger either in parallel or in series will be required when the plant is expanded. This may be the best option when the heat transfer involves condensation and subcooling. The exchanger can be designed to perform both functions initially, and then when the plant is expanded an aftercooler can be installed and the initial equipment can act only as a condenser.

Another approach that may be universally applied is to purchase an exchanger large enough to meet all projected expansion needs and then plug off a number of the tubes. This reduces the heat-transfer area and hence the rate of heat transfer. When the expansion occurs the plugs merely have to be removed. An alternative to plugging the tubes when flow rates are low would be to initially use the exchanger as a multipass exchanger and then later convert it to a single-pass exchanger. For a multipass exchanger the amount of energy transferred per unit area is always less than for a single-pass countercurrent exchanger.[39] These last two methods result in lower total capital, installation, and piping expenses than when parallel or series equipment is added. However, when the time value of money is considered this may not be the most economical plan (see Chapter 10).

Air coolers often consist of two tube bundles in one frame with one set of fans. In this case one tube bundle may be adequate for the initial capacity, and the space where the other tube bundle would reside can be blocked off by sheet metal to prevent the air from bypassing the cooling section. The other tube bundle can then be purchased when more cooling capacity is needed.

**Pumps**

When the pressure increase across the centrifugal pump is less than 125 psi (9 kg/cm²) and the flow rate is less than 300 GPM (70 m³/hr) the pump specified should be large enough to meet expansion needs.[39] This possibility should also be considered for other situations. A smaller impeller can be used to meet the lower initial requirements. Then when expansion occurs this can easily be replaced with the normal-sized one. The engineer must check to see that under both situations enough power is produced and there is an adequate NPSH.

When a spare pump is required, one alternative is to specify two pumps that can each adequately handle the initial needs. Then when the expansion occurs a third pump can be installed in parallel with the first two. Then two pumps can be run to meet the new conditions with one available as a spare.

For positive-displacement pumps the only variable is the operating speed. The only ways to change the capacity of these pumps are to use a variable-speed driver, use a variable-speed transmission (not usually recommended), or replace a given rpm motor by another. Since most motors run at 1,750 or 3,500 rpm, the last method may be used only if the speed is to be doubled.

**Compressors**

When multiple compressors are specified, only those required for the initial plant should be purchased. Adequate space must of course be allocated to meet expansion needs. If only one compressor is specified, one large enough to meet expansion needs should be specified. For centrifugal compressors the capacity can easily be varied over a wide range. The designer should, however, check to see that the initial requirements exceed its minimum capacity, which is called the surge point. In the case of reciprocal compressors, either a clearance volume can be added to the compressor cylinders or the suction valve can be removed. See reference 39 for a discussion of this.

**Steam Boilers and Cooling Towers**

Because steam is usually a critical item in a chemical plant, it is not advisable to rely on only one boiler. This means that boiler size is not a factor. Only enough generating capacity should be installed to meet the needs of the initial plant, and additional boilers should be installed to meet expansion needs. Cooling towers should also be designed to handle only the initial requirements.[39]

## LIGHTING

During the preliminary design of a chemical plant it is not necessary to give any details of the lighting system. All that is required is the power necessary for illuminating the plant. To calculate this, it is necessary to know how much light is needed in each area of the plant, the type of illumination, and the amount of time it will be used per year. Table 8-6 gives some recommended illumination levels for various situations.[40,41] The efficiencies of three different types of lamps are given in

Table 8-6

Recommended Illumination Levels for Various Locations or Tasks

| Location or Task | Recommended Illumination Level | |
| --- | --- | --- |
| | Ft-Candles | Hectolux |
| Detailed drafting | 200 | 22 |
| Regular office work | 100 | 11 |
| Control rooms | 50 | 6 |
| General chemical plant | 30 | 3 |
| Toilets and washrooms | 30 | 3 |
| Corridors, elevators, and stairways in office building | 20 | 2 |
| Locker rooms | 20 | 2 |
| Storage rooms or warehouses | | |
| Active bulk storage | 10 | 1 |
| Inactive | 5 | 0.5 |
| Emergency lighting | 3 | 0.3 |
| Street lighting | 1 | 0.1 |
| Parking areas | 1 | 0.1 |

Source:   Kaufman, J.E. (ed.): IES Lighting Handbook, Illuminating Engineering Society, New York, Ed. 4, 1966, p. 9-49.

Table 8-7

Average Efficiency and Life of Various Lamps

| Type of Lamp | Lumens/watt Initial | Rated Life, Hours | Lumens/Watt, Mean |
| --- | --- | --- | --- |
| Incandescent | 10-16 | 2,500 | 9-14 |
| Mercury vapor lamp | 35-60 | 1,600-24,000 | 30-45 |
| Fluorescent lamps | 50-70 | 7,500-12,000 | 45-60 |

Source:   Kaufman, J.E. (ed.): *IES Lighting Handbook,* Ed. 4, Illuminating Engineering Society, New York, 1966.

Table 8-7 as well as the efficiency at the mean life. The mean is the length of time the lamps are expected to burn.

The power required can be calculated from the following formula:

$$\text{Number of watts} = \frac{\text{foot-candles of illumination} \times \text{area lighted}}{(\text{lumens/watt}) \, (\text{coefficient of utilization}) \, (\text{maintenance factor})}.$$

The coefficient of utilization is determined by the amount of light reflected by the reflectors.[41] If more detailed information is lacking, this may be assumed[42] to be 0.6. The maintenance factor makes a correction for the normal operating condition of the lamps and the reflector.[41] It takes into account the age of the lamp and how dirty the reflector and lamp are. Since this is constantly changing, it is hard to estimate. A reasonable value to use[42] is 0.70.

For walkways, where 3 foot-candles is an adequate amount of illumination, two 175-watt R-40 mercury vapor lamps placed every 54 ft may be specified. For these and areas that are used continuously but have adequate windows it may be assumed that the lights are on 4,500 hours per year. (This is approximately half the time.) For some interior areas they may never be turned off. At the other extreme, Sarah Lee has an automated warehouse in which the only time lights are needed is when there is an equipment failure.

## VENTILATION, SPACE HEATING AND COOLING AND PERSONAL WATER REQUIREMENTS

The amount of ventilation required, like the amount of lighting, depends upon the circumstances. Some general rules are given in Table 8-8.[44,45] Note that there are several guidelines and that in some instances more than one may apply. When this occurs the requirements for each should be calculated and the maximum rate used. None of these guidelines applies to hazardous areas, where the air may have to be moved more rapidly. Some estimates are given in Table 8-9. Safety guidelines and OSHA (Occupational Safety and Health Act) rules should be consulted under these conditions.

To determine the fan or blower sizes necessary to move the air, a pressure drop of 3 in $H_2O$ (75 kg/m²) may be assumed.

Once the ventilation requirements have been determined, the energy requirements for the space heaters may be approximated. These must provide an adequate amount of heat even on the coldest days. A very rough estimate can be obtained by determining the energy required to heat all the air removed by the ventilation system from the average outside temperature on the coldest day to 70 or 80°F (21 or 27°C). Since this is for the coldest day and heating may only be needed 6 months of the year, the average power requirement will be approximately one-fourth of that calculated. This is obviously a very gross calculation, since it ignores the fact that warehouses may be kept at 50°F (10°C) and factories between 60 and 65°F (15 and 18°C[44]; the buildings may or may not be insulated; there is frequently heat transfer-

Table 8-8

Ventilation Guidelines

| Type of Area | Ventilation Requirements |
|---|---|
| General office | 15 $ft^3$/min/person (26 $m^3$/hr) |
| Private office | 25 $ft^3$/min/person (38 $m^3$/hr) |
| Factory | 10 $ft^3$/min/person (17 $m^3$/hr) or 0.10 $ft^3$/min $ft^2$ of floor area (0.03 $m^3$/min $m^2$ of floor area) or as governed by local codes |

Sources: Perry, J.H. (ed.): *Chemical Engineer's Handbook,* Ed. 4, McGraw-Hill, New York, 1963, Section 15.
Staniar, H.: *Plant Engineer's Handbook,* Ed. 2, McGraw-Hill, New York, 1959.

Table 8-9

Exhaust Air Requirements

| Operation | Ventilation Requirements |
|---|---|
| Bagging—paper bags | 100 cfm*/$ft^2$ open area (30 m/min) |
| Bagging—cloth bags | 200 cfm/$ft^2$ open area (60 m/min) |
| Drum filling or emptying by scoop | 100 cfm/$ft^2$ container cross-section (30 m/min) |
| Bins (closed top) | 150-200 fpm** at feed points (45-60 m/min) |
| Enclosed crushers and grinders | 200 fpm through openings (60 m/min) |
| Laboratory hood with door | 50-100 fpm (15-30 m/min) |
| Enclosed mixers | 100-200 fpm through feed and inspection openings (30-60 m/min) |
| Enclosed screens | 150-200 fpm through openings (45-60 m/min) but not less than 25-50 cfm/$ft^2$ of screen area (8-15 m/min) |
| Open surface tanks having canopy hoods used for: | |
| Plating | 175 cfm/$ft^2$ hood opening (55 m/min) |
| Salt baths | 125 cfm/$ft^2$ hood opening (40 m/min) |
| Hot water boiling | 175 cfm/$ft^2$ hood opening (55 m/min) |

* cfm = $ft^3$/min
** fpm = ft/min

Source:  *ASHRAE Guide and Data Book—Systems 1970,* American Society of Heating, Refrigerating, Air Conditioning Engineers, Inc., New York, 1970, chap 20.

red to the processing areas from the reaction vessels, steam lines, holding tanks, and so on; in the office areas, some of the air may be recycled to conserve energy (recycling is not done in processing or laboratory areas because the concentration of contaminants could be increased to a dangerous level); and other factors. A more precise method of performing these calculations and for determining air-conditioning loadings is given in reference 44. This should be used if ventilation costs appear to be a significant portion of the operating costs.

In most plants the only areas that are air-conditioned are the offices, laboratories, and control rooms. For offices, 3-6 volt amperes/ft² (32-65 volt amperes/m²) is a typical load.[46] For control rooms the load will be somewhat higher, and for laboratories, because of increased ventilation requirements, it may be much higher.

The amount of water consumption per employee per 8-hour day is normally 27-45 gal (0.1-0.17 m³). Of this amount the factory worker will consume 5 gal (0.019 m³) of hot water, while the office worker will only consume 2 gal (0.0075 m³). The hot water is assumed to be at 140°F (60°C).[46]

## UTILITY REQUIREMENTS

Once the material and energy balances have been completed, a summary must be made of all the utilities required. Since the amount needed varies with the season of the year, one summation is made for extreme winter conditions and another for extreme summer conditions. From these the peak loading can be determined. Then more common conditions are used and an average loading per year is estimated. This is expressed as the average amount used per pound of product produced. Care must be exercised in calculating these figures, since some equipment is used sporadically. For instance, a large unloading pump may be used only for a few hours per week; agitators installed on a batch reactor are operated only part of the time; and spare equipment is used only when there are operating difficulties with the on-line equipment. Adding up the horsepower for all the equipment would be a poor way of determining power requirements.

Under "utilities" should be included the costs for items that enter the plant but do not enter directly into the material formulation of the products or by-products. This includes coal, oil, gas, electricity, water, air, and inert gases. The amounts of cooling, process, and potable water should each be specified. Potable water is water that can be used for drinking and food preparation. It is usually purchased from a nearby municipality. The average electrical power required, peak power required, and demand power need to be calculated. Demand power is the number of kilowatts of energy that the utility company agrees to supply on an uninterrupted basis. A premium price is charged for this power.

If an auxiliary power supply is to be installed, its size must be determined. It should be large enough either to shut down all processes safely or to maintain them at a minimal operating level, assuming there is a complete failure of the incoming electrical supply.

Most companies maintain an elevated water supply for use in case of emergencies. This supply should be large enough to provide whatever cooling water or quenching is needed to shut down the plant safely, assuming a failure of all pumps and other water sources. It also serves as a source of water for fighting fires. It may be located on a nearby hillside or in an overhead tank.

As noted before, standard sizes of equipment are cheaper and are thus preferentially specified. Because the equipment must assuredly be large enough to produce the amount of product desired, undersized items are rarely specified. From this the reader may infer that the predicted utility requirements may be low, since for a preliminary design the standard items are usually not specified. Also, numerous small items of equipment have not been included, and possibly the designer may have mistakenly omitted at least one major item. Therefore, it is wise to increase the power requirements by 5%-10%.

## MANPOWER REQUIREMENTS

Before an economic balance can be completed, the number of men required to operate the plant must be estimated. A list of some of the types of employees is given in Table 8-10. Note that these categories include only those people involved directly in production and maintenance, and not purely administrative, research, and sales personnel. The number of salaried personnel varies with the size and complexity of the plant and can range from three or four to many times that number. Some plants are run at night and on weekends entirely by foremen, with the superintendent and engineers on 24-hour call to handle emergencies.

To determine the number of hourly personnel, the process engineer must decide what tasks must be performed and how many people each task will require per shift. Some petroleum refineries processing over 200,000 bbl/day require less than 6 operators per shift. A plant that is nearly completely automated may require essentially no one; however, for safety reasons two persons will be employed. Two

Table 8-10

Job Classifications Often Used in Chemical Plants

| Salaried Employees | Hourly Employees | Maintenance Personnel |
|---|---|---|
| Plant manager | Foremen | Security men |
| Process engineer(s) | Operators | Electricians |
| Chief chemist | Technicians | Plumbers |
| Chemist(s) | Packagers | Pipefitters |
| Supervisor(s) | Clerks | Millwrights |
| Bookkeeper | Janitors | Boilermen |
| Nurse | Groundskeepers | Instrument technicians |
| | Receiving and product shipping personnel | Helpers for the above |
| | Secretaries | Supplymen |

is a minimum requirement so that in case one is injured the other can call for help. A night watchman might be one of the two.

When operators are used around the clock, at least 4 men are needed for each post and an extra man is often required as a substitute for vacation periods, sickness, and other absences. This extra person may be used as a substitute for more than one position.

The typical manning estimate of a large grass-roots plant is given in reference 47.

Many tasks are only performed on the day shift, since this is when most supervisory personnel are employed and also no shift premiums are given. A shift premium of about 10% of an employee's hourly wage is paid to those nonsalaried personnel working the evening shift (evening hours—say 3:30 to 11:30 P.M.), and a 15% premium for those working the graveyard shift (early morning hours—say 11:30 P.M. to 7:30 A.M.). These tasks might include routine maintenance, packaging, and shipping operations. Storage hoppers large enough to hold all the product made over the weekend may be specified so that these packaging operations only need to be run during the day shift on weekdays. In case of machine failure or increased demand, an occasional extra shift can be run. Conversely, janitorial services are often performed during the evening hours when there are few people around to interfere with the operations.

Maintenance may be handled either by employees hired especially for that job or by an outside contractor. The latter is most economical when workers in most trades are needed only occasionally. The company does not need to hire these workers full time, but still has their talents available when they are needed. Even in a large plant where a number of employees skilled in each maintenance trade are needed full time, contract maintenance may be used as a supplement at times of turnarounds (a planned time when large continuous plants are shut down to do preventive maintenance, repairs, and inspections) or especially heavy demands. For instance, during the turnaround of a 140,000 bbl/day (22,000 m³/day) Tidewater Oil Company refinery, the number of maintenance personnel rose from a preshutdown level of 181 to 924 men.[48] Fewer men could have been used, but the downtime of the refinery would have been increased. In this instance the plant contracted for all of its maintenance.

Contract maintenance personnel cost more than plant personnel doing the same job. In 1967, it was estimated that the average contract person's services cost $4.71 per hour, while plant maintenance men cost $3.62 per hour.[49]

One way of avoiding estimating the number of maintenance personnel is to base the cost of maintenance on the total sales. In 1972, this cost ranged between 3 and 9% of total sales with the largest chemical companies ranging between 6.1 and 7.4%.[50] For the whole chemical industry, maintenance averaged 4.8% of the total fixed assets.

## RULES OF THUMB

As with the equipment list in Chapter 5, there are numerous rules of thumb that can speed up calculations. A few that have not already been given, and some ranges, follow.

**Heat Exchangers**[51]

1. Removable bundle heat exchangers do not exceed 10,000 ft² (930 m²).
2. Fixed tube heat exchangers are usually less than 50,000 ft² (4,650 m²).

**Thermosyphon Reboilers**

1. Heat transfer coefficients[52] for:
   Light hydrocarbons: 160-220 BTU/hr ft² °F (780-100 kcal/hr m² °C)
   Water or aqueous solutions: 220-330 BTU/hr ft² °F (1,100-1,600 kcal/hr m² °C)
   Heavy organics: 100-160 BTU/hr ft² °F (500-800 kcal/hr m² °C)
2. The maximum mean temperature difference between fluids[52,53] should be 90°F (50°C).
3. A maximum of 20% of the liquid should be vaporized per pass.[11]

**Forced Circulation Reboilers**

Forced circulation reboilers should be designed for velocities of 10-15 ft/sec (3-5 m/sec).[53]

**Evaporators**

1. Over-all heat transfer coefficients[54]
   Long tube evaporators
     Natural circulation: 200-600 BTU/hr ft² °F (1,000-3,000 kcal/hr m² °C)
     Forced circulation: 400-2,000 BTU/hr ft² °F (2,000-10,000 kcal/hr m² °C)
   Short tube evaporators
     Horizontal tube: 200-400 BTU/hr ft² °F (1,000-2,000 kcal/hr m² °C)
     Calandria tube: 150-500 BTU/hr ft² °F (750-2,500 kcal/hr m² °C)
   Coil evaporators: 200-400 BTU/hr ft² °F (1,000-2,000 kcal/hr m² °C)
2. Capacity of agitated thin film evaporators[55]
   Steam heated:
     Water evaporation: 50,000 BTU/hr ft² (135,000 kcal/hr m²)
     Organics distillation: 20,000 BTU/hr ft² (54,000 kcal/hr m²)
   Hot oil heated, organics distillation: 8,000 BTU/hr ft² (22,000 kcal/hr m²)

**Air-Cooled Heat Exchangers**

1. The average face velocity [56] is 675 ft/min (200 m/min).
2. The air pressure drop [57] is 0.25-1.0 in H₂O (6-25 kg/m²).
3. The fan efficiency [57] is 65%.
4. The fan power requirement is 0.04 hp/ft² (0.4 hp/m²) of tower cross-section.[33]

**Process Furnaces**

The amount of charring or decomposition is often related to the maximum rate of heat absorption, which is given below as a factor of the average rate of heat absorption.[58]

| Floor-fired vertical-circular heater | 1.2-1.8 times average |
| Horizontal heater, fired from one end | 1.3-1.8 times average |
| Horizontal heater, fired from both ends | 1.2-1.5 times average |
| Horizontal heater, floor-fired | 1.3-2.0 times average |
| Fire tube heater | 1.5-3.2 times average |

**Spray Equipment[59]**

The pressure drop required for various equipment where a fluid is sprayed into a gas is given below.

Spray drying of solids: 100-5,000 psi (7-350 kg/cm$^2$)

Humidification, gas and air washing: 10-100 psi (0.7-7 kg/cm$^2$)

Cooling, evaporation, and aeration: 7 psi (0.5 kg/cm$^2$)

**Flares [51]**

Flares cannot be used if a gas has a fuel value less than 150 BTU/SCF (1,350 kcal/m$^3$)

**Motors [51]**

Motors are generally limited to 10,000 hp.

**Dust Collection Equipment**

Table 8-11 gives the pressure drop and utility requirements for dust collection equipment.

### CASE STUDY: ENERGY BALANCE AND UTILITY ASSESSMENT FOR A 150,000,000 LB/YR POLYSTYRENE PLANT USING THE SUSPENSION PROCESS

An addendum to the equipment list is given in Table 8E-1. The basis of these specifications follows.

The energy requirements in summer and winter are different. Since the equipment must work at all times of the year, heat exchangers that are heating up raw materials must be designed to do their job during the winter, when the largest amount of energy must be transferred. This means that they will be overdesigned for summer conditions. The reverse is true for raw materials being refrigerated. The

Table 8-11

Pressure Drop and Utility Requirements
in Dust Collecting Equipment

| | Pressure Drop | | Utilities | |
|---|---|---|---|---|
| | in $H_2O$ | $kg/m^2$ | per 1,000 $ft^3$/min of gas | per 1,000 $m^3$/min of gas |
| **Dry Inertial Collectors** | | | | |
| Settling chambers | <0.2 | 5 | — | — |
| Baffle chamber | 0.1-0.5 | 2.5-13 | — | — |
| Skimming chamber | <0.1 | <2.5 | — | — |
| Louver | 0.5-2 | 13-50 | — | — |
| Cyclone | 0.5-3 | 13-75 | — | — |
| Multiple cyclone | 2-6 | 50-150 | — | — |
| Impingement | 1-2 | 25-50 | — | — |
| Dynamic | none | none | 1-2 hp | 30-70 hp |
| **Wet Scrubbers** | | | | |
| Gravity spray | <1 | <25 | 0.5-2 gpm | 0.7-2.7 $m^3$/min |
| Centrifugal | 2-6 | 50-150 | 1-10 gpm | 1.3-13 $m^3$/min |
| Impingement | 2-8 | 50-200 | 1-5 gpm | 1.3-7 $m^3$/min |
| Packed bed | 1-10 | 25-250 | 5-15 gpm | 7-20 $m^3$/min |
| Submerged nozzle | 2-6 | 50-150 | — | — |
| Dynamic | none | none | 1-5 gpm + 3-20 hp | 100-700 h.p. + 1.3-7$m^3$/min |
| Jet | none | none | 50-100 gpm | 65-130 $m^3$/min |
| Venturi | 10-30 | 250-750 | 3-10 gpm | 4-13 $m^3$/min |
| Fabric Filters | 2-6 | 50-150 | — | — |
| Electrostatic Precipitators | 0.2-1 | 5-25 | 0.1-0.6 kw | 3-20 kw |

Source: Sargent, G.D.: "Dust Collection Equipment," *Chemical Engineering*, Jan. 27, 1969, p. 130.

Table 8E-1

Equipment List Addendum

| Item No. | No. Reqd. | Description |
|---|---|---|
| P-101 & P-102 | 2 | Centrifugal Pump, cast iron. BHP = 38, GPM = 1,200 |
| P-103 & P-104 | 2 | Centrifugal Pump, cast iron. BHP = 18, GPM = 400 |
| P-105 & P-106 | 2 | Centrifugal Pump, cast iron. BHP = 1.2, GPM = 15 |
| P-201 & P-202 | 2 | Centrifugal Pump, cast iron. BHP = 5.4, GPM = 100 |
| E-201 to E-208 | 8 | Styrene Heat Exchanger, tubes of stainless steel. P = 200 psig, area = 386 ft$^2$ |
| D-205 & D-206 | 2 | Hot Water Storage Tanks; D = 11.5 ft, L = 13 ft;   aluminum, insulated |
| P-301 to P-309 | 9 | Centrifugal Pump, cast iron. BHP = 29, GPM = 475 |
| P-310 & P-311 | 2 | Centrifugal Pump, cast iron. BHP = 15, GPM = 140 |
| P-312 to P-314 | 3 | Centrifugal Pump, cast iron. BHP = 7, GPM = 340 |
| P-315 & P-316 | 2 | Centrifugal Pump, cast iron. BHP = 4, GPM = 140 |
| D-305 | 1 | Wash Water Tank; D = 11.5 ft, L = 18.75 ft, V = 14,600 gal; aluminum |
| B-401 to B-403 | 3 | Air Blower, BHP = 19.25 |
| P-401 to P-405 | 5 | Centrifugal Pump, cast iron. BHP = 2, GPM = 100.   A variable-speed motor is required |
| P-406 to P-408 | 3 | Centrifugal Pump, cast iron. BHP = 1, GMP = 20 |
| P-409 to P-411 | 3 | Centrifugal Pump, cast iron. BHP = 0.45, GPM = 20 |
| E-401 & E-402 | 2 | Air Heat Exchanger, carbon steel. P = 200 psig,   area = 2,810 ft$^2$ |
| P-412 & P-413 | 2 | Centrifugal Pump, stainless steel. BHP = 0.17, GPM = 133 |
| P-414 & P-415 | 2 | Centrifugal Pump, stainless steel, BHP = 0.17, GPM = 4.3 |
| P-501 to P-510 | 10 | Centrifugal Pump, cast iron. BHP = 1.30, GPM = 22 |
| Fi-501 & Fi-502 | 2 | Air Filters, dry throwaway type, 400 in$^2$ |
| B-501 & B-502 | 2 | Blowers, cast iron. BHP = 8.4 |
| RV-501 to RV-511 | 11 | 10 in Rotary Valves, stainless steel |
| CY-501 to CY-509 | 9 | Cyclone Separators, capacity 706 ft$^3$/min |
| Fi-503 & Fi-504 | 2 | Bag Filters, 450 ft$^2$, to process 706 ft$^3$/min |
| Fi-604 | 1 | Bag Filter, 900 ft$^2$, to process 1,375 ft$^3$/min |
| Fi-605 | 1 | Bag Filter, 1,800 ft$^2$, to process 1,800 ft$^3$/min |
| Fi-606 | 1 | Bag Filter, 3,600 ft$^2$, to process 3,600 ft$^3$/min |
| CV-603 | | Aluminum Ductwork for pneumatic conveying systems.   ~ 2000 ft |
| B-601 & B-602 | 2 | Blowers, cast iron. BHP = 15 |
| B-603 & B-604 | 2 | Blowers, cast iron. BHP = 40 |
| B-605 & B-606 | 2 | Blowers, cast iron. BHP = 125 |
| RV-601 | 1 | 12 in Rotary Valve, stainless steel |
| RV-602 | 1 | 15 in Rotary Valve, stainless steel |
| D-625 to D-637 | 13 | Receiver, carbon steel D = 5 ft, L = 8 ft |
| Fi-601 | 1 | Air Filter, disposable, 800 in$^2$ |
| Fi-602 | 1 | Air Filter, disposable, 1,200 in$^2$ |
| Fi-603 | 1 | Air Filter, disposable, 2,400 in$^2$ |
| CV-604 | 1 | Screw Conveyor. L = 70 ft, D = 10 in, 18,050 lb/hr, 2 hp |
| CV-605 & CV-606 | 2 | Screw Conveyor. L = 60 ft, D = 12 in, 36,000 lb/hr, 3 hp |
| CV-607 | 1 | Screw Conveyor. L = 150 ft, D = 12 in, 36,000 lb/hr, 6 hp |
| P-701 to P-703 | 3 | Centrifugal Pump, cast iron. BHP = 15, GPM = 1,000 |
| P-704 & P-705 | 2 | Centrifugal Pump, cast iron. BHP = 3.6, GPM = 400 |
| P-706 & P-707 | 2 | Centrifugal Pump, cast iron. BHP = 9.9, GPM = 400 |
| P-708 & P-709 | 2 | Centrifugal Pump, cast iron. BHP = 19, GPM = 1,600 |
| P-710 & P-711 | 2 | Centrifugal Pump, cast iron. BHP = 0.6, GPM = 40 |
| D-701 | 1 | Deionized Water Storage, wood. D = L = 38 ft |
| D-702 | 1 | Emergency Water Storage, wood. D = L = 38 ft |
| D-703 | 1 | Cooling Water Storage, wood. D = L = 24 ft |
| IE-701 | 1 | Ion Exchange System to handle 200,000 lb/hr = 400 GPM of  resin, 200 ft$^3$ |
| WTS-701 | 1 | Incoming Water Treatment System to handle 1,000,000 lb/hr = 2,000 GPM |
| SP-701 | 1 | Steam Plant to produce 20,000 lb/hr of 150 psi steam |
| P-801 | 1 | Centrifugal Pump, cast iron. BHP = 8, GPM = 80 |
| WTS-801 | 1 | Waste Water Treatment System, 200,000 lb/hr = 400 GPM |
| AP-801 | 1 | Auxiliary Power Unit, 400 kw |

design temperature to use is the one that is not exceeded more than 1% of the time in the summer or the one that the temperature is below less than 1% of the time in the winter. During the winter in Martins Ferry, Ohio, 99% of the time the air temperature is above 5°F and during the summer 99% of the time it is below 91°F.[6]

Estimating from this, it will be assumed that the temperature of the inlet water during the winter is 40°F and that during the summer it reaches a high of 85°F.

Steam is needed to heat up the reactants, air, and water. In none of these cases does the material need to be heated above 300°F. Therefore only 150 psig steam will be needed.

### Styrene Heat Exchanger E-201

When GPPS* is made, all but 238 pounds of styrene are heated to 200°F. For the other products less is heated up per batch, because less is used. The time to charge the reactor shall be set at 5 minutes. Steam at 150 psig will be used as the heating medium.

$$Q_s = m_s \, Cp_s \, \triangle t_s = U_s \, A \, \triangle t_{ss}$$

$Q_s$ = rate of heat transfer

$m_s$ = flow rate of styrene through exchanger

=

$$\frac{(1.032 \text{ lb Styrene/lb P.S.}) \times (18{,}050 \text{ lb P.S./hr}) \times (5.5 \text{ hr/batch/7 reactors}) - 238 \text{ lb}}{(5/60 \text{ hrs})}$$

        = 173,000 lb/hr

$Cp_s$ = heat capacity of styrene = 0.43 BTU/lb°F (average)[60].

$\triangle t_s$ = temperature difference of styrene entering and leaving exchanger (in winter)

        = $(200 - 5)°F = 195°F$

$Q_s$ = $173{,}000 \times 0.43 \times 195 = 14{,}500{,}000$ BTU/hr

$A$ = area of the heat exchanger, ft$^2$

$U_s$ = over-all heat transfer coefficient[61] $\sim$ 150 BTU/hr ft$^2$ °F

$\triangle t_{ss}$ = log mean temperature difference across exchanger

        = $\dfrac{(365 - 5) - (365 - 200)}{\ln(360/165)} = 250°F$

$A$ = 386 ft$^2$

---

\* General Purpose Polystyrene

One of these will be needed for each reactor, because they must be positioned vertically above the reactors. This is to prevent any hot styrene from remaining in the exchanger or the piping, where it might polymerize.

The average steam rate = $m'_s Cp_s \Delta t_s / \lambda$ = 18,350 × 0.43 × 195/857 = 1,800 lb/hr (in winter)

where $\lambda$ = latent heat of vaporization of steam = 857 BTU/lb
$m'_s$ = average flow rate of styrene through exchanger
= 1.032 × (18,050 – 238) = 18,350 lb/hr

The maximum steam rate = $\dfrac{1,800 \text{ lb/hr} \times 5.5 \text{ hr/batch/7 reactors}}{(5/60) \text{ hours}}$ = 16,950 lb/hr

## Air Heat Exchanger E-401

The air is to be heated to 300°F (see Chapter 5) using 150 psig steam (T = 365°F).

The amount of energy required = $Q_A = m_a C_p \Delta t_a$

where $m_a$ = flow rate of air = 32,700 lb/hr (see Chapter 5)
$Cp$ = heat capacity of air = 0.25 BTU/lb°F
$\Delta t_a$ = temperature difference of air entering and leaving exchanger
= 300 – 5 = 295°F
$Q_A$ = 32,700 (0.25)(295) = 2,410,000 BTU/hr

This assumes a 10% heat loss.
The area of the heat exchanger = $A = Q_A / U_A \Delta t_a$

where $U_A$ = over-all heat transfer coefficient
$\Delta t_a$ = log mean temperature difference across exchanger
= $\dfrac{(365 - 5) - (365 - 300)}{\ln (360/65)}$ = 172.5°F
$U_A \sim \dfrac{1}{1/h_o + 1/h_i}$
$h_o$ = heat transfer coefficient of condensing steam[23] $\sim$ 2,000 BTU/hr ft²°F
$h_i$ = heat transfer coefficient of air[1] = 5 BTU/hr ft²°F
$U_A$ = 5 BTU/hr ft² °F
$A$ = 2,410,000/5(172.5) = 2,800 ft² of surface area
Amount of steam required = $Q_A / \lambda$ = 2,410,000/857 = 2,810 lb/hr

**Water Bath D-501**

Amount of heat removed by the water = $Q = mCp\Delta T$
$$= \; 18{,}050 \text{ lb/hr} \times 0.32 \text{ BTU/lb}°\text{F} \, (400 - 150)°\text{F}$$
$$= \; 1{,}450{,}000 \text{ BTU/hr}$$

If the water leaving is maintained at $100°\text{F}$, then in winter the water needed

$$= \; \frac{1{,}450{,}000 \text{ BTU/hr}}{1.0 \text{ BTU/lb}°\text{F} \times (100 - 40)°\text{F}} = 24{,}200 \text{ lb/hr}$$

In summer the water needed $= \dfrac{1{,}450{,}000 \text{ BTU/hr}}{1.0 \text{ BTU/lb}°\text{F} \times (100 - 85)°\text{F}} = 96{,}500 \text{ lb/hr}$

**Preheating Water**

There are at least two possible ways to preheat the water before it enters the reactors. One would be to use a heat exchanger similar to that designed for heating the styrene. A second would be to use a sparger in a tank. In the latter case the steam is intimately mixed with the water.

The incoming water should be pure deionized water. It could contain condensate from the various heat exchangers, or it could come from the water baths following the extruders. Still another source might be the wash water from the centrifuge. In this case the effect of recycled material upon the reaction would need to be determined. Since this would involve laboratory testing, the use of wash water will not be considered at this time.

Before a final decision is made, an economic evaluation of the various possibilities should be run. Until this is done it will be assumed the water will come from the water baths (D-501 - D-509) together with the condensed steam from the air heat exchanger and the styrene heat exchanger. This will require using aluminum or plastic lines and extra collection equipment.

A sparger will be used to heat the water to 200°F.

The amount of 150-psig steam required is calculated below for conditions occurring in winter.

Amount of water supplied by the air heat exchangers = 5,620 lb/hr
Excess energy supplied by this water = $Q_{ED}$ = 5,620 lb/hr $(\Delta H_A)$

where  $\Delta H_A$  = enthalpy change of water between $365°$ and $200°\text{F}$
$$= \; (337 - 168) = 169 \text{ BTU/lb*}$$
$$Q_{ED} \; = \; 950{,}000 \text{ BTU/hr}$$

---

\* Enthalpy from steam tables.

Amount of water supplied by the styrene heat exchangers = 1,800 lb/hr
Excess energy supplied by this water = 1,800 $(\Delta H_A)$ = 304,000 BTU/hr
Amount of water available from the water baths = 24,200 lb/hr
Amount of water required from elsewhere

$$= 18,050 \frac{\text{lb P.S.}}{\text{hr}} \times \frac{2.0 \text{ lb } H_2O}{\text{lb P.S.}} - 5,620 - 1,800 - 24,200 \text{ lb/hr} = 4,480 \text{ lb/hr}$$

This means that the water baths cannot supply all the remaining water at $100°$F.

Let $X$ = pounds of steam required by sparger

Amount of energy required = (24,200 lb/hr)(1 BTU/lb°F)($200°$F – $100°$F) – 950,000
$$- 304,000 + (4,480 - X)(200°F - 40°F)$$
$$= 1,884,000 \text{ BTU/hr} - 160X = \Delta H_s X$$

where $\quad \Delta H_s$ = enthalphy change of steam = (1,193 - 168)* = 1,025 BTU/lb

$\quad\quad X = 1,590$ lb/hr

Amount of water needed from the ion exchanger directly = 4,480 – 1,590 = 2,990 lb/hr.

The same calculation carried out under summer conditions gives $X = 2,020$ lb/hr. These should both be increased by 10% to account for heat losses.


## Steam Requirements

The average process steam requirements are given in Table 8E-2. The peak rate occurs when styrene is charged (16,950 lb/hr in winter). At this point the demand is temporarily increased by 15,150 lb/hr. The total peak demand is then 24,319 lb/hr during the winter. The steam plant could be designed to supply this amount, but it would operate at less than one-third of its capacity most of the time. Therefore the system will be designed for 20,000 lb/hr of 150-psi steam. It will mean that during the winter months 10 minutes will be needed for charging the styrene to the reactor rather than the 5 previously specified.

This system will be large enough to accommodate the proposed expansion when it occurs.

## Wash Water

The amount of deionized water required as wash water for the wash tanks and centrifuge is

$$18,050 \frac{\text{lb P.S.}}{\text{hr}} \frac{3.0 \text{ lb } H_2O}{\text{lb P.S.}} = 54,150 \text{ lb/hr}$$

In the summer this water can come from the cooling baths; however, in the winter all of this is used as feed to the reactors. In winter it must come directly from the ion exchanger.

## Process Water Requirements

The amount of deionized water required is given in Table 8E-3.

The deionization system will be designed to handle the proposed expansion. It should be large enough that the ion exchange unit can replenish storage tanks while the plant is running at full capacity.

The ion exchange system will be capable of processing 200,000 lb/hr. A two-bed ion exchange unit will be specified. The minimum resin volume is 200 ft$^3$. This was calculated on the basis of 2 GPM/ft$^3$ resin.[62] During the summer the extra wash water will be sent directly to the river. It does not appear to be warm enough to justify a cooling tower and should contain no pollutants.

Table 8E-2

Process Steam Requirements (lb/hr)

| Purpose | Summer | Winter |
|---|---|---|
| To heat styrene | 1,100 | 1,800 |
| To heat water | 2,222 | 1,749 |
| To heat air | 4,000 | 5,620 |
| Total | 7,322 | 9,169 |

Table 8E-3

Process Water Requirements (lb/hr)

(Deionized Water)

| Destination | Summer | Winter |
|---|---|---|
| Water baths | 96,500 | 24,200 |
| Wash water | 0 | 54,150 |
| Steam plant | 7,322 | 9,169 |
| Reactors (direct) | 0 | 2,990 |
| Total | 103,822 | 90,509 |

## Reactor Cooling System

A tempered water system has been specified for the reactor. It will be assumed that a 10°F rise in cooling water temperature is typical. Using data from the Case Study section of Chapter 5.

$$\text{Average energy removed per hour} = 2{,}015\, D^3 \times \frac{4 \text{ hr reaction}}{5.5 \text{ hr cycle}} \times 7 \text{ reactors}$$
$$= 5{,}900{,}000 \text{ BTU/hr.}$$

This is an average of 590,000 lb/hr of cooling water or 1,180 gal/min. The peak rate for any reactor is double the average rate. Assume the pumps must be able to handle 1,600 gal/min.

To temper this water assume that 30% of the water must be heated to 140°F. Since it does not need to be deionized water, this will be done in a separate tank. This tank will be designed to have a 0.5 hr holdup. The vessel will have steam coils to heat the water. Assume the coils occupy 10% of the space, and the tank is 90% full.

$$\text{Volume of tank} = \frac{590{,}000 \text{ lb/hr} \times 1/3 \times 0.5 \text{ hr (holdup)}}{62.3 \text{ lb/ft}^3 \times 0.81}$$
$$= 1{,}940 \text{ ft}^3 = 14{,}500 \text{ gal}$$

If the tank has a diameter of 11.5 ft, the height = 18.7 ft

The amount of steam required (winter)

$$= \frac{590{,}000 \text{ lb/hr} \times 1/3 \times 1.0 \text{ BTU/lb}^\circ\text{F} \times (140 - 40)^\circ\text{F}}{857 \text{ BTU/lb}}$$
$$= 22{,}950 \text{ lb/hr}$$

The amount of steam required seems absurd, since this is more than the total amount required for the rest of the plant. If only 20% of the water were heated and the upper temperature were reduced to 120°F, the steam requirements would be reduced by 48%. This is still large. However, under these conditions it could be handled by the proposed steam system, since the temperature of the heated water is not critical and during the peak steam-loading periods the steam could be shut off to the tempered water system. Nevertheless, it appears to be very expensive. Therefore, although it is less desirable, the control system on the reactor will be changed to a flow control system. Should this not work, a tempered system can be installed rather rapidly, since the boiler system is large enough and there is adequate space to install a hot-water tank outside the reactor building.

### Water Purification WTS-701

A rough screen to remove large items such as tree limbs, tin cans, and bones will be installed near the intake. A sand filter [63] will be the only large piece of equipment needed for the cooling water.

$$\text{Area required} = 2{,}000 \text{ GPM}/0.05 \text{ GPM}/\text{ft}^2 = 40{,}000 \text{ ft}^2$$
$$\text{Volume required} = 2{,}000 \text{ GPM}/0.01 \text{ GPM}/\text{ft}^3 = 200{,}000 \text{ ft}^3$$

### Hot water Storage D-205

Two tanks will be designed to hold enough water for a total of 4 batches.

$$\text{Volume} = \frac{18{,}050}{0.9} \frac{\text{lb P.S.}}{\text{hr}} \times \frac{2.0 \text{ lb H}_2\text{O}}{\text{lb P.S.}} \times \frac{4}{60.13 \text{ lb/ft}^3} = 2{,}670 \text{ ft}^3$$

Assume the diameter is 11.5 ft; then the length will be 13 ft.
These tanks will be constructed of aluminum and will be insulated.

### Waste Treatment System WTS-801

The design of a waste treatment system requires data that was not available to me. A secondary system to reduce the biological oxygen demand and a tertiary system to at least remove most of the phosphates are required. Whether styrene is biodegradable was not known. If it is not, then the tertiary system will have to be designed to remove it also.

### Wash Water Tank D-305

This will be designed to hold a 2 hr supply.

$$V = \frac{54{,}150 \text{ lb/hr} \times 2 \text{ hr}}{62.0 \text{ lb/ft}^3 \times 0.9} = 1{,}950 \text{ ft}^3$$
$$D = 11.5 \text{ ft}, L = 18.75 \text{ ft}$$

### Deionized Water Storage D-701

A 1-day supply will be stored.

$$V = \frac{104{,}000 \text{ lb/hr} \times 24 \text{ hr}}{0.9 \times 62.4 \text{ lb/ft}^3} = 44{,}500 \text{ ft}^3$$

If the length and diameter are equal then they will be 38.4 ft.
This is to be a wooden tank. It will contain a sparger for cold winter days.

## Emergency Water Storage D-702

A 4 hr storage capacity of cooling water will be maintained in an elevated tank on a nearby hill. This will provide enough water to properly cool all the reactors in case of a power failure.

$$\text{Volume} = \frac{590,000}{0.9} \frac{\text{lb}}{\text{hr}} \times \frac{4.0 \text{ hr}}{62.4 \text{ lb/ft}^3} = 42,000 \text{ ft}^3$$

$$L = D = 38 \text{ ft}$$

## Cooling Water Storage D-703

A 1 hr storage supply will be maintained in the plant area in a wooden tank.

$$\text{Volume} = \frac{590,000 \text{ lb/hr}}{62.4 \text{ lb/ft}^3 \times 0.9} = 10,500 \text{ ft}^3$$

$$L = D = 24 \text{ ft}$$

## Pumps

Table 8E-4 gives the information needed to specify the pumps for the plant and the average power required to operate them. A detailed calculation for the barge unloading pump follows.

## Barge Unloading Pump P-101

Assume that a barge is to be unloaded in 2 hr. It is expected that the styrene can be raised from the lowest level of the river to the top of the storage tank in that time. Assume the ground level is 40 ft above the low-water mark, the barge is 10 ft deep, and the height of the storage tank is 40 ft. This means the styrene may have to be raised 90 ft.

The maximum distance the styrene must travel horizontally is around 300 ft. The total distance traveled inside pipes is 390 ft. If a pressure drop of 2 psi per 100 ft is assumed, the pressure drop due to friction will be about 8 psi (20.5 ft of styrene) and that due to elevation is 90 ft. The total pressure drop is 110.5 ft. Usually for normal flow rates the pressure drop due to the change of velocity is ignored.

The flow rate =

$$\frac{1,000}{2 \text{ hr}} \frac{\text{tons}}{\text{barge}} \times \frac{2,000 \text{ lb}}{\text{ton}} \times \frac{1 \text{ ft}^3}{56.3 \text{ lb}} \times \frac{7.48 \text{ gal}}{\text{ft}^3} \times \frac{1 \text{ hr}}{60 \text{ min}} = 2,220 \text{ GPM}$$

The pump will be placed on a floating dock that is just a few feet above the river level, so that the net positive suction head (NPSH) will be adequate. The maximum height from the bottom of the barge to the pump will be 20 ft.

The styrene temperature should not exceed 88°F, at which the vapor pressure is

Table 8E-4

Pump Calculations

| Pump | Use | Mat'l. Pumped | Max Flow Rate lb/hr | Density lb/ft³ | Q Max GPM | Distance Hgt. Change ΔL,ft | Material Travels ΔZ,ft | Head Reqd. ΔP,ft | LHP | Pump Effi-ciency | BHP | Motor Effi-ciency | Ave. Power Per Pump kw | Total Ave. Power Reqd. kw | Notes |
|---|---|---|---|---|---|---|---|---|---|---|---|---|---|---|---|
| P-101 & P-102 | Barge unloading | Styrene | 500,000 | 56.3 | 1,200 | 90 | 390 | 110 | 30 | .80 | 37.5 | .89 | 1.2 | 2.4 | F |
| P-103 & P-104 | Storage to reactor | Styrene | 173,000 | 56.3 | 383 | 74 | 650 | 144 | 12.6 | .70 | 18 | .87 | 1.64 | 1.64 | AG |
| P-105 & P-106 | Storage to mix tank | Styrene | 6,000 | 56.3 | 13.3 | 74 | 650 | 118 | 0.36 | .30 | 1.2 | .80 | 0.06 | 0.06 | E |
| P-201 & P-202 | Water bath to storage | Hot Water | 50,000 | 62.0 | 100 | 80 | 580 | 107 | 2.7 | .50 | 5.4 | .85 | 3.42 | 3.42 | B |
| P-301 to P-309 | Water to reactor jacket | Water | 231,000 | 62.4 | 462 | 60 | 500 | 166 | 20 | .70 | 28.3 | .88 | 8.9 | 62.2 | I |
| P-310 & P-311 | Water bath to wash hold tank | Water | 70,000 | 62.0 | 140 | 20 | 40 | 22 | 0.78 | .55 | 1.42 | .81 | 1.01 | 1.01 | B |
| P-312 to P-314 | Wash hold tank to wash tank | Water | 170,000 | 62.0 | 340 | 40 | 560 | 56 | 4.8 | .70 | 6.85 | .86 | 1.26 | 2.52 | C |
| P-315 & P-316 | Deionized water to wash hold tank | Water | 70,000 | 62.4 | 140 | 40 | 460 | 61 | 2.2 | .55 | 4.0 | .84 | 2.75 | 2.75 | B |
| P-401 to P-405 | Slurry to Centrifuge | P.S. Sol'n | 46,000 | 62.4 | 92 | 20 | 470 | 42 | 0.98 | .50 | 1.96 | .82 | 1.78 | — | |
| P-406 to P-408 | Deionized water to centrifuge | Water | 9,025 | 62.4 | 18 | 20 | 730 | 54 | 0.25 | .30 | .835 | .79 | 0.79 | 1.58 | |
| P-409 to P-411 | Wash hold tank to centrifuge | Water | 9,025 | 62.0 | 18 | 20 | 140 | 27 | 0.13 | .30 | .435 | .70 | 0.465 | 0.93 | |
| P-412 & P-413 | HCl unloading to storage | HCl | 76,000 | 71.4 | 133 | 30 | 60 | 33 | 1.3 | .55 | 2.4 | .82 | 0.008 | 0.008 | |
| P-414 & P-415 | HCl storage to wash tank | HCl | 2,450 | 71.4 | 4.3 | 35 | 135 | 41 | 0.051 | .30 | .17 | .65 | 0.021 | 0.021 | |
| P-501 to P-510 | Deionized water to water bath | Water | 11,000 | 62.4 | 22 | 0 | 760 | 70 | 0.39 | .30 | 1.30 | .80 | 1.26 | 11.4 | I |
| P-701 to P-703 | River to filter | Water | 500,000 | 62.4 | 1,000 | 40 | 150 | 47 | 11.9 | .80 | 14.9 | .87 | 9.24 | 18.5 | B |
| P-704 & P-705 | Filter to ion exchange | Water | 200,000 | 62.4 | 400 | 20 | 100 | 25 | 2.53 | .70 | 3.6 | .84 | 1.71 | 1.71 | HDB |
| P-706 & P-707 | Ion exchange to storage | Water | 200,000 | 62.4 | 400 | 40 | 100 | 68 | 6.9 | .70 | 9.9 | .86 | 3.9 | 3.9 | HDB |
| P-708 & P-709 | Filter to cooling water storage | Water | 800,000 | 62.4 | 1,600 | 30 | 150 | 37 | 15 | .80 | 18.8 | .88 | 11.8 | 11.8 | B |
| P-710 & P-711 | Deionized water to steam plant | Water | 20,000 | 62.4 | 40 | 0 | 500 | 23 | 0.223 | .40 | .585 | .71 | 0.26 | 0.26 | BD |
| P-801 | Water to emergency water storage | Water | 40,000 | 62.4 | 80 | 150 | 5,000 | 280 | 5.65 | .70 | 8.1 | .71 | — | — | |

The following pumps are spares: P-104, P-106, P-202, P-309, P-311, P-314, P-316, P-403, P-405, P-408, P-411, P-413, P-415, P-510, P-703, P-705, P-707, P-709, and P-711.

Notes

A  Styrene to be charged in 5 minutes.

B  Pump must be able to fill storage tank more rapidly than it is emptied.

C  Wash water is to be charged to wash tank in 10 minutes.

D  Pump designed to handle expansion.

E  Styrene is to be charged in 2.5 minutes.

F  Barge is to be unloaded in 2 hours.

G  Pressure drop includes 10 psi for flow through heat exchanger.

H  Pressure drop includes 10 psi for flow through ion exchange unit.

I  Control valve included, which doubles the pressure drop.

10 mm Hg. The maximum distance to the pump from the intake on the barge is approximately 100 ft, for a pressure drop of 2 psi.

$$\text{NPSH} = 14.7 \text{ psi} - 2 \text{ psi} - 10 \text{ mm Hg} \times \frac{14.7 \text{ psi}}{760 \text{ mm Hg}} - \frac{20 \text{ ft} \times 56.3 \text{ lb/ft}^3}{144 \text{ in}^2/\text{ft}^2}$$

$$= 4.7 \text{ psi} = 12.0 \text{ ft}$$

Two pumps will be purchased and will be installed in parallel. Each will be specified to handle 1,200 GPM at a pressure drop of 110 ft.

$$\text{The liquid horsepower} = \text{LHP} = \frac{QH}{3,960} \text{ (S.G.)} = 30 \text{ hp}$$

where $Q$ = flow rate = 1,200 GPM
$H$ = head developed (ft) = 110 ft
S.G. = specific gravity = 56.3/62.4 = 0.902

A pump efficiency of 80% is obtained from Figure 8-5. This gives a brake horse-power (BHP) of 37.50. A motor efficiency of 89% is obtained from Figure 8-6. The average power per day used by these pumps is:

$$P = \frac{37.5 \text{ hp}}{0.89} \times 2 \text{ hr} \times \frac{1 \text{ day}}{24 \text{ hr}} \times \frac{1 \text{ barge}}{4.4 \text{ days}} \times 2 \text{ pumps} \times \frac{0.745 \text{ kw}}{\text{hp}} = 1.2 \text{ kw}$$

## Blower B-401

The blower calculations are similar to those for the pumps. The optimum velocity for air is around 75 ft/sec and the pressure drop is about 0.2 psi per 100 ft of piping. At this velocity the kinetic energy term in the pressure-drop equation cannot be ignored. The pressure drop can be approximated if a 14 in. duct is specified and Figure 8-7b is used.

Air flow rate = 32,700 lb/hr (see Chapter 5)
Pressure drop through the air filter[64] = 0.5 in $H_2O$
Pressure drop through the dryer[65] = 3 in $H_2O$
Pressure drop through the heat exchanger and piping = 4 in $H_2O$
Pressure drop through the bag filter[66] = 3 in $H_2O$
Total pressure drop = 10.5 in $H_2O$
Change in kinetic energy = $\dfrac{V^2}{2g_c} = \dfrac{(75)^2}{64.4} = 87.5 \dfrac{\text{ft lb}_f}{\text{lb}_m}$

Blower horsepower = 19.25
(This assumes an over-all efficiency of 70%)
If the motor efficiency is 89% then the power required =

$$\frac{19.25 \text{ hp}}{0.89} \times \frac{0.746 \text{ kw}}{\text{hp}} = 16.1 \text{ kw}$$

**Pneumatic Conveying System**

The minimum velocity is the air velocity that must be maintained to prevent the solids from dropping out onto the sides of the pipe.

$$\text{Minimum velocity}^{67} = V_{\text{min}} = 910 \left( \frac{\rho_s}{\rho_s + 62.3} \right) D_s^{0.60}$$

where   $\rho_s$  =  density of polystyrene = 65.5 lb/ft$^3$
        $D_s$  =  diameter of largest particle = 1/8 in = 0.0104 ft
        $V_{\text{min}}$ = 30.8 ft/sec

A velocity of 70 ft/sec will be used because it is close to the optimal velocity for air alone. The ratio of solids to air varies between 1 : 1 and 10 : 1.[68] A ratio of 3 : 1 will be assumed. If a pressure drop of 0.2 psi/100 ft for air without any solids is assumed, when solids are present the pressure drop in vertical ducts will be 2.2 times that amount. For horizontal ducts the factor is 1.55 (see Fig. 8-11).[69]

**Pneumatic Conveying from the Cutters (Cu-501)**
**to the Test Hoppers (D-620)**

A vacuum system like that shown in Figure 4-3 will be used,[36] since delivery to a single receiver is desired. For this system there will be 100 ft of horizontal pipe and 35 ft of vertical pipe. The pressure drop across the air filter at the air inlet should be about 0.5 in. H$_2$O and that across the receiver should be negligible. The pressure drop across the fabric filter is 4.0 in. H$_2$O and across the receiver is 3 in H$_2$O.

Frictional pressure drop = $(0.2 \text{ psi } (1.55) + 0.2 \text{ psi } \dfrac{35 \text{ ft}}{100 \text{ ft}} \, 2.2) \, 1.25 = 0.580$ psi

Total pressure drop = $0.580 + 7.5$ in H$_2$O $\times$ 14.7 psi/407 in H$_2$O = 0.85 psi

Air flow rate = $18,050 \dfrac{\text{lb P.S.}}{\text{hr}} \times 1/3 = 6,017 \dfrac{\text{lb air}}{\text{hr}} = 1,375$ ft$^3$/min.

LHP = $\dfrac{1,375 \text{ ft}^3/\text{min} \times 0.85 \text{ lb/in}^2 \times 144 \text{ in}^2/\text{ft}^2}{33,000 \text{ ft lb/min hp}} = 5.1$

This value must be increased to account for air leakage and for possible blockage of pipelines. If it is doubled and the blower has an efficiency of 70% the required BHP will be around 15. If the motor has an efficiency of 90% the power required for blowers B-601 and B-602 will be 12.5 kw.

Another way of determining horsepower is to use a generalized correlation. This rate gives 23 hp.[70] Still another correlation gives a brake horsepower of 18.[71]

**Pneumatic Conveying from the Test Hoppers (D-620)**
**to the Bulk Storage Hoppers (D-601)**

The test hoppers must be able to be emptied at a more rapid rate than they are filled, so that should any short testing delays occur they will not cause the process to be shut down. A rate twice that of the production rate will be specified. The length of vertical pipe is 60 ft and that of horizontal pipe is 300 ft. The results for blowers B-603 and B-604 are: LHP = 14; BHP = 40; and the power = 16.8 kw.

### Pneumatic Conveying from the Bulk Storage Hoppers (D-601) to the Packaging Storage Hoppers (D-611) and the Final Bulk Shipment Hoppers (D-614)

A positive-pressure system will be used because the delivery is to a number of different points.

Since the packaging units will only operate about 35 hours a week, the system must be able to handle at least 0.60 (168 hr/35 hr) × 18,050 = 52,000 lb/hr. Assume it must handle 70,000 lb/hr. This rate will also allow the bulk shipment hoppers to be filled in 3 hr, which seems reasonable. The largest pressure drop will occur in transferring material to the railroad storage hoppers. The horizontal length will be a maximum of 500 ft and the vertical distance 100 ft. A pressure system will be used. The results of the calculations for blowers B-605 and B-606 are LHP = 45; BHP = 125; and the power = 26.4 kw.

### Pneumatic Conveying from the Dryers (DR-401) to the Extruder Feed Tanks (D-510)

This will be a positive-pressure system, since each dryer will feed 5 extruders. It will be designed to run at the same rate as each dryer, 9,210 lb P.S./hr. The vertical rise is 30 ft and the maximum distance traveled is 130 ft. Because powder is being conveyed a cyclone separator will be specified for each of the extruders. The blowers should be designed to have an LHP of 7.2 and a BHP of 8.4. The power required to operate both is 12.6 kw.

### Rotary Valves

Rotary valves have been specified for the receivers of the vacuum pneumatic conveying systems (RV-601 and RV-602) and for the feeders (RV-401) and receivers (RV-501) of the positive-pressure system transferring powder from the dryers to the extruders. Perry[72] gives some recommended sizes for rotary valves. The power required by all these valves will probably not exceed 2 kw and will be ignored.

### Receivers D-620

When ⅛ in cylinders are conveyed, a simple vertical tank will be used instead of a cyclone for separating the solids from the air. It is assumed a 4 ft × 8 ft aluminum vessel will suffice.

### Bag Filters and Air Filters

See Chapter 5 for how these should be sized.

### Screw Conveyor CV-604

The conveyor will extend over the tops of tanks D-620 through D-624 and distribute the material received to the proper tank. The material will enter near the top of the middle tank. It will travel a maximum of 34 ft (assuming a spacing of 5 ft between tanks). However, the conveyor will be 70 ft long. The flow rate is 18,050 lb/hr. A 2 hp motor will be needed.[73]

### Screw Conveyors CV-605 to CV-607

Screw conveyors CV-605 will normally handle only HIPS* and will empty into D-601, D-602, and D-603. Screw conveyor CV-606 will normally handle only MIPS** and will supply tanks D-603, D-604, and D-605. Screw conveyor CV-607 will normally convey only GPPS and will feed tanks D-605 through D-610. The calculations are similar to those for CV-604.

### Natural Gas Requirements

Natural gas will be used in the steam boilers to obtain steam. It is assumed that the thermal efficiency of a steam-generating plant is around 80%.[74]

$$\text{Number of BTU required} = 1,160 \text{ BTU/lb} \ \frac{9032 + 7134}{2} \ \frac{\text{lb}}{\text{hr}} = 9,360,000 \text{ BTU/hr}$$

$$\text{Amount of gas required}^{43} = 9,360,000/(0.80) \times (1,050 \text{ BTU/ft}^3) = 12,300 \text{ ft}^3/\text{hr}$$

### Manpower Requirements

An estimate of this is given in Table 8E-5. The total number of hourly personnel on each shift will be 6.

### Ventilation and Heating Requirements

The ventilation requirements come from Tables 8-9 and 8-10. The largest estimate should be used when more than one is obtained.

Product Warehouse

   Est 1 $Q_w$ = 18 men $\times$ 10 ft$^3$/min man = 180 ft$^3$/min

   Est 2 $Q_w$ = 1,356,000 ft$^3$ $\times$ 1/2 change/hr $\times \frac{1 \text{ hr}}{60 \text{ min}}$ = 11,400 ft$^3$/min

   Est 3 $Q_w$ = 68,250 ft$^2$ $\times$ 0.10 ft$^3$/min ft$^2$ = 6,825 ft$^3$/min

Raw Material Warehouse

   Est 1 = 120,000 ft$^3$ $\times$ 1/2 change/hr $\times \frac{1 \text{ hr}}{60 \text{ min}}$ = 1,000 ft$^3$/min

   Est 2 = 6,000 ft$^2$ $\times$ 0.10 ft$^3$/min ft$^2$ = 600 ft$^3$/min

Reactor Building

   $Q_R$ = 12,800 ft$^2$ $\times$ 0.10 ft$^3$/min ft$^2$ = 1,280 ft$^3$/min

Centrifuge and Drying Building

   $Q_{CD}$ = 2,400 ft$^2$ $\times$ 0.10 ft$^3$/min ft$^2$ = 240 ft$^3$/min

Extruder Building

   $Q_E$ = 9,600 ft$^2$ $\times$ 0.10 ft$^3$/min ft$^2$ = 960 ft$^3$/min

Offices, Laboratories, etc.

   Est 1 = 15 ft$^3$/min person $\times$ 8 = 120 ft$^3$/min

   Est 2 = 3,800 ft$^2$ $\times$ 0.10 ft$^3$/min ft$^2$ = 380 ft$^3$/min

Total ventilation requirement = 15,260 ft$^3$/min

---

\* High Impact Polystyrene
\*\* Medium Impact Polystyrene

Table 8E-5

Manpower Requirements

### Hourly Personnel

| | |
|---|---|
| Power plant | 1 man per shift |
| Utilities | 1 man on day shift |
| Reactor area | 2 men per shift |
| Centrifuge drying area | 1 man per shift |
| Extruder area | 1 man per shift |
| Automatic bagging area | 2 men on lifts on day shift |
| | 1 operator on day shift |
| Carton filling area | 2 men on lifts on day shift |
| | 1 operator on day shift |
| Drum filling area | 2 men on lifts on day shift |
| | 1 operator on day shift |
| | 1 pallet loader on day shift |
| Loading railcars & trucks | 6 men on lifts on day shift |
| | 2 men on bulk shipments on day shift |
| Janitors | 2 men on night shift |
| Warehouse foremen | 2 men on day shift |
| Yardman | 1 man on day shift |
| Extra men for vacations, sickness, etc. | 3 men total |

### Salaried Personnel

| | |
|---|---|
| Plant manager | 1 man |
| Process engineer | 1 man |
| Supervisors | 1 man per shift plus one |
| Bookkeeper (in charge of shipments) | 1 man |
| Secretary | 1 man |
| Chief chemist | 1 man |
| Chemists | 1 man per shift plus one |

The amount needed for the ion-exchange and steam-generating building will be insignificant. Since we are dealing with pellets, not powders, no special hoods are needed. The fluid power required to move air, assuming a pressure drop of 3 in $H_2O$, is 7.3 hp. Less than 7 kilowatts are required to power fans.

To heat this air from 5°F to 70°F during winter will require the following amount of power:

$$P_h = 15,260 \text{ ft}^3/\text{min} \times 0.0750 \text{ lb/ft}^3 \times 60 \text{ min/hr} \times 0.25 \text{ BTU/lb}°\text{F} \times 65°\text{F}$$
$$= 1,116,000 \text{ BTU/hr} = 330 \text{ kw}$$

This may be high, since in most of the processing areas large amounts of heat will be released from the equipment and the temperature in the warehouse need not exceed 50°F. Over the whole year the power requirement for heating and ventilating will be around 90 kw.

## Lighting

The lighting specifications are given in Table 8-7. A coefficient of utilization of 0.60 and a maintenance factor of 0.70 have been assumed. Fluorescent lighting is to be used. The power requirements are calculated below.

Product Warehouse

$$P_W = \frac{68{,}250 \text{ ft}^2 \times 7 \text{ ft-candles}}{50 \text{ lumens/watt } (0.60)\,(0.70)} = 22{,}800 \text{ watts}$$

Raw Material Warehouse

$$P_W = \frac{6{,}000 \text{ ft}^2 \times 7 \text{ ft-candles}}{50 \text{ lumens/watt } (0.60)\,(0.70)} = 2{,}000 \text{ watts}$$

Reactor Building

$$P_R = \frac{12{,}800 \text{ ft}^2 \times 30 \text{ ft-candles}}{50 \text{ lumens/watt } (0.60)\,(0.70)} = 18{,}300 \text{ watts}$$

Centrifuge and Drying Building

$$P_D = \frac{2{,}400 \text{ ft}^2 \times 30 \text{ ft-candles}}{50 \text{ lumens/watt } (0.60)\,(0.70)} = 3{,}400 \text{ watts}$$

Extruder Building

$$P_E = \frac{9{,}600 \text{ ft}^2 \times 30 \text{ ft-candles}}{50 \text{ lumens/watt } (0.60)\,(0.70)} = 13{,}800 \text{ watts}$$

Offices, Laboratories, etc.

$$P_L = \frac{3{,}800 \text{ ft}^2 \times 100 \text{ ft-candles}}{50 \text{ lumens/watt } (0.60)\,(0.70)} = 18{,}000 \text{ watts}$$

## Outside Lighting

175 w R-40 mercury vapor lamps are to be used for walkways and should be 26 ft apart.[42] The number of feet of walkways will be approximately 3,000 ft. This means that a total of 116 lights are needed. The total power used for walkways will be 20 kw. For the parking lot:

$$P_P = \frac{16{,}000 \text{ ft}^2 \times 1 \text{ ft-candle}}{50 \text{ lumens/watt } (0.60)\,(0.70)} = 2{,}800 \text{ watts}$$

This gives a total of 95 kw for peak lighting requirements.

The warehouses are designed to be operated only 35 hours per week and the outside lights should only be on about half the time. The other power will be on most of the time. This gives an average power of 67 kw.

### Power Requirements

The power requirements are given in Table 8E-6. The power stations should, however, be designed to handle the proposed expansion. Their specifications should be based on an average demand of 3,500 kw. Since this does not include peaks, a main transformer of 5,000 kva and three secondary transformers rated at 2,000 kva will be specified.

Table 8E-6

Average Power Requirements

| | | |
|---|---|---|
| All pumps | = | 130 kw |
| Air blowers | = | 44.8 kw |
| Airveying system | = | 55.7 kw |
| Reactor agitators | = | 316 kw |
| Hold tank agitators | = | 150 kw |
| Centrifuge | = | 90 kw |
| Dryer | = | 18 kw |
| Extruder | = | 1,040 kw |
| Ventilating, heating, and lighting | = | 170 kw |
| Waste disposal system, bagging, palletizing, steam generation, and Waste disposal | = | 200 kw (estimated) |
| All other items | < | 100 kw |
| Total Power | = | 2,302 kw |

### CHANGE OF SCOPE

Scope and process changes and additions:
1. The tempered water system on the reactors will be replaced by a simple flow control system.
2. It will take 5 min to charge a reactor.
3. It will take 2 hr to unload a barge containing styrene.

## References

1. "The Energy Crisis," *Chemical Week,* Sept. 20, 1972, p. 29.
2. Miller, R., Jr.: "Process Energy Systems," *Chemical Engineering* May 20, 1968, p. 130.
3. "Turned off by the Gas Shortage," *Chemical Week,* May 31, 1972, p. 10.
4. Gambs, G.C., Rauth, A.A.: "The Energy Crisis," *Chemical Engineering,* May 31, 1971, p. 64.
5. Lord, R.C., Minton, P.E., Slusser, R.P.: "Design of Heat Exchangers," *Chemical Engineering,* Jan. 26, 1970, p. 96.
6. *Evaluated Weather Data for Cooling Equipment Design. Addendum No. 1, Winter and Summer Data,* Fluor Products, Santa Rosa, Calif. 1964.
7. Winton, J.: "Plant sites '67," *Chemical Week,* Oct. 28, 1967, p. 88.
8. Perry, J.H. (ed.): *Chemical Engineers' Handbook,* Ed. 4, McGraw-Hill, New York, 1963, Section 9, p. 52.
9. Seifert, W.F., Jackson, L.L., Sech, C.E.: "Organic Fluids for High Temperature Heat Transfer Systems," *Chemical Engineering,* Oct. 30, 1972, p. 96.
10. Fried, J.R.: "Heat Transfer Agents for High Temperature Systems," *Chemical Engineering,* May 28, 1973, p. 89.
11. Buckley, P.S.: "Some Practical Aspects of Distillation Control," paper presented at Chemical Engineering Seminar, Ohio University, Nov. 20, 1967.
12. Brooke, M.: "Process Plant Utilities—Water," *Chemical Engineering,* Dec. 14, 1970, p. 135.
13. Thompson, A.R.: "Cooling Towers," *Chemical Engineering,* Oct. 14, 1968, p. 100.
14. Silverstein, R.M., Curtis, S.D.: "Cooling Water," *Chemical Engineering,* Aug. 9, 1971, p. 84.
15. Woods, G.A.: "Bacteria: Friends or Foes?" *Chemical Engineering,* Mar. 5, 1973, p. 81.
16. Troscinski, E.S., Watson, R.G.: "Controlling Deposits in Cooling Water Systems," *Chemical Engineering,* Mar. 9, 1970, p. 125.
17. Gazzi, L., Pasero, R.: "Process Cooling Systems: Selections," *Hydrocarbon Processing,* Oct. 1970, p. 83.
18. Wigham, I.: "Designing Optimum Cooling Systems," *Chemical Engineering,* Aug. 9, 1971, p. 95.
19. Kals, W.: "Wet Surface Aircoolers," *Chemical Engineering,* July 26, 1971, p. 90.
20. Perry, J.H.: *op. cit.,* Section 12.
21. Tanzer, E.K.: "Comparing Refrigeration Systems," *Chemical Engineering,* June 10, 1963, p. 215.
22. Perry, J.H.: *op. cit.,* Sections 10 and 11.
23. McCabe, W., Smith, J.C.: *Unit Operations of Chemical Engineering,* McGraw-Hill, New York, 1967, p. 316.
24. Ludwig, E.E.: *Applied Process Design for Chemical and Petrochemical Plants,* Gulf, Houston, 1965, p. 61.
25. Lord, R.C., Minton, P.E., Slusser, R.G.: "Guide to Trouble-Free Heat Exchangers," *Chemical Engineering,* June 1, 1970, p. 153.
26. Thurlow, C.E., III: "Centrifugal Pumps," *Chemical Engineering,* Oct. 11, 1971, p. 29.
27. Younger, A.H., Ruiter, J.L.: "Selecting Pumps and Compressors," *Chemical Engineering,* June 26, 1961, p. 117.
28. Stindt, W.H.: "Pump Selection," *Chemical Engineering,* Oct. 11, 1971, p. 43.
29. Bresler, A.: "Guide to Trouble-Free Compressors," *Chemical Engineering,* June 1, 1970, p. 161.
30. Simpson, L.L.: "Sizing, Piping for Process Plants," *Chemical Engineering,* June 17, 1968, p. 192.
31. Gallant, R.W.: "Sizing Pipe for Liquids and Vapors," *Chemical Engineering,* Feb. 24, 1969, p. 96.
32. Perry, J.H.: *op. cit.,* Section 5.
33. Edmister, W.C.: "Applied Hydrocarbon Thermodynamics," *Petroleum Refiner* 38:4, Apr. 1959, p. 195. (see Miller, R., Jr.: "Process Energy Systems," *Chemical Engineering,* May 20, 1968, p. 131)
34. *ASHRAE Guide and Data Book,* American Society of Heating, Refrigerating and Air Conditioning Engineers, New York, 1969, p. 61.
35. White, M.H.: "Surge Control For Centrifugal Compressors," *Chemical Engineering,* Dec. 25, 1972, p. 54.
36. Kraus, M.N.: "Pneumatic Conveyors," *Chemical Engineering,* Oct. 13, 1969, p. 59.
37. Munson, J.S.: "Dry Mechanical Collections," *Chemical Engineering,* Oct. 14, 1968, p. 147.

38. Aude, T.C., Cowper, N.T., Thompson, T.L., Wasp, E.L.: "Slurry Piping Systems," *Chemical Engineering,* June 28, 1971, p. 74; Feb. 7, 1972, p. 58.
39. Robertson, J.M.: "Design for Expansion," *Chemical Engineering,* May 6, 1968, p. 187.
40. Silverman, D.: "Lighting," *Chemical Engineering,* July 6, 1964, p. 121.
41. Kaufman, J.E., Christensen, J.F.: *IES Lighting Handbook,* Ed. 5, Illuminating Engineering Society, New York, 1972.
42. Mixon, G.M.: "Chemical Plant Lighting,'' *Chemical Engineering,* June 5, 1967, p. 113.
43. Constance, J.D.: "Estimating Exhaust—Air Requirements For Processes," *Chemical Engineering,* Aug. 10, 1970, p. 116.
44. Perry, J.H.: op. cit., Section 15, p. 25.
45. Staniar, H., "Plant Engineers' Handbook," 2nd Edition, McGraw-Hill, 1959, New York, N.Y.
46. Perry, R.H.: *Engineering Manual,* Ed. 2, McGraw-Hill, New York, 1967, pp. 7-68, 4-59, 4-61, 4-70, 5-27.
47. Jenckes, L.C.: "How to Estimate Operating Costs and Depreciation," *Chemical Engineering,* Dec. 14, 1970, p. 168.
48. "Contract Maintenance—How Far to Go?" *Chemical Engineering,* Apr. 3, 1961, p. 170.
49. Jordan, J.H.: "How to Evaluate the Advantages of Contract Maintenance," *Chemical Engineering,* Mar. 25, 1968, p. 124.
50. "It Costs More Than Ever to Run a Plant," *Chemical Week,* July 18, 1973, p. 33.
51. Axelrod, L., Daze, R.E., Wickham, H.P.: "Practical Aspects of the Large Plant Concept," paper presented at the 60th Annual Meeting of AIChE., New York, Nov. 28, 1967.
52. Frank, O., Prickett, R.D.: "Designing Thermosyphon Reboilers," *Chemical Engineering,* Sept. 3, 1973, p. 107.
53. Lord, R.C., Minton, P.E., Slusser, R.P.: "Design Parameters for Condensers and Reboilers," *Chemical Engineering,* Mar. 23, 1970, p. 127.
54. Coates, J., Pressburg, B.S.: "How Heat Transfer Occurs in Evaporators," *Chemical Engineering,* Feb. 22, 1960, p. 139.
55. Parker, N.: "Agitated Thin-Film Evaporators—Equipment and Economics," *Chemical Engineering,* Sept. 13, 1965, p. 179.
56. Lohrisch, F.N.: "What Are Optimum Conditions for Air-Cooled Heat Exchangers," *Chemical Engineering,* June 1966, p. 131.
57. Rubin, F.L.: "Design of Air-Cooled Heat Exchangers," *Chemical Engineering,* Oct. 31, 1960, p. 91.
58. Ellwood, P., Danatos, S.: "Process Furnaces," *Chemical Engineering,* Apr. 11, 1966, p. 151.
59. Tate, R.W.: "Sprays and Spraying for Process Use—Part II," *Chemical Engineering,* Aug. 2, 1965, p. 111.
60. Perry, J.H.:*op. cit.,* Section 11, p. 22.
61. *Kirk-Othmer Encyclopedia of Chemical Technology*, Ed. 2, Wiley, New York, vol. 19, p. 57.
62. Downing, D.G.: "Calculating Minimum Cost Ion-Exchange Units," *Chemical Engineering*, Dec. 6, 1965, p. 170.
63. "Sand Filter Saves Space," *Chemical Engineering*, Apr. 21, 1970, p. 112.
64. Perry, J.H.: *op. cit.,* Section 20, p. 95.
65. *Ibid.,* Section 20, p. 16.
66. Sargent, G.D.: "Dust Collection Equipment," *Chemical Engineering*, Jan. 27, 1969, p. 147.
67. Perry, J.H.: *op. cit.,* Section 5, p. 43.
68. Smith, W.M.: *Manufacture of Plastics*, vol. 1, Reinhold, New York, 1964, p. 45.
69. Constance, J.D.: "Calculating Pressure Drops in Pneumatic Conveying Lines," *Chemical Engineering,* Mar. 15, 1965, p. 200.
70. Mills, H.E.: "Costs of Process Equipment," *Chemical Engineering*, Mar. 16, 1964, p. 133.
71. Gluck, S.E.: "Design Tips for Pneumatic Conveying," *Hydrocarbon Processing*, Oct. 1968, p. 88.
72. Perry, J.H.: *op. cit.,* Section 7, p. 33.
73. Perry, J.H.: *op. cit.,* Section 7, p. 5.
74. "Fuel Costs for Steam Plants," *Chemical Engineering*, June 22, 1964, p. 164.

## Additional References

### General
Ludwig, E.E.: *Applied Process Design for Chemical and Petrochemical Plants*, vol. 1, 2, 3. Gulf, Houston, 1964, 1965.
Perry, J.H.: *Chemical Engineers' Handbook,* Ed. 5, McGraw-Hill, New York, 1974.
McCabe, W.L., Smith, J.C.: *Unit Operations of Chemical Engineering*, McGraw Hill, New York, 1967.
Kuong, J.F.: *Applied Nomography*, vol. 1, 2, 3, Gulf, Houston, 1965, 1968, 1969.
Bourton, K.: *Chemical and Process Engineering Unit Operations*, Plenum, New York, 1967. A bibliographic guide to reference media and specific unit operations.
Desk Book Issues of *Chemical Engineering*:
   "Liquids Handling" Apr. 14, 1969.
   "Solids Handling" Oct. 13, 1969.
   "Pump and Valve Selector" Oct. 11, 1971.
   "Plant and Maintenance Engineering" Feb. 26, 1973.
   (lists equipment manufacturers)
Gartmann, H. (ed.): *De Laval Engineering Handbook*, McGraw Hill, New York, 1971.

### Pumps, Fans, Blowers, and Compressors
Abraham, R.W.: "Reliability of Rotating Equipment," *Chemical Engineering*, Oct. 15, 1973, p. 96.
Neerken, R.F.: "Pump Selection for the Chemical Process Industries," *Chemical Engineering*, Feb. 18, 1974, p. 104.
Karassik, I.J.: "Pump Performance Characteristics," *Hydrocarbon Processing*, June 1972, p. 101.
Doyle, H.E.: "Highlights of API 610 Pump Standard," *Hydrocarbon Processing*, June 1972, p. 85.
Polcak, R.: "Selecting Fans and Blowers," *Chemical Engineering*, Jan. 22, 1973, p. 86.
"How to Specify Centrifugal Compressors," a series of articles in *Hydrocarbon Processing*, Oct. 1971, p. 64, onward.
Cavaliere, G.F., Gyepes, R.A.: "Evaluate Air Compressor Intercoolers," *Hydrocarbon Processing*, Oct. 1973, p. 107.

### Power and Power Recovery Equipment
Braun, S.S.: "Power Recovery Pays off at Shell Oil," *Oil and Gas Journal*, May 21, 1973, p. 129.
Abadie, V.H.: "Turboexpanders Recover Energy," *Hydrocarbon Processing*, July 1973, p. 93.
Casto, L.V.: "Practical Tips on Designing Turbine-Mixer Systems," *Chemical Engineering*, Jan. 10, 1972, p. 97.
Jewett, E.:"Hydraulic Power Recovery Systems," *Chemical Engineering*, Apr. 8, 1968, p. 159.
Farrow, J.F.: "User Guide to Steam Turbines," *Hydrocarbon Processing*, Mar. 1971, p. 71.

### Heat Transfer Equipment
Doyle, O.T., Benkly, G.J.: "Use Fanless Air Coolers," *Hydrocarbon Processing*, July 1973, p. 8.
Fair, J.R.: "Designing Direct Contact Cooler/Condensers," *Chemical Engineering*, June 12, 1972, p. 91.
Minton, P.E.: "Designing Spiral Tube Heat Exchangers," *Chemical Engineering*, May 4, 1970, p. 103; May 18, 1970, p. 145.
Thompson, J.W., "How Not to Buy Heat Exchangers," *Hydrocarbon Processing*, Dec. 1972, p. 83.
Chapman, F.S., Holland, E.A.: "Heat Transfer Correlations in Jacketed Vessels," *Chemical Engineering*, Feb. 15, 1965, p. 175.
Holt, Arthur D.: "Heating and Cooling of Solids," *Chemical Engineering*, Oct. 23, 1967, p. 145.
Brown, C.L., Kraus, M.: "Air Preheaters Cut Costs and Increase Efficiency," *Oil and Gas Journal*, Oct. 22, 1973, p. 77.
Standiford, F.C., Jr.: "Evaporation," *Chemical Engineering*, Dec. 9, 1963, p. 161.

Mutzenburg, A.B.: "Agitated Thin Film Evaporators—Thin Film Technology," *Chemical Engineering*, Sept. 13, 1965, p. 175.

Parker, N.: "Agitated Thin Film Evaporators—Equipment and Economics," *Chemical Engineering*, Sept. 13, 1965, p. 179.

Fischer, R.: "Agitated Thin Film Evaporators—Process Applications," *Chemical Engineering*, Sept. 13, 1965, p. 186.

McCarthy, A.J., Hopkins, M.E.: "Simplify Refrigeration Estimating," *Hydrocarbon Processing*, July 1971, p. 105.

Williams, V.C.: "Cryogenics," *Chemical Engineering*, Nov. 16, 1970, p. 72.


**Heat Loss Determinations**

Bisi, F., Menicatti, S.: "How to Calculate Tank Heat Losses," *Hydrocarbon Processing*, Feb. 1967, p. 145.

Gibbons, E.J.: "Chart Finds Heat Loss from Objects in Still Air," *Chemical Engineering*, Mar. 19, 1962, p. 192.

Thomas, B.L.: "For Heated Ponds and Thickeners—How to Calculate Heat and Water Losses," *Chemical Engineering*, Aug. 8, 1960, p. 129.


**Water Treatment**

Yost, W.H.: "Microbiological Control in Recirculating Water Systems Avoids Fouling," *Oil and Gas Journal*, Apr. 16, 1973, p. 107.

Walko, J.F.: "Controlling Biological Fouling in Cooling Systems— Part I," *Chemical Engineering*, Oct. 30, 1972, p. 128.

Curtis, S.D., Silverstein, R.M.: "Corrosion and Fouling Control of Cooling Waters," *Chemical Engineering Progress*, July 1971, p. 39.

Askew, T.: "Selecting Economic Boiler-Water Pretreatment Equipment," *Chemical Engineering*, Apr. 16, 1973, p. 114.

Ahlgren, R.M.: *Economic Production of Boiler Water through Membrane Methods*, pamphlet published by Aqua-Chem Inc., Waukesha, 1970.

Michalson, A.W: "Ion Exchange," *Chemical Engineering*, Mar. 18, 1963, p. 163-182.

Gilwood, M.E.: "Saving Capital and Chemicals with Countercurrent Ion Exchange," *Chemical Engineering*, Dec. 18, 1967, p. 83.

Seamster, A.H., Wheaton, R.M.: "Ion Exchange Becomes a Powerful Tool," *Chemical Engineering*, Aug. 22, 1960, p. 115.


**Miscellaneous**

Rowe, G.D.: "Essentials of Good Industrial Lighting, Part I," *Chemical Engineering*, Dec. 10, 1973, p. 113.

McAllister, D.G.: "Air," *Chemical Engineering,* Dec. 14, 1970, p. 138.

Price, H.A., McAllister, D.G.: "Inert Gas," *Chemical Engineering* , Dec. 14, 1970, p. 142.

Loeb, M.B.: "New Graphs for Solving Compressible Flow Problems," *Chemical Engineering*, May 19, 1969, p. 179.

Kleppe, C.A.: "Chart for Compression and Expansion Temperatures," *Chemical Engineering*, Sept. 19, 1960, p. 213.

Anderson, R.J., Russell, T.W.F.: "Designing for Two Phase Flow," *Chemical Engineering*, Dec. 6, 1965, p. 139; Dec. 20, 1965, p. 99.

Mottram, R.A.: "Mean Temperature Difference Found Quickly, Accurately," *Chemical Engineering*, June 16, 1969, p. 116.

Tate, R.W.: "Sprays and Spraying For Process Use—Part II," *Chemical Engineering*, Aug. 2, 1965, p. 111.

Hallas, R.S.: "Auxiliary Equipment," *Society of Plastics Engineers Journal*, Dec. 1972, p. 19.

# CHAPTER 9

# Cost Estimation

There are a number of different ways of estimating the cost of constructing a chemical plant. Some require very little information and some require a complete listing of every item, from pipe fittings to storage tanks and electrical sockets to generators. All assume a normal schedule and normal conditions. A normal schedule implies that the contractor and engineers will be allowed to operate in the most efficient way. Any attempt to complete the plant sooner will result in increased investment costs. This is discussed in Chapter 13.

Normal conditions mean that only minor amounts of overtime are involved, that an adequate number of competent tradesmen can be found, and that scheduled delivery times for equipment and supplies will be met. No provisions are made for work stoppages or slowdowns due to labor unrest.

## COST INDEXES

The effect of time on building costs is given by several indexes. The Chemical Engineering Plant Cost Index[1,2] (Table 9-1) and Nelson Refinery Construction Index[3] are the two most useful for estimating plant costs. The Engineering News Record Construction Cost Index[4] mainly measures the cost of civil structures such as dams, and buildings. The Marshall and Swift Equipment Cost Index[5] (Table 9-1) gives the index for the cost of equipment by industry and the cost of buildings by type and geographical region.

The indexes are used to determine the costs in year A if they are known in year B. This can be done by using the following formula.

$$\text{Cost in year A} = \frac{(\text{Index in year A})}{(\text{Index in year B})} (\text{cost in year B}).$$

### Example 9-1

The average Marshall and Swift index for 1959 was 234.5 and that for 1968 was 273.1. If a piece of equipment cost \$15,000 in 1959, what would be the expected price in 1968?

$$\text{Expected cost in 1968} = \$15,000 \times \frac{273.1}{234.5} = \$17,400.$$

This is a 16% increase in 9 years.

Table 9-1

The Annual Chemical Engineering Plant Cost Index
and
The Marshall and Swift Equipment Cost Index

| Year | | | | | 1953 | 1954 | 1955 | 1956 |
|---|---|---|---|---|---|---|---|---|
| Chemical Engineering Plant Cost Index | | | | | 84.7 | 86.1 | 88.3 | 93.9 |
| Marshall and Stevens Equipment Cost Index | | | | | 182.5 | 184.6 | 190.6 | 208.8 |
| 1957 | 1958 | 1959 | 1960 | 1961 | 1962 | 1963 | 1964 | |
| 98.5 | 99.7 | 101.8 | 102.0 | 101.5 | 102 | 102.4 | 103.3 | |
| 225.1 | 229.2 | 234.5 | 237.7 | 237.2 | 238.5 | 239.2 | 241.8 | |
| 1965 | 1966 | 1967 | 1968 | 1969 | 1970 | 1971 | 1972 | 1973 |
| 104.2 | 107.2 | 109.7 | 113.6 | 119.0 | 125.7 | 132.2 | 137.2 | 144.1 |
| 244.9 | 252.5 | 262.9 | 273.1 | 285.0 | 303.3 | 321.3 | 332.0 | 344.1 |

Source: Thorsen, D. R.: "The Seven-Year Surge in the CE Cost Indexes," *Chemical Engineering*, Nov. 13, 1972, p. 170 (and later issues).

Each index is based on a year or an average of several years for which the costs of the items on the list are given a value of 100. Then each succeeding year the prices are compared with what they were in the base year; the ratio times 100 is the index for that year.

The year chosen as a base is one that is close to normal. War years and periods of inflation or depression are avoided. The Marshall and Swift Index uses 1926 as a base. The Chemical Engineering Plant Cost Index uses an average of 1957-1959, and the Construction Cost Index uses 1913.

The Chemical Engineering Plant Cost Index is based on four major components, which are weighted as follows:

| | |
|---|---|
| Equipment, machinery, and supports | 61% |
| Erection and installation labor | 22% |
| Building materials and labor | 7% |
| Engineering and supervision manpower | 10% |
| | 100% |

Each of these items is further broken down into subcomponents, which are based mainly on figures supplied by the United States Bureau of Labor Statistics. Anyone who feels this index does not best represent his situation can make up his own index by weighing the components and/or subcomponents differently.

The following example shows that a different weighting of components gives a different index.

**Example 9-2**

In June, 1969, the indexes for the major components in the Chemical Engineering Plant Design Index were:

| | |
|---|---|
| Equipment, machinery, and supports | 115.9 |
| Construction labor | 127.0 |
| Buildings | 121.5 |
| Engineering and supervision | 109.8 |

Suppose the proposed plant costs consisted mainly of warehouses and construction labor. Assume the ratios are as follows:

| | |
|---|---|
| Equipment, machinery, and labor | 35% |
| Construction labor | 35% |
| Buildings | 20% |
| Engineering and supervision manpower | 10% |

The cost index would be
$$0.35(115.9) + 0.35(127.0) + 0.20(121.5) + 0.10(109.8) = 120.3$$

If the weightings used by the Chemical Engineering Plant Design Index are employed, the result would be 118.2. This is about a 2% difference.

Even if the indexes are correct, there is still a major problem. The plant being estimated will not be built for a number of years. What will be the index two or three years from now? In general, the rate of increase of the index is around 3% per year. However, there are wide variations from this. In 1954-1956 it increased at 6% per year while between 1959-1962 there was no noticeable increase. If no other information is available, a 3% increase per year should be used to estimate future values of the index.

A number of indexes that should be of interest to those in the chemical process industries are listed in Table 9-2 along with their sources.

## HOW CAPACITY AFFECTS COSTS

As a chemical plant increases in size, its cost also increases. However, there is not a linear relationship between capacity and cost. If the size doubles the cost will not increase two-fold. There are many reasons why this is so.

First, equipment costs do not increase linearly with size, since the amount of metal used is more closely related to the area than the volume of the vessel, and also the fabrication of a larger piece of equipment usually involves the same operations as a smaller piece, but each operation does not take twice as long. Second, actual construction costs are not twice as much. The piping of a 2 in. line does not take

Table 9-2

Indexes Useful to Chemical Engineers and Their Sources

| | Sources |
|---|---|
| Chemical Engineering Plant Cost | *Chemical Engineering* |
| Nelson Refinery Construction Cost | *Oil and Gas Journal* |
| Engineering and News Record Construction Cost | *Engineering News Record* |
| Marshall and Swift Equipment Cost (formerly Marshall and Stevens Index) | *Chemical Engineering* |
| Chemical Process Industries Output (production rate) | *Chemical Engineering* |
| Chemical Process Industries Output Value (sales) | *Chemical Engineering* |
| Chemical Process Industries Operating Rate | *Chemical Engineering* |
| Wholesale Prices of Industrial Chemicals | *Chemical Engineering* |
| Hourly Earnings for Employees in Chemical and Allied Products Industry | *Chemical Engineering* |
| Productivity of Chemical and Allied Products Industries | *Chemical Engineering* |
| Chemical Price | *Chemical Week* |
| Consumer Price | U.S. Bureau of Labor Statistics* |
| Wholesale Price (2,200 different items) | U.S. Bureau of Labor Statistics |
| Individual Equipment Costs | *Oil and Gas Journal* |
| Operating Costs for Specific Industries | *Oil and Gas Journal* |
| Engineering and News Record Building Cost | *Engineering News Record* |

* * The Bureau of Labor Statistics in the U.S. Department of Labor also publishes the average earnings of hourly and salaried employees for many categories.

twice as long as a 1 in. line. The wiring of a 10 hp pump does not take much longer than the wiring of a 5 hp pump. Third, the costs for engineering, drawing, ordering, and so on do not increase much with size. The cost of ordering 5 tons or 10 tons of steel is the same. The calculation of the stresses on a reactor takes the same amount of time regardless of the size.

Figure 9-1 (curve form=0.88) gives the cost of Ammonia Plants in 1960 versus capacity. This is a typical curve and it can be expressed in equation form as:

$$C_a = \left(\frac{A}{B}\right)^m C_b \qquad (1)$$

where $C_a$ = cost in dollars of a plant of capacity A
$C_b$ = cost in dollars of a plant of capacity B
$m$ = exponent
(The units of A and B must be the same)

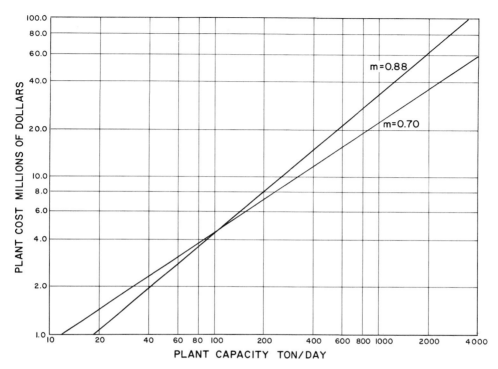

Figure 9-1    Plant cost as a function of capacity for two different cost exponents. The curve for $m = 0.88$ is the same as that given by Berk, J. M. and Haselbarth, J. E. for Ammonia Plants in "Cost Capacity Data IV" *Chemical Engineering,* Mar. 20, 1961, p. 186.

The values of m, $C_b$, and B have been obtained by John Haselbarth[6] for 60 different types of chemical plants and are given in Table 9-3.

The use of Equation 1 together with the engineering indexes is the easiest way of estimating plant costs.

### Example 9-3

Calculate the cost of building a 1,000 ton/day ammonia plant in 1969. From Table 9-3, which gives the values of $C_b$, m, and B for the year 1967, the following figures were obtained:

$$C_b = \$16,000,000$$
$$B = 500,000 \text{ tons/yr}$$
$$m = 0.70$$

Assuming the plant runs 98% of the time,

$$B_1 = 500,000/365(0.98) = 1,400 \text{ tons/day}.$$

The 1967 estimated plant investment is

$$\$16,000,000 \left(\frac{1,000}{1,400}\right)^{0.70} = \$12,630,000.$$

Table 9-3

1967 Capital-Cost Data for Processing Plants

| Compound | Source or Route | Typical Plant Size, Tons/Yr. | Investment Cost, $ | Investment, $ per Annual Ton | Size Factor m | Remarks |
|---|---|---|---|---|---|---|
| Acetaldehyde | Ethylene | 50,000 | 3,500,000 | 70 | 0.70 | Metallic catalyst required |
| Acetylene | Natural gas | 75,000 | 9,500,000 | 127 | 0.70 | High purity |
| Alumina | Bauxite | 100,000 | 9,000,000 | 90 | | |
| Aluminum sulfate | | 75,000 | 2,000,000 | 27 | | |
| Ammonia | | 500,000 | 16,000,000 | 32 | 0.70 | |
| Ammonium phosphate | | 250,000 | 2,500,000 | 10 | 0.68 | Fertilizer grade |
| Ammonium sulfate | | 140,000 | 1,200,000 | 9 | 0.68 | |
| Carbon black | | 30,000 | 3,000,000 | 100 | | |
| Carbon dioxide | | 200,000 | 2,400,000 | 12 | | |
| Carbon tetrachloride | | 30,000 | 2,500,000 | 85 | | |
| Butadiene | Butane | 100,000 | 50,000,000 | 500 | 0.70 | |
| Butadiene | Butylenes | 200,000 | 70,000,000 | 350 | 0.70 | |
| Chlorine/caustic | Cl₂: NaOH: | 70,000 | 13,000,000 | | | |
| Cyclohexane | | 78,000 | 750,000 | 8 | 0.69 | Does not include hydrogen plant |
| | | 100,000 | | | 0.70 | |
| Diphenylamine | | 10,000 | 2,400,000 | 240 | | |
| Ethanolamine | | 25,000 | 1,750,000 | 70 | | |
| Ethyl alcohol | From ethylene by direct hydration of via ethyl sulfuric acid | 75,000 | 3,750,000 | 50 | 0.72 | Manufacturing costs are lower in the direct hydration process |
| Ethylbenzene } Paraxylene | | 20,000 | 1,800,000 | | 0.71 | These chemicals are produced simultaneously |
| | | 8,500 | 1,100,000 | | | |
| Ethyl chloride | | 15,000 | 3,000,000 | 200 | | |
| Ethyl ether | | 35,000 | 1,200,000 | 35 | | |
| Ethylene | Refinery gases or hydrocarbons | 300,000 | 15,000,000 | 50 | 0.71 | |
| Ethylene dichloride | | 25,000 | 3,200,000 | 127 | 0.71 | |
| Ethylene oxide | Direct oxidation of ethylene | 100,000 | 9,000,000 | 90 | 0.67 | Cost also includes conversion to ethylene glycol as needed |
| 37% Formaldehyde | Hydrocarbons | 100,000 | 13,000,000 | 130 | | |
| Glycerin (synthetic) | | 35,000 | 5,500,000 | 157 | 0.67 | |
| Hydrofluoric acid | | 15,000 | 2,600,000 | 175 | | |
| Hydrogen | | 60,000 | 6,500,000 | 108 | | |
| Isopropyl alcohol | | 150,000 | 7,500,000 | 50 | 0.80 | |
| Maleic anhydride | | 50,000 | 18,000,000 | 360 | | |
| Melamine | | 70,000 | 11,500,000 | 164 | | |

| | | | | | | |
|---|---|---|---|---|---|---|
| Methanol | Natural gases | 210,000 | 9,000,000 | 43 | 0.71 | |
| Methyl chloride | Methanol | 10,000 | 500,000 | 50 | 0.72 | |
| Methyl ethyl ketone | | 35,000 | 3,750,000 | 107 | | |
| Methyl isobutyl ketone | | 25,000 | 1,250,000 | 50 | | |
| Methyl isobutyl carbonal | | 10,000 | 750,000 | 75 | | |
| Nitric acid | | 50,000 | 5,000,000 | 100 | | |
| Oxygen plants | | 150,000 | 2,250,000 | 15 | 0.71 | |
| Phenol | | 45,000 | 9,000,000 | 200 | | |
| Phosphoric acid (as $P_2O_5$) | | 100,000 | 2,400,000 | 24 | 0.66 | Wet process—contains 30% $P_2O_5$ |
| Cis-polybutadiene | | 50,000 | 12,000,000 | 240 | 0.67 | |
| Polyethylene (high-pressure) | | 200,000 | 14,000,000 | 70 | 0.70 | |
| Polyethylene (low-pressure) | | 50,000 | 22,000,000 | 440 | 0.70 | High-purity ethylene required |
| Polyisoprene (includes manufacture of the monomer) | | 30,000 | 5,000,000 | 320 | 0.74 | |
| Soda ash | Natural brine | 400,000 | 34,000,000 | 85 | | No synthetics plants built since 1934 |
| Sodium metal | | 20,000 | 7,000,000 | 350 | | |
| Styrene | | 20,000 | 8,500,000 | 425 | 0.67 | |
| Sulfuric acid | Contact process | 280,000 | 2,100,000 | 8 | | |
| Sulfur recovery | Refinery gases | 15,000 | 1,500,000 | 100 | | |
| Toluene diisocyanate | | 12,500 | 7,500,000 | 600 | | |
| Urea | | 140,000 | 4,300,000 | 31 | | |
| Vinyl acetate | | 40,000 | 7,000,000 | 175 | | |
| Vinyl chloride monomer | | 100,000 | 2,000,000 | 20 | | |
| Refinery Products | | (Bbl./Day) | | | | |
| Alkylation units ($H_2SO_4$ or HF) | From reformer streams; e.g. Udex | 10,000 | 7,750,000 | 775 | | |
| BTX extraction | | 10,000 | 3,400,000 | 340 | 0.70 | |
| Cat. cracker (fluid) | Cost based on fresh feed | 35,000 | 14,000,000 | 400 | | Includes vapor recovery and CO boiler |
| Cat. reformer | | 23,000 | 7,500,000 | 375 | | |
| Crude distillation units | | 100,000 | 4,700,000 | 47 | | |
| Delayed coker | | 14,000 | 5,000,000 | 357 | | |
| Hydrocracker | | 28,000 | 21,000,000 | 750 | | |
| Wax plants | | 7,500 | 900,000 | 120 | | |
| Gas absorption and dehydration plants | | 50 MM cfd. | 2,000,000 | | | |

Source: Haselbarth, J.E.: "Updated Investment Costs for 60 Types of Chemical Plants," *Chemical Engineering*, Dec. 4, 1967, p. 214.

243

The Chemical Engineering Plant Cost Index (CEPI or C.E. Index) in June, 1969, was 118.2; in 1967 it was 109.7.
The estimated investment cost for 1969 is

$$\$12,630,000 \times \frac{118.2}{109.7} = \$13,600,000.$$

The use of this method implies that all plants producing a given chemical are basically the same. In simpler language, this means the plant scopes must be similar in all aspects except for the time of construction and capacity. Since this is frequently not true, this estimate cannot be very accurate.

The possible errors that can result when this method is used blindly can be illustrated by considering the case of ammonia plants. Suppose we recalculate the cost of the plant given in Example 9-3 by using Figure 9-1 (m=0.88). An investment cost of $35,000,000 for 1960 can be obtained from extrapolating that curve. The CEPI in 1960 was 102, so the estimated cost for 1969 is

$$\$35,000,000 \times \frac{118.2}{102} = \$40,500,000.$$

Note that the estimates differ by a factor of three. What is the true cost of building such a plant, and why do these estimates differ by $26,900,000? To answer these questions the actual cost of building ammonia plants must be determined.

Table 9-4 gives the capital costs for six ammonia plants that were built between 1959 and 1969. When plant no. 5 is compared with the three other plants that have a capacity of 1,000 tons/day, it appears that its reported cost is in error. This could be a misprint, or the plant might be producing urea, nitric acid, and/or ammonium nitrate as well as ammonia. The reader must always be careful, since errors occur frequently in printed material. This is why care should be used when the cost of a plant is estimated from only one piece of information.

It should also be noted that plants 1 and 2 fall very near the values predicted by Figure 9-1 (m = 0.88). This would indicate that the reported costs are reliable.

The plant costs given in Table 9-4 must now be put on a common basis in order to compare them with our estimates. Haselbarth, in 1967, gave the exponent, m, in equation 9-1 as 0.70. Gallagher,[7] in the same year, stated that m = 0.74. From Figure 9-1 a value of m = 0.88 can be calculated.

The project cost of a 1,000 ton/day ammonia plant that was built in 1969 has been obtained, using Equation 9-1 with an m of 0.70 and 0.88 and the CEPI. The results appear in Table 9-5. The effect of the exponential factor is very evident for plants 1 and 2. This effect does not occur for the other plants because their rated capacity was the desired 1,000 tons/day. Exponential factors are only used when capacity extrapolations must be made. This illustrates how a difference of 0.18 in the exponential factor (m) can have a profound effect on the projected cost if the scale-up factor is large. This can be further demonstrated by drawing lines of these two slopes on log-log paper (Fig. 9-1). As the lines get farther away from the base

Table 9-4

Reported Price for Construction of Ammonia Plants

| Plant Number | Year Completed | Capacity | Cost | Reference |
|---|---|---|---|---|
| 1 | 1960 | 80 ton/day | $ 4,000,000 | 1 |
| 2 | 1962 | 100,000 ton/yr | $11,000,000 | 2 |
| 3 | 1967 | 1,000 ton/day | $14,000,000 | 3 |
| 4 | 1968 | 1,000 ton/day | $18,000,000 | 4 |
| 5 | 1969 | 1,000 ton/day | $38,000,000 | 5 |
| 6 | 1969 | 1,000 ton/day | $18,000,000 | 6 |

Sources:
1. "1960 Inventory of New Plants & Facilities," *Chemical Engineering*, Apr. 18, 1960, p. 176.
2. "Semi-Annual Inventory of New Plants and Facilities," *Chemical Engineering*, Oct. 16, 1961, p. 191.
3. Axelrod, L., Daze, R.E., Wickham, H.P.: "Technology—The Large Plant Concept," *Chemical Engineering Progress*, July 1968, p. 17.
4. "New Plants & Facilities," *Chemical Engineering*, Apr. 6, 1968, p. 148.
5. "New Plants & Facilities," *Chemical Engineering*, Oct. 7, 1968, p. 167.
6. "New Plants & Facilities," *Chemical Engineering*, Apr. 7, 1969, p. 141.

point, where the two lines cross, the percentage difference of their ordinate values at a given abscissa value increases rapidly. This should also be kept in mind when scaling up equipment costs.

The disparity between plant costs obtained before 1963 (Runs 1 and 2 in Table 9-5 and Fig. 9-1, with m = 0.88) and those obtained after 1966 (other runs and Fig. 9-1 with m = 0.70) can be explained by advances in technology. This was discussed in Chapter 3. The use of an exponential factor to scale up size assumes that a similar plant will be built. This was not true for ammonia plants. In fact, if a company is doing developmental research it should never be true. Each plant should be better than the last.

Table 9-5

The Scaled-up Cost of a 1,000 Ton per Day Ammonia Plant in 1969

| Basis* | m = 0.70 | m = 0.88 |
|---|---|---|
| 1 | $27,000,000 | $43,000,000 |
| 2 | 31,000,000 | 38,000,000 |
| 3 | 15,100,000 | 15,100,000 |
| 4 | 18,750,000 | 18,750,000 |
| 5 | 38,000,000 | 38,000,000 |
| 6 | 18,000,000 | 18,000,000 |

* The numbers refer to the plants listed in Table 9-4.

If Plant 5 in Table 9-5 is ignored, the cost estimates of Example 9-3 and those for the three other plants of Table 9-5 built in the late 1960s are within 17% of the average $16,400,000. This is very good for this type of estimate, since this method does not take into account any differences in the scope. It assumes everything is similar, which is obviously false. This is why this method cannot be expected to provide anything better than a ballpark estimate.

The exponential factors usually given for total plants should be used for grass-root plants, and not for new units constructed at a developed site or expansions to existing plants. Grass-roots plants are those built in a location that has not been previously developed. They cost more than plants built in an area where the company has other plants. When other plants are nearby, access roads, railroad sidings, sewers, and water lines may only need to be extended a short distance. Dock and steam generation facilities may be available, and office, lunchroom, medical, and change-room space may already be adequate.

The cheapest plant is usually the enlargement of an existing unit. This is especially true if a provision was made for expansion when the original plant was designed.

## FACTORED COST ESTIMATE

For the factored estimate a list is made of all pieces of equipment, and the delivered cost of each item is obtained. This could be determined by inquiring from manufacturers, from past records, or from published data. The delivered cost of all the equipment is summed and multiplied by an appropriate factor. According to Lang[8] this factor would be 3.10 for a solids process plant, 3.63 for a solid-fluid plant, and 4.74 for a fluid plant. These factors are referred to as Lang factors. This estimate is often used in the preliminary stages of engineering, but is not extremely accurate.

The major advantage of this method is that the cost of equipment is readily available. In the 1960s *Chemical Engineering* magazine alone published three extensive lists[9,10,11] for predicting process equipment costs. A list of other sources is given at the end of this chapter (see also Appendix B). These costs can be updated by using the Marshall and Swift Equipment Index.

Nearly all published costs are Free on Board (F.O.B.) prices at the manufacturer's plant. This means the manufacturer pays loading charges but not freight or unloading charges. These are assumed by the buyer. The Lang factor is based on delivered equipment charges. To obtain these costs from published information, the freight charges must be estimated. These are roughly 5% of the published costs for eastern, midwestern, and southern sites and 7% for western locations. A sales tax should also be added in those states for which it is applicable. When items are purchased for direct shipment to another state, only the sales taxes in the state to which the equipment is delivered need to be paid. Imported items often carry import duties of 10-15%, and the freight charges may be 10-12% of the purchase price.

The factored estimate has one obvious drawback. It is very easy in the early phases of a process engineering study to forget some items. Since this will always result in a low estimate, a contingency allowance is usually added. This contingency allowance will vary between 10 and 50% of the factored estimate. It is based on the estimator's feeling about the completeness of the information. If the piping and instrument diagrams (see Chapter 12) are complete, he may use a 15% contingency allowance. If there is merely a rough flow sheet, it will be nearer 50%.

To use the Lang factors the engineer must define what type of plant is being built. This is important, since the largest factor is 50% greater than the smallest. It is sometimes difficult, however, because there is a continuum of chemical plants between the two extremes. A coal-briquetting plant is obviously a solids processing plant. Methanol and ammonia plants are fluids plants. Plants that extract chemicals from solids fall between.

The Lang factor includes everything that is involved in the design and construction of the plant, from the engineering costs and contractor's profits to the erection, piping, insulating, and wiring of the equipment. It includes the cost of buildings, lunchrooms, roads, landscaping, and site preparation. A list of these items is given in Table 9-6.

### Example 9-4

Determine the cost of a fluid plant whose equipment costs are given in Table 9-7 for the year 1975. The Lang factors are to be used. From Table 9-7, the total F.O.B. equipment cost is $1,310,000. Let us assume a 4% sales tax and an average freight charge of 5%. The total cost of the delivered equipment is

$$\$1,310,000 \ (1.04) \ (1.05) = \$1,430,000.$$

The total estimated cost of the plant in 1972 is
$$\$1,430,000 \ (4.74) = \$6,800,000.$$

If a 3% increase in prices per year is expected, the cost of the plant in 1975 is
$$\$6,800,000 \ (1.03)^3 = \$7,440,000.$$

The premise upon which the Lang factors are based is that the equipment costs are a certain fraction of the total cost of the plant. Or conversely, it can be said that the costs of piping, insulation, wiring, site preparation, and so on are a function of the cost of the equipment. However, the cost of equipment is very dependent on the materials used to make it, while most of the other items listed in Table 9-6 are not dependent on materials.

For instance, equipment made of monel generally costs 6.5 times as much as the same item constructed of carbon steel. If 25% of the equipment purchased for a plant were made of monel, this would increase the equipment costs by 237%, and the factor cost estimate would be 2.37 times that for a plant constructed of carbon steel. This is unreasonable, since the cost of buildings, roads, wiring, piping utilities, insulating, and instrumentation are independent of the materials of construction. In fact the only major changes would be in the process piping, which,

Table 9-6

Items Included in the Capital Estimate
(Costs of Necessary Labor, Materials, and Equipment)

Site Preparation

Surveying
Clearing and grading
Soil testing
Landscaping
Pilings
Sewers
Roads and walks
Fencing
Drainage
Excavation

Processing Areas

Process equipment
Foundations
Supports
Structures and/or building
Erection
Piping
Electrical
Instruments and installations
Insulating
Steam tracing
Painting
Sprinkler system
Ventilating systems
Construction equipment

Storage Areas

Tanks
Warehouse
Piping to
Steam Tracing
Painting
Lift trucks
Foundations
Erection
Insulating
Docks
Construction equipment

Miscellaneous

Flares
Blowdown tanks
Piping to those areas

Utilities

Electrical substations
Water distribution systems
Electrical distribution systems
Steam plants and distribution systems
Air compressors
Gas distribution systems
Refrigeration plants and distribution systems
Light and heating systems

Services

Lunchrooms
Offices
Laboratories
Gatehouse
Change rooms
Toilets
Medical facilities
Firefighting equipment
Recreational facilities
Parking lots

Indirect Costs

Drafting
Engineering costs
Small tools used by craftsmen
Taxes
Contractor's overhead
Inspectors salaries
Company overhead
Contractor's profit
Equipment procurement costs
Insurance
A contingency to account
for inadequate scopes
and forgotten items

Table 9-7

Equipment Costs (1972) Used in Examples 9-4 and 9-5

| Equipment | Number of Pieces | Total Cost (F.O.B.) |
|---|---|---|
| Process pumps | 30 | $   15,000 |
| Compressors | 2 | 150,000 |
| Towers | 4 | 445,000 |
| Exchangers | 10 | 50,000 |
| Filters | 2 | 25,000 |
| Storage tanks | 10 | 200,000 |
| Process tanks | 8 | 175,000 |
| Reactors | 6 | 250,000 |
| | 72 | $1,310,000 |

because it probably would be monel- or glass-lined, would cost more and take more time to install. One way to avoid this problem is to base all costs other than equipment on the equivalent piece of carbon-steel equipment.

Similar arguments can be made for high-pressure and low-temperature apparatus. It can also be shown the Lang factor is dependent on the size of the equipment. As the size increases the proportionate cost of all other items decreases, and therefore the Lang factor should be smaller. The net result of the above inaccuracies is that the factored estimate as previously discussed cannot be very accurate. In 1951 W. T. Nichols[12] estimated that the error could easily be as great as 60% of the cost of the plant.

## IMPROVEMENTS ON THE FACTORED ESTIMATE

To make the factored estimate more accurate, the Lang factor has been broken into parts.[13] This allows the engineer to construct his own Lang factor for each project. Miller[14] expands this idea and suggests that picking a single value for each subfactor is difficult, so the engineer should determine a low, probable, and high value for each part. From this he can develop a minimum, most likely, and maximum plant investment cost. He also noted that large equipment, high-pressure equipment, and equipment made of expensive alloys are all very expensive per item of equipment and their over-all Lang factors are low. He therefore developed his factors as a function of the average cost of a *main plant item* (MPI). This is obtained by adding the cost of all such items on the equipment list and dividing by the number of items. The equipment costs are based on delivered prices, excluding sales taxes, in the year 1958.

Miller divides the plant into four areas: Battery Limits, Storage and Handling, Utilities, and Services. The last three can often be estimated separately. They are a function of size of the facility and do not depend on the type of product being made. For instance, a steam plant, a warehouse, or an office building are the same

Table 9-8

BATTERY LIMIT COSTS
RANGE OF FACTORS IN PERCENT OF BASIC EQUIPMENT

| | | AVERAGE UNIT COST OF M.P.I. IN 1968 IN DOLLARS | | | | | | |
|---|---|---|---|---|---|---|---|---|
| | | UNDER $3,000 | 3,000 to 5,000 | 5,000 to 7,000 | 7,000 to 10,000 | 10,000 to 13,000 | 13,000 to 17,000 | OVER 17,000 |
| **BASIC EQUIPMENT** Delivered to site, excluding sales taxes & catalyst | M.P.I. (Main plant Items) | X | X | X | X | X | X | X |
| | M.U.E. (Miscellaneous unlisted items) Early flowsheet stage Scope of work well defined | 20 to 10% of M.P.I.'s in all categories 10 to 1% of M.P.I.'s in all categories | | | | | | |
| | NOTE: Top of ranges: Complicated processes Many process steps Bottom of ranges: Simple processes Few process steps | | | | | | | |
| | BASIC EQUIPMENT = M.P.I. + M.U.E. | 100 | 100 | 100 | 100 | 100 | 100 | 100 |
| **FIELD ERECTION OF BASIC EQUIPMENT** | High percentage of equipment involving high field labour | 23/18 18/12.5 | 21/7 17/11.5 | 19.5/16 16/10.8 | 18.5/15 15/10 | 17.5/14.2 14.2/8.5 | 16.5/13.5 13.5/9.5 | 15.5/13 13/8 |
| | AVERAGE (Mild steel equipment) High percentage of corrosion materials and other high unit cost equipment involving little field erection | 12.5/7.5 | 11.5/6.7 | 10.8/6 | 10/5.5 | 9.2/5.2 | 8.5/5 | 8/4.8 |
| **EQUIPMENT FOUNDATIONS AND STRUCTURAL SUPPORTS** | HIGH-Predominance of compressors or mild steel equipment re- quiring heavy fdns. | | | 17/12 | 15/10 | 14/9 | 12/8 | 10.5/6 |
| | AVERAGE-For mild steel fabri- cated equipment solids | | | 12.5/7 | 11/6 | 9.5/5 | 8/4 | 7/3 |
| | AVERAGE-For predominance of alloy and other high unit price fabricated equipment | 7/3 | 8/3 | 8.5/3 | 7.5/3 | 6.5/2.5 | 5.5/2 | 4.5/1.5 |
| | LOW-Equipment more or less sitting on floor | 5/0 | 4/0 | 3/0 | 2.5/0 | 2/0 | 1.5/0 | 1/0 |
| | PILING OR ROCK EXCAVATION | Increase above values by 25 to 100% | | | | | | |
| **PIPING** includes ductwork excludes insulation | HIGH-Gases and liquids, petrochem- icals, plants with substantial ductwork | 105/65 | 90/58 | 80/48 | 70/40 | 58/34 | 50/30 | 42/25 |
| | AVERAGE FOR CHEMICAL PLANTS Liquids, electrolytic plants | 65/33 | 58/27 | 48/22 | 40/16 | 34/12 | 30/10 | 25/9 |
| | LIQUIDS AND SOLIDS | 33/13 | 27/10 | 22/8 | 16/6 | 12/5 | 10/4 | 9/3 |
| | LOW-Solids | 13/5 | 10/4 | 8/3 | 6/2 | 5/1 | 4/0 | 3/0 |
| **INSULATION OF EQUIPMENT ONLY** | VERY HIGH-Substantial mild steel equipment requiring lagging and very low temperatures | 13/10 | 11.5/8.5 | 10/7.4 | 9/6.2 | 7.8/5.3 | 6.8/4.5 | 5.8/3.5 |
| | HIGH-Substantial equipment re- quiring lagging and high temperatures (Petrochemicals) | 10.3/7.5 | 9/6.3 | 7.8/5.2 | 6.7/4.2 | 5.7/3.4 | 4.7/2.8 | 4.8/2.5 |
| | AVERAGE FOR CHEMICAL PLANTS | 7.8/3.4 | 6.5/2.6 | 5.5/2.1 | 4.5/1.7 | 3.6/1.4 | 2.9/1.1 | 2.2/8 |
| | LOW | 3.5/0 | 2.7/0 | 2.2/0 | 1.8/0 | 1.5/0 | 1.2/0 | 1/0 |
| **INSULATION OF PIPING ONLY** | VERY HIGH-Substantial mild steel piping requiring lagging and very low tempera- tures | 22/16 | 19/13 | 16/11 | 14/9 | 12/7 | 9/5 | 6/3.5 |
| | HIGH-Substantial piping requiring lagging and high temperatures (Petrochemicals) | 18/14 | 15/12 | 13/10 | 11/8 | 9/6 | 7/4 | 4.5/2.5 |
| | AVERAGE FOR CHEMICAL PLANTS | 16/12 | 14/10 | 12/8 | 10/6 | 8/4 | 6/2 | 4/2 |
| | LOW | 14/8 | 12/6 | 10/5 | 8/4 | 6/3 | 4/2 | 2/1 |
| **ALL ELECTRICAL** except building lighting and instrumentation | ELECTROLYTIC PLANTS (includes rectification equipment | | 55/42 | 50/38 | 45/33 | 40/30 | 35/26 | |
| | Plants with mild steel equipment, heavy drives; solids | 26/17 | 22.5/15 | 19.5/12.5 | 17/10 | 14/8.5 | 12/7 | 10/6 |
| | Plants with alloy or high unit cost equipment, chemical and petrochemical plants | 18/9.5 | 15.5/8.5 | 13/6.5 | 11/5.5 | 9/4.5 | 7.3/3.5 | 6/2.5 |
| | NOTE: Above figures include 1 to 3% for B/L outside lighting which is not covered in Building Services | | | | | | | |

| | | AVERAGE UNIT COST OF M.P.I. IN 1958 DOLLARS | | | | | | |
|---|---|---|---|---|---|---|---|---|
| | | UNDER $3,000 | 3,000 to 5,000 | 5,000 to 7,000 | 7,000 to 10,000 | 10,000 to 13,000 | 13,000 to 17,000 | OVER 17,000 |
| INSTRUMENT-ATION | Substential instrumentation, central control panels, petrochemicals<br>MISCELLANEOUS CHEMICAL PLANTS<br>Little instrumentation, solids<br><br>NOTE: Total instrumentation costs does not vary a great deal with size and hence is not readily calculated as a per cent of Basic Equipment. This is particularly true for distillation systems. If in doubt, detailed estimates should be made | | 58/31<br>32/13<br>21/9 | 46/24<br>26/10<br>17/7 | 37/18<br>20/7<br>13/4 | 29/13<br>15/5<br>10/3 | 23/10<br>11/3<br>7/2 | 18/7<br>8/2<br>5/1 |

| MISCELLANEOUS<br>Includes site preparation, painting & other items not accounted for above | Top of range - large complicated processes<br><br>Bottom of range - Smaller, simple processes | RANGE FOR ALL VALUES OF BASIC EQUIPMENT<br>6 to 1% |
|---|---|---|

| BUILDINGS -<br>ARCHITECTURAL<br>&<br>STRUCTURAL<br>(excludes bldg. services) | NOTE: When building specifications and dimensions are known a high speed building cost estimate is recommended especially if buildings are a significant item of cost. If a separate estimate is not possible, evaluate the buildings as follows before selecting the factors. |
|---|---|

BUILDING EVALUATION
when most of process units are located inside buildings

| | HIGH<br>Brick and Steel | MEDIUM | LOW<br>Economical | EVALUATION |
|---|---|---|---|---|
| QUALITY OF CONST. | +4 | +2 | 0 | |
| | VERY HIGH UNIT COST EQUIPMENT | MOSTLY ALLOY STEEL | MIXED MATERIALS | MOSTLY CARB. STEEL | |
| TYPE OF EQUIPMENT | -3 | -2 | -1 | 0 | |
| | VERY HIGH | INTERMED. | ATMOS. | |
| OPERATING PRESSURES | -2 | -1 | 0 | |

BUILDING CLASS = ALGEBRAIC SUM =

| | | AVERAGE UNIT COST OF M.P.I. IN 1958 DOLLARS | | | | | | |
|---|---|---|---|---|---|---|---|---|
| | BLDG. CLASS | UNDER $3,000 | 3,000 to 5,000 | 5,000 to 7,000 | 7,000 to 10,000 | 10,000 to 13,000 | 13,000 17,000 | OVER 17,000 |
| Most of Process UNITS INSIDE BUILDINGS | +2<br>+1 to -1<br>-2 | 92/68<br>72/49<br>50/37 | 82/61<br>62/43<br>44/33 | 74/56<br>56/38<br>40/29 | 67/49<br>51/33<br>35/25 | 59/44<br>45/29<br>30/21 | 52/39<br>41/26<br>27/18 | 46/33<br>36/21<br>23/15 |
| OPEN AIR PLANTS WITH MINOR BUILDINGS | | 37/16 | 32/13 | 28/11 | 24/8 | 20/6 | 17/4 | 14/2 |

| BUILDING SERVICES | NOTE: The following factors are Battery Limit (process) buildings only and are expressed in per cent of the Building-Architectural & Structural cost. They are not related to the Basic Equipment cost. |
|---|---|

| | HIGH | NORMAL | LOW |
|---|---|---|---|
| Compressed air for general service only | 4 | 1½ | 5 |
| Electric lighting | 18 | 9 | 5 |
| Sprinklers | 10 | 6 | 3 |
| Plumbing | 20 | 12 | 3 |
| Heating | 25 | 16 | 8 |
| Ventilation: without air conditioning | 18 | 8 | 0 |
| with air conditioning | 45 | 35 | 25 |
| TOTAL OVERALL AVERAGE* | 85 | 55 | 20 |

The above factors apply to those items normally classified as building services. They do not include:
1. Services located outside the building such as sub-stations, outside sewers, outside water lines, etc., all of which are considered to be outside the Battery Limit, as well as outside the building.

2. Process services
*The totals provide the ranges for the type of building involved and are useful when the individual service requirements are not known. Note that the overall averages are not the sum of the individual colums.

Source: Miller, C.A.; "Factor Estimating Refined for Appropriation of Funds," *American Association of Cost Engineers Bulletin,* Sept. 1965, p. 92.

regardless of whether ammonia, soda ash, polystyrene, or gasoline is being produced. The cost of a steam plant depends only on the amount and quality of the
steam required; the size of a warehouse depends solely on how much material is to
be stored in it; the cost of an office building is dependent mainly on the size of the
office staff.

The Battery Limits (BL) is the area in which the raw materials are converted into
products. Anything in that area is included in the BL costs. The Storage and
Handling (S&H) area consists of all product and raw material storage as well as
loading and unloading facilities. It includes all pipelines to and from the BL area. It
does not include any intermediate storage. The Utilities area consists of the production, storage, and transfer of all energy outside of the BL area. It would include
drains, sewers, waste treatment plants, steam generators, compressed air plants,
electric substations, cooling towers, and the storage facilities for coal, oil, and other
raw materials used by the various utilities. Sometimes small refrigeration units or
air compressors may be considered part of the BL. The Service area consists of
offices, laboratories, shops and lunchrooms, change houses, gatehouses, roads,
ditches, railways, fences, communication systems, service equipment, truck
scales, and the like.

The cost of the BL plants can be obtained by using Table 9-8 (see Example 9-5). If
the S&H or utilities costs cannot be determined separately, they can be obtained
from Table 9-9 as a percentage of the BL costs. More details can be found in
reference 14. Table 9-9 also gives the services as a percentage of the sum of the costs
for the battery limits, storage and handling, and utilities. Since it is highly unlikely
that all the items will be at their low or high estimates, Miller suggests that after
calculating the battery limits costs the low estimate be increased 10% and the high
estimate be reduced 10%. He also notes that usually all capital cost estimates
contain some contingency allowance. Since the high estimate should take care of
these problems, it is suggested that the low estimate only, be increased by another
10%.

### Example 9-5

Solve the problem given in Example 9-4 using the Miller method. Let us assume
that 10 of the pumps and all the storage tanks are not within the battery limits. The
total estimated F.O.B. price of the equipment in 1972 is then $1,105,000. The cost of
it delivered, assuming a 5% freight charge (sales tax is not included), is

$$\$1,105,000 \ (1.05) = \$1,160,000.$$

The CEPI in 1972 was 137.2. In 1958 it was 99.7. The estimated equipment cost in
1958 is

$$\$1,160,000 \ \frac{99.7}{137.2} = \$844,000.$$

Table 9-9

Auxiliary Costs

| Item | Grass Roots Plant | Battery Limit Addition on Existing Site |
|---|---|---|
| Storage and Handling (S&H) (% of BL cost) | | |
| Low | 2 | 0 |
| Raw material by pipeline; Little warehouses space | | |
| Average | 15-20 | 2-6 |
| Average raw material storage and finished product ware-housing | | |
| High | 70 | 20 |
| Tank farm for raw materials; Substantial warehousing for finished product | | |
| Utilities (U) (% of BL Cost) | | |
| Low | 10 | 3 |
| Average | 20-30 | 6-14 |
| High | 50 | 30 |
| Services (S) (% of [BL + S&H + U]) | | |
| Low | 5 | 0 |
| Average | 10-16 | 2-6 |
| High | 20 | 15 |

Source: Miller, C.A.: "Factor Estimating Refined for Appropriation of Funds," *American Association of Cost Engineers Bulletin,* Sept. 1965, p. 92.

The price per main plant item is then

$$\frac{\$844,000}{(72-20)} = \$16,200.$$

Since we do not know any details about the plant, let us assume that the average values given in the various categories by Miller in Tables 9-8 and 9-9 are the most reasonable ones. The column to be used is the one for average costs between $13,000 and $17,000. Assume the scope is poorly defined and 15% of the items may be unlisted. The calculations are given in Table 9-10. Note that in calculating the storage and handling, utilities, and services costs only, the factor is multiplied by the most probable cost. Otherwise the low factor is multiplied by the low estimate, and so on.

NOTE: The Lang estimate obtained in Example 9-4 falls between the high and low estimates.

One advantage of the three estimates is that it allows an economic evaluation to be made for each. This will then show the effect the capital investment has on the project.

Miller claims that if the basic equipment estimate has an accuracy of $\pm$ 10%, the "most likely" plant estimate should have an accuracy of $\pm$ 14%. This is much better than the ratio or Lang estimates, and considerably more accurate than Nichols said was possible with this type of data.[12] He claimed that there is a direct correlation between the cost of an estimate and its probable accuracy. Ever since he stated this in 1951, cost engineers have been trying to prove him wrong.

## MODULE COST ESTIMATION

Instead of summing all equipment costs and then multiplying by a factor to obtain the preliminary cost estimate for a plant, Guthrie[15] divides the plant into modules. A module is a grouping of similar items such as heat exchangers, horizontal pressure vessels, steam generators, mechanical refrigerator units, docks, cafeterias, and/or laboratories. The 1968 price of each item in the module is then obtained. In this case it is the F.O.B. price at the suppliers' plants. A modified Lang factor is then calculated for each module. This module factor includes the freight costs and sales taxes. It further differs from the Lang factor in that it does not include site development or any structures. These are figured as separate modules. By summing the costs for each module, and adding a contingency factor (13-25%) to correct for any omissions, inadequate scope definitions, and contractors' fees, a capital cost estimate was obtained by Guthrie for a plant built in mid-1968 at a Gulf Coast site.

In this method the first step is to calculate the standard cost of each process module, assuming that it is made of carbon steel and there are no high pressures. A correction is then made if either or both of these assumptions is false. This is done by determining the difference in equipment costs between low-pressure, carbon-steel equipment and that which is specified. This difference is then added to the cost of the standard module to obtain the estimated cost of the process module.

This assumes that the only difference in cost between standard modules and those constructed of other than carbon steel or those capable of withstanding high pressures is the cost of the equipment itself.

It has already been noted that the only other item that might be affected is the process piping cost. If necessary, the contingency allowance can be increased to account for this.

Table 9-10

Miller Estimate for Example 9-5

| No. of M.P.I.'s | Cost Indexes 1958  Current | Factor or Accuracy | Low | Probable | High |
|---|---|---|---|---|---|
| | | | | | Date Dec., 1972 |
| 52 | 99.7    137.2 | | | | |

Average Unit Cost of M.P.I.'s
    in 1958 Dollars 16,200

| | | | Low | Probable | High |
|---|---|---|---|---|---|
| M.P.I. (Main Plant Items)           Estimated | | | | 844,000 | |
| M.U.E. (Miscellaneous Unlisted Items) 15% | | | | 126,000 | |
| Basic Equipment (M.P.I. + M.U.E.)           100 (Excluding sales taxes and catalyst)   +10-10% | | | 873,000 | 970,000 | 1,067,000 |
| Field erection of basic equipment | | 9/11/13 | | | |
| Equipment foundations & structural supports | | 2/ 3/ 5 | | | |
| Piping | | 15/20/25 | | | |
| Insulation | | | | | |
| Equipment | | 1.5/2/2.5 | | | |
| Piping | | 3/ 4/ 5 | | | |
| Electrical | | 4/5/6 | | | |
| Instrumentation | | 5/7/9 | | | |
| Miscellaneous | | 3/4/5 | | | |
| Buildings           Evaluation = 0 Architectural & Structural | | 30/35/40 | | | |
| Building services | | 5/10/15 | | | |
| Subtotal—factored items           77.5/101/125.5 Adjustments:  Lows + 10       Highs – 10 | | | | | |
| Total factored items adjusted | | 85/101/113 | 742,000 | 980,000 | 1,200,000 |
| Direct cost of B/L (Excluding taxes and catalyst) | | | 1,615,000 | 1,950,000 | 2,267,000 |
| Storage and Handling   Medium to High | | 30/40/50 | 585,000 | 780,000 | 975,000 |
| Utilities | | 20/25/30 | 390,000 | 487,000 | 585,000 |
| Subtotal | | | 2,580,000 | 3,217,000 | 3,827,000 |

*(Table 9-10 continued on following page.)*

Table 9-10 (*continued*)

| Services | | | | |
|---|---|---|---|---|
| (In % of B/L + S&H + U) | 10/13/16 | 322,000 | 419,000 | 515,000 |
| Total B/L + Auxiliaries | | 2,902,000 | 3,636,000 | 4,342,000 |
| Catalyst      None | | | | |
| Taxes  4% sales tax | 3% | 87,000 | 109,000 | 130,000 |
| (assume 75% of | | | | |
| above subject to) | | | | |
| Total Direct Cost | | 2,989,000 | 3,745,000 | 4,472,000 |
| Indirect costs | | | | |
| Field construction, overhead,  & Profit | 20% | | | |
| Royalties, licenses, & Patents | 0% | | | |
| Engineering | 10% | | | |
| Total indirect costs | 30% | 897,000 | 1,120,000 | 1,340,000 |
| Total direct and indirect | | 3,886,000 | 4,865,000 | 5,812,000 |
| Contingencies (Including | 10% | 388,000 | 486,000 | 0 |
| contractors fee of 3%) | | | | |
| Total (1958) | | 4,274,000 | 5,351,000 | 5,812,000 |
| Total (1972) | | 5,900,000 | 7,370,000 | 8,000,000 |
| Total (1975)      above $\times$ $(1.03)^3$ | | 6,450,000 | 8,050,000 | 8,750,000 |

Guthrie makes no claims for the accuracy of his method. However, it should be more accurate than Miller's method. Appendix B gives some of Guthrie's data. Two examples using this information follow.

### Example 9-6

Determine the 1971 module cost of a 5,000 ft² U-tube heat exchanger constructed of carbon steel shell and monel tubes and rated at 500 psi.

F.O.B. equipment cost of a carbon steel exchanger rated at 150 psi is $23,000. See Appendix B, Fig. B-3. The bare module factor is 3.39 (only one exchanger; it would be 3.29 if there were two exchangers. See Table B-1).

$F_d =$ factor depending on the type of exchanger = 0.85
$F_p =$ factor relating to the design pressure = 0.52
$F_m =$ factor depending on size and the materials of construction = 3.75

The bare module cost for a carbon-steel exchanger rated at 150 psi is $23,000 × 3.39 = $78,000

The F.O.B. cost of the desired exchanger = $23,000 × $(F_d + F_p)$ × $F_m$

$$= (\$23,000) (0.85 + 0.52) (3.75) = \$118,000.$$

Bare module cost = $78,000 + ($118,000 - $23,000) = $173,000.

This is a 1968 cost. The CEPI for 1968 is 113.6. Assuming a 4% per year rise, the index for 1971 would be:

$$113.6 \times (1.04)^3 = 128.$$

The bare module in 1971 would cost approximately

$$\$173,000 \times \frac{128}{113.6} = \$195,000.$$

Add an 18% contingency. The module in 1971 would cost approximately $195,000 × (1.18) = $234,000

### Example 9-7

Determine the cost of constructing a 500,000 gal. stainless-steel vertical storage tank in mid-1968.
   From Table B-2 (Appendix B)
The cost of a carbon-steel tank is:
   $11.8(500,000)^{0.63} = \$45,900$
The cost of a stainless-steel tank is:
   $3.20 \times 45,900 = \$147,000$
The bare module cost of a carbon-steel tank is:
   $2.52 \times 45,900 = \$115,700$
The cost of installation is:
   $115,700 - $45,900 = $69,800
The bare module cost of a stainless-steel tank is:
   $147,000 + $69,800 = $216,800

   A quick estimate of the cost for a process equipment module can be obtained by using Figure B-9. This cannot be used for high pressure equipment. It is not as accurate as the data used in example 9-6 or 9-7.

### Example 9-8

   The total F.O.B. equipment cost of a processing section was $3,400,000 in 1958. What is the cost of the processing module in 1972? 40% of the equipment is monel.

Assume a 20% contingency, that the cost index will increase at 5% a year after 1969, and that monel equipment costs 5.8 times as much as carbon-steel equipment.

$$\text{CEPI in } 1958 = 99.7$$
$$\text{CEPI in } 1968 = 113.6$$
$$\text{CEPI in June } 1969 = 118.2$$
$$\text{Est. CEPI in } 1972 = 118.2 \, (1.05)^3 = 137$$

$$E = \text{F.O.B. cost } 1968 = 3{,}400{,}000 \, \frac{113.6}{99.7} = 3{,}870{,}000$$

Let $x$ = F.O.B. cost in 1968 of a totally carbon-steel module

$$0.6x + (0.4x) \, (5.8) = 3{,}870{,}000$$
$$x = \$1{,}320{,}000$$

$$\text{Alloy ratio} = \frac{\text{total equipment cost}}{\text{total equipment cost made of carbon steel}} = \frac{3{,}870{,}000}{1{,}320{,}000} = 2.93$$

From Figure B-9 (see Appendix B)

$F_r$ = ratio factor that depends on the material of construction = 2.25
$E \times F_r = 3{,}870{,}000 \, (2.25) = 8{,}710{,}000$
$F_m$ = magnitude factor that depends on the cost of the module = 0.980
Bare module cost in 1968 = $(E \times F_r) \times F_m = 8{,}710{,}000 \times 0.98 = \$8{,}550{,}000$

Bare module cost in 1972 = $8{,}550{,}000 \times \, (137/113.6) = \$10{,}300{,}000$

Contingency = $(0.20) \, (10{,}300{,}000) = \$2{,}060{,}000$
Total module cost = $\$10{,}300{,}000 + 2{,}060{,}000 = \$12{,}360{,}000$

## UNIT OPERATIONS ESTIMATE

The unit operations approach[16] is based on the premise that similar units have similar accessory equipment and should have similar costs. For instance, nearly all distillation columns would have a condenser, a condensate holding tank, and a reboiler. Why separate this into four different major items? By treating it as one item it is less likely that something will be forgotten. Forgetting a few items is more probable than omitting a whole processing step. Other unit operations would be ion exchangers, reactors, extractors, filters, and refrigeration units.

The costs for a specific unit operation would be a function of:

> size
> materials of construction
> pressure
> temperature

The importance of the first three of these factors has already been discussed. The temperature factor would include the cost of insulation plus the increase in metal thickness necessary to counteract the poorer structural properties of metals at high temperatures. Zevnik and Buchanan[17] have developed curves to obtain the average cost of a unit operation for a given fluid process. They base their method on the production capacity and the calculation of a *complexity factor*. The complexity factor is based on the maximum temperature (or minimum temperature if the process is a cryogenic one), the maximum pressure (or minimum pressure for vacuum systems) and the material of construction. It is calculated from Equation 2:

$$CF = 2 \times 10^{(F_{tz} + F_{pz} + F_{mz})} \tag{2}$$

where
$CF$ = complexity factor
$F_{tz}$ = temperature factor obtained from Figure 9-2
$F_{pz}$ = pressure factor obtained from Figure 9-3
$F_{mz}$ = material of construction factor obtained from Table 9-11

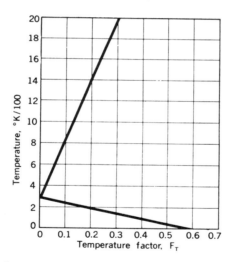

Figure 9-2 Temperature factor curve used in the Zevnik-Buchanan method.
Source: Zevnik, F.C., Buchanan, R.L.: "Generalized Correlation of Process Investment," *Chemical Engineering Progress*, Feb. 1963, p. 70.

The cost of a unit is then obtained from Figure 9-4. To obtain the cost of the plant this figure needs to be multiplied by the number of units, a factor (1.33) to account for utilities and general facilities, and the current or projected value of the Engineering News Record Chemical Cost Index. The accuracy of this method is unknown but should be better than the Lang factor and not as good as Miller's.

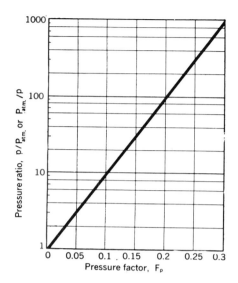

Figure 9-3    Pressure factor curve used in the Zevnik-Buchanan method.
Source: Zevnik, F.C., Buchanan, R.L.: "Generalized Correlation of Process Investment,"
*Chemical Engineering Progress,* Feb. 1963, p. 71.

Table 9-11

Material of Construction Factor for
Unit Operations Estimate

| $F_{mz}$ | Construction Material |
|---|---|
| 0 | Cast iron, carbon steel, wood |
| 0.1 | Aluminum, copper, brass, stainless steel (400 Series) |
| 0.2 | Monel, nickel, inconel, stainless steel (300 Series) |
| 0.3 | Hastelloy, etc. |
| 0.4 | Precious metals |

Source:    Zevnik, F.C., Buchanan, R.L.: "Generalized Correlation of Process Investment," *Chemical
Engineering Progress,* Feb. 1963, p. 70.

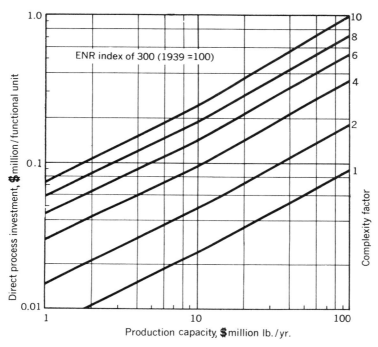

Figure 9-4   Process investment versus process capacity correlation.
Source: Zevnik, F.C., Buchanan, R.L.: "Generalized Correlation of Process Investment,"
*Chemical Engineering Progress,* Feb. 1963, p. 71

## Example 9-9

Estimate the cost of a 30,000 metric ton/yr isopropanol plant, based on the process given in Figure 9-5 in 1971. Use the method of Zevnik and Buchanan. The highest pressure is 4,260 psia and the maximum temperature is 554°F. It will be assumed that carbon steel can be used.

From Figure 9-5 the following six unit operations can be obtained:

1. Reactor                    4. Purification (light end) column
2. Gas separator              5. Dehydration column
3. Azeotrope column           6. Recovery column

To this will be added one unit for storage and shipping, giving a total of seven units.

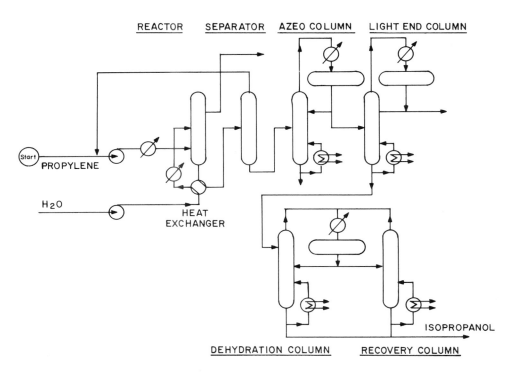

Figure 9-5   Isopropanol production flow sheet.
            Source: "Isopropanol," *Hydrocarbon Processing*, Nov. 1971, p. 172.

Maximum temperature of the process = 563°K
Temperature factor = 0.05 (from Fig. 9-2)
Pressure ratio = 4,260/14.7 = 290
Pressure factor = 0.24 (from Fig. 9-3)
Material factor = 0 (from Table 9-11)

Complexity factor = $2 \times 10^{(0.05 + 0.24)}$ = 3.90 (Equation 2)
Capacity of plant = 66,200,000 lb/yr
Cost of a functional unit for ENR Index of 300 = $260,000 (from Fig. 9-4)
Cost of plant for ENR Index of 300 = 7 × 260,000 × 1.33 = $2,420,000
In 1960 the ENR Index,[17] on the basis that it was 100 in 1939, was 350. The C.E. Index
will be used to extrapolate this to 1971.

C.E. Index in 1960 = 102.0
C.E. Index in 1971 = 132.2

$$\text{Cost of plant in 1971} = \$2,420,000 \times \frac{350}{300} \times \frac{132.2}{102} = \$3,660,000$$

The estimated cost for a battery limits plant in 1971 was $2,200,000.[20]

This difference of 40% could partially be due to different scopes. If storage and shipping were not considered then the cost of the plant would be reduced to $3,140,000. The other possibility could be that the estimate is low. An error of this size, however, need not be surprising because very little information is needed to obtain the estimate.

Hensley[16] suggests that a different factor should be developed for each unit operation and that another factor, the number of similar units, should be included. This is important because of the so-called learning factor.[19] The first time a 16-year-old tunes his car, it may take him a full day. Initially he may not know what a carburetor is or even where it is and may not know what a choke is for. The second time he at least knows where things are. Eventually he will be able to do this in less than an hour. Although the craftsman does not start at time zero like the teenager, he still must read and interpret the engineers' blueprints and the vendors' instructions, and sometimes both are vague. Since all problems must be resolved before the first item is installed, putting together the second unit will take less time. Also, he will develop labor-saving techniques if he has many duplicate operations to perform. Besides the learning factor, there are certain things that only need to be done once no matter how many items are desired. For instance, the engineering time spent in designing 10 identical reactors is the same as that spent in designing one. This means that the design cost per reactor in the former case is one-tenth that of the latter.

The major disadvantage of this method is that no cost information has been published and so anyone wanting to use it must develop his own data. This is time-consumming and hence expensive. Once data are developed, this method could produce more accurate results than any except the detailed cost estimate.

## DETAILED COST ESTIMATE

The detailed cost estimate requires more data than any of the estimates that have been discussed. For this estimate each item is estimated separately. For instance, the cost of piping is estimated by determining the exact footage of each size pipe, and the number of tees, elbows, valves, flanges, unions, reducers, bushings, and so forth. The material cost can be obtained exactly. The time for attaching each fitting and for laying the pipe is determined from tables. The total time obtained is then multiplied by the cost of labor per hour and divided by an efficiency factor to obtain the cost for installing the pipe. For installing a tank, every operation involved is analyzed for what equipment is necessary (for example, crane, helicopter, lift truck), what type of labor is required (for example, millwrights, boilermakers, electricians), and how many hours each is needed. This estimate may cost more than $80,000. This is, however, accurate to within ± 5% or better.

This type of estimating is used by contractors who are bidding for a given job. If they are wrong they will not make any profit. It is also used by companies who want to keep a close control on costs during the construction phase.

## ACCURACY OF ESTIMATES

The accuracy of any estimate will depend on the competence of the people involved, the completeness of the design information, the proper feedback of actual costs, and the subsequent updating of all cost data. It has been noted before that it takes a while for an engineer to get a qualitative and quantitative understanding of any process. This is also true for a cost estimator. For this reason, many companies have the process engineers do the capital cost estimating at the preliminary design stages. This saves time, since a cost estimator does not need to go through this gestation period. However, the process engineer must then spend some time keeping up with the advances in cost engineering. He must also receive feedback about the accuracy of his estimates.

In any profession it is important that the people involved see the results of their handiwork. If a mistake is made and the person making it never hears about it, he will probably make it again. Presumably he did what he felt was right the first time. Why should he change his ways unless he is corrected? For this reason it is important that each company have a procedure for gathering cost data while the plant is being built. At its completion an analysis of the costs should be compared with all the cost estimates made. The estimator should then be apprised of the results, and some method should be used to update the factors involved in the cost estimates. Only in this way can accurate cost estimating occur. Good cost estimations don't just happen. They require careful work and skilled judgment, and they cost money.

## CASE STUDY: CAPITAL COST ESTIMATION
## FOR A 150,000,000 LB/YR POLYSTYRENE PLANT
## USING THE SUSPENSION PROCESS

No information on the cost of a specific polystyrene plant could be found in the literature. One 1969 source[18], however, listed the average cost of a polystyrene plant as between $100 and $205 per annual ton. This will be extrapolated to 1974, using the Chemical Engineering Plant Design Index and an assumed inflation rate.

The Chemical Engineering Plant Cost Index (CEPI) in 1969 was 119.0. The CEPI in March, 1971, was 129.9. It will be assumed that for a plant to be completed by October, 1974, on the average all costs can be figured on an index for March, 1974.

If costs increase at 3% per year, the CEPI in March, 1974, will be 129.9 $\times$ (1.09) = 142. This may be low, since 1971 was an inflationary year, with costs rising nearly 6%. To account for this, the plant index will be increased to 147.

The capital cost for the proposed plant should range between $124 and $253 per annual ton, or between $9,300,000 and $19,000,000. It is expected to be nearer the lower figure, because it is being designed as a large, economical plant.

## Cost of Equipment

Table 9E-1 lists the F.O.B. cost of each item on the equipment list. These costs were gathered from a large number of different sources. At times these sources gave widely different answers. Table 9E-2 gives the costs of three different centrifugal pumps as obtained from three different authors. These costs were converted to the year 1968 by using the Marshall and Swift Index. Consistently, the costs predicted from Guthrie's figures were higher. This was true for all the pumps on the equipment list. It would be understandable if he was including something the others were not. However, by their statements, all that is missing is the driver (motor). The driver costs are also given in Table 9E-2 for purposes of comparison. If an explosion-proof motor is specified, an extra $200 should be added to each of those estimates.

The values given by Guthrie are average values, and he notes that the actual cost may vary from this by as much as 35%. This range almost includes the other two values. For all future calculations, the pump costs determined from Guthrie's figures are used because they are the most recent.

Table 9E-3 gives three different estimated costs for the blowers specified in Chapter 8. In this case, the estimates based on Guthrie's figures are usually below the others. Since he gives the same values for fans and blowers, this may mean he has assumed a pressure drop that is much lower than was calculated in Chapter 8. Under normal circumstances, it would therefore be wise to use the estimate obtained from Peter's book,[21] since it is more recent than Chilton's 1949 figures.[22] However, Simonds[23] cites the cost of an airveying system as $13,500 in 1959 ($16,400 in 1968). If either Peters's or Chilton's estimates were used, the airveying costs would greatly exceed this. So, again, the values obtained using Guthrie's tables are assumed to be right. Should the estimate obtained from Peter's graphs prove to be correct, this would mean the capital cost estimate is low by approximately $350,000.

The estimates from Guthrie's tables[15] for hoppers seemed extremely low, so the figures given by Wroth[22] are used. The material and labor factor also seemed low, so the factor for storage vessels is used instead. This will probably result in the estimate of hopper costs being somewhat high.

No specific costs could be found for the cutter, Cu-201; an anchor-type agitator, Ag-301; an epoxy-lined tank, D-101; and the waste treatment system, WTS-801. Further, the F.O.B. costs of the sand filter, WTS-701, and the extruders, EXT-501, had to be calculated from the installed costs.

One way to estimate prices of items when they cannot be found is to compare them with other items that might be similar or to items that might be expected to cost about the same. To be conservative for the cutter, it was assumed that a blender would cost at least as much and probably more. Its price was equated to that of a similar size blender ($5,000). In place of an anchor-type agitator, the costs for a propeller-type agitator was used. Similarly, it was assumed that an epoxy-lined tank would cost the same as a rubber-lined tank.

To calculate the cost of pollution control, it was decided to use an average cost for the industry as a whole. The capital cost for pollution control in 1971 was predicted

Table 9E-1

| Item No. | Number Reqd. | Cost per Item | Year Cost Estimate Was Made | Total Purchased Cost in 1968 | Bare Module Factor | Total Bare Module Cost $ | Ref. |
|---|---|---|---|---|---|---|---|
| D-101 to D-103 | 3 | $ 62,000 | 1968 | $186,000 | 2.52 | 377,000 | 6 |
| P-101 & P-102 | 2 | $ 2,600 | 1968 | $ 5,200 | 3.38 | 17,500 | 6 |
| P-103 & P-104 | 2 | $ 1,700 | 1968 | $ 3,400 | 3.38 | 11,500 | 6 |
| P-105 & P–106 | 2 | $ 520 | 1968 | $ 1,040 | 3.38 | 3,500 | 6 |
| Total 100 | 9 | | | $195,640 | | 409,500 | |
| D-201 & D-202 | 2 | $ 1,000 | 1968 | $ 2,000 | 4.23 | 4,900 | 6 |
| Ag-201 & Ag-202 | 2 | $ 350 | 1968 | $ 700 | 2.17 | 15,200 | 6 |
| D-203 to D-206 | 4 | $ 5,630 | 1968 | $ 22,520 | 4.23 | 54,800 | 6 |
| Ag-203 to Ag-206 | 4 | $ 350 | 1968 | $ 1,400 | 2.17 | 3,040 | 6 |
| Cu-201 | 1 | | | $ 5,000 | 2.08 | 10,400 | 9 |
| P-201 & P-202 | 2 | $ 820 | 1968 | $ 1,640 | 3.38 | 5,500 | 6 |
| E-201 to E-208 | 8 | $ 12,400 | 1968 | $100,000 | 3.29 | 191,000 | 6 |
| D-205 & D-206 | 2 | $ 13,500 | 1968 | $ 27,000 | 4.23 | 85,000 | 6 |
| Total 200 | 25 | | | $ 65,300 | | 369,840 | |
| R-301- to R-308 | 8 | $ 27,500 | 1967 | $226,000 | 4.23 | 425,000 | 1 |
| Ag-301 to Ag-308 | 8 | $ 3,850 | 1968 | $ 30,800 | 2.17 | 67,000 | 6 |
| D-301 to D-304 | 4 | $ 27,000 | 1968 | $108,000 | 4.23 | 203,000 | 6 |
| Ag-309 to Ag-312 | 4 | $ 2,500 | 1968 | $ 10,000 | 2.17 | 21,700 | 6 |
| P-301 to P-309 | 9 | $ 2,000 | 1968 | $ 18,000 | 3.38 | 60,800 | 6 |
| P-310 & P-311 | 2 | $ 600 | 1968 | $ 1,200 | 3.38 | 4,100 | 6 |
| P-312 to P-314 | 3 | $ 1,050 | 1968 | $ 3,150 | 3.38 | 10,700 | 6 |
| P-315 & P-316 | 2 | $ 800 | 1968 | $ 1,600 | 3.38 | 5,400 | 6 |
| D-305 | 1 | $ 21,000 | 1968 | $ 21,000 | 4.34 | 68,000 | 6 |
| Total 300 | 41 | | | $419,750 | | 865,700 | |
| D-401 | 1 | $ 66,000 | 1968 | $ 66,000 | 4.34 | 126,000 | 6 |
| S-401 & S-402 | 2 | $ 38,000 | 1968 | $ 76,000 | 2.16 | 164,000 | 6 |
| Dr-401 & Dr-402 | 2 | $ 39,000 | 1965 | $ 87,000 | 2.33 | 203,000 | 2 |
| Fi-401 & Fi-402 | 2 | $ 66 | 1960 | $ 142 | 2.26 | 320 | 10 |
| Fi-403 & Fi-404 | 2 | $ 12,100 | 1968 | $ 24,200 | 2.18 | 52,800 | 6 |
| B-401 to B-403 | 3 | $ 3,110 | 1968 | $ 9,330 | 2.13 | 19,920 | 6 |
| P-401 to P-405 | 5 | $ 1,100 | 1968 | $ 5,500 | 3.38 | 18,600 | 6 |
| P-406 to P-408 | 3 | $ 500 | 1968 | $ 1,500 | 3.38 | 5,100 | 6 |
| P-409 to P-411 | 3 | $ 480 | 1968 | $ 1,440 | 3.38 | 4,900 | 6 |
| P-412 & P-413 | 2 | $ 1,310 | 1968 | $ 2,620 | 3.38 | 5,860 | 6 |
| P-414 & P-415 | 2 | $ 830 | 1968 | $ 1,660 | 3.38 | 3,720 | 6 |
| E-401 & E-402 | 2 | $ 15,200 | 1968 | $ 30,400 | 3.29 | 108,600 | 6 |
| Total 400 | 28 | | | $305,792 | | 712,820 | |
| Ext-501 to Ext-509 | 9 ⎫ | | | | | | |
| D-501 to D-509 | 9 ⎬ | $ 54,500 | 1964 | $540,000 | 3.25 | 1,180,000 | 8 |
| Cu-501 to Cu-509 | 9 ⎭ | | | | | | |
| D-510 to D-518 | 9 | $ 800 | 1959 | $ 8,400 | 2.52 | 17,000 | 11 |
| Fi-501 & Fi-502 | 2 | $ 8 | 1960 | $ 18 | 2.26 | 40 | 10 |
| P-501 to P-510 | 10 | $ 520 | 1968 | $ 5,200 | 3.38 | 17,600 | 6 |
| B-501 & B-502 | 2 | $ 616 | 1968 | $ 1,232 | 2.05 | 2,530 | 6 |
| RV-501 to RV-511 | 11 | $ 2,800 | 1962 | $ 35,000 | 2.0 | 50,400 | 7 |
| CY-501 to CY-509 | 9 | $ 580 | 1968 | $ 5,220 | 2.18 | 11,400 | 6 |
| Fi-503 & Fi-504 | 2 | $ 2,200 | 1968 | $ 4,400 | 2.18 | 9,600 | 6 |
| Total 500 | 72 | | | $599,270 | | 1,288,570 | |
| D-601 to D-610 | 10 | $ 15,400 | 1959 | $179,000 | 1.96 | 290,000 | 11 |
| Bg-601 | 1 | $ 27,000 | 1971 | $ 23,000 | 2.06 | 47,500 | 4 |
| D-611 | 1 | $ 1,700 | 1959 | $ 1,980 | 2.52 | 4,000 | 11 |
| D-612 | 1 | $ 1,200 | 1959 | $ 1,400 | 2.52 | 2,800 | 11 |
| Dr-601 | 1 | $ 6,000 | 1971 | $ 5,200 | 1.87 | 9,700 | 5 |
| D-613 | 1 | $ 1,200 | 1959 | $ 1,400 | 2.52 | 2,800 | 6 |
| Cr-601 | 1 | $ 6,000 | 1971 | $ 5,200 | 1.87 | 9,700 | 5 |
| Tr-601 to Tr-612 | 12 | $ 11,000 | 1968 | $132,000 | 1.29 | 170,000 | 6 |
| W-601 | 1 | $ 1,500 | 1968 | $ 1,500 | 1.29 | 1,930 | 6 |

(Table 9E-1 continued on following page.)

| | | | | | | | |
|---|---|---|---|---|---|---|---|
| W-602 | 1 | $ 1,500 | 1968 | $ 1,500 | 1.29 | 1,930 | 6 |
| Cv-601 | 1 | $ 215 | 1968 | $ 215 | 2.13 | 460 | 6 |
| Cv-602 | 1 | $ 400 | 1968 | $ 400 | 2.13 | 850 | 6 |
| D-614- to D-619 | 6 | $ 7,000 | 1959 | $ 49,000 | 1.96 | 80,400 | 11 |
| Cv-603 | | | | | 1.29 | 460 | 6 |
| (all ductwork) | | | | | | | |
| B-601 & B-602 | 2 | $ 945 | 1968 | $ 1,890 | 2.05 | 3,870 | 6 |
| B-603 & B-604 | 2 | $ 1,525 | 1968 | $ 3,050 | 2.05 | 6,250 | 6 |
| B-605 & B-606 | 2 | $ 2,100 | 1968 | $ 4,200 | 2.05 | 8,600 | 6 |
| Rv-601 | 1 | $ 3,200 | 1962 | $ 3,650 | 2.0 | 5,200 | 7 |
| Rv-602 | 1 | $ 4,300 | 1962 | $ 4,900 | 2.0 | 7,100 | 7 |
| D-625 to D-637 | 13 | $ 375 | 1968 | $ 4,860 | 4.23 | 15,400 | 6 |
| Fi-601 | 1 | $ 13 | 1960 | $ 15 | 2.26 | 34 | 10 |
| Fi-602 | 1 | $ 26 | 1960 | $ 30 | 2.26 | 68 | 10 |
| Fi-603 | 1 | $ 56 | 1960 | $ 65 | 2.26 | 147 | 10 |
| Cv-604 | 1 | $ 8,100 | 1968 | $ 8,100 | 2.05 | 16,600 | 6 |
| Cv-605 & Cv-606 | 2 | $ 7,200 | 1968 | $ 14,400 | 2.05 | 29,500 | 6 |
| Cv-607 | 1 | $ 14,900 | 1968 | $ 14,900 | 2.05 | 30,500 | 6 |
| Vf-601 to Vf-633 | 33 | $ 6,000 | 1971 | $171,500 | 2.0 | 343,000 | 5 |
| D-620 to D-624 | 5 | $ 2,900 | 1959 | $ 17,000 | 2.52 | 34,000 | 11 |
| Fi-604 | 1 | $ 3,400 | 1968 | $ 3,400 | 2.18 | 7,400 | 6 |
| Fi-605 | 1 | $ 5,500 | 1968 | $ 5,500 | 2.18 | 12,000 | 6 |
| Fi-606 | 1 | $ 8,750 | 1968 | $ 8,750 | 2.18 | 19,000 | 6 |
| **Total 600** | 108 | | | $617,675 | | 1,161,699 | |
| D-701 | 1 | $ 18,000 | 1962 | $ 20,600 | 1.34 | 34,600 | 7 |
| D-702 | 1 | $ 18,000 | 1962 | $ 20,600 | 1.34 | 34,600 | 7 |
| D-703 | 1 | $ 5,500 | 1962 | $ 6,300 | 1.34 | 10,600 | 7 |
| IE-701 | 1 | $ 11,200 | 1965 | $ 12,500 | 1.83 | 22,900 | 3 |
| WTS-701 | 1 | | | $ 5,000 | | 27,300 | 9 |
| Sp-701 | 1 | $ 30,000 | 1968 | $ 30,000 | 1.83 | 55,000 | 6 |
| P-701 to P-703 | 3 | $ 1,600 | 1968 | $ 4,800 | 3.38 | 16,200 | 6 |
| P-704 & P-705 | 2 | $ 820 | 1968 | $ 1,640 | 3.38 | 5,555 | 6 |
| P-706 & P-707 | 2 | $ 1,200 | 1968 | $ 2,400 | 3.38 | 8,100 | 6 |
| P-708 & P-709 | 2 | $ 1,700 | 1968 | $ 3,400 | 3.38 | 11,500 | 6 |
| P-710 & P-711 | 2 | $ 500 | 1968 | $ 1,000 | 3.38 | 3,400 | 6 |
| **Total 700** | 17 | | | $108,240 | | 229,755 | |
| Ap-801 | 1 | $280,000 | 1968 | $280,000 | 1.46 | 409,000 | 6 |
| E1-801 | 1 | $ 21,500 | 1968 | $ 21,500 | 1.34 | 28,800 | 6 |
| D-801 | 1 | $ 9,000 | 1968 | $ 9,000 | 1.96 | 17,600 | 6 |
| P-801 | 1 | $ 1,100 | 1968 | $ 1,100 | 3.48 | 3,830 | 6 |
| LEQ-801* | 1 | | | $100,000 | 1.29 | 129,000 | 9 |
| WTS-801 | 1 | | | | | | |
| **Total 800** | 6 | | | $411,600 + WTS-801 | | 588,230 + WTS-801 | |
| **Grand Total** | | | | | | 5,630,000 + WTS-801 | |

---

\* Special laboratory equipment listed in scope

1. Derrick, G.C.: "Estimating the Cost of Jacketed Agitated and Baffled Reactors," *Chemical Engineering,* Oct. 9, 1967, p. 272.

2. "Estimating Costs of Process Dryers," *Chemical Enginnering,* Jan. 31, 1966, p. 101.

3. "Calculating Minimum-Cost Ion Exchange Units," *Chemical Engineering,* Dec. 6, 1965, p. 170.

4. Personal communication.

5. "Balky Materials Surrender to Good Vibrations," *Chemical Week,* Aug. 4, 1971, p. 35.

6. Guthrie, K.M.: "Capital Cost Estimating," *Chemical Engineering,* Mar. 24, 1969, p. 114 (see Appendix B).

7. Mills, H.E.: "Costs of Process Equipment," *Chemical Engineering,* Mar. 16, 1964, p. 133.

8. Carley, J.F.: "Introduction to Plastics Extrusion," *Chemical Engineering Progress* Symposium Series no. 49, vol. 60, 1964, p. 38.

9. Personal estimate.

10. Perry, J.H. (ed.): *Chemical Engineer's Handbook*," Ed. 4, McGraw-Hill, New York, 1963, Section 20, p. 95.

11. Wroth, W.F.: "Storage and Process Tanks," *Chemical Engineering,* Oct. 31, 1960, p. 124.

to be around 3.2% of the total capital expenditures for the chemical and allied products industry.[24] For the petroleum industry it was between 9% and 13%.[24] One consultant felt that the figure given for the chemical industry was low, and that because of government pressure it would at least double, and might rise by a factor of 13.[24] The higher factor might be correct over a short period of time, if industry is forced to meet rigid pollution-abatement standards. It was assumed that in the future the chemical industry will need to spend at least double what is currently being spent, and that in many cases this will exceed 10% of the capital costs. For this plant the cost of the waste-treatment system was estimated to be 8% of the total capital cost.

Guthrie[15] listed the cost of a water-treatment system. However, it is not known what this included. From his data the cost of the complete water-treatment system, installed, was estimated as $27,300. Since the equipment costs are needed for certain estimates, they had to be obtained from the installed cost. It was assumed that the major cost in constructing a sand filter (see Chapter 16) would be on-site labor. This gives an F.O.B. price of around $5,000 for the sand filter.

In 1964 the installed cost for each extruder and its accessory equipment was given as $90,000.[25] The indirect costs increase the cost for solids-handling equipment an average of 29%. This gives a bare module cost of $1,180,000 for all 9 extruders and accessories in 1968. No bare module factors are given for extruders. However, an extruder is really a polymer pump, and for pumps and compressors the bare module factor is around 3.25 and the material factor for stainless steel is 2.0. A back calculation gives a 1968 F.O.B. cost of $540,000.

In some cases a material factor was not given for aluminum. In these cases, a material factor of 1.50 was assumed. Where two figures were given for offsite calculations, the lower one was used, since this is to be a relatively small plant.

Table 9E-2

Cost of Centrifugal Pumps & Motors (1968)

| Pump No. | Cost of | | | |
| --- | --- | --- | --- | --- |
|  | Pump[1] | Pump[2] | Motor[2] | Pump & Motor[3] |
| P-101 | $865 | $1,020 | $800 | $2,600 |
| P-103 | $605 | $ 640 | $435 | $1,700 |
| P-201 | $378 | $ 577 | $206 | $ 820 |

1. Chapman, F.S., Holland, F.A.: "New Cost Data for Centrifugal Pumps," *Chemical Engineering*, July 18, 1966, p. 200.

2. Mills, H.E.: "Cost of Process Equipment," *Chemical Engineering*, Mar. 16, 1964, p. 133.

3. Guthrie, K.M.: "Capital Cost Estimating," *Chemical Engineering*, Mar. 24, 1969, p. 115 (see Appendix B).

Table 9E-3

Cost of Blowers (1968)

| Blower No. | Cost Basis | | | |
|---|---|---|---|---|
| | F.O.B.[1] | Installed[2] | F.O.B.[3] | Bare Module[1] |
| B-401 | $3,110 | $ 5,350 | $21,900 | $6,640 |
| B-601 | $ 945 | $ 4,590 | $ 7,800 | $1,935 |
| B-603 | $1,525 | $ 7,650 | $12,500 | $3,125 |
| B-605 | $2,100 | $15,300 | $16,600 | $4,300 |

1. Guthrie, K.M.: "Capital Cost Estimating," *Chemical Engineering,* Mar. 24, 1969, p. 115 (see Appendix B).
2. Chilton, C.H.: "Cost Data Correlated," *Chemical Engineering,* June 1949, p. 97.
3. Peters, M.S., Timmerhaus, K.D.: *Plant Design and Economics for Chemical Engineers,*" Ed. 2, McGraw-Hill, New York, 1968, p. 470.

**Factored Cost Estimate**

For this estimate, the prices of all items on the equipment list except those in the 700 and 800 categories were summed to obtain a 1968 F.O.B. cost of $2,203,000. This was increased by the Ohio sales tax of 4% and then by another 5%, to cover freight costs. This results in a delivered equipment cost of $2,400,000. The Lang factor is between that of a fluids plant and a solid-fluids plant. The value selected was 4.3. This was updated to 1974, and an 8% factor for waste treatment was added. The result is a factored cost estimate of $14,400,000.

**Miller Estimates**

Miller's method of estimating[14] is based on the costs of all items inside the battery limits. This does not include storage units nor the materials handling devices between the storage units and the battery limits. To obtain the F.O.B. cost, all items in the 200, 300, 400, and 500 categories were included except for the following 27 items: P-301 to P-309, P-315, P-316, D-401, P-406 to P-408, P-412, P-413 and P-501 to P-510. The calculations for this method are given in Table 9E-4. For details, see Miller's article.[14] Ordinarily, the costs for wastewater treatment would be included in the battery limits. However, in this case, since they were not available, an 8% surcharge was added to the final estimate. This was the same procedure used for the factored cost estimate.

**Module Estimate**

The bare module factors for each individual item are given in Table 9E-1. The offsite costs are given in Table 9E-5, the site development costs in Table 9E-6, and the cost of industrial buildings in Table 9E-7. The various costs are combined in Table 9E-8 and then updated to 1974. Much of the information is from Appendix B.

Table 9E-4

Miller Estimate

| No. of M.P.I.'s | Cost Indexes | | Factor or Accuracy | Low | Probable | High |
|---|---|---|---|---|---|---|
| | 1958 | Current | | | | |
| 113 | 99.7 | 147 | | | | |

Proj. or Study No. 1    Title    Polystyrene Plant    Date
Requested by  WDB    Capacity 150,000,000 lb/yr    Aug. 1971

Average Unit Cost of M.P.I.'s
       In 1958 Dollars  9,290

| | | | | | |
|---|---|---|---|---|---|
| M.P.I. (Main plant items) | | | Estimated | 1,190,000 | |
| M.U.E. (Miscellaneous unlisted items) | | | 10% | 119,000 | |
| Basic Equipment (M.P.I. + M.U.E.) | 100 | | 1,178,000 | 1,309,000 | 1,440,000 |
| (Excluding sales taxes and catalyst) +10 – 10 | | | | | |

Remarks

| | | |
|---|---|---|
| Field erection of basic equipment | Slightly under Average | 8/10/12 |
| Equipment foundations & Structural supports | Average | 3/5/7 |
| Piping | Liquids & solids | 6/8/12 |
| Insulation | | |
|   Equipment | Average | 2/3/4 |
|   Piping | Below average | 4/5/6 |
| Electrical | Average | 6/8/10 |
| Instrumentation | Below average | 8/10/13 |
| Miscellaneous | | 2/3/4 |
| Buildings     Evaluation = –1 to –2 | | 30/35/40 |
|   Architectural & Structural | | |
| Building services | | |
|   (in % of Arch'l. & Struct.) | | |

| | |
|---|---|
| Compressed air | 1 |
| Electrical lighting | 9 |
| Sprinklers | 6 |
| Plumbing | 12 |
| Heating | 16 |
| Vent. & air conditioning | 10 |
| Total services | 54 |

54%  16/19/22

(*Table 9E-4 continued on following page.*)

Table 9E-4 (*continued*)

| Subtotal—factored items | 85/106/130 | | | |
|---|---|---|---|---|
| Adjustments:  Lows + 10  Highs – 10 | | | | |
| Total factored items | 93/106/117 | 1,095,000 | 1,390,000 | 1,680,000 |
| Direct cost of B/L (Excluding taxes and catalyst) | | 2,273,000 | 2,699,000 | 3,120,000 |
| Storage and handling      High | 50/60/70 | 1,350,000 | 1,620,000 | 1,890,000 |
| Utilities  Water treatment             Steam             Ion exchange | 15/23/32 | 405,000 | 620,000 | 864,000 |
| Subtotal | | 4,028,000 | 4,939,000 | 5,974,000 |
| Services (in % of B/L + S&H + U) | 11/13/15 | 540,000 | 640,000 | 735,000 |
| Total B/L + Auxiliaries | | 4,568,000 | 5,579,000 | 6,709,000 |
| Catalyst      None | | | | |
| Taxes   4% Sales Tax (assume 75% of above subject to) | 3% | 137,000 | 167,000 | 201,000 |
| Total direct cost | | 4,705,000 | 5,746,000 | 6,910,000 |
| Indirect costs | | | | |
| Field construction, overhead, & profit | 21% | | | |
| Royalties, Licenses, & Patents | 0% | | | |
| Engineering | 10% | | | |
| Total indirect costs | 31% | 1,460,000 | 1,780,000 | 2,140,000 |
| Total direct and indirect | | 6,165,000 | 7,526,000 | 9,050,000 |
| Contingencies (including contractor's fee of 3%) | 13% for low 8% for prob. 3% for high | 802,000 | 602,000 | 271,000 |
| Total (1958) | | 6,967,000 | 8,128,000 | 9,321,000 |
| Total (1974) | | 10,300,000 | 12,000,000 | 13,700,000 |
| Wastewater treatment | 8% | 825,000 | 960,000 | 1,100,000 |
| Total (1974) | | 11,125,000 | 12,960,000 | 14,800,000 |

Table 9E-5

Offsite Costs

| | | | |
|---|---|---|---|
| Instrument air systems | $18,750 x 1.34* | = | $ 25,100 |
| Fireloop and hydrants | $12,580 x 1.34 | = | $ 16,850 |
| Drinking and service water | $2,500 x 1.34 | = | $ 3,350 |
| Main transformer station (5,000 kva) | $23(5,000)1.34 | = | $154,000 |
| Secondary transformer station (2,000 kva) | $3(19.3)(2,000)1.34 | = | $155,000 |
| Dock—20 ft wide, 215 ft long (3 in. medium construction) | $9.38(4,300)1.29** | = | $ 52,000 |
| Railroad facilities | | | |
|    Track—1,000 ft | $26.25(1,000)1.29 | = | $ 34,000 |
|    Turnout | $2,800 x 1.29 | = | $ 3,610 |
|    Bumper | $1,580 x 1.29 | = | $ 2,040 |
| Steam distribution | $14,100 x 1.34 | = | $ 18,900 |
| Yard lighting and communications | $18,750 x 1.34 | = | $ 25,100 |
| Total offsite piping—8,300 ft | $0.82(8,300)^{1.05}$ x 1.34 | = | $ 14,300 |
| Total offsite costs | | | $504,250 |

This is a small plant compared to many chemical complexes, so where lump sums were given the minimum price was used.

\*   1.34 is an average indirect factor.
\*\* 1.29 is an indirect factor for solids operations.

Source of Cost Data: Guthrie, K.M.: "Capital Cost Estimating," *Chemical Engineering,* Mar. 24, 1969, p. 114 (see Appendix B).

Table 9E-6

Site Development Costs

| | | | |
|---|---|---|---|
| Drainage ditch around perimeter—10,000 ft | $0.85(10,000)1.29* | = | $ 11,000 |
| Fencing (chain link) around perimeter—5,800 ft | 5.93(5,800)1.29 | = | $ 44,400 |
| Gates (chain)—4 | 4 x $128 x 1.29 | = | $ 660 |
| Landscaping—30,000 yd | 1.70 x 30,000 x 1.29 | = | $ 66,000 |
| Roads—2 in. thick asphalt top, 4 in subbase—2,000 ft | 4.68(2,000)1.29 | = | $ 12,000 |
| Roads—2 in. thick gravel surface—9,000 ft | 0.58(9,000)1.29 | = | $ 6,700 |
| Parking lots—blacktop, space for 50 cars 160 ft x 100 ft—1,780 yd² | 6.25(1,780)1.29 | = | $ 14,400 |
| Sewer facilities (36 in) around all buildings—1,500 ft | 15.96(1,500)1.29 | = | $ 30,900 |
| Site clearing, grading, and leveling | 0.90(227,000 yd²)1.29 | = | $264,000 |
| Total site development costs | | | $450,060 |

\*   1.29 is an indirect factor.
Sources of Cost Data: Guthrie, K.M.: "Capital Cost Estimating," *Chemical Engineering,* Mar. 24, 1969, p. 114 (see Appendix B).

Table 9E-7

Industrial Buildings

---

Product warehouse — 180 ft x 380 ft = 68,250 ft$^2$
    Cost = 68,250 ft$^2$ (4.64 + 1.17 + 1.30 + 1.43)
        = \$583,000
Raw material warehouse — 60 ft x 100 ft = 6,000 ft$^2$
    Cost = 6,000 ft$^2$ (4.64 + 1.17 + 1.30 + 1.43)
        = \$51,200
Office, change room, lunchroom — 2,000 ft$^2$
    Cost = 2000 ft$^2$ (8.80 + 6.89 + 3.25 + 2.60 + 2.83 + 1.69 + 9.10)
        = \$70,500
Laboratory, office — 1,800 ft$^2$
    Cost = 1,800 ft$^2$ (9.96 + 6.89 + 3.25 + 2.60 + 2.83 + 1.69 + 20.80)
        = \$86,500
Reactor building — 80 ft x 40 ft = 3,200 ft$^2$ (4 stories high)
    Cost = 3,200 ft$^2$ $\left[(5.30 \times 2.5) + (2.27 + 1.95 + 2.83 + 1.69)\right]$
        = \$70,400
Drying extruder building — 9,600 ft$^2$
    Cost = 9,600 ft$^2$ (5.30 + 2.27 + 1.95 + 2.21 + 1.43)
        = \$124,600
Buildings for utilities and maintenance — assume 4,000 ft$^2$
    Cost = 4,000 ft$^2$ (5.31 + 2.27 + 1.95 + 2.21 + 1.43)
        = \$52,700
Total cost for industrial buildings = \$1,040,300

---

Source of Cost Data:  Guthrie, K.M.: "Capital Cost Estimating," *Chemical Engineering*, Mar. 24, 1969, p. 114 (see Appendix B).

Table 9E-8

Total Cost of Plant (Guthrie Method)

| | |
|---|---:|
| Equipment installed | \$ 5,630,000 |
| Offsite costs | 504,250 |
| Site development costs | 450,000 |
| Piping costs | 168,000 |
| Industrial buildings | 1,040,300 |
| | \$ 7,792,550 |
| Wastewater treatment system | \$ 623,000 |
| (8% of above) | |
| Bare module cost | \$ 8,416,000 |
| Contingency plus contractor's fee (18%) | \$ 1,515,000 |
| Total cost of plant in 1968 = | \$ 9,931,000 |
| Total capital cost of plant in 1974 = | \$13,000,000 |

Sources of Cost Data: Guthrie, K.M.: "Capital Cost Estimating," *Chemical Engineering*, Mar. 24, 1969, p. 114 (see Appendix B).

Piping between units is difficult to estimate. From Figure 8E-4 there are approximately 20,000 ft of piping. Assuming that the average size piping is 3 in, the installed cost is $6.00 per foot (including valves; partially stainless steel), and the bare module factor is 1.40, the cost in 1968 is

$$20,000 \times \$6.00 \times 1.40 = \$168,000.$$

In the Miller method, a 5% contingency was used, plus a 10% allowance for miscellaneous unlisted equipment. This is equivalent to the 15% contingency used in the module method.

### Method of Zevnik and Buchanan

Table 9E-9 lists unit operations in the polystyrene plant. The highest temperature is 400°F, in the extruder. From this and Figure 9-5, a temperature factor of 0.04 is obtained. There are no high pressures except in the extruder, and its value is unknown. The pressure factor will be assumed to be zero. Stainless steel is used, so the material factor is 0.2. From Equation 2 a complexity factor of 3.48 can be calculated. A direct process investment cost of $350,000 per functional unit is obtained from Figure 9-7. This means that the cost of constructing the plant when the Engineering News Record Construction Index (ENRCI) is 300 would be $3,150,000. This will be updated to 1960 when the ENRCI was 350, and then the CEPI will be used to obtain the cost in 1974. The resultant cost in 1974 is

$$\$3,150,000 \times \frac{350}{300} \times \frac{147}{102} = \$5,500,000.$$

This cost is very low, compared to other estimates. This may be due to the fact that Zevnik and Buchanan assume a single-train plant, whereas the plant just designed has two trains. If this assumption is made, the direct investment cost of a functional unit for a 75,000,000 lb/yr plant would be $280,000. Making the same calculations as before, the cost of a 75,000,000 lb/yr plant would be $4,400,000. The cost of two such plants would be $8,800,000. This is still low.

Table 9E-9

The Unit Operations in a Polystyrene Plant

| | |
|---|---|
| 1. Raw material storage | 5. Centrifuging |
| 2. Feed preparation (including deionization) | 6. Drying |
| | 7. Extrusion |
| 3. Reaction | 8. Product storage |
| 4. Washing | 9. Waste treatment |

**Comparison**

From the summary of estimates given in Table 9E-10, it can be seen that all estimates except for the one obtained using the Zevnik and Buchanan method are in the same ballpark and are surprisingly close. The factored estimate is probably greater than the others because too high a factor was used. The estimate of Zevnick and Buchanan[17] may be low because it applies mainly to continuous fluid processes. This plant is partly batch and partly a solids-handling facility. The estimate obtained using the module method will be used in all future calculations.

Table 9E-10

Summary of Capital Costs (1974)

| Type of Estimate | Estimate |
|---|---|
| From past plants | High $19,000,000<br>Low $9,300,000 |
| Factored estimate | $14,400,000 |
| Miller estimate | Low $11,125,000<br>Probable $12,960,000<br>High $14,800,000 |
| Module estimate | $13,000,000 |
| Zevnik and Buchanan estimate | $ 8,800,000 |

**Possible Improvements**

The most expensive items, in order, are the extruders, wastewater treatment, product warehouse, reactors, emergency power generating facility, styrene storage, vibrating feeders, transformer stations, bulk product storage, site preparation, reactor building, dryers, and hold tanks. Together they account for 67% of the bare module cost. These are the items that should be further investigated to see if any substantial savings can be obtained. Some money could be saved if something other than vibrating feeders were used. Substantial savings could be made if the storage space for product or styrene could be reduced. This would involve a change of scope.

### References

1. Arnold, T.H., Chilton, C.H.: "New Index Shows Plant Cost Trends," *Chemical Engineering*, Feb. 18, 1963, p. 143.
2. Thorsen, D.R.: "The Seven-Year Surge in CE Cost Indexes," *Chemical Engineering*, Nov. 13, 1972, p. 168.
3. Nelson, N.L.: "Tabulated Values of Construction Cost Index," *Oil and Gas Journal*, Mar. 6, 1967, p. 110; also see Jan. 29, 1973, p.. 114.
4. "ENR Building Construction Cost Index Histories," *Engineering News Record*, Mar. 22, 1973, p. 79 (history in March issue each year).

5. Stevens, R.W.: "Equipment Cost Indexes for Process Industries," *Chemical Engineering*, Nov. 1947, p. 124.
6. Haselbarth, J.: "Updated Investment Costs for 60 Types of Chemical Plants," *Chemical Engineering*, Dec. 4, 1967, p. 214.
7. Gallagher, J.T.: "Rapid Estimation of Plant Costs," *Chemical Engineering*, Dec. 18, 1967, p. 89-96.
8. Lang, H.J.: "Simplified Approach to Preliminary Cost Estimates," *Chemical Engineering*, June, 1948, p. 112.
9. Walus, S.M., Nofsinger, C.W.: "Process Equipment Nomography," *Chemical Engineering*, Mar. 21, 1960, pp. 173-176.
10. Winfield, M.D., Dryden, C.E.: "Chart Gives Equipment Plant Costs," *Chemical Engineering*, Dec. 24, 1962, pp. 100-104.
11. Mills, H.E.: "Costs of Process Equipment," *Chemical Engineering*, Mar. 16, 1964, pp.133-156.
12. Nichols, W.T.: "Next Time Give the Boss a Precise Capital Cost Estimate," *Chemical Engineering*, June, 1951, p. 248.
13. Bach, N.G.: "More Accurate Cost Estimates," *Chemical Engineering*, Sept. 22, 1958, p. 155.
14. Miller, C.A.: "Factor Estimating Refined for Appropriation of Funds," *American Association of Cost Engineers Bulletin*, vol. 7, no. 3, Sept. 1965, pp. 92-118.
15. Guthrie, K.M.: "Captial Cost Estimating," *Chemical Engineering*, Mar. 24, 1969, pp. 114-142.
16. Hensley, E.F.: "The Unit Operations Approach," a paper presented at the American Association of Cost Engineers Annual Meeting, 1967.
17. Zevnik, F.C., Buchanan, R.L.: "Generalized Correlation of Process Investment," *Chemical Engineering Progress*, Feb. 1963, p. 70.
18. "Petro Plastics '69" (chart), Chemical Construction Corp., New York, 1969.
19. Hirschmann, W.B.: "Profit from the Learning Curve," *Harvard Business Review*, Jan-Feb 1964, pp. 125-139.
20. "Isopropanol," *Hydrocarbon Processing*, Nov. 1971, p. 172.
21. Peters, M.S. Timmerhaus, K.D.: *Plant Design and Economics for Chemical Engineers*, Ed. 2, McGraw-Hill, New York, 1968, p. 470.
22. Chilton, C.H.: "Cost Data Correlated," *Chemical Engineering*, June 1949, p. 97.
23. Simonds, H.R.: *The Encyclopedia of Plastics Equipment*, Reinhold, New York, 1964, p. 176.
24. Rushton, J.D.: "Capital Spending," *Chemical Engineering*, June 21, 1971, p. 161.
25. Carley, J.F.: "Introduction to Plastics Extrusion," *Chemical Engineering Progress Symposium*, Series no. 49, vol. 60, 1964, p. 38.

## Additional References

Popper, H.: *Modern Cost-Engineering Techniques*, McGraw-Hill, New York, 1970.

Chilton, C. (ed.): *Cost Engineering in the Process Industries*, McGraw-Hill, New York, 1960.

Bauman, H.C.: *Fundamentals of Cost Engineering in the Chemical Industries*, Reinhold, New York, 1964.

*Bibliography of Investment and Operating Cost for Chemical and Petroleum Plants, January-December 1969*, Bureau of Mines, Circ. 8478, 1970.

### Publications of the Environmental Protection Agency

*Estimating Costs and Manpower Requirement for Conventional Wastewater Treatment Facilities*, 17090, DAN 10/71, Oct. 1971.

*Capital and Operating Costs of Pollution Control Equipment Modules–Volume II. Data Manual*, R5-73-023b, July 1973.

*Cost Analysis of Water Pollution Control: An Annotated Bibliography*, R5-73-017, Apr. 1973.

*Projected Wastewater Treatment Costs in the Organic Chemical Industry*, 12020 GND 07/71, July 1971.

*State of the Art Review: Water Pollution Control Benefits and Costs-Volume 1*, 600/5-73-008a, Oct. 1973.


**Articles from** *Chemical Engineering*

Dybdahl, E.C.: "Subject Index of CE Cost Files 1958-1966," (subject list of 119 articles concerning cost which appeared in *Chemical Engineering), Dec. 5, 1966, p. 166.

Guthrie, K.M.: "Capital and Operating Costs for 54 Chemical Processes," June 15, 1970, p. 140.

Sommerville, R.F.: "Estimating Mill Costs at Low Production Rates," Apr. 6, 1970, p. 148.

Chase, J.D.: "Plant Cost vs. Capacity: New Way to Use Exponents," Apr. 6, 1970, p. 113.

Drew, J.W., Grinder, A.F.: "How to Estimate the Cost of Pilot-Plant Equipment," Feb. 9, 1970, p. 100.

Liptak, B.G.: "Costs of Process Instruments," Sept. 7, 1970, p. 60; Sept. 21, p. 175.

Liptak, B.G.: "Control-Panel Costs," Oct. 5, 1970, p. 83.

Marshall, S.P., Brandt, J.E.: "Installed Cost of Corrosion-Resistant Piping," Aug. 23, 1971, p. 69.

Sommerville, R.F.: "New Method Gives Quick Accurate Estimate of Distillation Costs," May 1, 1972, p. 71.

Guthrie, K.M.: "Estimating the Cost of High-Pressure Equipment," Dec. 2, 1968, p. 144.

Naundorf, C.H.: "Estimate High Vacuum Costs Graphically," Oct. 2, 1961, p. 107.

Mills, H.E.: "Estimating Bucket Elevator Costs," Nov. 28, 1960, p. 117.

Kirk, M.M.: "Cost of Mist Eliminators," Oct. 10, 1960, p. 240.

Bromberg, I.: "A Look at an Engineering Contractor's Overhead Costs," Sept. 12, 1966, p. 218.

Sokullu, E.S.: "Estimating Piping Costs from Process Flowsheets," Feb. 10, 1969, p. 148.

Bosworth, D.A.: "Installed Costs of Outside Piping," Mar. 25, 1968, p. 132.

McGovern, J.G.: "Inplant Wastewater Control", May 14, 1973, p. 137.

Hutchins, R.A.: "Activated Carbon Systems," Aug. 20, 1973, p. 133.

Imhof, H.: "Protecting Process Plants from Power Failures," Apr. 2, 1973, p. 56.

Guthrie, K.M.: "Field-Labor Predictions for Conceptual Projects," Apr. 7, 1969, p. 170.

Kirk, M.M.: "Cranes, Hoists and Trolleys," Feb. 27, 1967, p. 168.

Gallagher, J.T., "Cost of Direct-Fired Heaters," July 17, 1967, p. 232.

"Crystallizer Costs for Fertilizer and Fine Chemicals," June 20, 1966, p. 246.

Sargent, G.D.: "Dust Collection Equipment," Jan. 27, 1969, p. 130.

Belcher, D.W., Smith, D.A., Cook, E.M.: "Design and Use of Spray Dryers," Sept. 30, 1963, p. 83; Oct. 14, 1963, p. 201.

Harmer, D.E., Ballantine, D.S.: "Applying Radiation to Chemical Processing," May 3, 1971, p. 91.

Hyland, J.: "Teflon Tank Linings," June 11, 1973, p. 124.

Standiford, F.C., Jr.: "Evaporator Economics and Capital Costs," Dec. 9, 1963, p. 170.

Clark, F.D., Terni, S.P.: "Thick Wall Pressure Vessels," Apr. 3, 1972, p. 112.

Considine, D.M.: "Weight Proportioning and Batching," Sept. 14, 1964, p. 199.


**Articles from** *Cost Engineering*

Zimmerman, O.T.: "Elements of Capital Cost Estimating," Oct. 1968, p. 4.

Zimmerman, O.T.: "Miscellaneous Cost Data," Apr. 1969, p. 17 (desalination and process plants).

Zimmerman, O.T., Lavine, I.: "Balanced Air XQ Series Cyclone Dust Collectors," Apr. 1969, p. 13.

Zimmerman, O.T., Lavine, I.: "Process Plant Costs," Oct. 1969, p. 16; and Apr. 1972, p. 18.

Zimmerman, O.T., Lavine, I.: "Miscellaneous Cost Data—Process Equipment," July 1969, p. 18.

Zimmerman, O.T.: "Process and Analytical Instruments," Jan. 1973, p. 4.

Zimmerman, O.T.: "Wastewater Treatment," Oct. 1971, p. 11.

Zimmerman, O.T.: "Ribbon Blenders," Apr. 1971, p. 13.

Zimmerman, O.T.: "Process Equipment Cost Data," Apr. 1973, p. 13.

Zimmerman, O.T.: "Dust Collections," Jan. 1972, p. 4.

Zimmerman, O.T., Lavine, I.: "Peabody Gas Scrubbers," Apr. 1968, p. 13.

**Articles from** *Hydrocarbon Processing*

Swaney, J.B.: "Preliminary Cost Estimating," Apr. 1973, p.. 167 (for refinery unit).

Grigsby, E.K., Mills, E.W., Collins, D.C.: "What Will Future Refineries Cost?" May, 1973, p. 133.

Bishop, R.B.: "Find Polystyrene Plant Costs," Nov. 1972, p. 137 (impact polystyrene and crystal P.S.).

Epstein, L.D.: "What Do Jacketed Reactors Cost," Dec. 1972, p. 102.

**Articles from Miscellaneous Journals**

Nelson, N.L.: "Costs of Aromatics Plants," *Oil and Gas Journal,* Apr. 30, 1973, p. 188.

Dyer, J.P.: "Two Views on Ways for Extrusion Plants to Implement Their Health & Safety Programs," *Society of Plastics Engineers Journal,* May, 1973, p. 26 (specific costs for replacement ventilation, lighting systems, etc.).

"Gas Liquid: Move It or Cool It," *Chemical Week,* Mar. 22, 1969, p. 99.

Cook, E.M.: "Estimating Spray Drying Costs," *Chemical Engineering Progress,* June 1966, p. 93.

Block, L.J.: "Estimating Machinery and Equipment, Erection Costs," *Oil and Gas Journal,* Aug. 6, 1973, p. 77..

# CHAPTER 10

# Economics

This chapter and the next are concerned with determining how much it costs to produce a pound of product and whether the expected profit is enough to justify building a plant. First, all the costs that are involved in producing and selling a chemical will be investigated. Next, the amount of money that must be invested to produce the product will be determined. Then a number of different ways of evaluating the profitability of a plant will be considered. There are as many profitability measures, as there are ways to determine the cost of building a plant. Only some of them will be presented. No one measure can be definitely said to be best, although the discounted cash flow methods are better than the others. However, some companies still use the other methods, and therefore an engineer should be familiar with them.

Before determining the cost of producing a chemical, let us look at the cost of operating an automobile, since the two have many similarities. To be determined is how much it costs per mile to drive a car.

Suppose a person buys a car for $3,500. Three years later, when he buys a new car, the trade-in value of his old car is $1,500. This means $2,000 total or $667 per year is spent merely to own the vehicle. If the car is driven 10,000 miles per year, the cost per mile is 6.67¢.* This cost is called depreciation.

To run the car requires fuel. If the cost of gasoline is 50 cents per gallon and the car gets 16 miles per gallon, then the cost of fuel is 3.13¢ per mile. Maintenance for new cars that are factory-guaranteed is small, say $67 per year or 0.67¢ per mile. Insurance, if there are no teenage drivers, is about $200 per year or 2¢ per mile. To this should be added the cost of maintaining a garage and driveway—say $50 per year or 0.50¢ per mile; driver's license and city and state car taxes—$20 per year or 0.20¢ per mile; any interest payment on the loan for buying the car—8% per year of $3,000 (assuming $500 paid in cash) or 2.4¢ per mile, and any tolls paid on highways or for crossing bridges—say 0.5¢/mile. This is a total of 16.07¢ per mile (see Table 10-1).

---

*Those familiar with the metric system should replace mile with kilometer in this example.

Table 10-1

## Cost per Mile of Operating an Automobile in Cents per Mile

|  | Miles Driven per Year | |
|---|---|---|
|  | 10,000 | 20,000 |
| Gas | 3.13 | 3.13 |
| Maintenance | 0.67 | 1.33 |
| Tolls | 0.50 | 0.50 |
| Depreciation | 6.67 | 3.33 |
| Garage depreciation | 0.50 | 0.25 |
| Licences and taxes | 0.20 | 0.10 |
| Insurance | 2.00 | 1.00 |
| Interest on loan | 2.40 | 1.20 |
| Total | 16.07 | 10.84 |

Table 10-2

## How to Estimate the Approximate Costs of Producing a Product*

| Cost Category | Method of Cost Estimation |
|---|---|
| Raw materials | Unit ratio material balance & chemical prices |
| Conversion costs | |
|     Labor | Manpower estimate and average salaries |
|     Utilities | Energy balance and energy costs |
|     Supplies | 15–20% of maintenance costs |
|     Maintenance | 6.1% to 7.4% of total sales |
|  | or 5% of cost of constructing the plant |
|     Waste treatment | See Chapter 16 |
|     Royalties | $5/1,000 lb of product when applicable |
|     Packaging | See reference 5 |
| Depreciation | 9% of cost of constructing the plant |
| Sales | 5–50% of total sales (ave. 10%) |
| Research | 3-4% of total sales |
| Taxes and insurance | 2-3% of cost of constructing the plant |
| General administrative costs | 5% of total sales |

*   These are approximations and where better estimates are available they should be used.

One of the major factors that affects this cost is the number of miles driven per year. If the car were driven 20,000 miles per year, all the costs per mile would be cut in half except those for fuel, maintenance, and tolls. Those for fuel and tolls would be expected to remain the same, assuming the same percentage of driving is done on toll roads. That for maintenance would be expected to be higher. The greater the mileage, the greater the wear and tear on the car. Suppose the maintenance costs per mile doubled. The total cost of operating the car would then be 10.84¢ per mile.

Thus there are two types of expenses: those that are independent of the amount of usage and those that depend on how much the car is driven. The former are known as fixed charges and include depreciation, insurance, interest costs, garage expenses, driver's license, and city and state taxes.

## COST OF PRODUCING A CHEMICAL

A chemical plant has every type of cost associated with the operation of a car and many more. The major categories are raw materials, conversion costs, depreciation, sales, research, taxes and insurance, and general administration costs. A method for determining the magnitude of each of these follows. A summary is given in Table 10-2.

### Raw Material Costs

These are the costs delivered to the plant of all the various chemicals that are used in producing the products. They include those that react as well as those others that are not recovered, such as chelating compounds, suspending agents, and bleaching compounds. The prices of the raw materials can be obtained from the manufacturer or from the *Chemical Marketing Reporter*. A list of manufacturers of industrial chemicals is given annually by *Chemical Week* in a supplement called "The Buyer's Guide." The amount of each chemical required is given by the unit ratio material balance.

### Conversion Costs

The conversion costs are those that occur between the time the raw materials enter the plant and the time the product leaves, and that are a direct result of processing. These consist of labor, utilities, supplies, maintenance, waste treatment charges, royalties, and packaging. Labor costs include those for the operators of equipment, packagers, foremen, clerks involved in shipping and plant administrative details, supervisors, managers, laboratory personnel connected with quality control, and usually at least one process engineer directly assigned to the plant.

The average wage of the nonsalaried laborers can be obtained from *Chemical Week's* annual plant site issue (see references for Chapter 2) or the U. S. Department of Labor.[1] This average wage does not include retirement benefits, social security, workman's compensation, company health insurance contributions, stock options, holidays and vacations, and other fringe benefits. These benefits amount to between 24% and 37% of a man's wage.[2,3] The average in 1972 was 35%. For salaried employees the average pay for supervisory and nonsupervisory engineers is given yearly in nearly every chemical engineering publication as a function of years after receiving the bachelor's degree. These must be increased by around 26% to cover fringe benefits. The number of employees is obtained from the manpower estimate.

The utilities are those provided by facilities outside the specific plant boundaries. These costs may be obtained from publications issued by the Federal Power Commission[4] or *Chemical Week*'s plant site issue. The amounts of each needed can be obtained from the energy balance and/or the unit ratio material balance.

The supplies may include anything from gaskets to toilet paper or from lubricating oil to instrument charts. They are items that are needed in operating the plant. These costs usually amount to around 15%-20% of the maintenance costs.

Maintenance costs are those involved in keeping the plant equipment in operating order. This cost involves equipment and labor. In Chapter 8 it was noted that the maintenance costs may be estimated as between 6.1% and 7.4% of the total sales or as approximately 5% of the cost of building the plant.

Waste treatment costs are those involved with disposing of by-products or off-grade products when these tasks are performed by facilities not directly connected with the plant. These costs are negotiated and vary widely. Sometimes long-term contracts with a guaranteed minimum are demanded. This is especially true if the contractor must build new facilities or expand his present operations in order to process the waste material. Some costs are given in Chapter 16.

Royalties are the costs paid to the owners of patents for using their inventions or processes. These are agreed upon in advance and usually amount to around $5/1,000 lb (500 kg) of product produced. There are also lump-sum royalties, where for a stated amount, which is paid only once, all rights to use the invention or process are given to the leasee.

The packaging costs are those for obtaining disposable containers needed for shipping the product to the customers. (See reference 5 for other details.) For products handled in nondisposable containers no separate packaging cost is listed.

### Depreciation

Depreciation is the means by which the capital cost involved in constructing the plant is prorated over its prospective life. This is discussed in detail in Chapter 11. A rough estimate is 9% of the cost of constructing the plant.

### Sales Costs

The sales costs are those involved with contacting the customer and convincing him to buy. Included are such items as warehouses throughout the country for quick distribution, sales offices, salaries and expenses, advertising, and technical service departments.

A chemical salesman is a highly trained man, and his job cannot be likened to a Fuller Brush man or an Avon lady. For instance, in 1968 the cost of an average chemical salesman's call was $49.71. The range was $5.00-$310.00 per call. The average chemical salesman made 20 calls per week and the average order was $6,000 per call.[6]

In 1968 the average selling costs were 10% of the sales dollar. There is, however, a large variance in this percentage. For basic bulk chemicals such as sulfuric acid, caustic, and soda ash the sales costs are less than 5% of the selling price, while for specialty chemicals having a low sales volume the sales costs may exceed 50% of the selling price.

Consumer products require expensive magazine advertisements and television commercials. This can amount to 30% or 40% of the selling price. If a new laundry product is introduced, a large amount of money must be spent to convince super-markets and/or mass distribution outlets to grant it shelf space. They must be convinced it will sell before they will put it on their shelves. They are only interested in items that will move. In 1968 Colgate introduced its new pre-soak, Axion®. In the first 9 months it spent over $3,000,000 for advertisements alone—$640,000 for newspaper ads, $865,000 for network television, $886,000 for spot local radio commercials, $800,000 for spot local television commercials, and $199,000 for magazine ads.[6] Airwick spent over $7,000,000 to introduce a new disinfectant-deodorant spray in 1973.[7]

### Research Costs

Research expenses are those associated with the administration and running of research projects. Research expenses average 3-4% of sales for the chemical industry. For the large pharmaceutical companies this is nearly 10%.[8]

### Taxes and Insurance

This category includes property and franchise taxes and all insurance costs. They depend on the value of the physical plant. They may be assumed to be between 2 and 3% of the cost of building the plant.

### General Administrative Costs

These are the overhead costs that it takes to operate a company. The item consists of all the administrative costs that cannot be assigned to a given project. It includes the expenses and salaries of the president, board of directors, treasurer, division managers, long-range planners and accountants. These costs average about 4% of the total sales.

# CAPITAL

A certain amount of money must be invested if any product is to be produced. This is referred to as capital. The capital is made up of the fixed capital needed to construct the plant and the working capital needed to operate it. The fixed capital is the cost of building and equipping the plant and all its peripheral buildings and operations. Chapter 9 was devoted to methods for estimating the fixed capital investment.

The working capital is made up of all items not included in the fixed capital. It consists of accounts payable, raw materials inventory, work in progress, and product and by-product inventories.

*Accounts payable* is similar to a charge account at a local department store. The buyer selects an item and says, "Charge it." He does not pay for it until sometime after he receives a billing near the end of the month. Meanwhile, the department store has paid its supplier but has not received payment from the customer. The store has money invested in the item that it cannot use elsewhere. In a chemical plant, the amount of capital tied up in accounts payable is the dollar value of the products or by-products that have been shipped to customers but for which no payment has been received. This can be calculated for each item by multiplying the sales price times the average number of days between the date of shipping the item and receiving payment times the production rate per day. Generally it takes 30-45 days after shipment before payment is received.

The maximum *raw materials inventory* is given by the scope. A rough guess for the average would be two-thirds of the maximum capacity. A more accurate average figure could be obtained by subtracting from the maximum capacity one-half the average size shipment. The working capital involved in raw materials can be obtained by multiplying the cost of each raw material by the average inventory. This assumes that the company pays its bills on receipt of the raw materials, something its customers do not do. Realistically, the company might subtract 15 days' storage from the average before performing the calculation.

*Work in process* includes all the materials that are sitting in intermediate storage, are currently being processed, or are needed to perform conversions. In the reforming step of a petroleum refinery a platinum catalyst is used. Its value for a large plant can exceed $1,000,000. For catalysts, ion exchange resins, solvents that are recycled, and other such items the number of pounds needed is multiplied by the purchase price to obtain the working capital. One way to determine the amount of materials in process is to determine the average time it takes to convert the raw material into the finished product and multiply it by the flow rate. Where large amounts of recycle are involved, this is difficult. Another way is to determine the average amount contained in all processing and intermediate storage vessels at any given time. This may range from a couple of hours' storage to a couple of days. Since all other items in this category are very imprecise, an estimate to within one day is adequate. The cost assigned to each pound in process should be the cost of the weighted average of the raw materials, plus half the conversion costs.

The *finished goods inventory* is the average amount of product that is ready for shipment to customers. The maximum storage for each product or by-product is given by the scope. The average is usually one-half to three-quarters of the total. The value of the finished goods inventory is obtained by multiplying the average amount of each product or by-product that is stored by its dollar value and then summing. Since it has not been sold, its value is merely the sum of the raw material and conversion costs.

## ELEMENTARY PROFITABILITY MEASURES

A company wants to make the largest profit it can from the money it invests. It wants to be able to compare the prospective earnings it may get from different ventures. Two of the simplest measures are the return on investment and the payout time.

### Return on Investment (R.O.I.)

The return on investment is the expected profit divided by the total capital invested. This is the percentage return that an investor may expect to eventually earn on his money. Since the federal corporate income tax rate is around 48% on all profits, it must be stated whether the profit is the before- or after-tax earnings.

It is the after-tax R.O.I. that the company or individual must compare with the earnings from savings accounts, capital bonds, and other projects to determine whether this is a good project in which to invest. A return on investment of at least 15% figured on after-tax earnings or 30% based on pretax earnings is usually expected. These numbers would be greatly increased if there were large risks involved.

Example 10-1 shows how to determine the minimum selling price for a product, assuming a 30% pretax R.O.I. If the product does not have an excellent chance of selling at that price or better, the project should be terminated.

### Example 10-1

The investment cost for a 120,000,000 lb/yr plant is estimated at $15,000,000. The working capital is $3,000,000. From the unit ratio material balance, the energy balance, and an estimation of the labor, the following costs per pound of product were determined:

| | | |
|---|---|---|
| Raw materials | 8¢/lb | (1) |
| Utilities | 1.2¢/lb | (2) |
| Labor | 1.5¢/lb | (3) |
| Packaging | none | (shipped in bulk quantities) |
| Royalties | none | (designed by company engineers) |
| Waste treatment charges | none | (pollution included in capital & operating costs) |

What must be the selling price if the return on the investment before taxes of the fully operating plant is to be 30%? Assume that the plant is to be fully depreciated in 12 years using a straight-line method.

$$\text{Depreciation per year} = \frac{\$15,000,000}{12} = \$1,250,000$$

$$\text{Depreciation per pound of product} = \frac{\$1,250,000}{120,000,000} = \$0.0104/\text{lb} = 1.04\cent/\text{lb} \qquad (4)$$

Let $X$ = the selling price per pound of the product

$$\text{Maintenance} \left(@ \ 6.5\% \text{ of fixed capital/year}\right) = \frac{(0.065)\,(\$15,000,000)}{(120,000,000 \ \text{lb/yr})} = \$0.00811/\text{lb}$$

$$= 0.811\cent/\text{lb} \qquad (5)$$

$$\text{Supplies} \ (@ \ 18\% \text{ of maintenance}) = \$0.00146/\text{lb} = 0.146\cent/\text{lb} \qquad (6)$$

$$\text{Administrative expense} \ (@ \ 4\% \text{ of sales}) = 0.04X \qquad (7)$$

$$\text{Research \& development} \ (@ \ 5\% \text{ of sales}) = 0.05X \qquad (8)$$

$$\text{Sales cost} \ (@ \ 10\% \text{ sales}) = 0.10X \qquad (9)$$

$$\text{Taxes \& insurance} \ (@ \ 3\% \text{ of fixed capital/yr}) = \frac{(0.03)\,(\$15,000,000)}{(120,000,000 \ \text{lb/yr})}$$

$$= \$.00375 = 0.375\cent/\text{lb} \qquad (10)$$

The total costs equal the sum of (1) through (10) = $12.7\cent + 0.19X$

A 30% return on the investment before taxes means a profit before taxes =

$$(0.30)\,(\$15,000,000 + 3,000,000) = \$5,400,000/\text{yr}$$

This is a profit per pound of $\dfrac{\$5,400,000/\text{year}}{120,000,000 \ \text{lb/yr}} = \$0.0450/\text{lb} = 4.5\cent/\text{lb}$

Selling price per pound = $X = 4.5 + 12.7 + 0.19X$

$$X = 21.2\cent/\text{lb}$$

The problem with the economic analysis as presented in Example 10-1 is that it considers the plant to be operating at full capacity (a mature plant). Often it takes a couple of years, after the plant begins producing, for the sales volume to equal the plant capacity. During this time the return on the investment is less than that calculated for the mature plant. This is shown in Example 10-2.

## Example 10-2

For Example 10-1 calculate the return on the investment if the plant is running at 60% of capacity and the selling price is 21.2¢/lb.
The raw material costs remain the same.

$$\text{Raw materials} = 8.0¢/\text{lb} \tag{1}$$

Utilities will increase slightly, since light and heat for the buildings are nearly independent of capacity. Assume this is negligible.

$$\text{Utilities} = 1.2¢/\text{lb} \tag{2}$$

The labor may be slightly more, since more shutdowns and startups are required. The indirect labor—laboratory personnel, bosses, engineers—does not vary much with capacity. Assume labor costs increase 20% per lb.

$$\text{Labor costs per lb} = 1.5 \times (1.2) = 1.8¢/\text{lb} \tag{3}$$

$$\text{Depreciation per lb} = \frac{1,250,000/\text{yr}}{120,000,000 \text{ lb/yr } (0.6)} = 1.735¢/\text{lb} \tag{4}$$

Assume that the maintenance costs are the same per pound but the total supplies budget is constant.

$$\text{Maintenance} = 0.811¢/\text{lb} \tag{5}$$

$$\text{Supplies} = (0.146)/0.6 = 0.243¢/\text{lb} \tag{6}$$

The administrative and R&D expenses are overhead costs that are not directly affected by the amount of product produced, even though for accounting purposes they are figured as a percentage of the gross sales. Assume these remain the same per pound of product.

$$\text{Administrative, R\&D expenses} = (0.09)(21.2¢/\text{lb}) = 1.9¢/\text{lb} \tag{7}$$

The sales costs are generally greater for plants not running at capacity than for mature plants. It is easier to keep a customer than it is to attract new business. Assume that the total sales costs remain the same, so the cost per pound increases.

$$\text{Sales cost} = (0.10)(21.2) \Big/ (0.6) = 3.54¢/\text{lb} \tag{8}$$

The property taxes and insurance are independent of how much product is produced.

$$\text{Property taxes and insurance costs} = 0.375/0.6 = 0.625¢/\text{lb} \tag{9}$$

The total costs equal the sum of (1) through (9) = 19.85¢/lb

The pretax profit is 21.2 – 19.85 = 1.35¢/lb.

The total pretax profit = $\dfrac{(\$0.0135)}{lb} \dfrac{(120{,}000{,}000\ lb)}{yr}(0.6) = \$972{,}000/yr$

Return on investment before taxes = $\dfrac{\$972{,}000}{18{,}000{,}000} \times 100 = 5.4\%$

This assumes the total capital remains constant.

Example 10-2 reillustrates the point made at the beginning of this chapter when the costs of operating a car were estimated. The more the plant (car) is used, the less it costs per pound (mile) to operate.

For the chemical plant the fixed charges, those that do not vary with the throughput, are depreciation, taxes, and insurance. The labor, utilities, supplies, and sales also cost more per pound of product if the plant is not running at full capacity, but their total cost is less than if the plant were running at full capacity.

The R.O.I. analysis fails to take into account any startup expenses or to consider how long it will take the plant to reach full capacity. It ignores the fact that research and development costs, marketing expenses, and engineering expenses, are all made at different points of time. It considers only the mature plant.

**Payout Time**

The payout time or payback period is the number of years from the time of startup it would take to recover all expenses involved in a project if all the pretax profits were used for this purpose.

In determining profits for this case, no depreciation is included in the expenses. It is not necessary to assume a mature plant or to ignore startup expenses when using this method. Example 10-3 shows how to calculate the payout period for a plant that does not reach full production until the fourth year of operation.

### Example 10-3

The total amount of money needed for research and development, engineering, marketing, and construction prior to the startup of a plant is $25,000,000. After the plant is completed it will take 3 years before the plant runs at full capacity. During these years the net proceeds before depreciation are estimated to be:

| Year | |
|------|------|
| 1 | $1,000,000 |
| 2 | 4,000,000 |
| 3 | 6,000,000 |
| 4 (onward) | 7,000,000/yr |

For the first three years the proceeds are $11,000,000. After that it is $7,000,000/year. The years for payout =

$$3 + \frac{\$25,000,000 - 11,000,000}{\$7,000,000} = 5 \text{ years}$$

Often, as in the case of the return on the investment, expenses not incurred directly in the design and construction of the plant are excluded when the payout period is calculated. If the only prestartup expense considered is the fixed capital investment, a payout time of 3-5 years is reasonable. A time longer than this is considered unacceptable.

Whenever an economic evaluation is made, past costs to develop the product process or markets should be ignored. These expenses have already been incurred. The object of an economic evaluation is to determine what is the best way to allocate a company's present and future resources. The stockholder wants to know how to obtain greatest profit at the smallest risk. To consider past expenditures would be equivalent to crying over spilt milk.

The problem with the payout period is that it does not consider the timing of the payments or the profits earned by the plant after the payout period is over. To illustrate the importance of the former, suppose a plant has the same prestartup expenses as the one in Example 10-3. Assume it has a profit of $5,000,000 per year for the first 5 years and from then on earns $7,000,000 per year. The payout time for this plant is 5 years, the same as for the plant in Example 10-3. The return on investment of the two mature plants is the same. Yet this proposed plant has a definite advantage over the one in Example 10-3. This is illustrated in the following example.

## Example 10-4

A company is considering which of two plants to build. Both plants cost $25,000,000. They will each earn pretax profits of $7,000,000 a year after 5 years of operation. The pretax profits (these do not include any depreciation expenses) of each plant are as follows:

| Year | Plant 1 | Plant 2 |
|------|---------|---------|
| 1 | $1,000,000 | $5,000,000 |
| 2 | $4,000,000 | $5,000,000 |
| 3 | $6,000,000 | $5,000,000 |
| 4 | $7,000,000 | $5,000,000 |
| 5 | $7,000,000 | $5,000,000 |

The company is borrowing money from a bank at 7% interest compounded annually. Determine which plant should be built and what are the pretax savings.

Suppose the profits are paid to the bank to reduce the loan at the end of each year. At the end of year 1, plant 2 would have earned \$4,000,000 more than plant 1. Its savings in interest payments for year 2 = (\$4,000,000) (0.07) = \$280,000. At the end of year 2, plant 2 has paid \$5,000,000 more to the bank than plant 1.

Savings in year 3 = (\$5,000,000) (0.07) = \$350,000

At the end of year 3, plant 2 has earned \$4,000,000 more than plant 1.

Savings in year 4 = (\$4,000,000) (0.07) = \$280,000

At the end of year 4, plant 2 has earned \$2,000,000 more than plant 1.

Savings in year 5 = (\$2,000,000) (0.07) = \$140,000

At the end of year 5 the plants are both paid off, and from there on the profits are the same.

Building plant 2 will result in a pretax saving of \$1,050,000. This is an amount that cannot be sneezed at.

The importance of profits after the payout period can be shown by making a small change in Example 10-4. Suppose that plant 2 made a profit in year 6 of \$5,000,000 instead of \$7,000,000 and everything else was the same. Then plant 1 would result in a pretax saving of \$950,000 in year 6 and would be preferable. Neither the payback period nor the return on investment method would indicate this.

**Proceeds per Dollar of Outlay**

A third method that is sometimes used to compare the economic value of prospective plants is the Proceeds per Dollar of Outlay. To obtain this indicator the total income (proceeds) over the life of the plant is divided by the total investment. Again, in determining expenses no depreciation charges are included. The plant having the highest value of the indicator is supposedly the best. The order in which the profits are received is not considered, nor is it important how long the plant must operate to make such a profit. In other words, this index cannot distinguish between the two plants in Example 10-4 if they both have the same prospective life. Nor could it distinguish between a plant having a 3% return on the investment that operated 100 years and a plant having a 30% return on the investment that ran only 10 years. The following related indicator was developed to correct for the latter deficiency.

**Annual Proceeds per Dollar of Outlay**

The Average Annual Proceeds per Dollar of Outlay is obtained by dividing the Proceeds per Dollar of Outlay by the number of years that it is estimated the plant will operate. Again, the plant having the highest indicator value is considered the best.

### Example 10-5

Using the last two indicators, compare the following two plants, each having a cash outlay of \$25,000,000. Plant 1 will earn a profit of \$10,000,000 for each of 4 years. Plant 2 will earn a profit of $2,000,000 for each of 25 years. (The expenses do not include depreciation charges.)

The Proceeds per Dollar of Outlay for:

$$\text{Plant } 1 = \frac{\$40,000,000}{\$25,000,000} = 1.6$$

$$\text{Plant } 2 = \frac{\$50,000,000}{\$25,000,000} = 2.0$$

The Average Annual Proceeds per Dollar of Outlay for:

$$\text{Plant } 1 = \frac{\$40,000,000}{\$25,000,000} \times \frac{1}{4} = 0.4$$

$$\text{Plant } 2 = \frac{\$50,000,000}{\$25,000,000} \times \frac{1}{25} = 0.08$$

The two indicators used in Example 10-5 give different answers. The Proceeds per Dollar of Outlay procedure is obviously wrong. It would take 12.5 years to pay off the investment for plant 1. If the $15,000, 000 pretax profit available at the end of year 4 were invested at 7% interest, the accumulated interest would have exceeded $10,800,000 before plant 1 was paid for. In other words, the total profit for plant 1 would exceed $25,800,000 before plant 2 showed a profit. This is more than the total profit made by plant 2 during its whole life. If the $15,000,000 profit of plant 1 were invested for 21 years at 7% compounded annually, the total interest would amount to $47,100,000 or the total earnings would be $62,100,000. If after plant 2 were paid off the profits were invested at 7% per year, the interest would amount to $8,900,000 or a total pretax profit of $33,900,000 25 years after the plant startup. This is $28,200,000 less than that earned in the same amount of time by plant 1. This shows that for this example the Annual Proceeds per Dollar of Outlay is a better economic indicator.

The Annual Proceeds per Dollar of Outlay, however, does not consider the order in which proceeds are received, and has an even worse failing in that it favors investments that have large initial proceeds and a short life.

### Example 10-6

Compare the following two plants using the Annual Proceeds per Dollar of Outlay. Plant 1 has a life of 2 years and plant 2 will last 5 years. The capital investment for each plant is $20,000,000. The proceeds (depreciation charges have not been included as an expense) are as follows:

|       | Proceeds for: | |
| :---: | :---: | :---: |
| Year  | Plant 1 | Plant 2 |
| 1     | $9,000,000 | $6,000,000 |
| 2     | $9,000,000 | $6,000,000 |
| 3     | 0 | $6,000,000 |
| 4     | 0 | $6,000,000 |
| 5     | 0 | $6,000,000 |

The Annual Proceeds per Dollar of Outlay for

$$\text{Plant 1} = \frac{(9,000,000) \times 2}{(20,000,000) \times 2} = 0.45$$

$$\text{Plant 2} = \frac{(6,000,000) \times 5}{(20,000,000) \times 5} = 0.30$$

This indicates that plant 1 would be a better economic choice than plant 2, which is wrong. Plant 1 never earns enough to recover the capital expended on it, whereas plant 2 returns a $10,000,000 profit.

### Average Income on Initial Cost

Still another economic indicator is the Average Income on Initial Cost. This is the average income divided by the fixed capital costs. It is usually expressed as a percentage. The plant having the largest indicator is considered the best. In calculating the income, the depreciation of the plant is considered as a cost. This is its major difference from the Annual Proceeds per Dollar of Outlay. This difference would prevent it from indicating an investment as best that does not recover its own capital investment, as in Example 10-6. However, it still has the same major failings of that indicator. It favors investments that have a high income for a short period and does not consider the timing of the proceeds. It does not show that plant 2 is better than plant 1 in Example 10-4.

All of the indicators mentioned so far have another major disadvantage. They can lead us to erroneous conclusions when a project having a small capital expenditure and a high return is compared with an alternative that has a high capital expenditure and only a good return. This is illustrated in Example 10-7.

### Example 10-7

Compare the following alternatives that have been proposed to improve an operating plant. Only one can be adopted.

| Alternative | 1 | 2 |
|---|---|---|
| Cash outlay | $3,000,000 | $300,000 |
| Fixed capital | $2,500,000 | $250,000 |
| Working capital | $500,000 | $50,000 |
| Cash proceeds | $1,000,000 for 5 years | $200,000 for 5 years |
| (depreciation included as expense) | | |
| Depreciation | $500,000/yr | $50,000/yr |

|  | Plant | |
|---|---|---|
|  | 1 | 2 |
| Pretax R.O.I. | | |

Pretax R.O.I.

$$\frac{\$1,000,000}{3,000,000} \times 100 = 33\% \qquad \frac{200,000}{300,000} \times 100 = 66\%$$

Payout period

$$\frac{3,000,000}{1,500,000} = 2 \text{ yr} \qquad \frac{300,000}{250,000} = 1.2 \text{ yr}$$

Proceeds per dollar

$$\frac{\$7,500,000}{3,000,000} = 2.5 \qquad \frac{1,250,000}{300,000} = 4.17$$

Average annual proceeds per dollar of outlay

$$\frac{2.5}{5} = 0.5 \qquad \frac{4.17}{5} = 0.83$$

Average income on initial cost $\frac{1,000,000}{3,000,000}(100) = 33.3 \quad \frac{200,000}{300,000}(100) = 66.6$

Dollars earned in 5 years     $5,000,000     $1,000,000

The five economic indicators suggest that plant 2 is best. However, the greatest proceeds and hence the best economic choice is plant 1. (Interest was not considered here, but if an interest rate of less than 14% is used plant 1 will still come out best. The reader should prove this to himself.)

The indicators used in Example 10-7 should not be used for comparing alternatives. However, they are useful in evaluating a research project when the goal of the study is to determine whether the project should be continued. They can also be used to determine whether a more complicated evaluation is warranted. Their major advantage is that they are easy to perform.

## TIME VALUE OF MONEY

The major problems with the indicators that have been presented are that they fail to take into account the timing of payments and receipts and the magnitude of the profits. They also can give incorrect results when comparing projects having different time durations. To resolve these problems a different economic indicator will be developed. Before this can be done the reader must have a quantitative understanding of the time value of money. To introduce this concept, consider the following situation.

Your roommate is going to come into a sizeable inheritance at age 25. He is now 18, and ever since he saw a James Bond movie he has felt he must have an Aston Martin. (This automobile costs $13,500 with accessories.) He currently has only $100 a month spending money and he needs the car immediately. You have access to the required money because of the death of a wealthy maiden aunt. He is willing to sign over a portion of his inheritance. What would be a reasonable amount to request from him when he reaches 25?

In 1969 this money could be placed in a savings and loan association where it would earn 5.25% interest compounded annually, with the principal guaranteed by the federal government. There is no risk in this investment. At the end of the first year the interest would be

$$13,500 \ (.0525) = \$709$$

The total interest plus principal is then

$$13,500 + 709 = \$14,209$$

The interest earned on this in the next year is

$$(14,209) \ (.0525) = \$746$$

At the beginning of the third year the accumulated amount is

$$14,209 + 746 = \$14,954$$

If this continued for 7 years the principal plus interest would amount to $17,900. This would obviously be the minimum amount to request.

The maximum amount your roommate would pay would be determined by his alternate sources of financing. If he could get a loan from a bank it would be at an interest rate of 8% or more. A loan from a finance agency would cost more. At 8% interest compounded annually $13,500 would be worth $23,100 after 7 years. To request that he pay you $23,100 on his twenty-fifth birthday in return for giving him $13,500 now would certainly be fair—especially if he let you use the car and if an ironclad lien could be placed on his inheritance.

Before closing the deal your alternatives should also be considered. Some mutual funds have at times earned 12% annually. These are based on a portfolio of stocks and there is no guarantee that there will not be a loss instead of a profit. If $13,500 is invested at 12% per annum, after 7 years the value of the principal plus interest is $29,800. To request this amount from your roommate in return for financing his car would still be reasonable. Other alternatives would be municipal or government bonds, which return 6-8% annually but are nontaxable; industrial bonds, which pay at least 9% and speculative stocks.

In a situation of this sort your roommate's personality, his reliability, the size of his inheritance, whether you were close friends, and other factors would also need to be considered before you could decide whether to loan him the money at all and if you did, at what rate.

## COMPOUND INTEREST

Compound interest was used in the above case. Compound interest means that the interest earned is figured not only on the principal but also on any previously earned interest. This is equivalent to increasing the principal by the amount of interest after each interest period. See Table 10-3 for the development of the following simple formula for compound interest.

Table 10-3

Compound Interest

| Period | Principal & Earnings at Beginning of Interest Period* | Interest at End of Interest Period |
|---|---|---|
| 1 | $P$ | $Pi$ |
| 2 | $(P + Pi) = P(1 + i)$ | $P(1 + i)i$ |
| 3 | $P(1 + i) + P(1 + i)i = P(1 + i)(1 + i) = P(1 + i)^2$ | $P(1 + i)^2 i$ |
| 4 | $P(1 + i)^2 + P(1 + i)^2 i = P(1 + i)^2(1 + i) = P(1 + i)^3$ | $P(1 + i)^3 i$ |
| 5 | $P(1 + i)^3 + P(1 + i)^3 i = P(1 + i)^3(1 + i) = P(1 + i)^4$ | $P(1 + i)^4 i$ |
| n | $P(1 + i)^{n-1}$ | $P(1 + i)^{n-1} i$ |
| n + 1 | $P(1 + i)^n$ | $P(1 + i)^n i$ |
| n + 2 | $P(1 + i)^{n+1}$ | $P(1 + i)^{n+1} i$ |

\* This is also the principal & earnings at the end of the previous period. Example: At the end of 9 years the principal & interest equal $P(1 + i)^9$

P = principal

i = interest rate per period

$$S = P(1 + i)^n$$
$S$ = principal plus earnings at the end of the $n^{\text{th}}$ period
$P$ = principal invested at the beginning of period one at
$i$ = interest rate per period

Table 10-4 and Table 10-5 give $(1+i)^n$ for various periods and interest rates.

Table 10-4

Compound Interest Factors

$(1 + i)^n$

| Period n | | | | | | | Interest Rate per Period, i | | | | | | |
|---|---|---|---|---|---|---|---|---|---|---|---|---|---|
| | 4 | 5 | 6 | 7 | 8 | 9 | 10 | 11 | 12 | 15 | 20 | 25 | 30 |
| 1 | 1.040 | 1.050 | 1.060 | 1.070 | 1.080 | 1.090 | 1.100 | 1.110 | 1.120 | 1.150 | 1.200 | 1.250 | 1.300 |
| 2 | 1.081 | 1.102 | 1.123 | 1.144 | 1.166 | 1.188 | 1.209 | 1.232 | 1.254 | 1.322 | 1.439 | 1.562 | 1.690 |
| 3 | 1.124 | 1.157 | 1.191 | 1.225 | 1.259 | 1.295 | 1.330 | 1.367 | 1.404 | 1.520 | 1.727 | 1.953 | 2.197 |
| 4 | 1.169 | 1.215 | 1.262 | 1.310 | 1.360 | 1.411 | 1.464 | 1.518 | 1.573 | 1.749 | 2.073 | 2.441 | 2.856 |
| 5 | 1.216 | 1.276 | 1.338 | 1.402 | 1.469 | 1.538 | 1.610 | 1.685 | 1.762 | 2.011 | 2.488 | 3.051 | 3.712 |
| 6 | 1.265 | 1.340 | 1.418 | 1.500 | 1.586 | 1.677 | 1.771 | 1.870 | 1.973 | 2.313 | 2.985 | 3.814 | 4.826 |
| 7 | 1.315 | 1.407 | 1.503 | 1.605 | 1.713 | 1.828 | 1.948 | 2.076 | 2.210 | 2.660 | 3.583 | 4.768 | 6.274 |
| 8 | 1.368 | 1.477 | 1.593 | 1.718 | 1.850 | 1.992 | 2.143 | 2.304 | 2.475 | 3.059 | 4.299 | 5.960 | 8.157 |
| 9 | 1.423 | 1.551 | 1.689 | 1.838 | 1.999 | 2.171 | 2.357 | 2.558 | 2.773 | 3.517 | 5.159 | 7.450 | 10.604 |
| 10 | 1.480 | 1.628 | 1.790 | 1.967 | 2.158 | 2.367 | 2.593 | 2.839 | 3.105 | 4.045 | 6.191 | 9.313 | 13.785 |
| 11 | 1.539 | 1.710 | 1.898 | 2.104 | 2.331 | 2.580 | 2.853 | 3.151 | 3.478 | 4.652 | 7.430 | 11.641 | 17.921 |
| 12 | 1.601 | 1.795 | 2.012 | 2.252 | 2.518 | 2.812 | 3.138 | 3.498 | 3.895 | 5.350 | 8.916 | 14.551 | 23.298 |
| 13 | 1.665 | 1.885 | 2.132 | 2.409 | 2.719 | 3.065 | 3.452 | 3.883 | 4.363 | 6.152 | 10.699 | 18.189 | 30.287 |
| 14 | 1.731 | 1.979 | 2.260 | 2.578 | 2.937 | 3.341 | 3.797 | 4.310 | 4.887 | 7.075 | 12.839 | 22.737 | 39.373 |
| 15 | 1.800 | 2.078 | 2.396 | 2.759 | 3.172 | 3.642 | 4.177 | 4.784 | 5.473 | 8.137 | 15.407 | 28.421 | 51.185 |
| 16 | 1.872 | 2.182 | 2.540 | 2.952 | 3.425 | 3.970 | 4.594 | 5.310 | 6.130 | 9.357 | 18.488 | 35.527 | 66.541 |
| 17 | 1.947 | 2.292 | 2.692 | 3.158 | 3.700 | 4.327 | 5.054 | 5.895 | 6.866 | 10.761 | 22.186 | 44.408 | 86.504 |
| 18 | 2.025 | 2.406 | 2.854 | 3.379 | 3.996 | 4.717 | 5.559 | 6.543 | 7.689 | 12.375 | 26.623 | 55.511 | 112.455 |
| 19 | 2.106 | 2.526 | 3.025 | 3.616 | 4.315 | 5.141 | 6.115 | 7.263 | 8.612 | 14.231 | 31.947 | 69.388 | 146.192 |
| 20 | 2.191 | 2.653 | 3.207 | 3.869 | 4.660 | 5.604 | 6.727 | 8.062 | 9.646 | 16.366 | 38.337 | 86.736 | 190.049 |
| 21 | 2.278 | 2.785 | 3.399 | 4.140 | 5.033 | 6.108 | 7.400 | 8.949 | 10.803 | 18.821 | 46.005 | 108.420 | 247.064 |
| 22 | 2.369 | 2.925 | 3.603 | 4.430 | 5.436 | 6.658 | 8.140 | 9.933 | 12.100 | 21.644 | 55.206 | 135.525 | 321.183 |
| 23 | 2.464 | 3.071 | 3.819 | 4.740 | 5.871 | 7.257 | 8.954 | 11.026 | 13.552 | 24.891 | 66.247 | 169.406 | 417.539 |
| 24 | 2.563 | 3.225 | 4.048 | 5.072 | 6.341 | 7.911 | 9.849 | 12.239 | 15.178 | 28.625 | 79.496 | 211.758 | 542.800 |
| 25 | 2.665 | 3.386 | 4.291 | 5.427 | 6.848 | 8.623 | 10.834 | 13.585 | 17.000 | 32.918 | 95.396 | 264.697 | 705.641 |
| 26 | 2.772 | 3.555 | 4.549 | 5.807 | 7.396 | 9.399 | 11.918 | 15.079 | 19.040 | 37.856 | 114.475 | 330.8 | 917.3 |
| 27 | 2.883 | 3.733 | 4.822 | 6.213 | 7.988 | 10.245 | 13.109 | 16.738 | 21.324 | 43.535 | 137.370 | 413.5 | 1192.5 |
| 28 | 2.998 | 3.920 | 5.111 | 6.648 | 8.627 | 11.167 | 14.420 | 18.579 | 23.883 | 50.065 | 164.844 | 516.9 | 1550.2 |
| 29 | 3.118 | 4.116 | 5.418 | 7.114 | 9.317 | 12.172 | 15.863 | 20.623 | 26.749 | 57.575 | 197.813 | 646.2 | 2015.3 |
| 30 | 3.243 | 4.321 | 5.743 | 7.612 | 10.062 | 13.267 | 17.449 | 22.892 | 29.959 | 66.211 | 237.376 | 807.7 | 2619.9 |
| 31 | 3.373 | 4.538 | 6.088 | 8.145 | 10.867 | 14.461 | 19.194 | 25.410 | 33.555 | 76.143 | 284.851 | 1009.7 | 3405.9 |
| 32 | 3.508 | 4.764 | 6.453 | 8.715 | 11.737 | 15.763 | 21.113 | 28.205 | 37.581 | 87.565 | 341.821 | 1262.1 | 4427.7 |
| 33 | 3.648 | 5.003 | 6.840 | 9.325 | 12.676 | 17.182 | 23.225 | 31.308 | 42.091 | 100.699 | 410.186 | 1577.7 | 5756.1 |
| 34 | 3.794 | 5.253 | 7.251 | 9.978 | 13.690 | 18.728 | 25.547 | 34.752 | 47.142 | 115.804 | 492.223 | 1972.1 | 7482.9· |
| 35 | 3.946 | 5.516 | 7.686 | 10.676 | 14.785 | 20.413 | 28.102 | 38.574 | 52.799 | 133.175 | 590.668 | 2465.1 | 9727.8 |
| 36 | 4.103 | 5.791 | 8.147 | 11.423 | 15.968 | 22.251 | 30.912 | 42.818 | 59.135 | 153.151 | 708.801 | 3081.4 | 12646.2 |
| 37 | 4.268 | 6.081 | 8.636 | 12.223 | 17.245 | 24.253 | 34.003 | 47.528 | 66.231 | 176.124 | 850.562 | 3851.8 | 16440.0 |
| 38 | 4.438 | 6.385 | 9.154 | 13.079 | 18.625 | 26.436 | 37.404 | 52.756 | 74.179 | 202.543 | 1020.674 | 4814.8 | 21372.1 |
| 39 | 4.616 | 6.704 | 9.703 | 13.994 | 20.115 | 28.815 | 41.144 | 58.559 | 83.081 | 232.924 | 1224.809 | 6018.5 | 27783.7 |
| 40 | 4.801 | 7.039 | 10.285 | 14.974 | 21.724 | 31.409 | 45.259 | 65.000 | 93.050 | 267.863 | 1469.771 | 7523.1 | 36118.8 |
| 41 | 4.993 | 7.391 | 10.902 | 16.022 | 23.462 | 34.236 | 49.785 | 72.150 | 104.217 | 308.043 | 1763.725 | 9403.9 | 46954.5 |
| 42 | 5.192 | 7.761 | 11.557 | 17.144 | 25.339 | 37.317 | 54.763 | 80.087 | 116.723 | 354.249 | 2116.470 | 11754.9 | 61040.8 |
| 43 | 5.400 | 8.149 | 12.250 | 18.344 | 27.366 | 40.676 | 60.240 | 88.897 | 130.729 | 407.386 | 2539.765 | 14693.6 | 79353.1 |
| 44 | 5.616 | 8.557 | 12.985 | 19.628 | 29.555 | 44.336 | 66.264 | 98.675 | 146.417 | 468.495 | 3047.718 | 18367.0 | 103159.0 |
| 45 | 5.841 | 8.985 | 13.764 | 21.002 | 31.920 | 48.327 | 72.890 | 109.530 | 163.987 | 538.769 | 3657.261 | 22958.8 | 134106.8 |
| 46 | 6.074 | 9.434 | 14.590 | 22.472 | 34.474 | 52.676 | 80.179 | 121.578 | 183.666 | 619.584 | 4388.714 | 28698.5 | 174338.8 |
| 47 | 6.317 | 9.905 | 15.465 | 24.045 | 37.232 | 57.417 | 88.197 | 134.952 | 205.706 | 712.522 | 5266.457 | 35873.2 | 226640.5 |
| 48 | 6.570 | 10.401 | 16.393 | 25.728 | 40.210 | 62.585 | 97.017 | 149.796 | 230.390 | 819.400 | 6319.748 | 44841.5 | 294632.6 |
| 49 | 6.833 | 10.921 | 17.377 | 27.529 | 43.427 | 68.217 | 106.718 | 166.274 | 258.037 | 942.310 | 7583.698 | 56051.9 | 383022.4 |
| 50 | 7.106 | 11.467 | 18.420 | 29.457 | 46.901 | 74.357 | 117.390 | 184.564 | 289.002 | 1083.657 | 9100.437 | 70064.9 | 497929.2 |

Table 10-5

Nominal Interest Rates Compounded Monthly

$$\left(1 + \frac{i}{12}\right)^n$$

Nominal Interest Rate, i

| Period, n | 4 | 5 | 6 | 7 | 8 | 9 | 10 | 11 | 12 | 15 | 20 | 25 | 30 |
|---|---|---|---|---|---|---|---|---|---|---|---|---|---|
| 12 | 1.0407 | 1.0511 | 1.0616 | 1.0722 | 1.0829 | 1.0938 | 1.1047 | 1.1157 | 1.1268 | 1.1607 | 1.2193 | 1.2807 | 1.3448 |
| 24 | 1.0831 | 1.1049 | 1.1271 | 1.1498 | 1.1728 | 1.1964 | 1.2203 | 1.2448 | 1.2697 | 1.3473 | 1.4869 | 1.6402 | 1.8087 |
| 36 | 1.1272 | 1.1614 | 1.1966 | 1.2329 | 1.2702 | 1.3086 | 1.3481 | 1.3888 | 1.4307 | 1.5639 | 1.8131 | 2.1007 | 2.4325 |
| 48 | 1.1731 | 1.2208 | 1.2704 | 1.3220 | 1.3756 | 1.4314 | 1.4893 | 1.5495 | 1.6122 | 1.8153 | 2.2109 | 2.6904 | 3.2714 |
| 60 | 1.2209 | 1.2833 | 1.3488 | 1.4176 | 1.4898 | 1.5656 | 1.6453 | 1.7289 | 1.8166 | 2.1071 | 2.6959 | 3.4458 | 4.3997 |
| 72 | 1.2707 | 1.3490 | 1.4320 | 1.5201 | 1.6135 | 1.7125 | 1.8175 | 1.9289 | 2.0470 | 2.4459 | 3.2874 | 4.4131 | 5.9172 |
| 84 | 1.3225 | 1.4180 | 1.5203 | 1.6299 | 1.7474 | 1.8732 | 2.0079 | 2.1522 | 2.3067 | 2.8391 | 4.0086 | 5.6520 | 7.9580 |
| 96 | 1.3763 | 1.4905 | 1.6141 | 1.7478 | 1.8924 | 2.0489 | 2.2181 | 2.4012 | 2.5992 | 3.2955 | 4.8881 | 7.2387 | 10.7026 |
| 108 | 1.4324 | 1.5668 | 1.7136 | 1.8741 | 2.0495 | 2.2411 | 2.4504 | 2.6791 | 2.9289 | 3.8252 | 5.9605 | 9.2709 | 14.3938 |
| 120 | 1.4908 | 1.6470 | 1.8193 | 2.0096 | 2.2196 | 2.4513 | 2.7070 | 2.9891 | 3.3003 | 4.4402 | 7.2682 | 11.8735 | 19.3581 |
| 132 | 1.5515 | 1.7312 | 1.9316 | 2.1549 | 2.4038 | 2.6813 | 2.9905 | 3.3350 | 3.7189 | 5.1539 | 8.8628 | 15.2068 | 26.0345 |
| 144 | 1.6147 | 1.8198 | 2.0507 | 2.3107 | 2.6033 | 2.9328 | 3.3036 | 3.7209 | 4.1906 | 5.9825 | 10.8072 | 19.4758 | 35.0135 |
| 156 | 1.6805 | 1.9129 | 2.1772 | 2.4777 | 2.8194 | 3.2079 | 3.6495 | 4.1515 | 4.7220 | 6.9442 | 13.1782 | 24.9433 | 47.0893 |
| 168 | 1.7490 | 2.0108 | 2.3115 | 2.6568 | 3.0534 | 3.5088 | 4.0317 | 4.6319 | 5.3209 | 8.0605 | 16.0694 | 31.9457 | 63.3299 |
| 180 | 1.8203 | 2.1137 | 2.4540 | 2.8489 | 3.3069 | 3.8380 | 4.4539 | 5.1679 | 5.9958 | 9.3563 | 19.5949 | 40.9139 | 85.1717 |
| 192 | 1.8944 | 2.2218 | 2.6054 | 3.0548 | 3.5813 | 4.1980 | 4.9203 | 5.7660 | 6.7562 | 10.8604 | 23.8939 | 52.3998 | 114.5465 |
| 204 | 1.9716 | 2.3355 | 2.7661 | 3.2757 | 3.8786 | 4.5918 | 5.4355 | 6.4332 | 7.6130 | 12.6062 | 29.1360 | 67.1101 | 154.0523 |
| 216 | 2.0519 | 2.4550 | 2.9367 | 3.5125 | 4.2005 | 5.0226 | 6.0046 | 7.1777 | 8.5786 | 14.6327 | 35.5282 | 85.9500 | 207.1833 |
| 228 | 2.1355 | 2.5806 | 3.1178 | 3.7664 | 4.5492 | 5.4937 | 6.6334 | 8.0083 | 9.6665 | 16.9850 | 43.3228 | 110.0789 | 278.6385 |
| 240 | 2.2225 | 2.7126 | 3.3102 | 4.0387 | 4.9268 | 6.0091 | 7.3280 | 8.9350 | 10.8925 | 19.7154 | 52.8275 | 140.9815 | 374.7378 |

297

There is often some confusion in the use of the term *interest rate*. For instance, some banks advertise a given interest rate compounded daily. What this means is that the actual interest rate is the rate given divided by 365 and the period is one day. This is called the *nominal interest rate*, the amount that would be earned per year if the interest were not compounded:

$$S = P + n(Pi) = P(1 + i \times n) = P(1 + i_N)$$
$i$ = interest rate per period
$i_N$ = nominal interest rate
$n$ = number of periods per year

The federal government requires that the nominal interest rate be stated by each lender. Example 10-8 shows how to determine an equivalent yearly interest rate when the period of compounding is less than a year.

### Example 10-8

A bank advertises that it pays 5.25% interest, compounded daily, on savings certificates. This is equivalent to what rate of interest compounded annually?

$$S = P(1 + i)^n = P(1 + i_A)$$
$i_A$ = rate of interest compounded annually
$$i_A = (1 + i)^n - 1 = \left(1 + \frac{5.25}{365}\right)^{365} - 1 = 5.39\%$$

Even when the interest terms are clearly stated, it often takes some calculations to determine which proposition is best. The next example illustrates the importance of the timing of interest payments.

### Example 10-9

A man wishes to borrow $2,400 so he can buy a new car. He can get a bank loan for which he would be required to pay 7% interest on the initial loan. The payments would be in equal monthly installments over a period of one year. He could also get a loan from the credit union where he works. The terms are that each month he would pay $200 to reduce the capital borrowed and 1% interest on the unpaid balance. What is the total interest paid for each loan?
For the bank loan:
Interest = $Pi$ = $2,400 (0.07) = $168

For the credit union loan:
At the end of the first month the interest = $2,400 (0.01) = $24
At the end of the second month the interest = ($2,400 − 200) (0.01) = $22

At the end of the third month the interest = ($2,200 − 200) (0.01) = $20 etc.
Interest = $24 + $22 + $20 + $18 + $\cdots$ + $4 + $2 = $156

Example 10-9 shows why banks like short-term loans that are paid in equal installments. The actual rate the bank is charging is almost double the stated rate. When closing costs (a charge made to cover the cost of administering the loan) are added to the interest charges, the cost to the borrower is often greater than double the stated rate.

## Annuities

An annuity is a series of equal periodic payments that last for a given length of time. This is the usual way an individual pays for the capital necessary to purchase a new home, or buys life insurance.

When an individual buys a house he generally cannot afford to pay back a certain percentage plus a rate of interest on the unpaid balance each month. Even when he can, he would prefer to pay the loan back in equal installments each month. A formula is developed in Table 10-6 for determining the value of the payment. The result is:

$$X = \frac{P(1+i)^n i}{(1+i)^n - 1} = \frac{P}{\frac{[1-(1+i)^{-n}]}{i}}$$

where  $P$ = principal borrowed
$X$ = payment per time period
$n$ = number of periods
$i$ = interest rate per period

Some values of the denominator are given in Tables 10-7 and 10-8. These are used in Example 10-10 to determine the yearly and monthly payments for a mortgage.

## Example 10-10

Calculate the monthly and yearly payments for obtaining a loan of $30,000 that is to be fully repaid in 20 years. Assume an interest rate of 8%.

For yearly payments:

$$X = \frac{\$30,000}{\frac{1-(1.08)^{-20}}{(0.08)}} = \$3,055.50 \text{ per year}$$

# Table 10-6

## Installment Buying
### (Annuities)

| Period | Amount Owed at Beginning of Period | Interest Due at End of Period | Payment at End of Period |
|---|---|---|---|
| 1 | $P$ | $Pi$ | $X$ |
| 2 | $P + Pi - X = P(1+i) - X$ | $[P(1+i) - X]i$ | $X$ |
| 3 | $P(1+i) - X + [P(1+i) - X]i - X = P(1+i)(1+i) - 2X - Xi$ | $[P(1+i)^2 - 2X - Xi]i$ | $X$ |
| 4 | $[P(1+i)^2 - 2X - Xi] + [P(1+i)^2 - 2X - Xi]i - X =$ <br> $P(1+i)^2(1+i) - 3X - 3Xi - Xi^2$ | $[P(1+i)^3 - 3X - 3Xi - xi^2]i$ | $X$ |
| 5 | $[P(1+i)^3 - 3Xi - 3Xi^2] + [P(1+i)^3 - 3X - 3Xi - Xi^2]i - X$ <br> $= P(1+i)^3(1+i) - 4X - 6Xi - 4Xi^2 - Xi^3$ | $[P(1+i)^4 - 4X - 6Xi - 4Xi^2 - Xi^3]i$ | $X$ |
| . | . | . | . |
| . | . | . | . |
| n | $P(1+i)^{n-1} - \dfrac{X(1+i)^{n-1}}{i} + \dfrac{X}{i}$ | $\left[P(1+i)^{n-1} - \dfrac{X(1+i)^{n-1}}{i} + \dfrac{X}{i}\right]i$ | $X$ |
| n+1* | $\left[P(1+i)^{n-1} - \dfrac{X(1+i)^{n-1}}{i} + \dfrac{X}{i}\right](1+i) - X$ <br> $= P(1+i)^n - \dfrac{X(1+i)^n}{i} + \dfrac{X}{i}$ | | |

\* After $n$ years or at the beginning of year $n + 1$ nothing is owed.  Therefore $P(1 + i)^n - [X(1 + i)^n/i] + (X/i) = 0$ or $X = [P(1 + i)^n i]/[(1 + i)^n - 1]$ ,

where $P$ = amount borrowed, $i$ = interest rate period, $X$ = payment made at the end of each period.

Example: if $20,000 is borrowed on a 25-year loan at 8% per year the payment per year is

$$X = \frac{P(1+i)^n i}{(1+i)^n - 1} = \frac{20,000(1 + 0.08)^{25}\,(0.08)}{(1 + 0.08)^{25} - 1} = \$1,875.$$

300

Table 10-7

Present Value of an Annuity (Yearly Payments)

$$\left[ 1 - (1 + i)^{-n} \right] / i$$

| Period *n* | Interest Rate per Period, i | | | | | | | | | | | | |
|---|---|---|---|---|---|---|---|---|---|---|---|---|---|
| | 4 | 5 | 6 | 7 | 8 | 9 | 10 | 11 | 12 | 15 | 20 | 25 | 30 |
| 1 | 0.9615 | 0.9523 | 0.9433 | 0.9345 | 0.9259 | 0.9174 | 0.9090 | 0.9009 | 0.8928 | 0.8695 | 0.8333 | 0.8000 | 0.7692 |
| 2 | 1.8860 | 1.8594 | 1.8333 | 1.8080 | 1.7832 | 1.7591 | 1.7355 | 1.7125 | 1.6900 | 1.6257 | 1.5277 | 1.4400 | 1.3609 |
| 3 | 2.7750 | 2.7232 | 2.6730 | 2.6243 | 2.5770 | 2.5312 | 2.4868 | 2.4437 | 2.4018 | 2.2832 | 2.1064 | 1.9520 | 1.8161 |
| 4 | 3.6298 | 3.5459 | 3.4651 | 3.3872 | 3.3121 | 3.2397 | 3.1698 | 3.1024 | 3.0373 | 2.8549 | 2.5887 | 2.3616 | 2.1662 |
| 5 | 4.4518 | 4.3294 | 4.2123 | 4.1001 | 3.9927 | 3.8896 | 3.7907 | 3.6958 | 3.6047 | 3.3521 | 2.9906 | 2.6892 | 2.4355 |
| 6 | 5.2421 | 5.0756 | 4.9173 | 4.7665 | 4.6228 | 4.4859 | 4.3552 | 4.2305 | 4.1114 | 3.7844 | 3.3255 | 2.9514 | 2.6427 |
| 7 | 6.0020 | 5.7863 | 5.5823 | 5.3892 | 5.2063 | 5.0329 | 4.8684 | 4.7121 | 4.5637 | 4.1604 | 3.6045 | 3.1611 | 2.8021 |
| 8 | 6.7327 | 6.4632 | 6.2097 | 5.9712 | 5.7466 | 5.5348 | 5.3349 | 5.1461 | 4.9676 | 4.4873 | 3.8371 | 3.3289 | 2.9247 |
| 9 | 7.4353 | 7.1078 | 6.8016 | 6.5152 | 6.2468 | 5.9952 | 5.7590 | 5.5370 | 5.3282 | 4.7715 | 4.0309 | 3.4631 | 3.0190 |
| 10 | 8.1108 | 7.7217 | 7.3600 | 7.0235 | 6.7100 | 6.4176 | 6.1445 | 5.8892 | 5.6502 | 5.0187 | 4.1924 | 3.5705 | 3.0915 |
| 11 | 8.7604 | 8.3064 | 7.8868 | 7.4986 | 7.1389 | 6.8051 | 6.4950 | 6.2065 | 5.9376 | 5.2337 | 4.3270 | 3.6564 | 3.1473 |
| 12 | 9.3850 | 8.8632 | 8.3838 | 7.9426 | 7.5360 | 7.1607 | 6.8136 | 6.4923 | 6.1943 | 5.4206 | 4.4392 | 3.7251 | 3.1902 |
| 13 | 9.9856 | 9.3935 | 8.8526 | 8.3576 | 7.9037 | 7.4869 | 7.1033 | 6.7498 | 6.4235 | 5.5831 | 4.5326 | 3.7800 | 3.2232 |
| 14 | 10.5631 | 9.8986 | 9.2949 | 8.7454 | 8.2442 | 7.7861 | 7.3666 | 6.9818 | 6.6281 | 5.7244 | 4.6105 | 3.8240 | 3.2486 |
| 15 | 11.1183 | 10.3796 | 9.7122 | 9.1079 | 8.5594 | 8.0606 | 7.6060 | 7.1908 | 6.8108 | 5.8473 | 4.6754 | 3.8592 | 3.2682 |
| 16 | 11.6522 | 10.8377 | 10.1058 | 9.4466 | 8.8513 | 8.3125 | 7.8237 | 7.3791 | 6.9739 | 5.9542 | 4.7295 | 3.8874 | 3.2832 |
| 17 | 12.1656 | 11.2740 | 10.4772 | 9.7632 | 9.1216 | 8.5436 | 8.0215 | 7.5487 | 7.1196 | 6.0471 | 4.7746 | 3.9099 | 3.2947 |
| 18 | 12.6592 | 11.6895 | 10.8276 | 10.0590 | 9.3718 | 8.7556 | 8.2014 | 7.7016 | 7.2496 | 6.1279 | 4.8121 | 3.9279 | 3.3036 |
| 19 | 13.1339 | 12.0853 | 11.1581 | 10.3355 | 9.6035 | 8.9501 | 8.3649 | 7.8392 | 7.3657 | 6.1982 | 4.8434 | 3.9423 | 3.3105 |
| 20 | 13.5903 | 12.4622 | 11.4699 | 10.5940 | 9.8181 | 9.1285 | 8.5135 | 7.9633 | 7.4694 | 6.2593 | 4.8695 | 3.9538 | 3.3157 |
| 21 | 14.0291 | 12.8211 | 11.7640 | 10.8355 | 10.0168 | 9.2922 | 8.6486 | 8.0750 | 7.5620 | 6.3124 | 4.8913 | 3.9631 | 3.3198 |
| 22 | 14.4511 | 13.1630 | 12.0415 | 11.0612 | 10.2007 | 9.4424 | 8.7715 | 8.1757 | 7.6446 | 6.3586 | 4.9094 | 3.9704 | 3.3229 |
| 23 | 14.8568 | 13.4885 | 12.3033 | 11.2721 | 10.3710 | 9.5802 | 8.8832 | 8.2664 | 7.7184 | 6.3988 | 4.9245 | 3.9763 | 3.3253 |
| 24 | 15.2469 | 13.7986 | 12.5503 | 11.4693 | 10.5287 | 9.7066 | 8.9847 | 8.3481 | 7.7843 | 6.4337 | 4.9371 | 3.9811 | 3.3271 |
| 25 | 15.6220 | 14.0939 | 12.7833 | 11.6535 | 10.6746 | 9.8225 | 9.0770 | 8.4217 | 7.8431 | 6.4641 | 4.9475 | 3.9848 | 3.3286 |
| 26 | 15.9827 | 14.3751 | 13.0031 | 11.8257 | 10.8099 | 9.9289 | 9.1609 | 8.4880 | 7.8956 | 6.4905 | 4.9563 | 3.9879 | 3.3296 |
| 27 | 16.3295 | 14.6430 | 13.2105 | 11.9867 | 10.9351 | 10.0265 | 9.2372 | 8.5478 | 7.9425 | 6.5135 | 4.9636 | 3.9903 | 3.3305 |
| 28 | 16.6630 | 14.8981 | 13.4061 | 12.1371 | 11.0510 | 10.1161 | 9.3065 | 8.6016 | 7.9844 | 6.5335 | 4.9696 | 3.9922 | 3.3311 |
| 29 | 16.9837 | 15.1410 | 13.5907 | 12.2776 | 11.1584 | 10.1982 | 9.3696 | 8.6501 | 8.0218 | 6.5508 | 4.9747 | 3.9938 | 3.3316 |
| 30 | 17.2920 | 15.3724 | 13.7648 | 12.4090 | 11.2577 | 10.2736 | 9.4269 | 8.6937 | 8.0551 | 6.5659 | 4.9789 | 3.9950 | 3.3320 |
| 31 | 17.5884 | 15.5928 | 13.9290 | 12.5318 | 11.3497 | 10.3428 | 9.4790 | 8.7331 | 8.0849 | 6.5791 | 4.9824 | 3.9960 | 3.3323 |
| 32 | 17.8735 | 15.8026 | 14.0840 | 12.6465 | 11.4349 | 10.4062 | 9.5263 | 8.7686 | 8.1115 | 6.5905 | 4.9853 | 3.9968 | 3.3325 |
| 33 | 18.1476 | 16.0025 | 14.2302 | 12.7537 | 11.5138 | 10.4644 | 9.5694 | 8.8005 | 8.1353 | 6.6004 | 4.9878 | 3.9974 | 3.3327 |
| 34 | 18.4111 | 16.1929 | 14.3681 | 12.8540 | 11.5869 | 10.5178 | 9.6085 | 8.8293 | 8.1565 | 6.6090 | 4.9898 | 3.9979 | 3.3328 |
| 35 | 18.6646 | 16.3741 | 14.4982 | 12.9476 | 11.6545 | 10.5668 | 9.6441 | 8.8552 | 8.1755 | 6.6166 | 4.9915 | 3.9983 | 3.3329 |
| 36 | 18.9082 | 16.5468 | 14.6209 | 13.0352 | 11.7171 | 10.6117 | 9.6765 | 8.8785 | 8.1924 | 6.6231 | 4.9929 | 3.9987 | 3.3330 |
| 37 | 19.1425 | 16.7112 | 14.7367 | 13.1170 | 11.7751 | 10.6529 | 9.7059 | 8.8996 | 8.2075 | 6.6288 | 4.9941 | 3.9989 | 3.3331 |
| 38 | 19.3678 | 16.8678 | 14.8460 | 13.1934 | 11.8288 | 10.6908 | 9.7326 | 8.9185 | 8.2209 | 6.6337 | 4.9951 | 3.9991 | 3.3331 |
| 39 | 19.5844 | 17.0170 | 14.9490 | 13.2649 | 11.8785 | 10.7255 | 9.7569 | 8.9356 | 8.2330 | 6.6380 | 4.9959 | 3.9993 | 3.3332 |
| 40 | 19.7927 | 17.1590 | 15.0462 | 13.3317 | 11.9246 | 10.7573 | 9.7790 | 8.9510 | 8.2437 | 6.6417 | 4.9965 | 3.9994 | 3.3332 |
| 41 | 19.9930 | 17.2943 | 15.1380 | 13.3941 | 11.9672 | 10.7865 | 9.7991 | 8.9649 | 8.2533 | 6.6450 | 4.9971 | 3.9995 | 3.3332 |
| 42 | 20.1856 | 17.4232 | 15.2245 | 13.4524 | 12.0066 | 10.8133 | 9.8173 | 8.9773 | 8.2619 | 6.6478 | 4.9976 | 3.9996 | 3.3332 |
| 43 | 20.3707 | 17.5459 | 15.3061 | 13.5069 | 12.0432 | 10.8379 | 9.8339 | 8.9886 | 8.2695 | 6.6503 | 4.9980 | 3.9997 | 3.3332 |
| 44 | 20.5488 | 17.6627 | 15.3831 | 13.5579 | 12.0770 | 10.8605 | 9.8490 | 8.9987 | 8.2764 | 6.6524 | 4.9983 | 3.9997 | 3.3333 |
| 45 | 20.7200 | 17.7740 | 15.4558 | 13.6055 | 12.1084 | 10.8811 | 9.8628 | 9.0079 | 8.2825 | 6.6542 | 4.9986 | 3.9998 | 3.3333 |
| 46 | 20.8846 | 17.8800 | 15.5243 | 13.6500 | 12.1374 | 10.9001 | 9.8752 | 9.0161 | 8.2879 | 6.6559 | 4.9988 | 3.9998 | 3.3333 |
| 47 | 21.0429 | 17.9810 | 15.5890 | 13.6916 | 12.1642 | 10.9175 | 9.8866 | 9.0235 | 8.2928 | 6.6573 | 4.9990 | 3.9998 | 3.3333 |
| 48 | 21.1951 | 18.0771 | 15.6500 | 13.7304 | 12.1891 | 10.9335 | 9.8969 | 9.0302 | 8.2971 | 6.6585 | 4.9992 | 3.9999 | 3.3333 |
| 49 | 21.3414 | 18.1687 | 15.7075 | 13.7667 | 12.2121 | 10.9482 | 9.9062 | 9.0362 | 8.3010 | 6.6595 | 4.9993 | 3.9999 | 3.3333 |
| 50 | 21.4821 | 18.2559 | 15.7618 | 13.8007 | 12.2334 | 10.9616 | 9.9148 | 9.0416 | 8.3044 | 6.6605 | 4.9994 | 3.9999 | 3.3333 |

Table 10-8

Present Value of an Annuity (Monthly Payments)

$$\left[\frac{1 - \left(1 + \dfrac{i}{12}\right)^{-n}}{i/12}\right]$$

Nominal Interest Rate per Period, i

| Period n | 4 | 5 | 6 | 7 | 8 | 9 | 10 | 11 | 12 | 15 | 20 | 25 | 30 |
|---|---|---|---|---|---|---|---|---|---|---|---|---|---|
| 12 | 11.743 | 11.681 | 11.618 | 11.557 | 11.495 | 11.434 | 11.374 | 11.314 | 11.255 | 11.079 | 10.795 | 10.521 | 10.257 |
| 24 | 23.028 | 22.793 | 22.562 | 22.335 | 22.110 | 21.889 | 21.670 | 21.455 | 21.243 | 20.624 | 19.647 | 18.736 | 17.884 |
| 36 | 33.870 | 33.365 | 32.871 | 32.386 | 31.911 | 31.446 | 30.991 | 30.544 | 30.107 | 28.847 | 26.908 | 25.151 | 23.556 |
| 48 | 44.288 | 43.422 | 42.580 | 41.760 | 40.961 | 40.184 | 39.428 | 38.691 | 37.973 | 35.931 | 32.861 | 30.159 | 27.773 |
| 60 | 54.299 | 52.990 | 51.725 | 50.501 | 49.318 | 48.173 | 47.065 | 45.993 | 44.955 | 42.034 | 37.744 | 34.070 | 30.908 |
| 72 | 63.917 | 62.092 | 60.339 | 58.654 | 57.034 | 55.476 | 53.978 | 52.537 | 51.150 | 47.292 | 41.748 | 37.123 | 33.240 |
| 84 | 73.159 | 70.751 | 68.453 | 66.257 | 64.159 | 62.153 | 60.236 | 58.402 | 56.648 | 51.822 | 45.032 | 39.507 | 34.973 |
| 96 | 82.039 | 78.989 | 76.095 | 73.347 | 70.737 | 68.258 | 65.901 | 63.660 | 61.527 | 55.724 | 47.725 | 41.369 | 36.262 |
| 108 | 90.571 | 86.826 | 83.293 | 79.959 | 76.812 | 73.839 | 71.029 | 68.372 | 65.857 | 59.086 | 49.933 | 42.822 | 37.221 |
| 120 | 98.770 | 94.281 | 90.073 | 86.126 | 82.421 | 78.941 | 75.671 | 72.595 | 69.700 | 61.982 | 51.744 | 43.957 | 37.933 |
| 132 | 106.647 | 101.373 | 96.459 | 91.877 | 87.600 | 83.606 | 79.872 | 76.380 | 73.110 | 64.478 | 53.230 | 44.843 | 38.463 |
| 144 | 114.216 | 108.120 | 102.474 | 97.240 | 92.382 | 87.871 | 83.676 | 79.773 | 76.137 | 66.627 | 54.448 | 45.535 | 38.857 |
| 156 | 121.489 | 114.539 | 108.140 | 102.241 | 96.798 | 91.770 | 87.119 | 82.813 | 78.822 | 68.479 | 55.407 | 46.075 | 39.150 |
| 168 | 128.477 | 120.646 | 113.476 | 106.906 | 100.875 | 95.334 | 90.236 | 85.539 | 81.206 | 70.075 | 56.266 | 46.497 | 39.368 |
| 180 | 135.192 | 126.455 | 118.503 | 111.255 | 104.640 | 98.593 | 93.057 | 87.981 | 83.321 | 71.449 | 56.937 | 46.826 | 39.530 |
| 192 | 141.643 | 131.981 | 123.238 | 115.312 | 108.116 | 101.572 | 95.611 | 90.171 | 85.198 | 72.633 | 57.488 | 47.083 | 39.650 |
| 204 | 147.842 | 137.239 | 127.697 | 119.095 | 111.326 | 104.296 | 97.923 | 92.133 | 86.864 | 73.653 | 57.940 | 47.284 | 39.740 |
| 216 | 153.799 | 142.240 | 131.897 | 122.623 | 114.290 | 106.786 | 100.015 | 93.892 | 88.343 | 74.532 | 58.311 | 47.441 | 39.806 |
| 228 | 159.522 | 146.998 | 135.854 | 125.914 | 117.027 | 109.063 | 101.909 | 95.468 | 89.655 | 75.289 | 58.615 | 47.563 | 39.856 |
| 240 | 165.021 | 151.525 | 139.580 | 128.982 | 119.554 | 111.144 | 103.624 | 96.881 | 90.819 | 75.942 | 58.864 | 47.659 | 39.893 |

For monthly payments at a nominal interest rate of 8%:

$$X = \frac{\$30,000}{\left[1 - (1 + \frac{0.08}{12})^{-240}\right] \Big/ \left[\frac{0.08}{12}\right]} = \$250.94 \text{ per month } = 3,011.72 \text{ per year}$$

In the life insurance annuity a person contributes equal amounts over a number of years, and then at a given age (assuming he has not died previously) he receives a lump sum of money or some other form of payment. To determine how this compares with other forms of investment, the investor must determine at what interest rate his money would need to be invested in order to earn that lump sum in the same period of time. The first payment would earn compound interest for $n$ periods. The second payment, which is made at the end of the first period, would earn interest for $(n - 1)$ periods. The general rule is that each payment earns interest for one less period than the preceeding one. This can be expressed as

$$S = X(1 + i)^n + X(1 + i)^{n-1} + X(1 + i)^{n-2} + \ldots + X(1 + i)^2 + X(1 + i)$$

where  $S$ = lump-sum payment to be made after $n$ years of equal payments are made
$i$ = interest rate per period
$X$ = amount paid at the beginning of each period

Multiply the above by $(1 + i)$ and subtract one equation from the other.

$$S - S(1 + i) = X(1 + i) - X(1 + i)^{n+1}$$

$$S = X\frac{(1 + i)^{n+1} - (1 + i)}{i} = X\frac{(1 + i)[(1 + i)^n - 1]}{i}$$

Example 10-11 shows that the interest rate paid by one big insurance company is less than 1%.

### Example 10-11

A large, reputable insurance company's table states that if a person pays annual premiums of $132.70 beginning at age 21 until he reaches age 65, they will give him $7,160.00 cash, plus insuring his life for $10,000 until age 65. If instead this money were invested at 3% compounded annually, what would be its accumulated value?

$n$ = 44 (the insured gets paid at the start of the year he becomes 66)
$X$ = \$132.70
$i$ = 0.03
$S$ = \$132.70 $\dfrac{(1.03)[(1.03)^{44} - 1]}{0.03}$
$S$ = \$12,150
(at an interest rate of 1% per year, $S$ = \$7,350)

All of the equations developed in this section assume that payments are made at the beginning of a period and interest is calculated at the end of the period. The bank in Example 10-9 would probably ask for the interest at the time the payment is made. This is not uncommon. If this is the case, different formulas must be developed for the various annuities.

## Present Value

Previously it was determined that if $13,500 is borrowed today and invested at 8% interest compounded annually, this will be worth $23,100 after seven years. This can also be reversed to say that something worth $23,100 in 1982 is worth only $13,500 in 1975. $13,500 is called the *present value*. It is the value today that is equivalent to some future or past profit or expense.

$$\text{Present value} = P/(1 + i)^n$$
$$P = \text{value of an item after } n \text{ periods}$$
$$i = \text{interest rate per period}$$

### Example 10-12

Calculate the present value (PV) of a $2,000 gift that will be received after 3 years have passed. Assume money is worth 6% per year.

$$PV = \frac{2,000}{(1.06)^3} = \$1,680$$

The present value of an income is the money that must be invested at the stated interest rate at time zero to yield the same amount of income at the same time. It is also the money that must be invested at time zero to yield the same total obtained if all the income is invested at the stated interest rate when it is received. For a practical situation, this latter interpretation implies that alternate investments are available that can earn income at this interest rate.

The present value of an annuity that has been established to repay a loan is the amount of money borrowed. From the previous discussion

$$PV = \frac{X}{(1 + i)} + \frac{X}{(1 + i)^2} + \frac{X}{(1 + i)^3} + \cdots + \frac{X}{(1 + i)^n} \tag{1}$$

where $PV$ = present value of an annuity
$X$ = payment per time period (made at end of period)
$i$ = rate of interest per period

The equation can be simplified.

Multiply Equation 1 by $(1 + i)^n$

$$PV(1 + i)^n = X[(1 + i)^{n-1} + (1 + i)^{n-2} + \cdots + 1] \tag{2}$$

Multiply Equation 2 by $(1 + i)$

$$PV(1 + i)^{n+1} = X[(1 + i)^n + (1 + i)^{n-1} + \cdots + (1 + i)] \tag{3}$$

Subtract Equation 2 from Equation 3

$$PV(1 + i)^n[1 + i - 1] = X[(1 + i)^n - 1]$$

$$PV = \frac{X[(1 + i)^n - 1]}{(1 + i)^n i}$$

$$PV = \frac{X[1 - (1 + i)^{-n}]}{i} \tag{4}$$

Another way of obtaining this is to consider a life insurance annuity. After $n$ periods the value of this annuity is

$$S = \frac{X(1 + i)[(1 + i)^n - 1]}{i}$$

The net present value is merely this divided by $(1 + i)^n$ or

$$PV = X\frac{(1 + i)[(1 + i)^n - 1]}{i(1 + i)^n} = \frac{X(1 + i)[1 - (1 + i)^{-n}]}{i} \tag{5}$$

The reason Equation 5 and Equation 4 differ is that Equation 5 assumes payments start at the beginning of the first period, whereas Equation 4 assumes the first payment is made at the end of the first period.

If $n$ is very large, the term $(1 + i)^{-n}$ may be assumed to approach zero, in which case Equations 4 and 5 become Equations 6 and 7, respectively.

$$PV_\infty = X/i \tag{6}$$

$$PV_\infty = X(1 + i)/i \tag{7}$$

These equations are very useful for determining the present value of costs that occur periodically, such as those for raw materials, catalysts, maintenance, and utilities. This is illustrated in Example 10-13.

## Example 10-13

The utilities for a chemical plant cost $2,000 per month. What is their present value if money is worth 15% compounded monthly? Do this for plants that last 10 years and 50 years.

Utilities bills are paid after the energy has been used; therefore, from Equation (4)

$$PV = \frac{X[1 - (1 + i)^{-n}]}{i} = \frac{2,000\left[1 - (1 + \frac{0.15}{12})^{-n}\right]}{\frac{0.15}{12}}$$

If the plant lasts 10 years

$$n = 12 \times 10 = 120$$

$$PV = 160,000 \ (0.776) = \$124,000$$

If the plant lasts 50 years

$$n = 12 \times 50 = 600$$

$$PV = 160,000 - 93 \sim \$160,000$$

If $n = \infty$ is used

$$PV_\infty = \$160,000$$

In the last example an average figure was given for utilities. This is usual even though it is realized that the amount of these items used varies with the seasons. Some economists like to compound daily, since utilities are constantly is use. This is not reasonable, because payments to utilities companies or fuel suppliers are usually made on a monthly basis. For plant design evaluations the accuracy of the figures usually does not warrant even compounding semiannually.

## Perpetuities

Closely related to the annuity is the perpetuity. This is an amount of money set aside at time zero that will provide something indefinitely. Care for a cemetery lot is frequently purchased as a perpetuity. The owners of the cemetery agree for a given initial fee to cut the grass, rake the leaves, trim the trees, and do whatever else is necessary to preserve the lot's appearance forever.

While no plants last forever, some last 40 or more years, but the pumps, heat exchangers, and other pieces may need to be replaced periodically because of corrosion. One way of looking at the cost of this is to consider what money must be invested today at a given interest rate so that when a replacement is needed the interest accumulated exactly equals the cost of the new item. Then the original amount is still available for earning interest, which eventually will be spent on another replacement, and so forth.

This can be expressed mathematically by

$$P + C = P(1 + i)^n$$

or

$$P = \frac{C}{(1 + i)^n - 1}$$

where  $P$ = amount of money that must be invested initially or the net present value of a perpetuity

$C$ = cost of replacing the item

$i$ = interest rate per period

$n$ = number of periods the item lasts

This formula purposely does not include the initial installed price of the item. This is because the initial installation cost of a piece of equipment is different from the cost of replacing the old item with a new one.

### Example 10-14

How much money (P) must be set aside at time zero to provide that a pump may be replaced every 3 years thereafter? The cost of purchasing a new pump and replacing the old pump with the new one will be $7,000. The money can be invested at 8% interest, which is compounded annually.

$$P = \frac{\$7,000}{(1 + 0.08)^3 - 1} = \frac{\$7,000}{0.260} = \$26,900$$

### NET PRESENT VALUE—A GOOD PROFITABILITY MEASURE

The net present value of a project is what is obtained when the sum of the present values of all expenditures is subtracted from the sum of the present values of all incomes. This places all costs and incomes on a comparable basis.

Whenever the net present value (NPV) is positive this means the project will yield more money (assuming all income is invested at the given interest rate when it is received) than if the money expended had been invested at the interest rate initially. A positive value, then, means the plant appears to be a winner. If the net present value is negative the project should be dropped. This assumes that the interest rate is realistic and there is an alternate project that can yield the stated interest rate. When alternatives are compared, the one with largest positive net present value is supposedly the best.

When determining the net present value, taxes should be included. To do this an understanding of depreciation is required. This is discussed in Chapter 11. For all further examples in this chapter it will be assumed that taxes and depreciation charges are included in the expenditures.

### Example 10-15

Determine the net present value of the two plants in Example 10-4 if the plants only operate 7 years and have no salvage value. Assume that money is worth 7% per year and the profits become available at the end of the year.

The construction costs were incurred before the plant began operating and are spread unevenly over a number of years. Assume that on the average they date from a year before the plant startup. Time zero is assumed to be the beginning of the initial plant startup.

The present value of the plant construction costs is

$$\frac{(25,000,000)}{(1.07)^{-1}} = \$26,800,000$$

The present value of the profits for plant 1 is

$$\frac{1,000,000}{(1.07)} + \frac{4,000,000}{(1.07)^2} + \frac{6,000,000}{(1.07)^3} + \frac{7,000,000}{(1.07)^4} + \frac{7,000,000}{(1.07)^5} + \frac{7,000,000}{(1.07)^6}$$

$$+ \frac{7,000,000}{(1.07)^7} = \$28,697,000$$

The present value of the profits for plant 2 is

$$\frac{5,000,000}{(1.07)} + \frac{5,000,000}{(1.07)^2} + \frac{5,000,000}{(1.07)^3} + \frac{5,000,000}{(1.07)^4} + \frac{5,000,000}{(1.07)^5} + \frac{7,000,000}{(1.07)^6}$$

$$+ \frac{7,000,000}{(1.07)^7} = \$29,521,000$$

NPV for plant 1 = \$28,697,000 − \$26,800,000 = \$1,897,000
NPV for plant 2 = \$29,521,000 − 26,800,000 = \$2,721,000

The NPV for plant 2 is better than that for plant 1 by \$824,000. This could also be calculated by finding the net present value of the savings in interest payments. These are given in Example 10-4.

NPV of difference $= \dfrac{280,000}{(1.07)^2} + \dfrac{350,000}{(1.07)^3} + \dfrac{280,000}{(1.07)^4} + \dfrac{140,000}{(1.07)^5} = \$824,000$

It should be noted that for Example 10-4 none of the evaluation methods that did not take into account the time value of money could differentiate between these plants, but the net present value method not only could differentiate but determined which was best.

When comparing alternative types of equipment that last different lengths of time and are then replaced, it is important that the period considered be the expected life of the project. If the project is expected to last indefinitely, either a time that is a multiple of both replacement time periods or an infinite amount of time should be considered (perpetuity).

### Example 10-16

An engineer is considering whether he should buy a stainless-steel pump that will cost $6,000 installed and will last 5 years or a carbon-steel pump that will cost $3,000 installed and last 2 years. The cost of replacing the pumps is $5,000 and $2,500 respectively for the stainless-steel and carbon-steel pumps. Money is worth 9% per annum. Which pump is best, considering the lift of the project to be (a) infinite; (b) 7 years.

(a-1) Consider that the length of project is 10 years (the lowest number that is a multiple of 5 and 2)

NPV for the stainless-steel pump $= -\$6,000 - \dfrac{\$5,000}{(1.09)^5}$

$$= -6,000 - 3,250 = -\$9,250$$

NPV for the carbon-steel pump $= -3,000 - \dfrac{2,500}{(1.09)^2} - \dfrac{2,500}{(1.09)^4} - \dfrac{2,500}{(1.09)^6}$

$$- \dfrac{2,500}{(1.09)^8} = -3,000 - 2,100 - 1,770 - 1,490 - 1,260 = -\$9,620$$

Therefore the stainless-steel pump is best.

(a-2) If the project is continued indefinitely (see perpetuity)

NPV for the stainless-steel pump $= -\$6,000 - \dfrac{\$5,000}{(1.09)^5 - 1} = -6,000 - 9,300$

$$= -\$15,300$$

NPV for the carbon-steel pump $= -\$3,000 - \dfrac{\$2,500}{(1.09)^2 - 1} = -3,000 - 13,300$

$$= -\$16,300$$

Again the stainless-steel pump is the best choice.

(b)    If the life of the project is 7 years

$$\text{NPV for the stainless-steel pump} = -6,000 - \frac{5,000}{(1.09)^5} = \$9,250$$

$$\text{NPV for the carbon-steel pump} = -3,000 - \frac{2,500}{(1.09)^2} - \frac{2,500}{(1.09)^4} - \frac{2,500}{(1.09)^6}$$

$$= -\$8,360$$

For this situation the carbon-steel pump should be selected. The reason for the difference is that no replacement carbon-steel pump would be installed in the eighth year. This is equivalent to saving an amount of money having a present value of $-\$1,260$.

One of the problems with the Net Present Value method of evaluation is that an interest rate must be chosen, and in cases where the timing of incomes and outlays differs greatly, different interest rates will result in different conclusions.

### Example 10-17

Compare the following two plants, which have a present value of $11,100,000 in year 1 at interest rates of 5% and 20%. Their profits are given below.

| Year | Plant 1 | Plant 2 |
|------|---------|---------|
| 1 | $1,000,000 | $4,000,000 |
| 2 | $3,000,000 | $4,000,000 |
| 3 | $5,000,000 | $4,000,000 |
| 4 | $6,000,000 | $4,000,000 |
| 5 | $6,000,000 | $4,000,000 |

The net present value of plant 1 is

$$-11,100,000 + \frac{1,000,000}{(1+i)} + \frac{3,000,000}{(1+i)^2} + \frac{5,000,000}{(1+i)^3} + \frac{6,000,000}{(1+i)^4} + \frac{6,000,000}{(1+i)^5}$$

The net present value of plant 2 is

$$-11,100,000 + 4,000,000 \frac{(1-(1+i)^{-5})}{i}$$

For $i = 0.05$

NPV of plant 1 = $6,520,000
NPV of plant 2 = $5,730,000
(or plant 1 appears best)

For $i = 0.20$

NPV of plant 1 = $0.00
NPV of plant 2 = $800,000
(or plant 2 appears best)

Two different answers can be obtained for Example 10-17 because as the interest rate increases the net present value of money earned in future years decreases. This means that high interest rates favor projects that have large initial incomes and low initial costs. Low interest rates favor projects with low initial earnings and high initial outlays.

## RATE OF RETURN—ANOTHER GOOD PROFITABILITY MEASURE

After the net present value has been determined, what does it mean? Is $800,000 a good, average, or poor value? If the net present value is positive we know the plant is making more profit than if the capital were invested at the interest rate used. But how much higher—1%—2%—10%? The only way of knowing is to determine the interest rate that gives a net present value of zero. This interest rate is known as the *rate of return*. Some companies use the rate of return to determine the optimum investment. The project with the largest return is considered the best. This is also known as the *yield on investment* method.

Where the calculation of the net present value was straightforward, the determination of the rate of return requires a trial-and-error procedure. An interest rate is chosen and then the net present value is determined. If it is not zero, another interest rate is chosen and the net present value is recalculated. This is continued until a zero net present value is obtained.

### Example 10-18

Calculate the expected rate of return for a plant that has a present value of $-\$18,000,000$ at startup. The proceeds are expected to be:

| Year | Proceeds |
|------|----------|
| 1 | $3,000,000 |
| 2 | $5,000,000 |
| 3 | $6,000,000 |
| 4 | $6,000,000 |
| 5 | $6,000,000 |
| 6 | $6,000,000 |

The net present value is

$$NPV = -18,000,000 + \frac{3,000,000}{(1+i)} + \frac{5,000,000}{(1+i)^2} + \frac{6,000,000}{(1+i)^3} + \frac{6,000,000}{(1+i)^4}$$

$$+ \frac{6,000,000}{(1+i)^5} + \frac{6,000,000}{(1+i)^6}$$

Let i = 10%

NPV = − 18,000,000 + 2,720,000 + 4,130,000 + 4,510,000 + 4,090,000 + 3,720,000 + 3,380,000 =
− 18,000,000 + 22,550,000 = + $4,550,000

Let i = 15%

NPV = − 18,000,000 + 2,610,000 + 3,780,000 + 3,940,000 + 3,430,000 + 2,980,000 + 2,590,000 =
− 18,000,000 + 19,350,000 = + $1,350,000

Raising the interest rate 5% decreases the net present value $3,200,000. If this were extrapolated
to get NPV = 0, i should be around 17% (this is a very rough guide).

Let i = 17%

NPV = − 18,000,000 + 2,560,000 + 3,650,000 + 3,740,000 + 3,200,000 + 2,740,000 + 2,340,000 =
− 18,000,000 + 18,230,000 = + $230,000

Not quite enough; try i = 18%

NPV = − 18,000,000 + 2,540,000 + 3,590,000 + 3,650,000 + 3,100,000 + 2,620,000 + 2,220,000 =
− 18,000,000 − 17,720,000 = − $280,000

The rate of return is between 17% and 18%. To attempt to come closer than 1% for a preliminary
estimate is not worth while, since the estimated values do not warrant that accuracy. The rate of return is thus 17%.


The Rate of Return method is able to distinguish between the two plants in
Example 10-4. It correctly predicts that plant 2 is preferable.

In Example 10-17 the net present value for plant 1 is zero at an interest rate of
20%. So this is the rate of return. The return for plant 2 is 23.5%. This indicates that
plant 2 is superior to plant 1. This method gives a single answer that does not require
the advance choice of an interest rate. This means, for Example 10-17, that the Net
Present Value method would give a different answer than the Rate of Return
method if an interest rate of 5% were used to determine the former.

Both methods assume that the money earned can be reinvested at the nominal
interest rate. Suppose the rates of return calculated are after tax returns and the
company is generally earning a 5% or 6% return on investment. Is it reasonable to
expect that all profits can be reinvested at 23% or even 20%? No, it isn't! Yet this is
what is assumed in the Rate of Return method. Sometimes the rate of return
may be as high as 50%, while a reasonable interest rate is less than 15%. Therefore if
a reasonable value for the interest rate has been chosen (this is discussed later in this
chapter) and the two methods differ, the results indicated by the Net Present Value
method should be accepted.

### A Modification for Comparing Alternatives

The Rate of Return method can be modified to give the same answer as the Net
Present Value method. This can be illustrated by considering Example 10-7. Table
10-9 gives the NPV for various interest rates for both alternatives of the example.

The values in this table can be interpreted to mean that if an interest rate of 20% is
reasonable both possibilities are acceptable, with alternative 2 having a much
higher net present value. However, if 14% is an acceptable interest rate then

Table 10-9

Net Present Value at Various Interest Rates for the Plants of Example 10-7*

| | Net Present Value | |
|---|---|---|
| Interest Rate | Alternative 1 | Alternative 2 |
| 10% | $691,000 | $458,000 |
| 14% | 433,000 | 387,000 |
| 15% | 352,000 | 370,000 |
| 20% | 000,000 | 300,000 |
| 60% | – – – | 000,000 |

* Assumes outlay given is net present value at time zero.

alternative 1 is the best. Let us look at this somewhat differently. For this example it can also be said that as a result of spending $2,700,000 more on alternative 1, an extra income of $800,000 per year is obtained. A plant having such an outlay and income would have a rate of return of just under 15%. This means that as a result of an extra expenditure of money a 15% rate of return on that expenditure can be expected. This says that if other projects are available that have better rates of return than 15%, alternative 2 should be constructed. Otherwise plant 1 should be constructed. This is the same answer the net present value method gives. Note that this rate of return is below that for either plant 1 or plant 2.

This method can be generalized. Two alternatives can be compared by determining the rate of return of the difference between the cash flows of each plant. The cash flow for any period of time is the income minus the expenses. This will give the crossover interest rate above which one plant is best and below which the other is best.

### Example 10-19

Using the Rate of Return method described above, compare the plants of Example 10-17.

The cash flow for the plants and the difference follow.

| | Cash Flow | | |
|---|---|---|---|
| Year | Plant 1 | Plant 2 | Difference (1 – 2) |
| 0* | -$ 11,100,000 | -$ 11,100,000 | 0 |
| 1 | 1,000,000 | 4,000,000 | - $3,000,000 |
| 2 | 3,000,000 | 4,000,000 | - 1,000,000 |
| 3 | 5,000,000 | 4,000,000 | +1,000,000 |
| 4 | 6,000,000 | 4,000,000 | +2,000,000 |
| 5 | 6,000,000 | 4,000,000 | +2,000,000 |

*A negative cash flow is a payment instead of an income.

The NPV of the difference for various interest rates is

| $i$ | NPV |
|------|------|
| 10% | -$190,000,000 |
| 9% | -$150,000,000 |
| 8% | - $13,000,000 |
| 7% | + $93,000,000 |

This says plant 1 is best if less than an 8% interest rate is acceptable. Otherwise plant 2 is best. The Net Present Value method gives the same answer. At an interest rate of just below 8% the net present values of the two plants are equal.

The disadvantage of this method is that if three or more alternatives are being compared the process is time-consuming unless a digital computer is used. First any two projects are compared, then the best is compared with one of those remaining, and this process is continued until all have been considered and only the best remains. Each comparison involves trial-and-error calculations. On the other hand, the Net Present Value method requires only one calculation for each project.

**Two or More Rates of Return**

The Rate of Return method may give two answers. This can happen whenever large capital expenditures occur after startup. See Example 10-20.

**Example 10-20**

A company has sold bonds to finance a new chemical plant. The bonds are to be redeemed in 4 years for $16,000,000. The plant makes a profit of $2,000,000 per year. All interest and financing charges have been included as outlays. Assume the plant will run for 8 years. What is the rate of return?

$$\text{Net present value} = 0 = \frac{-16,000,000}{(1+i)^4} + \$2,000,000 \frac{1-(1+i)^{-8}}{i}$$

The rate of return is 0% and 22%

Whenever two or more rates of return are obtained, the net present value should be determined over a range of reasonable interest rates. Table 10-10 gives these for Example 10-20.

What this says is that if the reasonable interest rate does not exceed 22% the project should be abandoned. In other words, scrap this plan unless all profits can be invested to yield better than 22%. It should be noted that this is the exact opposite of the usual situation (see Example 10-15) where the project should be pursued if the rate of return is greater than the reasonable interest rate.

Table 10-10

Net Present Value versus Interest Rate for Example 10-20

| Interest Rate | Net Present Value |
|---|---|
| 0% | 0 |
| 5% | -$260,000 |
| 7% | -$270,000 |
| 10% | -$240,000 |
| 15% | -$170,000 |
| 20% | -$70,000 |
| 22% | 0 |
| 25% | +$100,000 |
| 45% | +$520,000 |

To see why two rates of return occur, consider three values: first, the present value of the proceeds received before the bonds were redeemed; second, the present value of the bonds, and third, the present value of the proceeds received after the bonds were redeemed.

Present value of the proceeds before the bonds were redeemed (PVPB) =

$$\$2,000,000 \, \frac{1 - (1 + i)^{-4}}{i}$$

Present value of the bonds (PVB) $= \dfrac{\$16,000,000}{(1 + i)^4}$

Present value of the proceeds after the bonds were redeemed (PVPA)

$$= \$2,000,000 \left[ \frac{1 - (1 + i)^{-8}}{i} - \frac{1 - (1 + i)^{-4}}{i} \right]$$

These values are given in table 10-11 as a function of the interest rate. Also included in parentheses is the present value divided by a constant to give a value of 100 to the first row. This indicates how each present value changes as the interest rate increases. The PVPB changes much less than the PVB, which changes much less than the PVPA as the interest rate increases. In fact, at an interest rate of 49% the PVPB and PVB are equal, even though the bonds are worth twice as much if interest is not considered. For this same rate the PVPB is 5 times the PVPA. If the time value of money were ignored they would be equal. If the interest rate were plotted versus the present values it would be seen that at low interest rates PVB changes less than the sum of PVPA and PVPB, thereby causing the net present value to become negative. At around 7% this reverses and the PVB changes more rapidly than the sum. As a result, the net present value increases until it becomes zero around 22%.

Table 10-11

Interest vs. Present Value for Example 10-20

| Interest Rate | Proceeds before Bond Redemption (PVPB) | Present Value of Bonds (PVB) | Proceeds after Bond Redemption (PVPA) |
|---|---|---|---|
| 0 | $8,000,000(100)* | -$16,000,000(100) | $8,000,000(100) |
| 5% | 7,260,000(91) | - 13,180,000(82) | 5,660,000(71) |
| 10% | 6,350,000(79) | - 10,920,000(68) | 4,230,000(53) |
| 15% | 5,710,000(71) | - 9,150,000(57) | 3,270,000(41) |
| 20% | 5,190,000(65) | - 7,730,000(48) | 2,470,000(31) |
| 22% | 4,990,000(62) | - 7,230,000(45) | 2,240,000(28) |
| 25% | 4,730,000(59) | - 6,550,000(41) | 1,820,000(23) |
| 45% | 3,440,000(43) | - 3,620,000(23) | 700,000(8.8) |
| 49% | 3,250,000(41) | - 3,240,000(20) | 650,000(8.1) |

\*   In parentheses is given the PV divided by PV at zero interest rates times 100.

This again illustrates that the earlier the income is received and the later the outlays occur, the greater their value. It can also be concluded that regardless of the size of the outlays, whenever income occurs before outlays some interest rate will give a positive net present value. However, this rate of return may not be a reasonable one.

Upon decreasing the bond sales in Example 10-20 by $100,000, two interest rates greater than zero would be obtained for which the net present value is zero. This would indicate that as long as the interest rate is not between the two values the project should be accepted. Other examples could be constructed where only if the interest rate is between the limits should it be accepted. There also could be more than two points, or no points, where the net present value is zero. If there are incomes and expenses over 8 years the equation that determines the rate of return has 8 roots. Theoretically all, some, or none of these could be positive.

## COMPARISON OF NET PRESENT VALUE AND RATE OF RETURN METHODS

Some writers claim that an advantage of the Rate of Return method is that no interest rate needs to be chosen. It has already been illustrated that this is not true. The difference in the two measures is when, not whether, a reasonable interest rate is chosen. With the Rate of Return method it does not need to be chosen until after the rate of return is obtained. If it is too high and two processes are both found acceptable, then the net present value must be calculated to determine which is best. In doing this the interest rates for proceeds and outlays may be different.

Various other evaluation schemes based on the concept of time value of money are also sometimes used. These, together with the Net Present Value and Rate of Return methods, are all grouped together under the title of *discounted cash flow* methods.

## PROPER INTEREST RATES

The key to the discounted cash flow methods is the determination of a proper interest rate. For this, two factors must be known. One is: how much does it cost to obtain money? The second is: what is a reasonable amount of profit to expect from a plant? The first depends on the source of money. This can be corporation earnings, the sale of stock, the issuance of bonds, the selling of assets, or borrowing from some outside source. The second depends on economic conditions.

### Corporation Earnings

The corporation earnings that are left after the dividends have been distributed to the stockholders can be used for capital improvements. This money costs nothing to the corporation. It usually amounts to about 5% of net sales. Capital expenditures for the large chemical companies generally are between 8 and 20% of the net sales. It is obvious that money from other sources is usually required. While undistributed profits cost nothing, they should not be thought of as valueless. Unless they can produce an increase of stock value and/or future dividends greater than the immediate cash value to the stockholder, these funds should be distributed as dividends. The profits can be considered to have the same value as stock.

Depreciation charges, which can be used in the same way as profits, should be treated in a similar way (see Chap. 11).

### Stock

Stocks are certificates that represent an individual's investment in a company. They indicate that he is an owner of the corporation. People buy stocks to make money either from the dividends that are paid or because of an expected increase in the selling price of the stock. This price can be found in most large metropolitan papers.

If a company plans to finance a large project by issuing new stock, it means that the percentage of the company's ownership represented by one share of stock decreases. As a result, the stock price often goes down after a large new block of stocks is issued. Because of this, when a large new block of stock is issued these stocks must be sold at a price below that currently listed on the stock exchange.

New stock can be issued in a number of different ways. Dow, in 1966, offered its employees the opportunity to buy new stock at 80% of its value on a given day. The employees could buy an amount not exceeding a certain percentage of their salary, and could not sell it for 6 months after it was received. Other companies give similar options to stockholders. Here, the amount that can be purchased is a percentage of the individual's current stock holdings. Still other companies offer blocks of stock for sale on the open market.

Stock can be likened to an annuity. For a certain amount paid to the corporation it pays a yearly dividend. It differs from many annuities in that the amount of the dividend is not guaranteed, and hopefully the dividends will increase in size and the stock in value. The rate of return that must be paid as a result of issuing stock can be estimated by assuming the dividends will increase at a given rate:

Let   $P$ = the sale price per share of stock
       $D$ = current dividend paid yearly
       $r$ = rate of return (to be determined)
       $a$ = expected percentage increase per year in the dividend rate

$$\text{NPV} = O = P - \frac{D}{(1+r)} - \frac{D(1+a)}{(1+r)^2} - \frac{D(1+a)^2}{(1+r)^3} - \frac{D(1+a)^3}{(1+r)^4} - \cdots - \frac{D(1+a)^n}{(1+r)^{n-1}}$$

$$= P - \frac{D}{(1+r)}\left[ 1 + \frac{1+a}{1+r} + \left(\frac{1+a}{1+r}\right)^2 + \cdots + \left(\frac{1+a}{1+r}\right)^n + \cdots \right]$$

The terms in brackets form a power series, and if $a < r$

$$P = \frac{D}{(1+r)}\left[ \frac{1}{1 - \frac{(1+a)}{(1+r)}} \right] = \frac{D}{(r-a)}$$

$$r = \frac{D}{P} + a \qquad (\text{since } \frac{D}{P} \text{ is positive, } a \text{ is less than } r)$$

As discussed above, the value of $P$ used above will be less than the value listed on the stock exchange. It is equal to the amount received per share minus the cost per share of issuing and selling the stock. $P$ is usually about 80% of the current market value. In general, for the chemical industry the dividends are about 3% of the market price of the stock. *Chemical Week* lists the dividends paid in each issue.

### Example 10-21

A stock that sells for $84 has annual dividends of $2.80. It is expected that the value of the stock and the amount of the dividend will increase 5% per year. Calculate the rate of return paid by the company for issuing stock.

$$P = 0.8\ (\$84) = \$67$$
$$r = \frac{\$2.80}{\$67} + 0.05 = 0.042 + 0.05 = 0.092$$

### Bonds

Bonds are promissory notes that can be issued by corporations or governmental bodies. They are contracts in which the seller agrees to pay the owner the face value (called *par value*) of the bond at a certain date (*maturity* or *due date*) and a given amount of interest at various stated times. Bonds can be sold to individuals, but frequently they are sold in blocks to banks, insurance companies, and others. They usually have a par value of $1,000 and are issued for 10-30 years. If a company becomes bankrupt the bonds must be redeemed before any money is paid to the stockholders. In 1968 over $96,000,000,000 was raised by this method. This includes government and corporation bonds. Since the risk is less, the rate of return is also less than for issuing stock.

On Nov. 15, 1969, B. F. Goodrich[9] offered $100,000,000 in bonds at 8.25% interest, to mature on Nov. 15, 1994. It was selling them at 100% of the par value. This means it expected to get $100,000,000 for them. Some bonds are sold at below par. This is especially true if the interest rate is below average. In January, 1970, Standard Oil (Ohio)[10] sold $150,000,000 worth of bonds that paid an interest rate of 8.5%. The bonds would mature in 1994. In September, 1969, Tenneco[11] sold $50,000,000 of 20-year bonds yielding 9% interest. These were sold at 100.5% of par.

The reason why companies offer different interest rates is because their bond ratings differ. With every major issue of bonds both Moody's[12] and Standard and Poor[13] issue a rating based on the financial standing of the company and the specific plans the company has for repaying the bondholders. Table 10-12 gives the average interest rates for three different bond ratings.

**Selling Assets**

A company may get capital by selling a subsidiary or a portion of its business. In 1969 Monsanto decided that it would quit the low-density polyethylene business. It sold its plant at Texas City, Tex., plus its research findings and technology concerning low-density polyethylene to Northern Petrochemical.[14] In the same year Atlantic Richfield paid $1,800,000 for a 30% interest in a mining lease on about 8,560 acres of oilshale reserves in eastern Utah.[15] W. R. Grace & Co. sold its 53% interest in the Miller Brewing Company to Philip Morris for $130,000,000 and made an after-tax profit of $54,000,000.[16] In 1969 it also sold its steamship business.

**Loans**

Loans are usually money borrowed from financial institutions. The *prime interest rate* is the minimum interest rate that leading banks charge their most credit-worthy customers on large loans. In 1973 this reached a high of 10% per year (Table 10-12). When it was 7.5% per year one big chemical company was reported[17] to be paying 10% interest per year. Usually 0.5-2% more than the prime interest rate is a reasonable value.

**Comparison of Financial Sources**

One advantage to loans and bonds is that the interest paid can be deducted from profits before taxes are figured, whereas the dividend paid stockholders is an after-tax expense. In 1971 the income tax rate was 48% of all earnings. If the interest rate at that time was 10%, then after taxes it would equivalently be 5.2%. In other words, for each dollar spent on interest the profits would be reduced $1.00 and the

Table 10-12

Prime Rate and Bond Yields

| Bond Rating | Moody's Average Industrial Bond Yields (Per Cent) | | | Prime Rate on Jan. 1% |
|---|---|---|---|---|
| | Aaa | Aa | Baa | |
| 1950 | 2.55 | 2.59 | 2.86 | 2.00 |
| 1951 | 2.78 | 2.82 | 3.04 | 2.25 |
| 1952 | 2.88 | 2.93 | 3.20 | 3.00 |
| 1953 | 3.12 | 3.23 | 3.55 | 3.00 |
| 1954 | 2.82 | 3.02 | 3.40 | 3.25 |
| 1955 | 3.00 | 3.11 | 3.47 | 3.00 |
| 1956 | 3.30 | 3.39 | 3.84 | 3.50 |
| 1957 | 3.76 | 3.89 | 4.79 | 4.00 |
| 1958 | 3.61 | 3.78 | 4.59 | 4.50 |
| 1959 | 4.27 | 4.36 | 4.91 | 4.00 |
| 1960 | 4.28 | 4.39 | 5.11 | 5.00 |
| 1961 | 4.21 | 4.33 | 5.10 | 4.50 |
| 1962 | 4.18 | 4.30 | 4.98 | 4.50 |
| 1963 | 4.14 | 4.29 | 4.90 | 4.50 |
| 1964 | 4.32 | 4.41 | 4.87 | 4.50 |
| 1965 | 4.45 | 4.50 | 4.92 | 4.50 |
| 1966 | 5.12 | 5.15 | 5.68 | 5.00 |
| 1967 | 5.49 | 5.55 | 6.21 | 6.00 |
| 1968 | 6.12 | 6.24 | 6.90 | 6.00 |
| 1969 | 6.93 | 7.05 | 7.76 | 6.75 |
| 1970 | 7.77 | 7.94 | 9.00 | 8.50 |
| 1971 | 7.05 | 7.23 | 8.37 | 6.34 |
| 1972 | 6.97 | 7.11 | 7.99 | 5.25 |
| 1973 | 7.28 | 7.40 | 8.07 | 6.00 |
| 1974 | | | | 10.00 |

Sources: *Mood's Industrial Manual,* Mood's Investor Service Inc., New York, 1973.

*Federal Reserve Bulletin,* Board of Governors of the Federal Reserve System, Feb. 1974, A32; Nov. 1972, p. 34.

amount paid the federal government in income taxes would be reduced 48¢. This would make the actual cost 52¢ for each dollar of interest paid.

Since common stock (or equity) costs more than bonds, why do companies issue stock? This is because a company is only able to borrow money at a good rate if the lender is 100% sure he can get his money back plus interest. The less sure he is that he can get his investment back, the higher the interest rate. When a bank is asked to finance the building of a private home it will rarely lend more than 85% of the cost of construction. It also insists that the house be insured, with the bank having the first lien. This means that in case of a catastrophe the bank loan is repaid first. Then the other creditors and the owners receive what is left. A lien's significance is that it takes precedence over all other debts. Even under this arrangement an individual is likely to pay at least 1% more than the prime interest rate. If the amount borrowed is only 35% of the cost of construction, the interest rate may be lowered 0.5%, since the risks are reduced. A homeowner taking out a mortgage always pays a higher interest rate than a corporation with a good credit rating. This is because the costs of bookkeeping and the like are as much as for a small loan as for a large loan, hence the costs per dollar of interest returned are more.

For corporations the same reasoning applies. To offer the prime interest rate the lender must be sure he can get his capital back plus interest. This means that the borrower's total assets must be considerably greater than the current liabilities and debts. Consider the simplified balance sheet given in Table 10-13. By *current assets* is meant cash and everything involved in working capital—feedstocks, unsold product, plus all the product that has been shipped but for which no payment has

Table 10-13

| Assets (in millions of dollars) | | Liabilities (in millions of dollars) | |
|---|---|---|---|
| Current Assets | 400 | Current liabilities | 200 |
| Plant | 600 | Capital | |
| | | Bonds | 300 |
| | | Stock | 500 |
| Total | $1,000 | Total | $1,000 |

been received. *Current liabilities* refers to all bills that are outstanding, such as shipping charges, feed costs, and rental expenses. This does not include any long-term debts such as stocks or bonds. The *plant assets* are the current value of the buildings and equipment. A lender looking at this balance sheet would try to determine what is the probability that he will get his money back. Suppose there is a major disaster and the company goes bankrupt. First all the current liabilities would be paid. Then all the bonds would be redeemed, the ones that were issued first being redeemed first. If anything were left it would then be distributed among the stockholders.

The value of the assets is their value to the company. If they were to be sold on the open market they could not be sold at this price. For instance, if a person won a $3,600 new car in a sweepstakes contest and decided to sell it immediately, he probably could not get over $3,000 for the car. Very few people would want all the optional extras on the car; they generally would want to trade in an old car that he would not want; they expect the seller to repair it if something is faulty, something he ordinarily could not do; and most people would not trust him as much as the dealer who has been in business 25 years.

In 1965, Carling started up a $11,000,000 continuous brewery that was a failure. In 1967 it sold it for $5,500,000.[18] If the costs of startup and other expenses were included the loss would be greater than is indicated. In other words, a company might have to sell its plant assets at less than 50% of the listed value. Its working capital would also be worth less than that listed.

If the assets for the company whose balance sheet is given in Table 10-13 can only be sold at half their listed value, then after all the current liabilities and bonds have been paid off, there would be nothing left for the stockholders. In fact, some of the bondholders might not be totally reimbursed, since it would cost something to liquidate the company's assets. This company could not get a loan at prime interest rates. It would have a better chance of getting a good interest rate if its balance sheet resembled that given in Table 10-14.

Table 10-14

| Assets (in millions of dollars) | | Liabilities (in millions of dollars) | |
|---|---|---|---|
| Current Assets | 400 | Current Liabilities | 200 |
| Plant | 600 | Capital | |
| | | Bonds | 100 |
| | | Stock | 700 |
| Total | $1,000 | Total | $1,000 |

This would also look better to stockholders. They hold a larger share in the company and would at least get something back in case of a disaster. Therefore the stock would sell for more (and have a higher present value to the company) than the stock of the company whose balance sheet was given in Table 10-13 (assuming each had the same number of shares outstanding). This indicates that there is a balance that must be maintained between the total value of the loans, bonds, and stocks outstanding.

In 1973 the companies specializing in the production of industrial chemicals and synthetic materials needed to raise about $6,000,000,000. It was expected that about 80% of that would be obtained from net income and depreciation charges. About 10% would be borrowed and the remaining 10% would come from the other categories.[19]

## Average Cost of Raising Funds

To determine the average cost of money, the cost for each source must be multiplied by the fraction of money obtained from that source. All rates must be either on a before-tax or after-tax basis. Example 10-22 illustrates how this can be done.

**Example 10-22**

A company needs $300,000,000 for capital expansions in the coming year. This will be obtained from the following sources at the following cost of money:

Retained earnings and depreciation allowances $140,000,000
Bonds @ 9% before taxes 50,000,000
Common stock @ 11% after taxes 110,000,000

50% of all earnings will be taken by income taxes.
Assume the retained earnings are worth 10% after taxes. (Issuing new stock has administrative costs and it must be sold for less than the market value.)
Cost of money after taxes =

$$0.10 \left(\frac{140,000,000}{300,000,000}\right) + 0.09\,(0.50) \left(\frac{50,000,000}{300,000,000}\right)$$

$$+ 0.11 \left(\frac{110,000,000}{300,000,000}\right) = 0.095 \text{ or } 9.5\%$$

For any company the earnings, depreciation allowance, bonds outstanding, and stock value can be obtained from its annual report to the stockholders.

## EXPECTED RETURN ON THE INVESTMENT

The profit expected back on any project must be greater than that for projects having fewer risks. In 1974 some banks were offering certificates of deposit that returned a 7.5% profit per year. This investment and some Aaa bonds (see Table 10-12) or government bonds are probably minimum-risk situations.

Most chemical companies aim at making about a 10% profit on all sales. This means that the pretax profits must be around 20%. Since new products involve a large amount of risk and cost uncertainty, a new process is generally not considered unless at least 30% average profit is expected before paying federal corporate income taxes. From this a reasonable rate of return or interest rate may be obtained if some other information on the plants is available. This is done in the following example.

**Example 10-23**

A 100,000 ton/yr polyethylene plant costs $8,500,000 to construct.[20] If polyethylene sells for 0.08¢ a pound, what is the return on investment for this plant if a 10% profit is to be expected after taxes?

$$\text{Profit per year} = 0.10\,(100,000 \text{ tons/yr} \times 2,000 \text{ lb/ton} \times \$0.08/\text{lb})$$

$$= \$1,600,000/\text{yr}$$

Assume the working capital is 25% of the fixed capital

$$\text{R.O.I.} = \frac{1,600,000}{8,500,000 \times 1.25} = 15\%$$

This is the after-taxes return on the investment and assumes a mature plant.

One source[21] states that a 20% return after taxes is a minimum for some companies. Obviously the expansion of a well-established chemical line like sulfuric acid or chlorine has less risk and hence will require a lower rate of return.

Since different values are usually obtained for the cost of money and the minimum return on the investment, it is desirable to use both. The former is used to determine the present value of all outlays and the latter to determine the present value of all incomes. If a positive net present value is obtained, this is an acceptable project. Then if a modified rate of return is determined, an idea can be obtained of how much better than merely acceptable it appears. This can be obtained by using the cost of money to determine the present value for all outlays and determining the interest rate for proceeds that will make the net present value zero. The advantage of the modified rate of return is that it evaluates outlays at a realistic rate.

## CASE STUDY: ECONOMIC EVALUATION
## FOR A 150,000,000 LB/YR POLYSTYRENE PLANT
## USING THE SUSPENSION PROCESS

### Costs

The major costs are summarized by category as follows.

### Raw Materials

The raw materials costs are given in Table 10E-1. The amounts of styrene and rubber used per pound of product were obtained by multiplying the pounds of substance used in producing GPPS* per pound of polystyrene by the fraction of GPPS produced and adding to it a similar product obtained for MIPS and HIPS.*

Pounds of styrene used per pound of composite product =

1.032 (0.60) + 0.982 (0.20) + 0.912 (0.20) = 0.998 lb/lb P.S.

Pounds of polybutadiene used per pound of composite product =

0.05 (0.20) + 0.12 (0.20) = 0.034 lb/lb P.S.

---

\* GPPS = General Purpose Polystyrene
\* MIPS = Medium Impact Polystyrene
\* HIPS = High Impact Polystyrene

The list price for chemicals is often not the price they are selling at. For instance, in 1968 the cost of styrene in tank cars was listed as 8.5¢/lb but it was sold at about

Table 10E-1

Raw Material Costs[1,2]

| | |
|---|---|
| Styrene — 0.988 lb/lb P.S. x $0.065/lb = | $0.0650[3] |
| Polybutadiene — 0.034 lb/lb P.S. x $0.25/lb x 1.05 = | 0.00894 |
| Tricalcium phosphate — 0.005 lb/lb P.S. x $0.1255/lb x 1.05 = | 0.00066 |
| Dodecylbenzene sulfonate — 0.00006 lb/lb P.S. x $0.1125/lb x 1.05 = | 0.0000071 |
| Benzoyl peroxide — 0.0025 lb/lb P.S. x $1.00/lb x 1.05 = | 0.00262 |
| Hydrochloric acid — 0.004 lb/lb P.S. x $0.024/lb x 1.05 = | 0.00010 |
| Total = | $0.07733 |

1. A 5% charge for freight costs was added to all items except styrene, which is quoted as a delivered price.
2. The prices for raw materials were obtained from the *Oil, Paint and Drug Reporter,* Aug. 9, 1971, and *Chemical Week,* Mar. 18, 1970, except as noted in text.
3. List price is 8¢/lb.

7.5¢/lb.[22] No listing is given for barge shipments obtained under a long-term contract. It will be assumed that styrene can be obtained at 1.50¢/lb less than listed.

Since the Wholesale Price Index for Industrial Chemicals changed only 1% between 1959 and 1971, no correction will be applied to obtain the 1974 prices.

**Manpower**

From Table 8E-5, there are 47 hourly employees; 24 of these work days only, and the others work all three shifts. Those who work the night shift generally get a 10% increase in pay and those on the graveyard shift get a 15% increase in pay. Four men are used for each shift position. This means that on the average each will work 42 hours per week. Since the standard week is 40 hours, the workers obtain time-and-a-half wages for 2 hours per week. When all these extras are considered, the average shift worker earns 10.5% more per hour than a man having the same base wage who only works days.

Table 10E-2 gives the expected average wages in 1971. It has been assumed that fringe benefits are about 26% of all wages. To obtain 1974 costs, the wages should be increased about 7% per year. This was the average increase between 1968 and 1971 of the hourly earnings index. Therefore in 1974 the wages and fringe benefits will total $970,000. If the plant is running at full capacity, this is equivalent to $0.00646/lb P.S.

Table 10E-2

Salaries of Plant Personnel (1970)

| Title | Number | Wage | Annual Total |
|---|---|---|---|
| Hourly day workers (2,080 hr worked/yr) | 24 | $3.76/hr* | $188,000 |
| Hourly shift workers (2,190 hr worked/yr) | 23 | $3.76/hr x 1.105 | $210.000 |
| Plant manager | 1 | $20,000/yr | $ 20,000 |
| Process engineer | 1 | $15,000/yr | $ 15,000 |
| Supervisors | 5 | $13,000/yr | $ 65,000 |
| Bookkeeper | 1 | $11,000/yr | $ 11,000 |
| Secretary | 1 | $8,000/yr | $ 8,000 |
| Chief chemist | 1 | $15,000/yr | $ 15,000 |
| Chemists | 5 | $11,000/yr | $ 55,000 |
| | | Subtotal | $587,000 |
| | | Fringe benefits (26%) | 153,000 |
| | | Total | $740,000 |

* "Plant Sites," *Chemical Week,* Aug. 19, 1970, p. 51.

## Utilities

Cost of electricity[23]

$$= 2,300 \text{ kw} \times \$0.002/\text{kwhr} \times 8,300 \text{ hr/yr} = \$38,200/\text{yr}$$
($0.000254/lb P.S.)

Cost of gas[23]

$$= 12,300 \text{ ft}^3/\text{hr} \times 8,300 \text{ hr/yr} \times \$0.000522/\text{ft}^3 = \$53,300/\text{yr}$$
($0.000355/lb P.S.)

Cost for an antifoulant[24] to be added to the cooling water

$$= \$2.00/\text{day} \times 365 = \$700/\text{yr}$$
($0.00000465/lb P.S.)

Operating costs for the ion exchanger[25]

$$= \$35,000/yr \times (1.03)^5 = \$40,600$$
$$(\$0.000270/lb \text{ P.S.})$$

Total utilities costs (1970) = \$132,000

$$(\$0.000884/lb \text{ P.S.})$$

Everyone is predicting that utility rates will rise, but how much is uncertain. In general they have not risen as much as the cost of living. A 3% rise per year will be assumed. Then the total utilities cost in 1974 will be \$0.00100/lb P.S.

### Packaging

The costs for bags and fiber drums[26] are given below.

| | |
|---|---|
| Bags with polyethylene liner, extensible | \$0.195 |
| Fiber drums with Polyethylene barrier (47 gal) | \$2.80 |

For cartons a cost of \$0.005/lb of product was assumed. The average price per pound of product is:

$$0.30 (0.195/50) + 0.15 (2.80/200) + 0.15 (0.005) = \$0.00402/lb$$

Note that a bag holds 50 lb and a carton, 200 lb.

Assume this cost increases 3% per year. Then in 1974 the average cost of packaging per pound of product will be \$0.00466/lb.

### Fixed Capital

This was estimated in Chapter 9 to be \$13,000,000 in 1974.

### Working Capital

It is assumed that all storage facilities are about two-thirds full. The value of the product will be rated at what it costs to make it, not its selling price. From Table 10E-5 it can be seen that this amounts to around \$0.105/lb. The value of the material in process will be taken as midway between the raw material costs and the product's value, or \$0.091/lb. Table 10E-3 gives the storage capacities and the approximate value of their contents. The company pays its bills an average of 15 days after the material is received. Therefore, 15 days were subtracted from the average storage capacity for raw materials. As explained before, the value of chemicals changes only slightly with time, so no inflationary factor is included.

### Polystyrene Selling Price

The price of polystyrene is given in Table 10E-4. To obtain the average price of a pound of product it was assumed that 50% of the bulk product (20% of total) would

Table 10E-3

Working Capital

| Raw Materials | Amount | Unit Value[1] | Total Value |
|---|---|---|---|
| Styrene | 0 lb | $0.065/lb | $    0.00 |
| Polybutadiene | 340,000 lb (25 days) | 0.2625/lb | 89,300.00 |
| Tricalcium phosphate | 50,000 lb (25 days) | 0.1318/lb | 6,590.00 |
| Dodecylbenzene sulfonate | 600 lb (25 days) | 0.1181/lb | 71.00 |
| Benzoyl peroxide | 0 lb | 1.05  /lb | 0.00 |
| Hydrochloric acid | 1,800 lb (1 day) | 0.0252/lb | 45.00 |
| Total | | | $  96,000.00 |
| **Product** | | | |
| Polystyrene | | | |
| Bulk (40%) | 2,580,000 lb (16.3 days) | $0.105/lb | $ 270,000.00 |
| Packaged (60%) | 9,520,000 lb (40 days) | 0.105/lb | 1,000,000.00 |
| Total | | | $1,270,000.00 |
| **Large Process Holdups** | | | |
| R-301 to R-308 (reactors) $(9 \times 814 \text{ ft}^3) \times 0.67 \times 62.4 \text{ lb/ft}^3 \times 0.33$ | | $0.091/lb | $    9,200.00 |
| D-301 to D-304 (hold tanks) $(4 \times 5500 \text{ ft}^3) \times 0.67 \times 62.4 \text{ lb/ft}^3 \times 0.20$ | | 0.091/lb | 16,800.00 |
| D-510 to D-518 (storage for extruders) 36,100 lb x 0.67 | | 0.091/lb | 2,200.00 |
| D-519 to D-523 (storage for testing) 397,000 lb x 0.67 | | 0.105/lb | 28,000.00 |
| Total | | | $  56,200.00 |
| Accounts Payable | 11,900,000 lb  (30 days) | $0.127/lb | $1,510,000.00 |
| **Other Items** | | | |
| Bags | 35,000    (15 days) | $0.195 ea | $    6,840.00 |
| Fiber drums | 4,500    (15 days) | 2.80  ea | 12,600.00 |
| Boxes | 900    (15 days) | 5.00  ea | 4,500.00 |
| Ion exchange resins[2] | 200 ft$^3$ | 62.95/ft$^3$ | 12,600.00 |
| Total | | | $  36,500.00 |
| Total Working Capital | | | $2,969,000.00 |

1. See Table 10E-1 for most of these.

2. Downing, D.G.: "Calculating Minimum Cost Ion-Exchange Units," *Chemical Engineering,* Dec. 6, 1966, p. 170.

be shipped by rail and the rest (20% of total) by truck. Using the same weighting scheme used for the feed, the composite selling price is:

$$0.20 \; [0.60 \,(13.5) + 0.2 \,(15.5) + 0.2 \,(16.5)]$$
$$+ \; 0.20 \; [0.60 \,(14.0) + 0.2 \,(16.0) + 0.2 \,(17.0)]$$
$$+ \; 0.60 \; [0.60 \,(14.5) + 0.2 \,(16.5) + 0.2 \,(17.5)] = \$0.1514/lb.$$

Table 10E-4

Polystyrene Selling Price*

| Shipment Category | GPPS | MIPS | HIPS |
|---|---|---|---|
| Truckload in bags | 14.5¢/lb | 16.5¢/lb | 17.5¢/lb |
| Bulk truck | 14.0¢/lb | 16.0¢/lb | 17.0¢/lb |
| Hopper Car | 13.5¢/lb | 15.5¢/lb | 16.5¢/lb |

* *Chemical Week*, Dec. 14, 1968, p. 51; *Chemical Week*, June 9, 1971, p. 30.

### Economic Indicators

The calculations of the return on the investment and payout period follow. Those for the Net Present Value and Rate of Return are given following Chapter 11.

#### Return on Investment

Table 10E-5 gives the total costs for the polystyrene plant. The pretax return on investment is

$$\frac{(\$0.1514 - 0.1278)/\text{lb} \times 150{,}000{,}000 \text{ lb/yr}}{13{,}000{,}000 + 2{,}969{,}000} \times 100 = 22.2\%$$

This is low for a new product or process. If the product storage were cut in half this would give the following savings. (The costs for the warehouse, items D-601 to D-610, and D-620 to D-634 are cut in half.)

Working capital saving = $635,000
Total capital saving = $1,290,000
Saving in depreciation = 0.00022/lb

The return on investment would be 24.3%.

#### Payout Time

It will be assumed that the plant runs at full capacity after January 1975. The profit made in the first 3 months of operation will be assumed to cover all startup expenses.

Net proceeds per year = $(0.1514 - 0.12785 + 0.00790)/lb × 150,000,000 lb/yr
= $4,730,000/yr

$$\text{Payout time} = \frac{\$13{,}000{,}000 + \$2{,}969{,}000}{\$4{,}730{,}000/\text{yr}} = 3.4 \text{ yr}$$

If only the fixed capital is considered

$$\text{Payout time} = \frac{\$13{,}000{,}000}{\$4{,}730{,}000/\text{yr}} = 2.7 \text{ yr}$$

Either of these times suggests that the plant should be built.

Table 10E-5

Costs for 150,000,000 lb/yr Polystyrene Plant

|                                                            | $/lb     |
|------------------------------------------------------------|----------|
| Raw materials                                              | 0.07733  |
| Utilities                                                  | 0.00100  |
| Labor                                                      | 0.00646  |
| Packaging                                                  | 0.00466  |
| Royalties                                                  | (none)   |
| Maintenance (6.5% of fixed capital/yr)                     | 0.00564  |
| Supplies (19% of maintenance)                              | 0.00107  |
| Administrative (4% of sales)                               | 0.00605  |
| Research and development (5% of sales)                     | 0.00757  |
| Sales (5% of sales)                                        | 0.00757  |
| Taxes and insurance (3% of fixed capital/yr)               | 0.00260  |
| Depreciation (straight-line assuming 11 yr life)           | 0.00790  |
| Total                                                      | 0.12785  |

## PROBLEMS

Problem 1.

Estimate the working capital and determine the annual proceeds per dollar of outlay and the payout time for a plant producing melamine. Use the figures given in *Chemical Week,* Nov. 25, 1967, p. 78. These are duplicated below. Melamine sells for 26.5¢/lb.

Problem 2.

You wish to borrow $20,000 to build a home. The bank offers you a 20-year mortgage at 8% compounded monthly. What is your monthly payment? What is the total interest that will be paid over 20 years? Repeat the calculation for a 10-year and 30-year mortgage (note that the above is standard wording for the nominal interest).

Problem 3.

What is the present value of:

      (a)  $1,000 earned 5 years from now

      (b)  A guaranteed income of $100 per month for life

      (c)  A payment of $500 per month for 5 years

(Do this problem assuming money is worth 5% and 10%. Discuss the results.)

Manufacturing Cost Worksheet for Melamine

| Cost Category | Item | Unit Consumption | Unit Price | Unit Cost |
|---|---|---|---|---|
| Raw materials | Urea | 3.3 tons/ton | $50/ton | $165/ton |
| | Ammonia, 99% | 0.1 tons/ton | $60/ton | 6 |
| By-product credit | Ammonia | 1.1 tons/ton | $30/ton | – 33 |
| Utilities | Steam, 400 psig | 14.5 tons/ton | $ 1/ton | 14.5 |
| | Electricity | 1,900 kwh/ton | 0.5¢/kwh | 9.5 |
| | Cooling water | 94,000 gal/ton | 2¢/1,000 gal | 2 |
| Labor | Operating & supervision | 4 men/shift | $4.00/hr/man + 150% | 25 |
| Fixed charges | Maintenance | 4% of capital yr | $240/ton | 9.5 |
| | Depreciation | 11% of capital/yr | $240/ton | 25.5 |
| | Insurance & taxes | 3% of capital/yr | $240/ton | 7 |
| Total estimated manufacturing cost, | | $232/ton 11.6¢/lb | | |

Basis: 25,000,000 lb*/yr battery-limits plant erected on Gulf Coast, requiring an investment of $3,000,000.

* 38 tons/day; 1.6 tons/hr.

Problem 4.

A company is considering purchasing a power-driven post-hole digger mounted on a line truck. The machine will cost $12,000 and have an 8-year life; 2 men will operate it and can dig 25 holes per day. It uses $10/day of fuel and oil. It will be used in place of a present line truck (value $500) when not digging. The line truck is only used 50% of the time.

It presently takes one man-day to dig a hole. About 350 holes per year are dug. A man works 240 days per year and other work is available when a man is not digging holes. Money is worth 8% and a man is paid $8,000 per year.

(a) Is the purchase justified?

(b) What is the minimum number of holes that must be dug
    per year to justify the truck?
(This problem was modified from one obtained from Oren Ross at the Dow Chemical Company.)

Problem 5.

A pipe containing 150 psi steam in an outdoor location loses 8,400,000 BTU per hour if not insulated. (Unrecovered energy loss = 1,000 BTU/lb vapor.) Assume steam costs 40¢/1,000 lb, and money is worth 8%. Determine whether it is economically justified to insulate the pipe, and if so, which of the thicknesses given below should be specified: assume the insulation will last 20 years.

| Thickness, Inches | Efficiency, Percent | Capital |
|---|---|---|
| 1 | 89.0 | $ 3,800 |
| 1.5 | 90.5 | 5,400 |
| 2 | 92.5 | 8,000 |
| 2.5 | 93.5 | 10,000 |
| 3 | 94.3 | 12,600 |

(This problem was modified from one obtained from Oren Ross at the Dow Chemical Co.)

Problem 6.

Calculate

(a) annual proceeds per dollar of outlay
(b) payout period
(c) net present value assuming money is worth 5%
(d) net present value assuming money is worth 30%
(e) rate of return

for the following six plants. (All costs are given in thousands of dollars.)

| Investment | Initial Cost at Year Zero (including working capital) | Net Cash Proceeds per Year | | |
|---|---|---|---|---|
| | | Year 1 | Year 2 | Year 3 |
| A | $30,000 | $30,000 | | |
| B | 30,000 | 10,000 | 10,000 | $10,000 |
| C | 30,000 | 3,000 | 10,000 | 20,000 |
| D | 30,000 | 17,000 | 10,000 | 3,000 |
| E | 30,000 | 20,000 | 5,000 | 5,000 |
| F | 30,000 | 14,000 | 14,000 | 2,000 |

Discuss your results.

Problem 7.

Determine whether it is more economical for a 150,000,000 lb/yr polystyrene plant to buy styrene in 3,000-ton or 1,000-ton shipments. The cost of shipping is 0.23¢/ton mile in the former case and 0.26¢/ton mile in the latter case. The distance to be shipped is 1,250 miles. Assume the former requires a 26-day storage capacity and the latter a 17-day storage capacity. (See example in Chapter 3.) The value of money is 10%. Use the tank sizes given in Table 5-2 only. The Net Present Value method should be used.

Problem 8.

A centrifugal pump is to be used to produce a 150-psi head at a flow rate of 80 GPM. The maximum pressure in the system will be 200 psi and the maximum temperature is 220°F. A corrosive fluid is being pumped, and it must be determined whether a stainless-steel or cast-iron pump is to be used. The cast-iron pump will last 3 years while the stainless-steel pump will last 7 years. Assume that in 1968 it cost $250 in labor costs to replace a pump. Assume overhead costs involved with obtaining a pump come to 15% of its cost. Assume money is worth 12%. (For other costs see Appendix B)

Using a Net Present Value method, determine which pump should be used if it is estimated that the plant will last

(a) 15 years
(b) an infinite period of time

Problem 9.

A finance company gives the following figures: For a loan of $1,000 a person must make 36 monthly payments of $38.62 per month. What is the rate of return? What is the nominal interest rate?

Problem 10.

The following figures were obtained from a brochure produced by the Boeing Company. From these figures determine whether it would have been a good deal for the American taxpayer to support the building of the Supersonic Transport (SST) in the winter of 1971. What other factors should be considered?

"Market is forecast on a 10% increase per year in free world air traffic between 1970 and 1990." The projection was that 540 SSTs will be needed by 1990. Repayment: "Government will recover all money appropriated plus $1.2 billion dollars through delivery of 540 airplanes. The first production SST is scheduled for 1978 with a continuing production rate to meet growing traffic demands.''

| Financing | |
|---|---|
| Government | $1,342,000,000 |
| BOEING | 215,000,000 |
| General Electric | 94,000,000 |
| Airline prepayments | 59,000,000 |
| Subcontractors | 25,000,000 |

**Problem 11.**

According to an AP report in the *Athens* (Ohio) *Messenger* (Feb. 11, 1974), George Washington was allowed 11¢/day to feed a soldier, 33¢/day to feed on officer, and $5.28/day for himself.

The average soldier received "one pound of fresh beef or one pound of salt-fish; three-fourths of a pound of pork or 20 ounces of salt beef; one loaf of bread and one pint of milk." (Hardly a balanced diet.)

In 1974 the cost of these victuals would be $3.00 per day. What is the average rate of inflation per year for these foods between 1776 and 1974?

**Problem 12.**

Calculate the net present value assuming money is worth 8% and rate of return for the following stocks in 1923 knowing the following data. Assume no dividends were paid (which is false).

| Company | Original Investment in 1923 | Value of Stock in 1973* |
|---|---|---|
| Allied Chemical | $1,000 | $ 4,400 |
| Dow Chemical | 1,000 | 511,500 |
| Du Pont | 1,000 | 88,600 |
| Union Carbide | 1,000 | 15,000 |

* Source: *Chemical and Engineering News*, Jan. 15, 1973.

**Problem 13.**

A large (120 ft x 48 ft) uninsulated tank had a heat loss through the roof of 71.4 BTU/hr ft² and a heat loss of 38.6 BTU/hr ft² through the walls. After insulating the tank with 2 in. of urethane foam, the heat loss through the roof was reduced to 3.0 BTU/hr ft² and that through the wall to 5.1 BTU/hr ft². The total cost of purchasing and installing the insulation was $24,500. If the tank had not been insulated the company would have had to repaint it at an expected cost of $6,000. The material in the tank is maintained at 130°F and the average yearly outside temperature is 50°F. The cost of fuel is $0.80/million BTU.

Determine

(a) rate of return
(b) return on the investment
(c) net present value (assuming money is worth 10%)
(d) payout period

The figures in this problem were obtained from Soderlind, C.: "Tank Insulation Can Pay Off," *Hydrocarbon Processing,* July 1973, p. 122.

Problem 14.

In 1972 a distillery in Scotland offered to sell casks of unaged Scotch grain whiskey for $300. Each cask contains approximately 50 original proof gallons. These casks are stored in government-bonded warehouses. Scotch increases in value as it ages. The average cost for storage and insurance is $6.00 per year per cask.

(a) How much must the Scotch sell for at the end of 3 and 6 years if the buyer wishes to make a 10% profit per year?

(b) What is the net present value (assuming money worth 10%) in 1958 and rate of return for the Scotch if (1) two casks purchased for $161 per cask in 1958 sell for $171.50 each in 1962, (2) two casks purchased for $161 per cask in 1958 sell for $600 each in 1973?

Problem 15.

A term insurance policy for $10,000 will cost a person the following amounts biannually between ages 25 and 69:

| | | | | | |
|---|---|---|---|---|---|
| 25-29 | $ 22.00 | 40-44 | $ 46.00 | 55-59 | $171.00 |
| 30-34 | $ 24.00 | 45-49 | $ 71.00 | 60-64 | $257.00 |
| 35-39 | $ 31.00 | 50-54 | $110.00 | 65-69 | $390.00 |

a) A regular life insurance policy taken out at age 25 will cost $149/year. The term insurance pays the policy owner only if he dies. The regular life policy pays back $5,923 at age 65. Which is a better buy at age 25?

(b) Suppose the term policy is taken out. If the difference between it and the regular policy payments were invested at 6% interest, how long would it take to accumulate $10,000?

Problem 16.

Which of the following options is best? Discuss.

(a) Purchase a regular life insurance policy at age 25 and pay $149.30 per year until age 65. At that time the cash value of the policy is $5,923.50.

(b) Purchase a regular life insurance policy at age 35 and pay $206.90 per year until age 65. At that time the cash value of the policy is $5,357.70.

(c) Purchase a regular life insurance policy at age 45 and pay $289.80 per year until age 65. At that time the cash value of the policy is $4,421.00.

(d) Beginning at age 25, place $149.30 per year in a savings account that pays 5% interest per year, until age 65.

Calculate the net present value, assuming money is worth 8%, and the rate of return for each option.

Problem 17.

For a plant producing 1,000,000,000 polystyrene dairy tubs per year, the following costs were given.

| | |
|---|---|
| Fixed capital | $1,075,000 |
| Working capital | 657,000 |
| Raw materials | 1,344,000/yr |
| Direct labor | 97,000/yr |
| Operating supplies | 6,000/yr |
| Maintenance, labor and materials | 72,000/yr |
| Electric power | 24,000/yr |
| Miscellaneous expenses | 30,000/yr |

(Source: Layman, L. R.: "How to Build a Plant for Thermo-forming Containers," Plastics Technology, Feb. 1971, p. 27.)

(a) What must a tub sell for if a 0.30 annual proceeds per dollar of outlay before taxes is desired?
(b) At the selling price found in (a), what is the net present value, assuming the plant will last 12 years and money is worth 8%?
(c) At the selling price found in (a), determine the rate of return assuming the plant will last 12 years.

Problem 18.

The Evangelical Covenant Church of America offered the following two debenture certificates for sale in 1972 to help finance capital expansions. Which would you advise buying? Determine the Net Present Value, assuming money is worth 10%, and Rate of Return for each:

Option A: A 10 year certificate that pays 6.5% per annum semiannually for 10 years. After 10 years the loan is repaid.

Option B: No semiannual payments are made, but the loan doubles in value after 10 years and 9 months.

Problem 19.

Determine the optimal amount of insulation for an electric water heater, using the net present value (assume money is worth 6%) and rate of return methods. Assume the heater will last 10 years.

| Thickness of Insulation | Cost to Buy & Install | Energy Saving |
|---|---|---|
| (in) | ($) | (kw hr/year) |
| 1 | 3.96 | 188 |
| 2 | 8.34 | 290 |
| 3 | 13.33 | 350 |
| 4 | 18.93 | 384 |

Do this problem for two different electrical rates: $0.02/kw hr and $0.04/kw hr.

Source of data: *Hearings before the Interior and Insular Affairs Committee of the U.S. Senate,* March 22, 23, 1973, p. 174.

Problem 20.

Suppose a person buys six $1000 bonds for $640 in 1974. These bonds have a nominal interest rate of 9 ½% and mature in 1988. What is the rate of return?

# References

1. *Employment and Earnings—States and Areas, 1939-1971,* Bureau of Labor Statistics, U. S. Government Printing Office, Washington, D. C., updated yearly.
2. Berenson, C.: "How Much Are Your Fringe Benefits Worth?," *Chemical Engineering,* Oct. 21, 1968, p. 156.
3. Winton, J.M.: "Plant Sites '74," *Chemical Week,* Oct. 17, 1973, p. 29.
4. *National Electric Rate Book,* Federal Power Commission, U. S. Government Printing Office, Washington, D. C., issued periodically.
5. Raymus, G.J.: "Evaluating the Options for Packaging Chemical Products," *Chemical Engineering,* Oct. 8, 1973, p. 67.
6. Terby, H.: "How to Court the Consumer," *Chemical Week,* Aug. 16, 1969, p. 59-70.
7. "Disinfectants and Fresheners Put on New Airs," *Chemical Week,* July 18, 1973, p. 23.
8. "Chalking up New Highs in R & D", *Chemical Week,* Dec. 14, 1968, p. 74.
9. "Rapid Wrap-up," *Chemical Week,* Nov. 19, 1969, p. 44.
10. "Business Newsletter," *Chemical Week,* Jan. 21, 1970, p. 21.
11. "Business Newsletter," *Chemical Week,* Sept. 27, 1969, p. 10.
12. *Moody's Bond Survey,* Moody's Investor Service, Inc., New York (published weekly; summary published yearly).
13. *Bond Guide,* Standard and Poor's Corp., New York (published monthly; summary published yearly).
14. "Monsanto Quits LDPE," *Chemical Week,* Sept. 20, 1969, p. 39.
15. "National Roundup," *Chemical Week,* May 10, 1969, p. 18.
16. Lurie, M.: "Grace Scrambles for Second Billion," *Chemical Week,* Sept. 13, 1969, p. 30-44.
17. "So That's Where the Money Goes," *Chemical Week,* Apr. 5, 1969, p. 25.
18. "Chementator," *Chemical Engineering,* Nov. 21, 1966, p. 61.
19. "Money Strategy in '73: Spend More, Borrow Less," *Chemical Week,* May 23, 1973, p. 17.
20. Guthrie,K.M. "Capital and Operating Costs for 54 Chemical Processes," *Chemical Engineering,* June 15, 1970, p. 140.
21. Childs, J.F.: "Should Your Pet Project Be Built? What Should the Profit Be?" *Chemical Engineering,* Feb. 26, 1968, p. 188-192.
22. "Market Newsletter," *Chemical Week,* June 15, 1968, p. 52.
23. "Plant Sites," *Chemical Week,* Aug. 19, 1970, p. 51.
24. Silverstein, R.M., Curtis, S.D.: "Cooling Water," *Chemical Engineering,* Aug. 9, 1971, p. 84.
25. Downing, D.G.: "Calculating Minimum-Cost Ion-Exchange Units," *Chemical Engineering,* Dec. 6, 1965, p. 170.
26. Uncles, R.F.: "Containers and Packaging," *Chemical Engineering,* Oct. 13, 1969, p. 87.

# Additional References

Bierman, H., Jr., Smidt, S.: *The Capital Budgeting Decision,* Macmillan, New York, 1960.

Ohsol, E.O.: "Estimating Marketing Costs," *Chemical Engineering,* May 3, 1971, p. 116.

Jenckes, L.C.: "How to Estimate Operating Costs and Depreciation," *Chemical Engineering,* Dec. 14, 1970, p. 168.

Jenckes, L.C.: "Developing and Evaluating a Manufacturing Cost Estimate," *Chemical Engineering,* Jan. 11, 1971, p. 168.

Massey, D.J., Black, J.H.: "Predicting Chemical Prices," *Chemical Engineering,* Oct. 20, 1969, p. 150.

# CHAPTER 11

# Depreciation, Amortization, Depletion and Investment Credit

Depreciation, amortization, depletion, and investment credit are all factors that affect the taxes a company must pay, and hence the profit that can be made. When the government wishes industry to change its direction, it can manipulate these factors to make certain options more profitable. The engineer must be aware of these changes, since they can be the deciding factor on whether a project should be continued.

### DEPRECIATION

Depreciation and amortization are means of recovering your investment in property that has a useful life of more than a year and is used in your trade or business or held for the production of income.[1]

When a corporation constructs a new plant, the firm expects that it will last for a number of years. It is expected that when the plant begins producing it will be worth the outlay of funds needed to construct it. However, as the plant runs it tends to wear out and/or become obsolete. Depreciation is the means by which this loss in value can be deducted as a business expense.

This is a bookkeeping operation. There is no physical exchange of money, as occurs for most other expenses. This means that the money listed as a depreciation expense actually is available to the company to spend as it pleases.

Depreciation is important for two reasons. First, it reduces federal income taxes, because the amount of depreciation occurring in any one year is considered as an expense. Second, it is a means whereby the stockholder can assess the physical value of a company.

For tax purposes it is best to depreciate property as rapidly as possible. This makes income taxes less for the first few years and greater in the last years of operation. Since the total depreciation, and hence tax deduction, is the same this is equivalent to having money available sooner. This, as discussed below, will result in a higher present value.

When the government wishes to encourage construction it can give permission for a fast tax write-off. This means the company can depreciate the plant much faster than it will wear out. This practice has been prevalent in war years when the government needed to have certain defense plants built. It has been proposed that this method might be used to encourage companies to install pollution-reducing systems.

Investors, however, like companies that have large tangible assets, because they think they have a better chance of getting their money back should the company become bankrupt. The tangible assets are the undepreciated assets of the company. So if a company is interested in selling bonds, it looks better if it has depreciated its assets slowly. As a result, some companies keep dual books—one for the public and the other for the Internal Revenue Service. There is nothing illegal about this. The capitalized cost minus the amount that has been depreciated is called the *book value* of the asset. This may be above, below, or the same as its resale value.

## Capitalized Costs

The U. S. government has ruled that none of the costs involved in the planning, construction, and testing of a plant can be deducted from the company's income as a business expense. These are known as capitalized costs. Depreciation is the only means by which these costs can be used to reduce income taxes. All the costs that occur from the time the preliminary process design is begun until the plant begins production fall into this category. They include nearly all the items listed in Table 9-7.

The cost of purchasing land is not subject to depreciation. Neither is the cost of clearing the land, grading, planting, or landscaping. These activities should, if properly done, permanently improve the value of the land. Hence there is no reason for allowing any depreciation expenses.

If a company paves streets and sidewalks, puts in sewers and water mains, and then gives these to a local government to run and maintain, this is not a depreciable item. It can be considered a business expense, since if a city had done this initially it would have assessed the company for the expense.

Inventories, automobiles used for pleasure, and buildings used only as residences are also not depreciable.

## Salvage Value

In determining the total amount of depreciation an estimate must be made of the value of the asset when it is taken out of service. This is called the salvage value. Usually before an item can be sold it must be disconnected and removed from the system. When the cost of these operations is subtracted from the salvage value, the net salvage value is obtained. The total amount to be depreciated is then the original cost minus net salvage value.

Since the net salvage value may be difficult to determine, a company may estimate the cost of removing and disposing of the used equipment as up to 10% of the capital cost. If this is done, the amount to be depreciated is the original cost minus the salvage value plus up to 10% of the capital cost. Either method may be used, depending on the practice of the company. In no case may more than the total capitalized cost be depreciated.

### Example 11-1

The capitalized cost of a plant is $12,000,000. Its estimated salvage value is $1,000,000 and its estimated net salvage value is $400,000. What is the total amount that can be depreciated?

(a) Assume the salvage value is used: 10% of $12,000,000 is $1,200,000, which is greater than the estimated salvage value, so the total capitalized cost, $12,000,000, can be depreciated.

(b) Assume the net salvage value is used: in this case the amount that can be depreciated is

$$\$12,000,000 - \$400,000 = \$11,600,000.$$

In this case, if the company has an option it would choose to use the salvage value, since this would result in a higher present value after taxes.

Table 11-1

Useful Life of Various Depreciable Items

| | |
|---|---|
| Transportation Equipment | |
| Aircraft | 6 years |
| Automobiles, including taxis | 3 years |
| Buses | 9 years |
| General-purpose trucks | |
| Light (actual unloaded weight less than 13,000 lbs) | 4 years |
| Heavy (actual unloaded weight 13,000 lb or more) | 6 years |
| Railroad cars | 15 years |
| Tractor units | 4 years |
| Trailers and trailer-mounted containers | 6 years |
| Vessels, barges, tugs, and similar water transportation equipment | 18 years |
| Land improvements | 20 years |
| Dwellings | 45 years |
| Factories | 45 years |
| Machine shops | 45 years |
| Office buildings | 45 years |
| Warehouses | 60 years |

Source:  *Depreciation Rules and Guidelines,* Publication No. 456, Internal Revenue Service, U.S. Government Printing Office, Washington, D.C., 1964.

Table 11-2

Useful Life for Capital Equipment and Special Structures
Used in the Following Industries

| | |
|---|---|
| Aerospace industry | 8 years |
| Apparel and fabricated textile products | 9 years |
| Cement manufacture | 20 years |
| Chemicals and allied products | 11 years |
| Electrical equipment | 12 years |
| Fabricated metal products | 12 years |
| Food and kindred products (except grain and grain mill products, sugar and sugar products, and vegetable oil products) | 12 years |
| Glass and glass products | 14 years |
| Mining industries | 10 years |
| Paper and allied products | |
|     Pulp and paper | 16 years |
|     Paper finishing and converting | 12 years |
| Petroleum and natural gas | |
|     Drilling, geophysical, and field services | 6 years |
|     Petroleum refining | 16 years |
|     Exploration, drilling, and production | 14 years |
|     Marketing | 16 years |
| Plastics products | 11 years |
| Primary metals | |
|     Ferrous metals | 18 years |
|     Nonferrous metals | 14 years |
| Rubber products | 14 years |
| Stone and clay products (except cement) | 15 years |
| Textile mill products (except knitwear) | |
|     Textile mill products, excluding finishing and dyeing | 14 years |
|     Finishing and dyeing | 12 years |
| Central steam production and distribution | 28 years |
| Electric utilities | |
|     Hydraulic production plant | 50 years |
|     Nuclear production plant | 20 years |
|     Steam production plant | 28 years |
|     Transmission and distribution facilities | 30 years |
| Gas utilities | |
|     Distribution facilities | 35 years |
|     Manufactured gas production plant | 30 years |
|     Natural gas production plant | 14 years |
|     Trunk pipelines and related storage facilities | 22 years |
| Pipeline transportation | 22 years |
| Water utilities | 50 years |

Source: *Depreciation Rules and Guidelines,* Publication No. 456, Internal Revenue Service, U.S. Government Printing Office, Washington, D.C., 1964.

## Useful Life

The amount of depreciation per year depends on how long it is expected the plant will operate. The average life for various assets has been set by the Internal Revenue Service. A selected group of classifications is given in Tables 11-1 and 11-2. These may be reduced by 20%.[2] If a faster depreciation schedule than this is to be used, it must be approved by the Internal Revenue Service and must be based on the current practices of the company.

## Depreciation Schemes

Three different depreciation methods are recognized by the Internal Revenue Service. They are:

      1. Straight line
      2. Declining balance
      3. Sum of the years—digits

Actually, any reasonable method that is used consistently is acceptable, provided it meets some very minimal guidelines.

## Straight Line

The straight-line depreciation method reduces the asset value by the same amount for each year of the plant's expected life. The amount can be determined by dividing the total amount that can be depreciated by the number of years the plant is expected to last. This is the easiest of the depreciation methods.

### Example 11-2

For Example 11-1 compute the depreciation rate per year. Assume the plant will last 11 years. Use the straight-line method of depreciation.

(a)   If the salvage value is used

$$\text{depreciation per year} = \frac{\$12,000,000}{11} = \$1,091,000.$$

(b)   If the net salvage is used

$$\text{depreciation per year} = \frac{\$11,600,000}{11} = \$1,050,000.$$

## Declining Balance

In this method the depreciation allowance is a percentage of the undepreciated capitalized costs. For new items this percentage can be as much as twice the straight-line depreciation rate. When twice the straight-line rate is used it is known as the *double declining balance* rate. For used items, the percentage cannot exceed 1.5 times the straight-line rate. The salvage value is not deducted from the capitalized costs when this method is used. It is permissible to change from this method to a straight-line depreciation method at any time. The usual scheme, in fact, does this when about half the expected life has been expended.

## Example 11-3

A plant costs $6,000,000 and is expected to last 6 years. Its salvage value is negligible. Calculate the rate of depreciation using the double declining balance method.

Amount to be depreciated = $6,000,000

Straight-line depreciation rate = 1/6 or 16 2/3%

Double declining balance rate = 2 × 16 2/3 = 33 1/3%

$1^{st}$ year depreciation × $6,000,000 × 1/3 = $2,000,000

 Book value of asset after first year = $6,000,000 - $2,000,000 = $4,000,000

$2^{nd}$ year depreciation = $4,000,000 × 1/3 = $1,333,000

 Book value of asset after second year = $4,000,000 - $1,333,000 = $2,667,000

$3^{rd}$ year depreciation = $2,667,000 × 1/3 = $889,000

 Book value after third year = $2,667,000 - $889,000 = $1,778,000

$4^{th}$ year depreciation = $1,778,000 × 1/3 = $593,000

$5^{th}$ year depreciation = ($1,778,000-$593,000)× 1/3 = $395,000

$6^{th}$ year depreciation = ($1,185,000-$395,000) × 1/3 = $263,000

 Book value after 6 years = $790,000-$263,000 = $527,000

Note that after 6 years, when the plant is supposedly valueless, it still has a book value of $527,000. The double declining balance can never fully depreciate a plant that has zero salvage value, just as the frog that jumps a third of the way to the well with each jump will never reach the well.

If, however, in the fourth year the straight-line depreciation method had been adopted, the full plant could be depreciated by the end of 6 years. At the beginning of the fourth year the book value was $1,778,000 and there were 3 years of expected life remaining. By the straight-line method this would mean the plant should be depreciated $593,000 in each of the last 3 years.

## Sum of the Years - Digits Method

This method may only be used on new acquisitions. There are two accepted variations: the total-life plan and the remaining-life plan.

In the total-life plan the amount of depreciation is obtained by multiplying the total amount that can be depreciated by a fraction. The numerator of this fraction is the number of years of useful life remaining. The denominator is the sum of the digits from one through the total estimated number of years of useful life. The denominator is a constant.

## Example 11-4

A plant costs $10,000,000 and has a net salvage value of $1,000,000. It has an expected useful life of 5 years. Calculate the amount of depreciation per year using the sum of the years - digits, total-life plan.

Amount to be depreciated = $10,000,000-$1,000,000 = $9,000,000
Sum of digits = 1 + 2 + 3 + 4 + 5 = 15.

$$\text{Depreciation in first year} = \frac{5}{15}(9,000,000) = \$3,000,000$$

$$\text{in second year} = \frac{4}{15}(9,000,000) = \$2,400,000$$

$$\text{in third year} = \frac{3}{15}(9,000,000) = \$1,800,000$$

$$\text{in fourth year} = \frac{2}{15}(9,000,000) = \$1,200,000$$

$$\text{in fifth year} = \frac{1}{15}(9,000,000) = \underline{\$600,000}$$

$$\text{Total depreciation} = \$9,000,000$$

In the remaining-life plan the depreciation for any year is the book value reduced by an acceptable salvage value times a fraction. The numerator of the fraction, as before, is the number of years of useful life remaining. The denominator is the sum of the digits from one to the number of years of useful life remaining. For this plan the denominator is not a constant.

## Example 11-5

Determine the depreciation per year for Example 11-4 using the sum of the years - digits remaining-life plan.

$$\text{Total amount to be depreciated} = \$9,000,000$$

For year 1

Sum of digits = 1 + 2 + 3 + 4 + 5 = 15
Depreciation = $\frac{5}{15}$ (9,000,000) = $3,000,000

For year 2

Sum of digits = 1 + 2 + 3 + 4 = 10
Amount left to be depreciated = $9,000,000 – $3,000,000 = $6,000,000
Depreciation = $\frac{4}{10}$ (6,000,000) = $2,400,000

For year 3

Sum of digits = 1 + 2 + 3 = 6
Amount left to be depreciated = $6,000,000 – $2,400,000 = $3,600,000
Depreciation = $\frac{3}{6}$ (3,600,000) = $1,800,000

For year 4

    Sum of digits = 1 + 2 = 3
    Amount left to be depreciated = \$3,600,000 – \$1,800,000 = \$1,800,000
    Depreciation = $\frac{2}{3}$ (\$1,800,000) = \$1,200,000

For year 5

    Sum of digits = 1
    Amount left to be depreciated = \$1,800,000 – \$1,200,000 = \$600,000
    Depreciation = $\frac{1}{1}$ (\$600,000) = \$600,000
Total depreciation = \$9,000,000

## Comparison of Depreciation Plans

The determination of the present value for the depreciation plans is one of the best ways of comparing depreciation plans. In calculating the present value it will be assumed that depreciation expenses remain in the company and effectively reduce income taxes. If the income tax rate on earnings is 48%, then the amount of income tax saved when depreciation expenses are increased by \$100 is \$48. Therefore, the net savings of including depreciation as an expense is 48% of all depreciation. If the net salvage value is less than the book value after depreciation, the difference is an income and is subject to taxation. Since the amount will be the same for each of the depreciation schemes, it will not be considered in comparing the different methods.

### Example 11-6

A plant costs \$14,000,000 and has a salvage value of \$2,000,000 and a net salvage value of \$1,000,000. The expected life of the plant is 8 years. Calculate the present value for the depreciation plans presented in this chapter. Assume that the interest rate is 10% and the income tax rate is 48% of all earnings.

The amount to be depreciated may be calculated by two means. If the net salvage value is used the amount that can be depreciated is

$$\$14,000,000 - \$1,000,000 = \$13,000,000$$

If the salvage value is used the amount that can be depreciated is

$$\$14,000,000 - \$2,000,000 + \$1,400,000 = \$13,400,000$$

Since a higher value is obtained using the salvage value, this figure will be used.

Straight-Line Depreciation

$$\text{Depreciation per year} = \frac{\$13,400,000}{8} = \$1,675,000$$

$$PV = \left[ \frac{\$1,675,000}{(1.10)} + \frac{\$1,675,000}{(1.10)^2} + \text{- - - -} + \frac{\$1,675,000}{(1.10)^8} \right] 0.48$$

$$= \$4,289,340$$

(assumes taxes are paid at the end of the year)

Double Declining Balance

| Year | Book Value | Rate | Depreciation |
|------|-----------|------|--------------|
| 1 | $14,000,000 | 25% | $3,500,000 |
| 2 | 10,500,000 | 25% | 2,625,000 |
| 3 | 7,875,000 | 25% | 1,969,000 |
| 4 | 5,906,000 | 25% | 1,476,000 |
| 5 | 4,430,000 | 25% | 1,107,000 |
| | (Switch to straight line depreciation) | | |
| 6* | 3,323,000 | 33% | 908,000 |
| 7* | 2,415,000 | 50% | 908,000 |
| 8* | 1,507,000 | 100% | 907,000 |
| | | Total | $13,400,000 |

* Depreciation figured on (book value – $600,000)

$$PV = \left[ \frac{\$3,500,000}{(1.10)} + \frac{\$2,650,000}{(1.10)^2} + \text{- - - -} + \frac{\$907,000}{(1.10)^8} \right] 0.48$$

$$= \$4,764,000$$

Sum of the Years – Digits: Total-Life Plan

| Year | Total Amount Depreciable | Fraction | Depreciation |
|------|-------------------------|----------|--------------|
| 1 | $13,400,000 | 8/36 | $2,978,000 |
| 2 | 13,400,000 | 7/36 | 2,605,000 |
| 3 | 13,400,000 | 6/36 | 2,233,000 |
| 4 | 13,400,000 | 5/36 | 1,861,000 |
| 5 | 13,400,000 | 4/36 | 1,489,000 |
| 6 | 13,400,000 | 3/36 | 1,117,000 |
| 7 | 13,400,000 | 2/36 | 745,000 |
| 8 | 13,400,000 | 1/36 | 372,000 |
| | | Total | $13,400,000 |

$$PV = \left[ \frac{\$2,978,000}{(1.10)} + \frac{\$2,605,000}{(1.10)^2} + \text{---} + \frac{\$372,000}{(1.10)^8} \right] 0.48$$

$$= \$4,761,000$$

Sum of the Years – Digits: Remaining-Life Plan

| Year | Amount Depreciable | Fraction | Depreciation |
|------|--------------------|----------|--------------|
| 1 | $13,400,000 | 8/36 | $2,978,000 |
| 2 | 10,422,000 | 7/28 | 2,605,000 |
| 3 | 7,817,000 | 6/21 | 2,233,000 |
| 4 | 5,584,000 | 5/15 | 1,861,000 |
| 5 | 3,723,000 | 4/10 | 1,489,000 |
| 6 | 2,234,000 | 3/6 | 1,117,000 |
| 7 | 1,117,000 | 2/3 | 745,000 |
| 8 | 372,000 | 1/1 | 372,000 |
| | | Total | $13,400,000 |

$$PV = \$4,761,000$$

The answers obtained in Example 11-6 are typical of those usually obtained. If earnings are the only consideration, the straight-line depreciation method is the worst plan to use. However, the present values for the other methods are generally so close that no obviously best one can be picked.

But earnings are not the only consideration. In 1968 the Union Carbide Corporation, along with many other companies, had a very bad year financially. In order to make its financial picture look better it switched from the double declining balance method of depreciation to the straight-line method. This reduced expenses (deprecitation being considered an expense) and hence increased profits. This is a semipermanent move, however, since a return to the double declining balance method would require approval of the Internal Revenue Service.

## AMORTIZATION

Amortization is basically the same as depreciation except that it applies to intangible property, such as franchises, designs, drawings, or research expenses. Generally straight-line depreciation methods must be used, and only certain items that are amortized can be deducted as expenditures for federal income tax purposes. The value of goodwill, trademarks, and trade names generally cannot be amortized.

## DEPLETION ALLOWANCE

If a company mines an ore, has an oil or gas well, or cuts timber, the company may be entitled to a depletion allowance on its income tax. A depletion allowance is in effect a negative tariff or import tax. It allows a company to deduct from its income a certain percentage of its gross income (income before deducting expenses) before computing income taxes. This amount cannot exceed 50% of the net income (income after deducting expenses). For instance, a company operating a gas or oil well in 1970 could deduct up to 22% of the gross income from its gas and oil property as the equivalent of an expense.

### Example 11-7

A company operating an oil well has a gross income of $100,000,000 and expenses totaling $28,000,000. Assuming a 48% income tax rate, calculate the income taxes paid by the company.

Maximum depletion allowance = 0. 22($100,000,000) = $22,000,000
Net income = $100,000,000 - $28,000,000 = $72,000,000
Maximum amount deductable as depletion allowance = 0.5($72,000,000) = $36,000,000
Since this is greater than $22,000,000 the total depletion allowance can be taken.
   Income tax paid = ($72,000,000 - $22,000,000)0.48 = $24,000,000
The savings in federal income tax by using the depletion allowance is ($22,000,000)0.48 = $10,570,000

Table 11-3

1970 Depletion Allowances

| | |
|---|---|
| Oil and gas | 22% |
| Sulfur and Uranium | 22% |
| Asbestos, lead, zinc, nickel, and mica (if from deposits in United States) | 22% |
| Coal and sodium chloride | 10% |
| Clay and shale used in making bricks or used as sintered or burned light-weight aggregates | 7.5% |
| Gravel, sand, and stone | 5% |
| Most other minerals and metallic ores | 15% |

Source:   *Depreciation, Investment Credit, Amortization, Depletion,* Internal Revenue Service Publication 534 (10-69), U.S. Government Printing Office, Washington, D.C., 1969.

The depletion allowance in 1970 for various substances is given in Table 11-3.

### INVESTMENT CREDIT

The investment credit system for reducing taxes was devised in order to encourage companies to expand. This credit applies to machinery and equipment and does not apply to buildings. For new or used items that are acquired and will last at least 8 years, a company's tax liability may be reduced by up to 7% of the investment. The amount of reduction cannot exceed the calculated tax shown before this credit is taken. It also cannot exceed $25,000 plus 50% of the tax liability in excess of $25,000. This tax relief is generally ignored when an engineer is considering whether a plant is profitable.

### Example 11-8

A company has paid $800,000 for new equipment. Its income tax is figured to be $44,000. What tax actually has to be paid?

Maximum investment credit $= 0.07(\$800,000) = \$56,000$

Limit on investment credit $= \$25,000 + (\$44,000 - \$25,000) 0.5 = \$34,500$
Income tax $= \$44,000 - \$34,500 = \$9,500$
(Under certain circumstances the remaining $21,500 investment credit can be taken in the following year).

### SPECIAL TAX RULES

Tax laws include various exceptions and special options that usually must be considered when corporate income taxes are figured. This is no different from the case when an individual uses the long form to figure his federal income tax. For instance, in 1970 a corporation could deduct 20% of the capitalized cost as an additional first-year depreciation, subject to the following restrictions: the costs of constructing buildings do not apply; the maximum amount deducted does not exceed $10,000; it is taken in the year the asset is purchased; and the taxpayer is not a trust.

These and other such items are generally ignored when an engineer runs an economic evaluation to determine the merits of a given project.

Those wishing to find out more about these items should consult the various publications of the Internal Revenue Service. They can most easily be located by using the *United States Government Publication Index*, which is available in most large libraries. The individual reports can be purchased from the Superintendent of Documents in Washington, D. C.

### CASE STUDY: THE NET PRESENT VALUE AND RATE OF RETURN FOR A 150,000,000 LB/YR POLYSTYRENE PLANT USING THE SUSPENSION PROCESS

#### Net Present Value (NPV)

An interest rate of 8% will be chosen. The fixed capital charges are assumed to have occurred a year before plant startup. The working capital costs occur during the first 3 months of operation. The salvage value of the plant is assumed to be negligible. Straight-line depreciation will be used.

Fixed capital estimate by Guthrie's method = $13,000,000
Working capital = $2,969,000
Profit per year = $(0.1514 - 0.12785)150,000,000$
$\qquad\qquad = \$3,540,000$
Income tax rate = 48% of profits

If the plant operates at full capacity for 11 years, the net present value in 1975 is:

$$-\frac{\$13,000,000}{(1.08)^{-1}} - \$2,969,000 + \left[\frac{\$3,540,000}{(1.08)} + \frac{\$3,540,000}{(1.08)^2} + \cdots + \frac{\$3,540,000}{(1.08)^{11}}\right] 0.52$$

$$+\frac{\$13,000,000}{11 \times 1.08} + \frac{\$13,000,000}{11 \times (1.08)^2} + \cdots + \frac{\$13,000,000}{11 \times (1.08)^{11}} + \frac{\$2,969,000}{(1.08)^{11}}$$

$$= \$5,845,000$$

**Rate of Return**

The same assumptions are made as those given above for the net present value calculations. The rate of return is 13.3% after taxes.

## PROBLEMS

Problem 1.  It has been stated that should any company desire to assume a longer period of useful life than that suggested by the Internal Revenue Service, it may do so, but if the assumed period is less it must be approved by the government. Why is this so?

Problem 2.  Why is the salvage value not considered when depreciation for the double declining balance method is calculated?

Problem 3.  Show that the sum of the digits between 1 and $X = \dfrac{X^2 + X}{2}$

Problem 4.  Some authors claim that in most cases it is better to destroy an object than give it a salvage value. Is this true for Example 11-5?

Problem 5.  Calculate the present value for the depreciation of a $16,000,000 plant. It has a salvage value of $2,000,000 and a net salvage value of $1,200,000. Assume the plant will last 7 years. Use (a) straight-line depreciation (b) double declining balance method (c) sum of the years-digits method

## References

1. *Depreciation, Investment Credit, Amortization, Depletion*, Internal Revenue Service Publication 534 (10-69), U. S. Government Printing Office, Washington, D.C., 1969.
2. "Depreciation Reform Will Ease the Squeeze, but How Much," *Chemical Week*, June 30, 1971, p. 11.

# CHAPTER 12

# Detailed Engineering, Construction, and Startup

With the completion of the economic analysis, the preliminary plant design is now finished. What happens next depends on the reason for obtaining the preliminary design. If this was done to determine the economic feasibility of a new process or the cost of producing a new chemical, the results will be returned to the area that initiated the study. That group will then determine if the research or development should be stopped, continued at its present level, or expanded. If the design was done to determine the impact a competitive compound or process might have on one of the company's proposed or present products, this project can now be completed by marketing experts. If the object of the preliminary design was to determine whether it is economically justifiable to build a plant, the evaluation will be forwarded to the proper authorities to determine whether the project should proceed, be shelved for the present, or be abandoned. For multimillion-dollar projects, the board of directors will perform the final review. In making its decision the board will consider factors besides the economic analysis. It will take into account the market analysis for the product and raw materials and the present and predicted financial condition of the company, the country, and the world. It will compare this project with a number of others, both present, past, planned, and see if it fits into the company's long-range policy. If the project is to be continued, a completion time will be set, and the project will be returned to the process engineering department. The rest of this chapter concerns itself with what happens if the project is approved (finally or tentatively) for detailed engineering.

## DETAILED ENGINEERING

After approval to proceed has been obtained, a project manager is assigned, if this has not been done previously. The responsibility of the project manager is to see that all the thousands of jobs that must be performed in order to design, build, and start up the chemical plant are performed in the most expeditious manner. He will keep track of costs, and if it appears that they will be much greater than projected he will alert the proper authorities. He may be likened to a shepherd. He is to make certain that everything gets done on time and that the total projected capital costs remain within 10% of those estimated in preliminary design phase.

The whole thrust of the preliminary plant design was to determine all the costs involved in producing a given product or products. In detailed engineering, the object is to make certain that no details that are important in the production of a salable product, in a safe manner, are omitted or ignored. This is the task of the process and project engineers. The process engineers have the responsibility of detailing everything that is important to the process. When this is completed the project engineers take this information and complete the detailed specifications. These are required for every item that is to be included in the plant. For instance, the amount and type of agitation occurring within a reactor may play an important part in obtaining a uniform, reproducible product. Then the process engineer may have to specify not only the size, shape, and materials of construction for a reactor, but also the size, shape, and position of the agitator and the baffles, plus the rpm of the shaft, the position and size of the entry and exit ports, and the size of the agitator motor. On the other hand, for a storage vessel the process engineer may give only an approximate capacity.

When the process engineers fail to note some important parameter, major operating problems can result. In one benzene plant the aromatics preheater had to be shut down for cleaning once a month, because of fouling on the tubes. The velocity within the tubes was increased from 1 to 10 ft/sec (0.3 to 3 m/sec); the heater had to be cleaned only once during the first year after the change was made.[1] In this case, the process engineer did not specify an important factor and the project engineer, not knowing that fouling could occur, probably designed the system to optimize the pressure drop.

In the detailed design stage, everything must be specified. Each phase of the preliminary design must now be done in much more detail. The flow sheets develop into piping and instrument diagrams. The duty requirements for a piece of equipment become a specification sheet. The layout drawings may be replaced by a scale model, and a construction bid or detailed cost estimate is obtained to verify the previous cost estimate.

For a multimillion dollar project, this obviously involves dozens of people, and it may cost hundreds of thousands of dollars before construction is begun.

## Piping and Instrument Diagrams (P&IDs)

From the detailed flow sheets, P&IDs are developed. In this diagram all the equipment is drawn to scale, and placed in its proper location within the plant. All piping, pipefittings, valves, strainers, bypasses, rupture discs, sample ports, and so on are included. Every item that needs to be included in the plant is shown on this diagram.

Each pipeline is coded to denote its size and material of construction. For ease in tracing lines, it is given a sequence number. There is a specification sheet for each type and size of pipe. A number for each pipeline is keyed to a table that gives the average composition, pressure, temperature, and flow rate of the material that will be transferred through it. The equipment code is the same as that used for the original flow sheets and the equipment list, although a hyphenated number is sometimes added to indicate on which P&ID drawing the item appears. For the chemical plant, usually only about three or four different sizes of pipe are specified. This reduces the number of fittings, valves, and pieces of pipe that must be kept on hand, so repairs can be quickly made. It also reduces the size of the supply room and the capital tied up in spare parts.

### Specification Sheets

There is a specification sheet for each item shown on a P&ID. Obviously, if two or more items are identical, they may refer to the same specification sheet. To make certain all duplicate equipment is ordered, the code number of each item that must meet the specifications is included on this sheet. An example of a specification sheet for a pump is given in Fig. 12-1. This is often supplemented by general specifications that would apply to perhaps all pumps. The Bechtel Corporation drew up 11 pages of detailed information entitled "General requirements for the design, fabrication, and testing of horizontal pumps." It included such items as what codes had to be met—for example, pipe threads must meet the American Standard for Pipe Threads ASA B 2.1; motors must pass National Electrical Manufacturers Association standards. It gave detailed specifications for such items as castings and connections. It said that "impellers shall be of one piece construction and have a solid hub, packing shall be graphite asbestos, shrink couplings are unacceptable, and bearings must be sealed against entry of foreign materials." It even stated where the name plate was to be located on the pump and what information was to be provided on it. These general specifications are periodically reviewed and updated. They are developed for use with any plant that is built.

When the company does not have the engineers to set and update specific information on centrifugal pumps, it may elect to use one of the two major standards.

The first is the American Petroleum Institute (API) specification number 610. Pumps manufactured according to its specification are the most widely used ones in the petroeum industry.[2] The second is the American National Standards Institute (ANSI) B-123 (previously called the American Volunteer Standard (AVS). It was developed for use in the chemical industry. A committee is charged with improving this standard and establishing criteria for pumps not now standard.[3] One advantage of using standard pumps is that fewer spare parts need to be kept on hand, since the standards are designed to make most parts for pumps of the same size interchangeable.

The specification sheets are generally made out by experts who know from experience what is important and what is not. For the neophyte, the best sources of information come from books and monographs devoted to the design of a single item, material supplied by manufacturers and various testing and standards associations, and articles appearing in journals. A large amount of information is available.

The leading organization in the development of standards in the United States is the American National Standards Institute (ANSI). It is the member body representing the United States in the International Organization for Standardization (ISO) and the Pan American Standards Commission (COPANT). ANSI does not write any standards. It promotes standardization, coordinates efforts toward standardization, and approves standards. It annually publishes a list of American

## HEAT EXCHANGER SPECIFICATION SHEET

| | |
|---|---|
| 1 | JOB NO. |
| 2 CUSTOMER | REFERENCE NO. |
| 3 ADDRESS | INQUIRY NO. |
| 4 PLANT LOCATION | DATE |
| 5 SERVICE OF UNIT | ITEM NO. |
| 6 SIZE           TYPE | (HORIZ.) (VERT.)   CONNECTED IN |
| 7 SQ. FT. SURF./UNIT (GROSS) (EFF.)     SHELLS/UNIT      SQ. FT. SURF./SHELL (GROSS) (EFF.) | |

### PERFORMANCE OF ONE UNIT

| | | SHELL SIDE | TUBE SIDE |
|---|---|---|---|
| 10 | FLUID CIRCULATED | | |
| 11 | TOTAL FLUID ENTERING | | |
| 12 | VAPOR | | |
| 13 | LIQUID | | |
| 14 | STEAM | | |
| 15 | NON-CONDENSABLES | | |
| 16 | FLUID VAPORIZED OR CONDENSED | | |
| 17 | STEAM CONDENSED | | |
| 18 | GRAVITY | | |
| 19 | VISCOSITY | | |
| 20 | MOLECULAR WEIGHT | | |
| 21 | SPECIFIC HEAT | BTU/LB-°F | BTU/LB-°F |
| 22 | THERMAL CONDUCTIVITY | BTU/HR-FT-°F | BTU/HR-FT-°F |
| 23 | LATENT HEAT | BTU/LB | BTU/LB |
| 24 | TEMPERATURE IN | °F | °F |
| 25 | TEMPERATURE OUT | °F | °F |
| 26 | OPERATING PRESSURE | PSIG | PSIG |
| 27 | NO. PASSES PER SHELL | | |
| 28 | VELOCITY | FT/SEC | FT/SEC |
| 29 | PRESSURE DROP | PSI | PSI |
| 30 | FOULING RESISTANCE (MIN.) | | |
| 31 | HEAT EXCHANGED-BTU/HR | MTD CORRECTED-°F | |
| 32 | TRANSFER RATE-SERVICE | CLEAN | |

### CONSTRUCTION OF ONE SHELL

| | | | |
|---|---|---|---|
| 34 | DESIGN PRESSURE | PSI | PSI |
| 35 | TEST PRESSURE | PSI | PSI |
| 36 | DESIGN TEMPERATURE | °F | °F |
| 37 | TUBES    NO.    O.D.    BWG.    LENGTH    PITCH | | |
| 38 | SHELL    I.D.    O.D. | SHELL COVER | (INTEG) (REMOV) |
| 39 | CHANNEL OR BONNET | CHANNEL COVER | |
| 40 | TUBESHEET-STATIONARY | TUBESHEET-FLOATING | |
| 41 | BAFFLES-CROSS    TYPE | FLOATING HEAD COVER | |
| 42 | BAFFLES-LONG    TYPE | IMPINGEMENT PROTECTION | |
| 43 | TUBE SUPPORTS | | |
| 44 | TUBE TO TUBESHEET JOINT | | |
| 45 | GASKETS | | |
| 46 | CONNECTIONS-SHELL SIDE   IN    OUT    RATING | | |
| 47 | CHANNEL SIDE   IN    OUT    RATING | | |
| 48 | CORROSION ALLOWANCE-SHELL SIDE    TUBE SIDE | | |
| 49 | CODE REQUIREMENTS     TEMA CLASS | | |
| 50 | WEIGHTS-EACH SHELL    BUNDLE    FULL OF WATER | | |
| 51 | NOTE: INDICATE AFTER EACH PART WHETHER STRESS RELIEVED (S.R.) AND WHETHER RADIOGRAPHED (X-R) | | |
| 52 | REMARKS: | | |

Figure 12-1    Heat transfer specification sheet. Courtesy of Heat Transfer Division of American Standard.

standards, as well as the standards themselves. These can be obtained from ANSI, 1430 Broadway, New York, N.Y. 10018.

The engineer can also obtain information directly from the organization that has written the standard. For instance, he might write to the American Society for Testing Materials (ASTM) and the American Society of Mechanical Engineers (ASME) to obtain standards for materials. The advantage of specifying that a material conform to these standards is that the engineer then knows how it will perform.

The ASME has also developed, and continuously updated, a set of boiler and pressure-vessel codes.[4] These are manuals that take the engineer step by step through the detailed design of pressure vessels and boilers. Usually the project engineer will merely give the requirement on the specification sheet that the vessel must meet the code. The manufacturer will then complete the detailed design. However, if the item is made of some expensive material such as hastelloy or monel, or the design has some unique aspects, the engineer may have to go through the calculations to determine where he can save on material and construction costs. These codes are available in most engineering libraries and can be purchased at a nominal charge from the sponsoring organizations. They are developed by engineers who donate their time as a public service.

There are also two notable research institutes offering information that may assist the engineer in making out the specification sheets. One of these is the Heat Transfer Research Institute (HTRI). Its staff performs tests to optimize the design of heat-transfer equipment. This information is then encoded in computer programs, which are distributed to the companies that support the research. This means that the engineer's company must purchase a membership to obtain these data. The Fractionation Research Institute (FRI) is a similar institution that deals with the separation processes.

Sometimes a company, either because it lacks certain design information or because its engineering department is so overloaded it cannot take on the project, will obtain the service of an outside firm to do the detailed engineering. When this happens, the same company will probably also be hired to construct the plant.

This company will take the packet of information provided by the process engineers, perform the detailed engineering, and oversee the construction of the plant. Since the company for which the plant is being built does not want the details of its process to become general knowledge, the raw materials and products are given code names and only the general properties needed by the contractor are revealed. Much of the detailed process information is also omitted, such as the time a batch remains in a reactor and the operating conditions for the reactor.

## Layout

The layout may be done totally on paper or with the use of a scale model. Figure 12-2 shows a scale model. The amount of detail put into a scale model depends on the use that will be made of it.

Figure 12-2    Scale Model used in the construction of an ethylene-propylene synthetic rubber plant. The completed plant is partially shown in the background. The model is built to a scale of 3/8 inch to 1 foot. After construction was completed, it was used to train technicians. Courtesy of Uniroyal.

Some companies use it only to obtain a rough three-dimensional view of the plant. Others include on it all pipes, valves, and electrical lines. In this case, a large number of orthographic drawings and layouts will be eliminated. In some cases the dimensions for some piping may be taken directly off the model, although usually a pipe sketch is also included (see Fig. 12-3) for each pipeline. This is sometimes

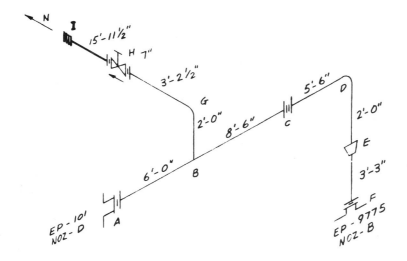

Figure 12-3 A portion of pipe sketch. At point A is a flange connecting nozzle D on a piece of equipment EP-101 with the line. Similarly, at point F a flange connects nozzle B of equipment EP-9775 to the line. H is a valve, E is a reducer, and C is a joint. The distances of each pipe segment are given in feet and inches. Courtesy of E. I. duPont de Nemours & Company.

known as a *spool* or *isometric*. It may be drawn by hand or by a computer from information obtained from the model. Other diagrams that will be developed are structural and architectural diagrams, which show all structural supports and foundation details, and electrical diagrams, which show the details of the electrical system in the same way the piping diagrams give piping details.

One advantage the model has over drawings is that it can easily be modified. A piping change or equipment relocation merely means physically reordering the pieces. When a change is made on drawings, it means replacing lines andor equipment on complicated diagrams. There is a good chance that in doing this errors will be made, since visualizing in two dimensions all the ramifications of a change is difficult. More than one plant has been designed with two different pipes going through the same space or a cleanout port that is inaccessible. Since changes may be made many times in an attempt to optimize the plant, often at least a rough model will be constructed to evaluate these changes. It also makes it easier for people not intimately familiar with the process to visualize what is happening and suggest improvements. The model may also be used after the plant is nearly complete, to help train the operators.

### Instrumention

The instrumentation details must be specific in the same way as the other pieces of equipment. The accuracy, range, and type of the various sensors must be specified. The position of the sensors, sampling ports, and control devices must be indicated on the PIDs. The type of controller must be specified as, for example, on-off, proportional, proportional integral, and derivative, etc. As usual, standard items should be selected whenever possible.

Often the position of the sensors or control valves will determine how easy it is to control the process. When this is felt to be the case, the location should be specified by the process engineer.

### Safety Review

After the detailed design is essentially complete, a safety review is held in which each area of the plant is thoroughly evaluated to make certain that no major damage would result if the two worst disturbances to the area occurred at the same time. One of these is usually the suspension of electrical service, and the other is often the failure of the plant's water supply. This review attempts to determine whether the plant can be shut down under these circumstances with no damage to equipment and only a small production of off-grade material. It must find out whether any plugged lines that may occur can be quickly reopened and any foreign material that entered the process equipment or storage vessels can easily be removed. It must also ascertain whether a dangerous spill to either a nearby body of water or the atmosphere might occur.

In order that different perspectives may be applied, members from research, process engineering, project engineering, construction, and production should be at the review meeting. Notes should be taken at this meeting on how to shut down the plant under any set of circumstances. Startup procedures should also be documented to see that they can be done easily and safely. It is reviews like this that help prevent the multimillion-dollar fires and explosions that have destroyed many plants and taken many lives in recent years.

### Cost Estimates and Construction Bids

When all of the above information has been obtained, the capital cost of the plant is again determined. If it is much higher than the preliminary estimate, the project will be thoroughly reviewed by the group or individual that approved it for detailed engineering. It could be dropped if the difference is much greater than 10%.

One way to obtain a reasonable assured cost estimate is to submit the package of information that has been accumulated to contractors for a firm construction bid.

These contractors will use a detailed cost estimate (see Chapter 9), to determine the costs. The cost of doing this is given in reference 5. Some companies, at the same time or previously, submit the same information to their own cost estimating group for a similar estimate. After the construction bids have been obtained, they are compared with this estimate. If the lowest bid is much lower than predicted by the company's cost estimators, that bidder may be asked to compare details with the company's estimate, to determine whether he has misunderstood some details and so underestimated some items. It is better to clear up differences before awarding the contract than after. The contractor might have been planning some shortcuts that will affect the operability of the plant and should not be allowed. It is better to know about this at the preliminary stage than to have delays as it is argued about later. Also, if the contractor runs into financial difficulties because of underestimating costs, the completion of the plant may be delayed. This could be a very expensive proposition for the company that engaged the contractor, even though it carries insurance against such calamities.

If the bids are much larger than estimated, again a conference may be held to see if there have been some misinterpretations. If the bids still remain high, the company may decide to build on a cost-plus basis, in hopes of reducing costs. For a cost-plus contract, the contractor keeps track of all costs he incurs in constructing the plant. He is then reimbursed for all these costs, and up to 10% in addition. This additional amount is for his overhead and profit. The exact percentage is negotiated in advance. There may also be contracted bonuses if the job is completed early or if costs are less than a given amount.

The contractor can not lose on a cost-plus contract. However, if he ever wants another contract with the company he had better make certain costs are controlled. When there is a lot of construction under way and all contractors are busy, they will increase their profit margin on any firm bids. Under these conditions, a cost-plus contract may be advantageous.

Some companies such as E.I. DuPont de Nemours and Co. have their own construction divisions. The construction division is usually not large enough to handle all the company's construction business. On certain jobs it may bid for the construction contract along with other firms. The lowest bidder is then awarded the job. In other cases the company may only contract for work that the construction division cannot handle.

## CONSTRUCTION

The construction of a chemical plant will usually take from 6 to 18 months if the total cost is under $1,000,000. It will take between 18 and 42 months if the cost exceeds $5,000,000.[6] Everything that has been detailed on the specification sheets and on the various diagrams is now put in place. Construction begins with preparing the site and laying foundations, and ends with the startup of the plant.

To detail all the jobs performed during the construction phase is beyond the scope of this book. However, one job that occurs near the end of the construction phase deserves special mention. It is the pressure testing of equipment and lines to detect any leaks. This is done using water or air wherever possible. The section to be tested is blocked off, and the fluid is pumped in until the proper pressure is reached. The system is then allowed to set for a number of hours. If the pressure loss is less than 2 psi/hr, it is considered acceptable.[7] If it is greater, the leak is repaired and the system is retested.

### Representative at the Construction Site

Before the construction begins, the project manager, in consultation with others, appoints a company representative who is charged with making certain that the plant is built as designed and that the contractor keeps the time schedule he submitted along with his bid. This schedule will be in the form of either a PERT or a CPM diagram (a discussion of these occurs in the next chapter). The representative will handle all communications between company and contractor. He will arrange for company inspectors to observe the required testing of equipment, to inspect the workmanship of the contractor, and to check all incoming equipment to see that it meets the specifications and was not damaged in transit.

During the construction phase many questions will arise, plus many minor and maybe some major problems. The company's representative must find the answers and resolve the problems as expeditiously as possible. For instance, in one case a large amount of control equipment was due to arrive at a northern construction site in midwinter. It would not be needed until spring. Because freezing and thawing would occur in the intervening months, some protective storage location had to be found. The equipment could be left on a freight car, but this would be expensive and it might be cheaper to temporarily house it in a warehouse, a nearby barn, or some inflatable structure. The company representative had to determine what was to be done and make all the necessary arrangements.

In certain cases he will have to contact the process or project engineer to determine whether something is acceptable. The problem may concern an item that has not been completely specified, or one that has been overspecified. It may even be due to an honest mistake on the part of the contractor.

When there is a cost-plus contractor, the representative tries to make certain the laborers are being properly used. He may run into such problems as a grievance over whether certain items such as piping may be prefabricated in a shop or whether they must be put to together in the field. In other cases, he may have to act as the mediator between the contractor and the union.

### Procurement of Equipment

The contractor may be in charge of procuring all the equipment, or the contracting company's purchasing department may have ordered all the major equipment. The latter may be necessary when the equipment delivery times are long, if the plant is to be completed on schedule. In November 1973, a buyer would have faced a wait of 78 weeks after he placed a firm order for a compressor before he received delivery. The average time for pumps was 52-68 weeks.[8] Chemical Week periodically publishes a survey of these delivery times.

Sometimes a set of items such as all the process control instruments or all piping is lumped together, and a number of suppliers are asked to bid on this package. At other times the company may have a contract with, say, a pump manufacturer that guarantees the company will receive a specified discount on the cost of all pumps, if it purchases more than a certain number per year. Then the pumps would be purchased from this vendor. There are even instances where the company may forbid the purchase of items from a given vendor. For instance, one company found that the pumps it purchased from a particular manufacturer cost approximately 5 times as much to maintain as the other pumps it had performing similar tasks. The company refused to buy any more pumps from that manufacturer.

### Changes to the Process after Construction Begins

During the construction phase, as various personnel review the process, desirable changes will often become apparent. If at all possible, these changes should be deferred until after the construction phase is complete. Any changes made when there is a firm contract will generally result in excessive overcharges and may greatly delay the process. The contractor can charge whatever he wants and the customer can do nothing except to forget about making the change then. Even if there is a cost-plus contract, there are so many people who must be coordinated and so many places the changes must be registered that even a simple revision can cost far more than it will after the contractor has completed his contract. In fact, if possible, the changes should be deferred until after the plant is in operation. This is because large numbers of extra people and services are involved in any startup, and any delay at that time can be very expensive.

## STARTUP

Startup is the time when the validity of all the approximations made in the design stage is tested in practice. It is also a test of the competency of the contractor and his crews, and of the ability of the equipment to meet its predicted performance levels. It is a time of stress when everyone is racking their brains trying to figure out why something is not responding as expected. It can also be a very expensive operation in terms of both time and money.

The costs of startup usually range from 5% to 20% of the total cost for design and construction.[9] The time required for startup varies greatly. The Continental Oil Company (Conoco) and its contractor, the Lummus Company, spent a large amount of extra money in preparing for the startup of 500,000,000 lb/yr ethylene plant. One of the reasons it did this was that the project team had estimated normal flaring costs for this plant during startup as $35,000 per day and extra staffing costs as $30,000/day. The extra money spent was considered worth it, because they were able to complete startup operations in 8 days, even though it was Conoco's first ethylene plant. Since they were so successful, examples of what they did will be used throughout this section.[20]

On the other hand, Union Carbide was plagued by all sorts of problems in starting up its chemical complex at Taft, La. It was reported that startup problems reduced corporation after-tax earnings by $30,000,000 and set the time schedule for full production back 18-24 months.[12]

Holroyd has categorized the causes of startup problems as equipment deficiencies (61%), design errors (10%), construction mistakes (16%), and operator errors (13%). This shows that the planning for startup cannot begin too soon. It must be considered during the design stage and actively worked on while construction is proceeding.

The person in charge of this phase of the operation is often the person designated to become the plant manager. Some companies, however, have a startup division that is placed in charge. Others contract for this function with the construction firm. In any case, the person in charge should have had prior experience with startups.

The choosing of the startup manager should be made early in the construction phase, or before if possible. The sooner this is done, the more chance he has to become familiar with all aspects of the process. He also needs time to schedule and oversee all the preparations for the startup. The supervisory team that will assist him will start from 3 to 6 months in advance of the scheduled startup date. Preferably, engineers who have been involved with the design of the plant should be chosen. This gives them the opportunity to determine whether their designs are correct. This is especially important for the process engineers. Unless they are on the site during startup, they will never know which portions of their design worked and which posed problems. If they do not get this feedback, they will keep making the same errors over and over again.

One of the things the process and project engineers on the startup team will do is to write the operating manual. This book not only tells how to start-up, operate, and shut down the plant, but explains what is happening in terms an operator with only a high-school diploma can understand. It will, in layman's terms, explain each unit operation, discuss the chemistry involved, show the reasoning that went into the process design, and state and give reasons for all safety precautions. The process and project engineers will also be responsible for making certain an adequate supply of raw materials is available to start-up the plant.

Maintenance engineers on the team write similar manuals discussing the repair, cleaning, lubrication, and operation of all the equipment. They will also decide what spare parts and tools must be available for startup, and make certain these are ordered by the purchasing department.

The team's laboratory specialist will make certain that all the equipment he needs is in operating order before startup. He will need to familiarize himself with not only the tests that will be regularly performed, but any others that may be useful in diagnosing startup problems. He may arrange for the use of a number of special analytical instruments, which will be returned after startup is complete. He will also need to assemble a manual for the technicians who will assist him.

Usually the operators begin their training somewhere between 4 and 8 weeks in advance of the startup. In some cases they will spend part of the time running similar plants owned by the company. When this is not possible, they may be placed in a simulated situation using either an analog or a digital computer. In other cases, only the lecture and discussion method will be used. The Conoco training program for operators began 6 months before startup. Because the company was inexperienced in operating this type of plant, management felt extra time should be spent familiarizing the men with their jobs.

The maintenance and laboratory technicians begin training at about the same time. The former may spend time in training schools conducted at the equipment manufacturers' plants. These teach them how to service the equipment. The latter may be indoctrinated, at least partially, in the company's main laboratories.

Some companies do not begin startup operations until the construction is completed. Other companies arrange to take over the plant and begin testing on a piecemeal basis as soon as the construction of various portions is completed.

One of the first operations in the startup phase is the flushing of all lines using air or water. Its purpose is to remove any foreign material that may have gotten into the lines. This includes welding rods, nails, plastic lunch bags, handkerchiefs, and the like. To do this, the water should flow at a velocity exceeding 12 ft/sec (3.6 m/sec) and the air at a velocity above 200 ft/sec (60 m/sec).

Before the individual pieces of equipment are started, they should be checked out. Conoco, as part of its prestartup procedure, sent observers to the manufacturer's plant to observe the fabrication and to test the equipment before it was shipped. They found, among other things, at least 5 pumps with defective seals. These errors were corrected by the vendor before the items were shipped. This checking was in addition to checks performed during the construction phase.

Troyan[13] suggests that all pumps, compressors, turbines, and other equipment should be checked internally before they are turned on. He notes that such an inspection of a large gas compressor revealed rust and scale from nearby pipes, a welding rod, and a small pipe wrench within the cylinder itself. Starting up the compressor before these items were removed could have done enough damage that whole sections would need replacement. This could have delayed the startup for weeks or even months.

After the flushing and visual checking is finished, the equipment is tested under simulated conditions. Again, safe fluids such as air, water, and steam are used wherever possible. Sometimes inexpensive compounds with similar boiling points to the process chemicals' are used to check distillation or other separative steps. For solids-handling equipment sometimes salt, the actual raw material, or product purchased for this test may be used. The purpose of these tests is to see whether every piece of equipment performs properly before more expensive and possibly more hazardous process compounds are charged to the system. It also gives the operators a chance to see how the equipment performs. This is a time when the engineer can begin to obtain dynamic data that will be useful in setting controllers and in determining optimal operating conditions. The testing is called a *dry run*. Often, for large expensive pieces of equipment since this is when the equipment is first run at near-processing conditions, it is wise to have a vendor's representative present to assist in solving any problems that may arise.

When a piece of equipment does not respond properly, sometimes the reason may be totally unexpected. In one packed column the pressure drop was many times the design estimate. After many things had been checked, the column was opened and it was found that the workmen had stuffed the boxes in which the packing had been received, into the column along with the packing. The predicted pressure drop was obtained after the boxes were removed.

Troyan[13] tells how to spot certain symptoms during the startup phase so that problems may be averted. Some examples follow. Noisy equipment or a higher-than-normal temperature may mean poor lubrication. This could be due to plugging because of dirt that accumulated during construction. Vibration may mean that the equipment should be anchored better. Excessive pressures or unexpected vacuums may be due to spray painting that clogged ventilators. In one case, a tank was filled with water and then spray-painted. When the tank was emptied, a vacuum was formed that collapsed it.

Either before the dry run begins or while it is progressing, the instrument engineers and technicians should calibrate all the instruments over their full range and make certain they are connected to the correct recording and indicating equipment.

During startup it is wise to check warehouses and other places where equipment was kept before it was assembled, to see if any pieces have been left out. One company installed a scrubbing tower to remove $SO_2$ and spent months trying to figure out why it was performing so poorly. When the tower was opened, they found the device that was supposed to distribute the water within the column was missing. The water was entering in the same way water leaves a faucet, and there was practically no contact between the gas stream containing $SO_2$ and the water. When the distributor was put in place, the column performed as had been expected.

If any cleaning of lines or equipment needs to be done to avoid contamination of products or reactants, this should be done following the dry run. Included with this would be the drying of lines and equipment that have had water in them, and the removal of grease and oil films when the presence of these items is undesirable. The solvents that are to be used in the plant are next charged, and the units are again tested for leaks and run at operating conditions. When this testing is complete, the plant is ready to be started up.

The first units to be started are the utilities. These are needed in many parts of the plant, and must be functioning properly before the rest of the plant can be brought up to capacity. Unless there are reasons for altering the procedure, the rest of the plant is generally brought on stream piece by piece, starting with the feed preparation and following in sequence the steps on the process flow sheets. The process is begun by running at a small percentage of the expected capacity until all units are operational, and then gradually increasing the throughputs.

Once the plant has been operated at the design capacity for a prescribed period of time, the startup phase is officially declared ended, and all the extra personnel who have assisted during this period leave. The plant is now a part of the production facilities of the company, and the rate at which it will operate depends on the orders obtained by the sales department. The process engineering, however, should not stop here. The plant should be thoroughly investigated to see how the operating expenses can be reduced and the product quality improved. Various techniques to assist the engineer in this task are given in Chapter 14.

## References

1. Lord, R.C., Minton, P.E., Slusser, R.P.: "Guide to Trouble-Free Heat Exchangers," *Chemical Engineering,* June 1, 1970, p. 153.
2. Rost, M., Visich, E.T.: "Pumps," *Chemical Engineering*, Apr. 14, 1969, p. 45.
3. Doolin, J.H.: "Updating Standards for Chemical Pumps," *Chemical Engineering,* June 11, 1973 p. 117.
4. *Unfired Pressure Vessels*, American Society of Mechanical Engineers, New York (updated regularly).
5. Loring, R.J.: "The Cost of Preparing Proposals," *Chemical Engineering*, Nov. 16, 1970, p. 126.
6. Perry, J.H. (ed.): *Chemical Engineer's Handbook*, Ed. 4, McGraw-Hill, New York, 1963, Section 26, p. 26.
7. Troyan, J.E.: "How to Prepare for Plant Startups in the Chemical Industries," *Chemical Engineering,* Nov. 3, 1969, p. 87.
8. "You'll Have to Wait Longer for Equipment," *Chemical Week,* Nov. 7, 1973, p. 69.
9. Matley, J.M.: "Keys to Successful Startups," *Chemical Engineering*, Sept. 8, 1969, p. 110.
10. "Preplanning Reaps Dividends for Giant Ethylene Plant," *Chemical Engineering*, July 29, 1968, p. 78.
11. Feldman R.P.: "Economics of Plant Startups," *Chemical Engineering*, Nov. 3, 1969, p. 87.
12. Holroyd, R.: "Ultra Large Single Stream Chemical Plants: Their Advantages, Disadvantages," *Chemistry and Industry*, Aug 5, 1967, p. 1310.
13. Troyan, J.E.: "How to Prepare for Plant Startups in the Chemical Industries," *Chemical Engineering*, Sept. 5, 1960, p.107.

### Additional Reference

Rase, H.F., Barrow, M.H.: *Project Engineering of Process Plants*, Wiley, New York, 1957

# CHAPTER 13

# Planning Tools—CPM and PERT

The importance of completing the plant design, construction, and startup on time has been noted many times in the previous chapters. Every delay means a loss of money. This is true if for no other reason than that some of the money for design, construction, and startup has been spent, and the rest has been committed. No return can be obtained on what has been spent, and a subnormal return will be received on the remainder until after the plant is producing a product. As stated in Chapter 10, time is money.

Delays can also escalate the cost of the plant. If the construction phase is behind schedule, expensive rented equipment and men may be sitting idle while they wait for something to be completed. The company, of course, receives no benefits from the idle labor and equipment, but it still must pay wages and rent for them.

Delays may also mean inability to meet sales commitments. The salesmen, working under the assumption that the new plant will be in operation, may sell greater quantities of the product than the company can produce in its other facilities. If the plant does not start as scheduled, the company may be forced to purchase a competitor's product at a premium price to meet the commitment. The result often is a net loss on each pound (kilogram) that must be purchased under these circumstances.

Two major planning tools to prevent delays, were developed by separate groups around 1957. One, the Critical Path Method (CPM), was developed by Morgan Walker of DuPont and J.E. Kelley, Jr., of Remington Rand. The other, the Program Evaluation and Review Technique (PERT), was developed by the U.S. Navy's Special Project Office along with the firm of Booz, Allen and Hamilton.[1] These techniques have proven to be so successful that either one, or some variation of them, is required for all large government projects and most industrial projects.

The PERT technique was developed in connection with the Navy's crash project to produce the Polaris submarine. It is given much of the credit for the completion of that program 18 months ahead of schedule. In another spectacular success, the turnaround time (the time necessary to shut down, repair, maintain, inspect, and start up the unit) for a methanol unit was cut from 12 to 9 days, with no increase in personnel, by using CPM. This is especially impressive since a similar turnaround had been done annually for 25 years,[2] and using the best methods available the turnaround time had never been less than 12 days.

To prove to himself the importance of using a scheduling technique, the reader should attempt to determine the shortest time necessary to install a scrubber in an existing process. The normal time it takes to perform each task connected with this project is given in Table 13-2. He should get 136 hours.

## CPM

To use CPM or PERT, the job must first be divided into a number of tasks, and the average time to perform each of the tasks must be estimated. These tasks are called *activities*. The average time is the usual time it would take to perform the task when doing it the most economical way. To illustrate this, consider a job we have all performed, changing the front tire on an automobile. A list of the activities involved and a time estimate for them is given in Table 13-1. Before looking at that table, the reader should develop his own list and estimate how long each task will take.

The next step is to determine which tasks must precede each activity. For instance, before the lug bolts can be removed, the wheel cover must have been removed, or D must occur before E. Before the tire can be removed from the hub, the car must be jacked up, the lug bolts must be removed, and, for safety reasons, the brake must be applied. That is, before H can be performed F, E, and A have to be completed. In fact, all the items between A and F must be finished before H can be started.

One useful way of depicting this process is by constructing an arrow diagram. See Figure 13-1. On this diagram each arrow represents one of the activities listed or a so-called dummy activity. The points at which the arrows begin and end are called *nodes*. The length of the arrow or the angle at which it leaves or enters the nodes is immaterial, but the arrow always points to the right.

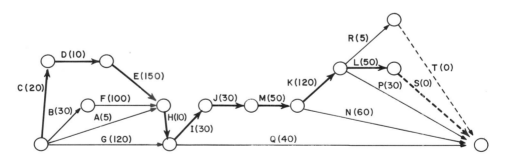

Figure 13-1    An arrow diagram for changing a tire.

The letters on Figure 13-1 refer to the activities given in Table 13-1, and the number in parentheses that follows each letter is the time necessary to complete the activity. Activities S and T are dummy activities that take no time to complete. For CPM and PERT diagrams, it is a rule that no two arrows can be connected to the same two nodes. Note that if it were not for the dummy activities, the arrows representing the tasks L, P, and R would go between the same two nodes. Dummy activities are shown using a dotted line.

Table 13-1

Changing a Tire

| Code | Activity | Time to Complete Activity |
|------|----------|---------------------------|
| A | Put on hand brake | 5 sec |
| B | Get jack from trunk | 30 sec |
| C | Get lug wrench and hub-cap remover from trunk | 20 sec |
| D | Remove wheel cover | 10 sec |
| E | Remove lug bolts | 150 sec |
| F | Jack up car | 100 sec |
| G | Get spare tire out of trunk | 120 sec |
| H | Remove tire from hub | 10 sec |
| I | Place spare tire on hub | 30 sec |
| J | Replace lug bolts | 30 sec |
| K | Tighten lug bolts | 120 sec |
| L | Replace wheel covers | 50 sec |
| M | Lower jack | 50 sec |
| N | Place jack in trunk | 60 sec |
| P | Place lug wrench and hub-cap remover in trunk | 30 sec |
| Q | Place tire that was removed in trunk | 40 sec |
| R | Release brake | 5 sec |
| | Total time | 850 sec |

When the tip of an activity arrow is at a node, this indicates that the activity must be completed before any activity designated by the arrows originating at that node can start. Thus, by going backward to the beginning of the diagram from the node at which an arrow begins, one can determine all the activities that must be completed before that activity can begin. To do this, every possible backward path must be followed. For instance, before Task I can begin all activities denoted by letters previous to it in the alphabet must have been completed. Similarly, before the lug bolts can be removed (E), the lug wrench and hub-cap remover must be available (C), and the wheel cover must have been removed (D). The reader should check to see that he understands all aspects of Figure 13-1.

If the times are correct, it would take 850 sec. to change a tire. Let us suppose the tire changer's son David is eager to help. Could the tire be changed in 425 seconds? If his other son, Daniel, and daughter, Nancy, agreed to help, could it be done in 213 seconds? To answer this, let us assume that the time required to perform each task is invariant. That is, it only takes one person to perform each task, and putting two people to work on any single activity will not reduce the time required to complete it. That means the children can be of assistance in reducing the total time to perform the job only when there are activities that can be done in parallel. For instance, theoretically Nancy could put on the hand brake while David was getting the jack, Danny was getting the hub-cap remover, and the father was getting the spare tire out of the trunk. However, none of them could remove the tire from the hub until the car had been jacked up.

What is the shortest time in which this task could be done? First, since there are at most five parallel activities (N, L, P, R, and Q can occur at the same time), a maximum of five people can be employed at any one time. Second, let us consider the path that gives us the longest time necessary to reach each node. This will be the minimum time that must elapse before the activity that follows can begin. For instance, before activity H can begin, 180 seconds must elapse. Each item on the path, C, D, and E, must be completed before H can begin, and these activities must be done in sequence.

When this procedure has been followed through the whole CPM diagram, the minimum time it takes to reach the final node will be determined. This will be the shortest time in which the project could be completed.

The path through the diagram that gives the maximum time to reach the final node is called the *critical path*. The path must proceed from left to right throughout the diagram. For our tire-changing case, the following activities lie on the critical path: C, D, E, H, I, J, M, K, L, and S. The sum of the times for these tasks is 470 seconds. Since they must be done in order, this is the minimum time it can take to change a tire.

Can the tire be changed in 470 seconds if only two people are working? Let us consider that the father performs every activity on the critical path. This fully occupies his time. One of his children must then be able to perform all the other activities, and reach every node at the same time or before he does. The activities that are not on the critical path and must be done before task I can begin are A, B, F, and G. The total time needed to complete these tasks is 255 seconds. To complete thethe job in 470 seconds, task I must be started 190 seconds (time to complete C, D, E, and H) after beginning the job. Therefore, two persons cannot hope to finish the job in 470 seconds. The reader should try to figure out the shortest time that two people would take to finish the job. Could three persons finish it in 470 seconds?

**Crash Time and Cost**

In constructing the critical path and determining the minimum time in which the tire could be changed, it was assumed that the times for each activity are inviolate. Obviously, this is not true. The lugs can be removed and tightened in seconds, using an electric drill with an impact wrench adapter. By changing the way the spare tire is stored, its removal from the trunk could be facilitated. If a hydraulic hoist were available, the time required for raising the car could be decreased. This can even be done without a jack or hoist. I remember that once, in my youth, a group of boys were in such a great rush that they changed a tire without using a jack. They manually supported the car while the one who was an auto mechanic changed the tire.

Let us assume that extra labor costs money. It is not free like the help given by David, Nancy, and Daniel. Then, all the above proposals to reduce the time it takes to change a tire cost money. If the use of a jack or hoist is to be avoided, more labor must be hired. If a hydraulic jack or power wrench is to be used, it must be purchased or rented. Finally, the remounting of the spare tire means spending extra money and effort in advance of the project.

Some of the activities cannot be speeded up. It takes a certain amount of time to open the front door of the car and engage the hand brake. There is no way of reducing this time. Similarly, the time it takes to remove the tire from the hub is unchangeable.

These generalizations can be applied to any job. There are some activities that cannot be speeded up. Others can be done faster, but in most cases this involves spending more money in obtaining special equipment, hiring more labor, paying overtime, working an extra shift, and so on. For all activities there is some minimum finite performance time. No matter what is done, it cannot be completed any faster. This is called the *crash time* of the activity, and the cost associated with it is the *crash cost*. The same is true for projects. The crash cost of a project is the price associated with finishing the project in the crash (minimum) time.

To illustrate this, let us consider the installation of a scrubber in an existing process. The activities are given in Table 13-2, along with the necessary time and cost data. For this example it will be assumed that for each activity there is a linear relationship between the cost and the time required. The slope of this line is called the *cost slope*. It is the cost for saving one unit of time. From the data given in Table 13-2, a CPM diagram can be constructed. The reader should do this independently, and then compare it with Figure 13-2. Next he should determine the critical path. This is indicated in the figure by the thickest line.

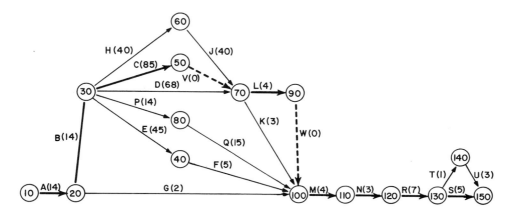

Figure 13-2   An arrow diagram for the installation of a scribber in an existing process. If the numbers in the circles are ignored, this diagram is typical of one used for a CPM problem; if the projects are identified by numbers and the normal or most likely times are replaced by the expected time, the diagram is typical of one used for a PERT problem.

Table 13-2

Cost and Time Data
for
The Installation of a Scrubber in an Existing Process

| Code | Activity | Normal Time (days) | Cost ($) | Crash Time (days) | Crash Cost ($) | Cost Slope $/day |
|------|----------|-------------------|----------|-------------------|----------------|------------------|
| A | Process design | 14 | 1,000 | 10 | 1,400 | 100 |
| B | Project design | 14 | 1,500 | 9 | 2,500 | 200 |
| C | Procure scrubber | 85 | 15,000 | 65 | 16,600 | 80 |
| D | Procure pump | 68 | 1,300 | 58 | 1,700 | 40 |
| E | Procure piping | 45 | 2,000 | 35 | 2,200 | 20 |
| F | Prefabricate pipe | 5 | 2,300 | 3 | 2,700 | 200 |
| G | Erect scaffolding | 2 | 500 | 1 | 700 | 200 |
| H | Site preparation | 40 | 6,000 | 30 | 7,000 | 100 |
| J | Pour concrete and allow to set | 40 | 3,000 | 40 | 3,000 | 0 |
| K | Install pump | 3 | 3,000 | 3 | 3,000 | 0 |
| L | Install scrubber | 4 | 3,500 | 2 | 4,500 | 500 |
| M | Fit up pipe | 4 | 400 | 2 | 600 | 100 |
| N | Weld pipe | 3 | 300 | 2 | 400 | 100 |
| P | Shut down process | 14 | 500 | 10 | 600 | 25 |
| Q | Remove some existing piping and make changes in existing equipment | 16 | 5,000 | 8 | 7,000 | 250 |
| R | Pressure test, flush, and make dry run | 7 | 3,000 | 5 | 4,000 | 500 |
| S | Start up process and test scrubber | 5 | 7,000 | 3 | 8,000 | 500 |
| T | Remove scaffolding | 1 | 150 | 1 | 150 | 0 |
| U | Clean up | 3 | 200 | 2 | 300 | 100 |
| | Total | | 55,650 | | 66,350 | |

If all the activities were crashed, the project cost would be $66,350. However, not all the activities need to be crashed to complete the project in the crash time.

To prove this, the crash time and minimum crash cost will now be determined. A cost-versus-completion-time curve will also be constructed. This curve increases the project manager's decision possibilities from two to many. With it he can determine whether it is most economical to get the job done as fast as possible, to proceed at a normal pace, or to choose some intermediate time.

For instance, a company might have a 50-day supply of product available, but be incurring a $250-a-day fine from the Environmental Protection Agency for each day it operates without a scrubber. Under these circumstances, it might be expedient not to crash those activities on the critical path whose slopes are greater than $250 per day.

To construct the cost-time curve, the planner must decide the order and amount by which he will speed up the various activities. It will be assumed that there are always enough men and an adequate amount of equipment available to perform all the activities that are not on the critical path in such a way that the completion of the project will not be delayed by them. If this is so, then speeding up any of the activities not on the critical path will not in any way produce an earlier completion date. Thus, only the times of activities on the critical path need to be shortened. Obviously the activity that should be speeded up first is the one with the smallest cost slope. This is because it will cost less for the time that is saved than any other task. From Table 13-2 it can be seen that C, procuring the scrubber in a more expeditious manner, fits the criteria. However, if we reduce this by more than 5 days, activity C will no longer be on the critical path. The critical path will go through H and J rather than through C and V. So reduce the time to complete C by 5 days. The increased cost is $400. Now H and J as well as C and V are on the critical path. There are now two parallel critical paths. The next cheapest change that can be made will cost $100 per day saved. This can be done by changing activities A, M, or N. As far as CPM is concerned, it does not matter which is changed first. However, the plant manager will undoubtedly have a very strong preference. After these activities are reduced to their crash times, the next logical single item to change would be B, at a cost of $200 per day. However, if a change is made in both items C and H, the cost would only be $180. Since these items are on separate parallel critical paths, an item on both paths must be changed to shorten the time required to complete the total job. If only one is changed, that activity merely leaves the critical path. A summary of all the changes, in order, is given in Table 13-3. It

Table 13-3

Optimal Change in Going from
Normal to Crash Times for the
Activities Given in Table 13-2

| Order in Which Changed | Activity Changed | Number of Days Saved | Cost of Change | Cost per Day Saved |
|---|---|---|---|---|
| 1 | C | 5 | $ 400 | $ 80 |
| 2 | A | 4 | $ 400 | $100 |
| 3 | M | 2 | $ 200 | $100 |
| 4 | N | 1 | $ 100 | $100 |
| 5 | C&H | 10 | $1,800 | $180 |
| 6 | B | 5 | $1,000 | $200 |
| 7 | L | 1 | $ 500 | $500 |
| 8 | R | 2 | $1,000 | $500 |
| 9 | S | 1 | $ 500 | $500 |
| 10 | S&U | 1 | $ 600 | $600 |
| Total | | 32 | $6,500 | $203 |

shows that a maximum of 32 days can be saved at a minimum cost of $6,500. The crash cost is $62,150. This is $4,200 less than if all the activities were crashed.

For the procedure just presented the assumption of linearity is traditional, even though it is frequently wrong. Sometimes a man can be hired for a day or shifted from some other job to speed up a task; then the assumption of linearity may be valid. However, if a new procedure is proposed, such as using a crane rather than a hoist, there is a quantum change. It is either one or the other—either the normal way is used or the crash procedure. To assume linearity in this case would be totally wrong.

### Planning Is Not Scheduling

CPM is known as a planning tool, not a scheduling device. The CPM diagram tells us what must be completed before an activity can begin. It also tells us that the activities on the critical path must be done in sequence, and must immediately follow each other if the job is to be completed in the minimum possible time. It cannot tell, when two persons are changing a tire, how each person should be scheduled to complete the job in the shortest time, nor does it tell how fast that job can be done. CPM also does not consider the interactions that may occur. In the tire-changing problem activities B, C, and D all involve removing an item from the trunk. If three different people tried to perform these acts at the same time, they would get in each other's way and, besides the cursing that would result, each of these activities would end up taking longer than predicted. This same problem arises when there is only one crane at a plant site and the CPM diagram indicates that two items need to be positioned using the crane at the same time. Obviously this cannot be done unless the contractor wishes to pay to have another crane transported to the plant site. Similarly, if a task requires five electricians, and only three are available, it cannot be done in the allotted time.

### MANPOWER AND EQUIPMENT LEVELING

CPM is an important scheduling aid, but from it alone a schedule cannot be devised. To schedule a project, after a CPM or PERT diagram has been constructed the planner must evaluate the work force and special equipment needed for each activity. Then he must devise a scheduling plan that will maintain a fairly even level of labor and assure the most efficient use of specialty items.

Suppose the CPM diagram for a job is given in Figure 13-3, and the men required for each task are given in Table 13-4. If each activity is scheduled at the earliest possible time it could begin, and it is desired to complete the job in the shortest possible time, number of men employed changes from 5 to 37 and fluctuates widely. (See figure 13-4) This is obviously an intolerable situation.

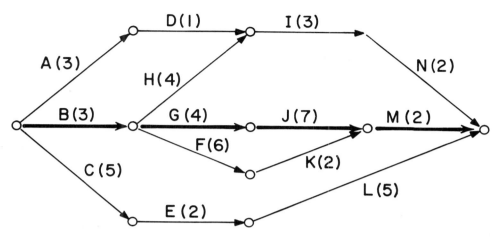

Figure 13-3   The arrow diagram for a project where manpower leveling is important.

Table 13-4

Manpower Required for Activities Diagrammed in Figure 13-3

| Activity Designation | A | B | C | D | E | F | G | H | I | J | K | L | M | N |
|---|---|---|---|---|---|---|---|---|---|---|---|---|---|---|
| Time to Complete Activity | 3 | 3 | 5 | 1 | 2 | 6 | 4 | 4 | 3 | 7 | 2 | 5 | 2 | 2 |
| Number of Men Required | 14 | 16 | 6 | 13 | 15 | 6 | 3 | 3 | 12 | 5 | 13 | 7 | 12 | 6 |

The clear part of the bar chart indicates the slack in each activity. This is called the *float*. A float of 3 days means the activity may begin on any of 4 consecutive days. The activities on the critical path have no float. All the other activities have some leeway in the time at which they may be started. The manager must consider all the possibilities in order to reduce to a minimum the fluctuations in manpower. Figure 13-5 shows the results of juggling the starting times for the activities. In doing this, it was assumed that once an activity had begun it would be pursued until it was finished, using the given crew size. No variance in the manpower or time to complete the task was considered. This is the same as assuming that the optimal number of men required and the time needed to complete each activity had been determined, and that any variance in the number of men would result in unacceptable inefficiencies.

For many jobs the classification of men merely as laborers may be inadequate. This procedure may need to be used for each trade (such as electricians, plumbers, millwrights). A similar type of procedure may be used for scheduling equipment such as electric welders and cranes.

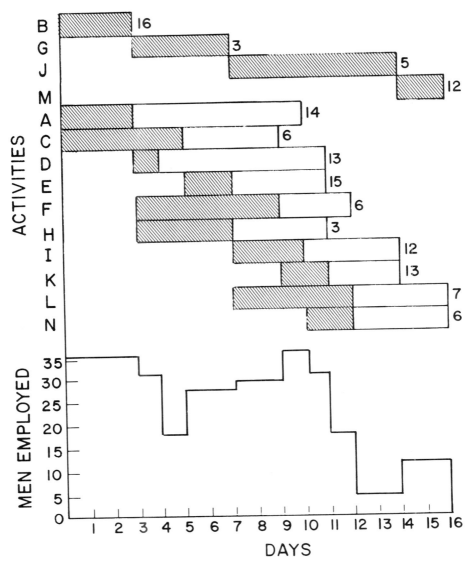

Figure 13-4   The number of men employed in each activity and totally if each activity for a project begins as soon as it can.

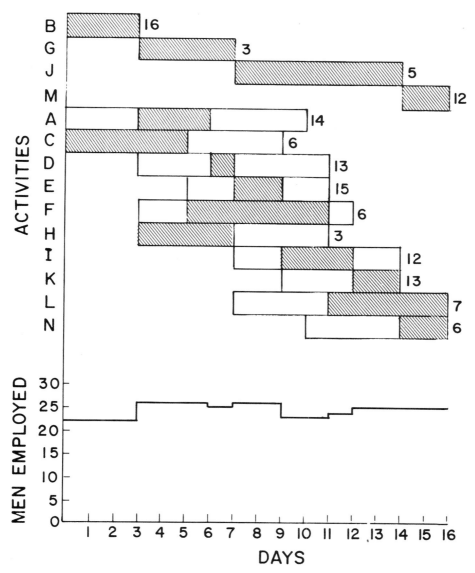

Figure 13-5  The number of men employed in each activity and totally if the starting times are varied in order to reduce the fluctuations in manpower that occurred in Figure 13-4. The job is still to be completed in the minimum amount of time possible assuming no activities are crashed.

## COST AND SCHEDULE CONTROL

Once a project has been approved, budgeted, and scheduled, it is important to make certain that the costs are within control limits and the project will meet its set completion date. From the schedule made using the CPM diagram, the project manager can tell where each activity should be, and compare this with where things are currently. This is done by having competent people estimate weekly or biweekly what percentage of each activity had been completed. This job could be assigned to the company representative or the contractor. An accountant can report how much money has been allocated, and this can be compared with the amount budgeted. By monitoring these items, the manager can tell when the costs are mounting too rapidly or the project is lagging behind schedule. Then he can make whatever decisions may be necessary to keep the price down or to speed up the project. Often both cannot be done. Even when this is the case, it is better for a competent manager to make the decision than to leave the decision to chance.

From the cost information, cost engineers can update their estimates so that future estimates will be more reliable. The planners can similarly update their time estimates. Even though this costs money, it is the best way a company can improve its predictive ability for future projects, as well as maintaining control over the current job.

Table 13-5

Estimated Activity Requirements for Process Plant Construction

| Cost of Plant Construction | Average Number of Activities |
| --- | --- |
| $ 1,000,000 | 900 |
| $ 5,000,000 | 3,500 |
| $ 8,000,000 | 4,100 |
| $12,000,000 | 6,600 |
| $20,000,000 | 8,700 |

Source:   Cosinuke, W.: *Critical Path Technique,* monograph published by Catalytical Construction Co., Philadelphia 1961. (I have been informed that since 1961 the number of activities listed above has increased. This is in part due to additional requirements imposed by clients, state and federal agencies, and labor unions, as well as an upgrading of the planning criteria).

## TIME FOR COMPLETING ACTIVITY

The number of activities into which a given job is divided depends on the size of the job. An estimate for process plants is given in Table 13-5. For any reasonably sized construction project, it is obvious that no one person can have all the information necessary to estimate the time for completing each activity. He must rely on many different people to get this information. In obtaining this information he must be aware that how a person responds to questioning depends on many factors.

Consider the way a politician and a job-seeker respond to questions. A politician may be very noncommital in public, but may be very open with his close friends. He wants to be reelected. At a job interview, most people are more conservative in their opinions than when they are talking to a sociologist or a close friend. This is because what the job-seeker says to an interviewer can affect his future. He wants the job, and feels that if he provokes anger he may not be recommended very highly.

The sequence of questions on a test is very important. If the first few questions are very difficult, the person being tested may get discouraged and do poorly throughout. The reverse may happen if the first few questions are easy. Even when there are no "correct" answers, the sequential order is important. The Minnesota Multiphasic Personality Inventory (MMPI) consists of over 500 questions. Many of these questions are not used in any evaluation. However, if these questions were eliminated, the responses obtained for the other questions would be different.

A manager seeking various time estimates must weigh all the responses he gets, since he can get different answers from the same person, depending on:

1. The order in which he asks for the information.
2. Whether the person feels the information is important.
3. Whether the person feels the activity is on the critical path.
4. Whether the person will be held responsible for a poor estimate.
5. Whether the person responsible for the estimate will be responsible for completing the activity.

The solution may be to use the *Delphi approach*. One asks a number of knowledgeable people to provide estimates. Then if there is a wide variance in answers, the experts are asked to evaluate their peers' responses. The manager must then use his experience to evaluate these.

## COMPUTERS

Most large computer companies have canned programs for PERT, CPM, and manpower leveling for sale. Such programs are a necessity for a project manager.

He must update the schedules at least monthly. This means obtaining a new CPM diagram based on current knowledge. He wants the results as soon as possible after the raw data have been collected, so he can make decisions while the information is current. Doing the calculations by hand would take too long for any large project.

The reason why he must continually revise his plans is that some activities will be completed more quickly than expected, and others will take longer. This may cause the critical path to change. Also, many unforeseen events will arise. In one instance, when a large Pfaulder vessel was delivered to the plant site the glass lining was found to be cracked. It had been scheduled to be installed shortly after its arrival. Instead, the vessel had to be shipped back to the manufacturer, repaired, and then returned to the plant site before it could be installed. This took over 3 months. Not only was this activity delayed, but a corridor from the railroad siding to the planned location had to be kept open. This delayed the installation of piping, wiring, and some other equipment. It changed the critical path and caused a major revision in scheduling.

## PERT

It has already been noted that PERT was developed to speed up the Polaris project. The Navy was dealing with hundreds of subcontractors, each constructing parts in a different location. These were then sent to an assembly point where the submarine was put together. Until the items arrived there, the activities were essentially independent. This is different from CPM, which was developed mainly for maintenance turnarounds at a given location. Under these conditions, everything was closely interrelated.

The Navy was interested in whether the individual contractors could meet a common schedule so that the final assembly would not be delayed. It developed the concept of getting three time estimates for each activity, and then using probability theory to determine whether a schedule could be met. One time estimate is the most likely time ($t_m$) it would take to finish the task. This is the same as the CPM time estimate. The second is the optimistic time estimate ($t_o$). This is the minimum time in which the activity could be completed if no problems arose. It can be assumed that at least 95% of the time the project will take longer than this to complete. The third is a pessimistic time estimate ($t_p$). If nothing went right, this is how long it would take to finish the project. The project will take longer than this less than 5% of the time. All of these estimates assume standard operations; they should not be confused with a crash time estimate, where more men might be used or a new procedure might be devised. A crash time estimate is the normal time to perform a task under different circumstances.It may or may not be faster than the optimistic time estimate.

From these three estimates a single value called the expected time ($t_e$) is obtained from the following formula;

$$t_e = \frac{t_o + 4t_m + t_p}{6}$$

This estimate is used in much the same way the normal time estimate was used in constructing a CPM diagram and determining the critical path. Then an estimate is made of the probability of completing the total job in a certain time period.

To illustrate how a computer might assist in these calculations, a numerical procedure that could be programmed will be used. Let us reconsider the problem of installing a scrubber on an existing process. Even though the procedure does not require a diagram, it will be useful in explaining some details. A PERT diagram does not differ greatly from a CPM diagram (see Fig. 13-2). However, the nodes on the PERT diagram are numbered in such a way that no matter which path is followed through the diagram, the number of each node is always greater than any previous one.The nodes are usually numbered initially by tens. This allows other activities, which may have been forgotten, to be inserted later without renumbering all the activities that succeed them. All the activities are coded by denoting the node at which they begin (P = Preceding node) and the node where they must be completed (S = Succeeding node). Activity A is symbolized as (10,20). These numbers are given in Table 13-6, along with the three time estimates and the expected time. In parentheses are the code letters used in Table 13-2. Note that if dummy activities were not inserted activities C and D, as well as L and K, would have the same code numbers. The code numbers would not then uniquely apply to a given activity, and the computer would be unable to distinguish them.

Table 13-6

Time Estimates (Days) for the Installation of a Scrubber in an Existing Process

| P | S | Activity | $t_o$ | $t_m$ | $t_p$ | $t_e$ |
|---|---|---|---|---|---|---|
| 10 | 20 | Process design (A) | 11 | 14 | 16 | 13.8 |
| 20 | 30 | Project design (B) | 12 | 14 | 17 | 14.2 |
| 30 | 50 | Procure scrubber (C) | 75 | 85 | 100 | 85.8 |
| 30 | 70 | Procure pump (D) | 58 | 68 | 78 | 68.0 |
| 30 | 40 | Procure piping (E) | 40 | 45 | 55 | 46.7 |
| 40 | 100 | Prefabricate pipe (F) | 4 | 5 | 6 | 5.0 |
| 20 | 100 | Set up scaffolding (G) | 2 | 2 | 2 | 2.0 |
| 30 | 60 | Site preparation (H) | 30 | 40 | 60 | 41.7 |
| 60 | 70 | Pour concrete and allow to set (J) | 40 | 40 | 41 | 40.2 |
| 70 | 100 | Install pump (K) | 3 | 3 | 4 | 3.2 |
| 70 | 90 | Install scrubber (L) | 3 | 4 | 5 | 4.0 |
| 100 | 110 | Fit up pipes and valves (M) | 3 | 4 | 5 | 4.0 |
| 110 | 120 | Weld pipe (N) | 3 | 3 | 4 | 3.2 |
| 30 | 80 | Preparation for and shutdown of Process (P) | 12 | 14 | 16 | 14.0 |
| 80 | 100 | Remove some existing piping and make changes in existing equipment (Q) | 12 | 16 | 20 | 16.0 |
| 120 | 130 | Pressure test, flush, and dry run (R) | 6 | 7 | 10 | 7.3 |
| 130 | 150 | Start up process and test scrubber (S) | 4 | 5 | 8 | 5.3 |
| 130 | 140 | Remove scaffolding (T) | 1 | 1 | 1 | 1.0 |
| 140 | 150 | Clean up (U) | 3 | 3 | 3 | 3.0 |
| 50 | 70 | Dummy (V) | 0 | 0 | 0 | 0 |
| 90 | 100 | Dummy (W) | 0 | 0 | 0 | 0 |

Beginning with the first activity in Table 13-7, the earliest time each activity could end ($T_E$ = Early completion) is determined. Remember, before any activity can begin all activities that end on its preceding node must be completed. The early completion time can then be determined by adding the expected time of the activity to the earliest time at which it can begin. If there is only one path from the beginning of the project to the start of the activity, the early completion time is the sum of the expected times ($t_e$) of the activity and all those activities that precede it. If there is more than one path, only the expected times of the activities on the path that gives the latest starting time should be added to the expected time of the activity. For instance, before the activities beginning at node 70 can begin, activities (30,70),

(50,70), and (60,70) must be completed. These are the only activities whose ending node is 70. There are therefore three paths to node 70. Since all must be completed, the one that is completed last is the one that determines the earliest time any activity beginning at 70 can start. In this instance, it is the path through (50,70). The earliest time at which (70,90) can be completed is the sum of the expected time of (10,20), (20,30), (30,50), (50,70), and (70,90), or 117.8 days. The best way of understanding the calculations is to follow them on a PERT or CPM diagram. Note that the activities in Table 13-7 have been rearranged so that the preceding nodes are in numerical order. This facilitates the calculating procedure.

Table 13-7

PERT Worksheet for the Installation of a Scrubber in an Existing Process

| P | S | Activity | $t_e$ | $T_E$ | $T_L$ | Slack $T_L - T_E$ |
|---|---|---|---|---|---|---|
| 10 | 20 | (A) | 13.8 | 13.8 | 11.2 | −2.6 |
| 20 | 30 | (B) | 14.2 | 28.0 | 25.4 | −2.6 |
| 20 | 100 | (G) | 2.0 | 15.8 | 115.2 | 99.4 |
| 30 | 40 | (E) | 46.7 | 64.7 | 110.2 | 45.5 |
| 30 | 50 | (C) | 85.8 | 113.8 | 111.2 | −2.6 |
| 30 | 60 | (H) | 41.7 | 69.7 | 71.0 | 1.3 |
| 30 | 70 | (D) | 68.0 | 96.0 | 111.2 | 15.2 |
| 30 | 80 | (P) | 14.0 | 42.0 | 99.2 | 57.2 |
| 40 | 100 | (F) | 5.0 | 69.7 | 115.2 | 45.5 |
| 50 | 70 | (V) | 0.0 | 113.8 | 111.2 | −2.6 |
| 60 | 70 | (J) | 40.2 | 109.9 | 111.2 | 1.3 |
| 70 | 90 | (L) | 4.0 | 117.8 | 115.2 | −2.6 |
| 70 | 100 | (K) | 3.2 | 117.0 | 115.2 | −1.8 |
| 80 | 100 | (Q) | 16.0 | 58.0 | 115.2 | 57.2 |
| 90 | 100 | (W) | 0.0 | 117.8 | 115.2 | −2.6 |
| 100 | 110 | (M) | 4.0 | 121.8 | 119.2 | −2.6 |
| 110 | 120 | (N) | 3.2 | 125.0 | 122.4 | −2.6 |
| 120 | 130 | (R) | 7.3 | 132.3 | 129.7 | −2.6 |
| 130 | 140 | (T) | 1.0 | 133.6 | 132 | −1.6 |
| 130 | 150 | (S) | 5.3 | 137.6 | 135 | −2.6 |
| 140 | 150 | (U) | 3.0 | 136.6 | 135 | −1.6 |

After calculating all the early completions, a time is chosen in which it is hoped the project will be completed. Assuming the time will be met, the planner now goes backward through the table in a similar way to obtain the latest time ($T_L$) at which each activity can be completed to allow the project to be finished at the desired time. Assume it is desired to complete the project in 135 days. For this to occur, activities

(130,150) and (140,150) must be completed by day 135. Activity (130,140) must be completed in time to finish (140,150), or 3 days prior to day 135. This is day 132. Two activities begin at node 130. Any activity ending at 130 must be completed in time for the longest of the two paths that follow to be completed. In this case, the expected time of the longest path through (130,150) is 5.3 days. Therefore, (120,130) must be completed, at the latest, 129.7 days after the job begins.

The *slack* is the difference between the latest time at which an activity must be completed to meet a schedule and the earliest time at which it can be completed. This would be the same as the float if the normal rather than the expected times are used and if the desired completion date is the minimal time required to complete the project. The items with the minimum slack are on the critical path. Thus, by following this procedure the items on the critical path are determined without the use of a diagram.

In calculating the probability that the project can be completed in a given length of time, all the functions that are not on the critical path will be ignored. This is the same as assuming that their slack is great enough that they will not slow down the completion of the project. Then the sum of the variances, of all the items on the critical path will be estimated using the following equation:

$$\sigma_{TE}^2 = \sum_{i=1}^{n} \left[ \frac{t_{pi} - t_{oi}}{6} \right]^2$$

where   $\sigma_{TE}^2$   =   a variance approximation

        $n$   =   number of items on the critical path

        $t_{pi}$   =   pessimistic time estimate of the $i^{th}$ activity on the critical path

        $t_{oi}$   =   optimistic time estimate of the $i^{th}$ activity on the critical path

Next a function that is assumed to be normally distributed with a mean of zero and a variance of one is created.

$$Z = \frac{(T_L - T_E)_{min}}{(\sigma_{TE}^2)^{1/2}}$$

where   $(T_L - T_E)_{min}$ = slack for items on critical path

        $Z$ = function with mean = 0 and variance = 1

By going to a table or graph of the normal distribution, the probability that $Z$ will not be exceeded can be determined. This is the probability that the project can be completed on time. The table can be found in nearly any elementary statistics book[3] and in many handbooks.

**Example 13-1**

Complete the example of the installation of a scrubber in an existing process.

$$\sigma_{TE}^2 = \left(\frac{16-11}{6}\right)^2 + \left(\frac{17-12}{6}\right)^2 + \left(\frac{100-75}{6}\right)^2 + \left(\frac{0-0}{6}\right)^2 + \left(\frac{5-3}{6}\right)^2$$

$$+ \left(\frac{0-0}{6}\right)^2 + \left(\frac{5-3}{6}\right)^2 + \left(\frac{4-3}{6}\right)^2 + \left(\frac{10-6}{6}\right)^2 + \left(\frac{8-4}{6}\right)^2$$

$$= \frac{716}{36} = 19.9$$

$$Z = \frac{-2.6}{(19.9)^{1\!/\!2}} = -0.582$$

The probability of completing the project in 135 days is 0.28. This means that 28% of the time the deadline will be met.

Traditionally the PERT diagram did not involve costs, but in the early 1960s a plan called PERT/Cost was developed that was similar to the cost-control concepts of CPM. Some authors claimed they were now essentially the same. PERT/Cost involves obtaining one cost for each activity. If the expected time were the same as the normal time, the procedure for cost control would be the same as for a CPM system. The PERT system can also be easily adapted to the method presented for speeding up projects (crashing). The remarks made previously about scheduling and obtaining time estimates apply equally well to PERT and CPM.

Both methods work well, although CPM and its many variations seem to be used more extensively in the chemical industry.

## PROBLEMS

Problem 1.

List all the activities involved in either making a float for the homecoming parade or decorating a house for homecoming. Assume both displays will be judged at night, so lighting is essential. Put a time on each activity and obtain the critical path. Which items can be and should be crashed to finish ahead of time?

## Problem 2.

Often certain pipes in chemical plants must be replaced because of corrosion and erosion. Table 13-P2 gives the costs and times of the activities which are involved in renewing a pipeline.

(a) Draw a CPM diagram of the project
(b) Determine the critical path and the minimum time required to complete the project
(c) Determine the minimum crash cost
(d) Determine the minimum crash time

Table 13-P2

Activities, Costs, & Times for
Renewing a Pipeline

| Job Code | Job | Sequential Code | Elapsed Time (max.) | Cost ($) | Elapsed Time (min.) | Cost ($) |
|---|---|---|---|---|---|---|
| Q | Lead time | 1,2 | 10 | — | 5 | — |
| R | Line available | 1,5 | 44 | — | 28 | — |
| A | Measure & sketch | 2,3 | 2 | 300 | 1 | 400 |
| B | Develop materials list | 3,4 | 1 | 100 | 1 | 100 |
| C | Procure pipe | 4,7 | 30 | 850 | 20 | 1,100 |
| D | Procure valves | 4,8 | 45 | 300 | 30 | 600 |
| E | Prefabricate sections | 7,9 | 5 | 1,200 | 3 | 2,000 |
| F | Deactivate line | 5,6 | 1 | 100 | 1 | 100 |
| G | Erect scaffold | 4,6 | 2 | 300 | 1 | 500 |
| H | Remove old pipe | 6,9 | 6 | 400 | 3 | 1,000 |
| I | Place new pipe | 9,10 | 6 | 800 | 2 | 2,000 |
| J | Weld pipe | 10,11 | 2 | 100 | 1 | 300 |
| K | Place valves | 8,11 | 1 | 100 | 0.5 | 250 |
| L | Fit up | 11,12 | 1 | 100 | 0.5 | 250 |
| M | Pressure test | 12,14 | 1 | 50 | 0.5 | 100 |
| N | Insulate | 11,13 | 4 | 300 | 2 | 700 |
| O | Remove scaffold | 13,14 | 1 | 100 | 0.5 | 200 |
| P | Clean up | 14,15 | 1 | 100 | 0.5 | 200 |
| | | | | $5,200 | | $9,800 |

Source: *Arrow Diagram Planning,* monograph by E. I. Du Pont de Nemours and Co., Wilmington, 1962.

Problem 3.

A company is interested in introducing a new product to the public. All technical problems have been solved and the economic prospect appears rosy. The various activities involved and their time estimates are given in Table 13-P3.

    (a) Draw a PERT diagram
    (b) Obtain the critical path

Table 13-P3
Activities and Times for Introducing a New Product

| Activity | Number | Optimistic, Wk. | Most Likely, Wk. | Pessimistic, Wk. |
|---|---|---|---|---|
| Large-scale engineering feasibility studies | 1-3 | 8 | 9 | 10 |
| Prepare mini-plant quantity of product | 1-2 | 6 | 7 | 9 |
| Conduct lab tests for government approval | 2-4 | 5 | 5 | 5 |
| Train sales force | 2-5 | 8 | 10 | 11 |
| Plant construction | 3-6 | 9 | 12 | 15 |
| Analyze shipping, storage, warehouse, etc. | 2-6 | 11 | 15 | 20 |
| Time for government to approve submitted tests | 4-5 | 3 | 4 | 6 |
| Install quality control and and special safeguards | 5-6 | 5 | 6 | 8 |
| Plant startup | 6-7 | 4 | 5 | 10 |
| Advertising campaign and initial customer contacts | 5-7 | 8 | 10 | 12 |

Source:   Klimpel, R.R.: "Operations Research: Decision Making Tool," *Chemical Engineering,* Apr. 30, 1973, p. 87.

Problem 4.

Table 13-P4 gives information about the manufacture of a unit of electronic equipment. From those data obtain

    (a) A PERT diagram
    (b) The minimum time required to complete the manufacture
    (c) The critical path
    (d) The probability that the manufacturing process can be completed in 26 days

Table 13-P4

Data for the Manufacture of a
Unit of Electronic Equipment

| Job Code | Activity | Time (days) | | | Cost ($) |
|---|---|---|---|---|---|
| | | $t_o$ | $t_m$ | $t_p$ | |
| 10, 20 | Release final design | 0 | 0 | 0 | 0 |
| 20, 30 | Prepare layouts | 3 | 4 | 5 | 325 |
| 30, 70 | Fabricate machine shop parts | 6 | 16 | 20 | 406 |
| 20, 40 | Engineering analysis | 1 | 2 | 3 | 123 |
| 40, 50 | Place outside supplier orders | 2 | 3 | 5 | 0 |
| 40, 60 | Place interworks orders | 2 | 3 | 5 | 0 |
| 50, 70 | Procure outside supplier parts | 6 | 8 | 10 | 2,638 |
| 60, 70 | Procure interworks parts | 12 | 16 | 20 | 2,240 |
| 70, 80 | Final assembly | 2 | 3 | 6 | 309 |
| 80, 90 | Inspect and test | 1 | 1 | 4 | 199 |

Source:    Frantz, R.A., Nothern, L.B.: "PERT/Cost," *The Western Electric Engineer*, July 1964, p. 25.

## Problem 5.

Determine the critical path and a reasonable schedule for the tasks given in Table 13-P5.

Table 13-P5

Data for Problem 5

| Job Code | Time | Number of Men of Specific Craft | |
|---|---|---|---|
| | | A | B |
| 1,2 | 2 | 2 | |
| 1,3 | 1 | | 2 |
| 2,3 | 4 | 2 | |
| 2,4 | 2 | 2 | |
| 3,4 | 2 | 2 | |
| 3,5 | 5 | | 2 |
| 4,5 | 1 | 2 | |
| 4,6 | 2 | | 2 |
| 5,6 | 3 | 2 | |

Source:    Martino, R.L.: "Plain Talk on Critical Path Method," *Chemical Engineering,* June 10, 1963, p. 221.

Problem 6.

Assume each activity in Table 13-P6 begins at its earliest starting time, and determine the critical path and the minimum and maximum manpower needs. Then manipulate the activities to minimize the variance in manpower requirements, assuming the project is to be completed in the shortest possible time.

Table 13-P6

Activities, Times, and Manpower Needs for Problem 6

| Preceding Node | 0 | 2 | 4 | 2 | 2 | 2 | 6 | 8 | 10 | 12 | 12 | 14 | 16 |
|---|---|---|---|---|---|---|---|---|---|---|---|---|---|
| Succeeding Node | 2 | 4 | 6 | 6 | 10 | 8 | 12 | 14 | 14 | 14 | 16 | 16 | 18 |
| Time | | 3 | 1 | 2 | 4 | 5 | 8 | 2 | 3 | 5 | 6 | 9 | 1 | 7 |
| Number of Men | | 6 | 4 | 3 | 5 | 3 | 2 | 5 | 6 | 7 | 2 | 3 | 1 | 5 |

Source of problem unknown.

## References

1. Zalokar, F.J.: *The Critical Path Method–A Presentation and Evaluation,* monograph published by General Electric Co., Schenectady, May 18, 1964.
2. Mauchly, J.W.; "Critical Path Scheduling," *Chemical Engineering*, Apr. 16, 1962, p. 141.
3. Brownlee, K.A.: *Statistical Theory and Methodology*, Ed. 2, Wiley, New York, 1965, p. 558.

# CHAPTER 14

# OPTIMIZATION TECHNIQUES

Optimization techniques are procedures to make something better. Some criteria must be established to determine whether something is better. The single criterion that determines the best among a number of alternatives is referred to as the *performance index* or the *objective function*. Economically, this is the expected profit for a plant design. It may be expressed as the net present value of the project.

Some conditions usually must be met. For the process engineer, the stipulations as given in Chapter 1 were that a given product meeting certain quality standards be produced safely. These are called the *constraints*. They limit the problem. The scope is a list of agreed-upon constraints.

To illustrate, consider an amateur fisherman who has only a 2-week vacation every year. He works to optimize the pleasure he will have during the period. His performance index may be the number of hours he spends with his rod and reel in a boat. However, he cannot spend 24 hours a day for 2 weeks in a boat fishing, since there are a number of constraints. First, he requires a certain amount of sleep each day to fully enjoy the time he spends fishing. Second, his wife and children, who are on vacation with him, want him to spend some time water-skiing with them each day. He and his wife are the only ones who can safely operate their only boat, and if she is to do any skiing he must be in command. If he ignores this request, their unhappiness will diminish his enjoyment. He could rent or buy another boat and hire someone else to run a boat while his wife water-skis, but he has a third constraint—he has a limited amount of money to spend. Fourth, he must spend some time cleaning and frying the fish he catches, since his wife flatly refuses to do this. Somehow our fisherman will attempt to optimize his vacation within these and probably many other constraints.

The performance index of a company or a chemical plant is its profit. It attempts to increase this while acknowledging ethical constraints, governmental constraints, and financial constraints. The ethical constraints are, or should be, to be honest in all dealings with employees and with others, and to produce a high-quality product or service. The governmental constraints are various laws governing pollution, hiring practices, minimum wages, safety regulations, union bargaining, rates and methods of depreciation, and the like. The general financial constraint is the amount of money the company can raise at reasonable interest rates.

An optimization procedure is a way of maximizing or minimizing the performance index. There are many different procedures, some of which will be discussed later on in this chapter. To determine the best optimization procedure, a performance index for the procedures must first be established. It could be the procedure that reaches a point within 5% of the optimum in the shortest time. It could be the one that requires the fewest steps or costs the least to reach that point. It could have constraints like a maximum cost or a time limit.

Say a farmer wishes to obtain the greatest crop yield at the least expense. However, in the temperate zones he can only raise one crop per year. This is equivalent to saying that he can only perform one experiment on each plot of ground each year. One optimization procedure he can follow is to heed the advice of the county agriculture agents. In doing this he relies upon the extensive research done by the state agriculture department and the state agricultural college. For the small farmer this is probably the best thing he can do.

Much of the information presented in the first eight chapters of this book consisted of guidelines that would help the process engineer to save time and money. What has been presented is an optimization procedure for obtaining a preliminary chemical plant design. Like the wise small farmer, the efficient process engineer relies heavily upon information that has been obtained by others. We do not need to reinvent the wheel every time we want to construct a new vehicle.

This attempt to use available information in the most efficient way is basic to all optimization procedures. They all begin in the same way by gathering as much information as possible. This is, of course, subject to the constraints of time and money. Then some starting point is chosen and one or a series of tests are made. From the results of these tests a decision will be made as to where future tests should be conducted. The resulting information from these tests will in turn be analyzed and used to choose the position of further tests, if any more are needed. Finally, some procedure, the *end game*, must be available to decide when the optimum has been reached.

## STARTING POINT

No matter which optimization technique is used, it must begin somewhere. If a good choice is made, even a poor optimization procedure will be fairly efficient. Unfortunately, there are no specific guidelines that will point to a good starting point. The best advice is: gather all the information available and use the best engineering judgment.

The engineer's preliminary chemical plant design can be considered the starting point for obtaining an optimal plant design. The point at which to begin operating a new chemical plant is at the operating conditions that were found to be best in the pilot plant for that process.

Our fisherman may begin his vacation by spending 4 hours water-skiing and 6 hours fishing each day. This is the starting point of his optimization procedure. If he had started with 1 hour of water-skiing his family would be unhappy. Contrarily, only 1 hour of fishing would have made him miserable. Luckily, the best times for fishing are at dawn and dusk, while water-skiing is most enjoyable during the middle of the day.

## ONE-AT-A-TIME PROCEDURE

In the one-at-a-time procedure a change is made in a single variable and the results are evaluated. As this procedure is continued, a change in only one variable is made at each step along the way. This is well suited to plant design.

In the Case Study following Chapter 5, the volume of styrene storage was given as 1,088,000 gallons and the size of the storage tanks had to be determined. To facilitate cleaning, the minimum number of tanks was set at 3. Standard tanks were to be used because it was known that specially designed tanks cost more. To determine the optimum size and number of tanks to be used, the initial point should be chosen as 3 tanks, because generally the installed cost per unit volume of large tanks is less than for smaller ones. The optimization procedure would be to choose the next smallest standard size tank and determine whether it is better. If it is, this procedure would be repeated. If not, the optimum would be assumed to be 3 tanks. In this case the size of the tanks was the single change made. No attempt was made to determine the cost savings that might arise if the total storage volume or the material of construction were changed.

After the preliminary design has been completed, the most expensive steps in the process should be investigated to see if some less expensive operation could be used. For instance, maybe an extraction system could be replaced by distillation or crystallization. Maybe two separations can be done in one unit, like a crude still, or possibly some step is unnecessary. Changing the sequence of processing steps may also reduce costs by reducing the size of equipment or the material of construction. Each of these involves comparing two alternatives and selecting the best for further investigations.

Once the basic unit operations and their sequence have been decided upon, each one should be investigated to see how it can be improved. Finally, the conditions at which each step is run should be scrutinized to determine whether they are optimal.

For this last stage, the one-at-a-time procedure may be a very poor choice. At Union Carbide, use of the one-at-a-time method increased the yield in one plant from 80 to 83% in 3 years. When one of the techniques, to be discussed later, was used in just 15 runs the yield was increased to 94%. To see why this might happen, consider a plug flow reactor where the only variables that can be manipulated are temperature and pressure. A possible response surface for this reactor is given in Figure 14-1. The response is the yield, which is also the objective function. It is plotted as a function of the two independent variables, temperature and pressure. The designer does not know the response surface. Often all he knows is the yield at point A. He wants to determine the optimum yield. The only way he usually has to obtain more information is to pick some combinations of temperature and pressure and then have a laboratory or pilot plant experimentally determine the yields at those conditions.

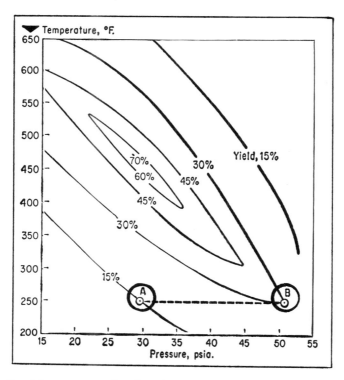

Figure 14-1    Possible response surface for a Chemical Reactor.
Courtesy of Baasel, W.D.: "Exploring Response Surfaces to Establish Optimum Conditions," *Chemical Engineering*, Oct. 25, 1965, p. 147.

The one-at-a-time method is to pick one of the variables and make a number of tests until the optimum for that variable is obtained at a constant level of the other variables. If the temperature is held constant at 250°F in Figure 14-1, the optimum value of the pressure would be 51 psia (point B).

One way to cut down on the number of tests is to approximate the response surface by a quadratic equation and from it to predict where the maximum will occur. The equation at constant $T$ would be

$$Y_T = C_0 + C_1 P + C_2 P^2 \tag{1}$$

where    $Y_T$    = yield at constant temperature

$P$    = pressure in psia

$C_0, C_1$, and $C_2$ are constants

Since there are three constants, we must have the results of three tests to evaluate them. The yield is known at point A (P=29 psia, Y=15%). If the second pressure were chosen as 35 psia the yield would be better (Y=20%). This would indicate that the third point should be taken at a higher pressure, since the yield appears to increase with pressure. It might be taken at 41 psia (Y=25%). If the yield for the second test had been less than for the first, the third experiment point should have been taken at a pressure less than that for the first test, such as 23 psia.

From these three points Equation 2 can be obtained:

$$Y = -23.3 + 1.725P - 0.139P^2 \tag{2}$$

If the first derivative of the yield with respect to pressure is set equal to zero, an approximation of the maximum will be obtained, provided the second derivative is negative. In this case the second derivative is negative and the predicted maximum is at 62 psia. This calculated value could be high because of experimental error or because the quadratic equation is a poor estimator of the shape of the true surface. In this case an error of a few tenths of a per cent in the yield may make a difference of 10 psia in the value of the maximum (see Problems 1 and 2). Once the value of the yield at a pressure of 62 psia is obtained, another approximation of the location of the maximum may be made using the three points that are nearest the highest experimental value. The procedure is then continued until the optimum is reached. The value of this variable (pressure) is now fixed at 51 psia and one of the other independent variables is changed until again the highest possible yield is obtained. In Figure 14-1, at a pressure of 51 psia the optimum temperature is 250°F. If there were other variables these would be changed in turn. After the yield has been maximized for all the variables individually, the procedure would begin again at the new conditions.

On Figure 14-1 there is no way to improve the yield using the one-at-a-time method if the experimenter is at point B. If the temperature alone is changed (a vertical line on the graph) the yield decreases. The same is true if only the pressure is changed (a horizontal line on the graph). But point B has a yield of 30%, and a 70% yield is possible. The only way that yield can be approached is to increase the temperature and decrease the pressure at the same time. This cannot be done using the one-at-a-time method, since it only allows one variable to change at a time.

The last example differs from the previous examples in this section in that they involved discrete variables, while pressure and temperature are continuous functions. The same problem could also arise in the discrete case. For instance, although the initial design might favor crystallization over extraction, if the sequence of processing steps were changed the extractive process might be preferable.

While there is always this possibility of a blind spot, for discrete variables the one-at-a-time procedure is still frequently used. For continuous variables other procedures, which follow, should be used.

## SINGLE VARIABLE OPTIMIZATIONS

When there is only one continuous independent variable, there are some well-developed theories that give the best method for sampling. This is adequately discussed in the literature. Since it is not a typical plant-design situation, it will not be discussed further here. Anyone who is interested should read Douglas Wilde's book *Optimum Seeking Methods*.[1]

## MULTIVARIABLE OPTIMIZATIONS

Many techniques can be used to solve multivariable optimizations. Unfortunately, there is no single best method that applies to every type of response surface. Therefore, I will give a number of different procedures, with the advantages and disadvantages of each one. The reader will then have to decide which one(s) he wishes to use.

There are two major approaches. One is a statistical method involving a factorial design. This will be briefly discussed. Most of the rest of this chapter will be devoted to the sequential methods.

### Factorial Design

The object of this procedure is to obtain enough experimental responses over the entire region being explored that the approximate shape of the response surface can be predicted. Consider the problem of locating the highest point in North America. The factorial design approach would be to place a two-dimensional grid over this area and to sample the heights at each node. Then statistically[2] an approximation of the surface would be made and the region(s) in which the maximum might occur would be chosen for further investigation in a similar way, using a smaller grid. This would be repeated until an approximate peak was identified. In this case we start with the area of North America, which is over 10,000,000 square miles. If the grid lines were a mile apart, millions of points would be required. If they were 100 miles apart, over 100,000 points would still need to be determined; there would be a chance that Mount McKinley might be inside the lines of the grid and the area it is in could be eliminated from further exploration. This is unlikely, since Mount McKinley is part of a mountain range and most probably some high points in the region would be indicated. However, if the highest point were a volcano like Mount Orizaba in Mexico, which stands by itself, the region it is in could easily be eliminated from further consideration in the first series of tests.

This method for optimizing a process parallels the method given in the mapping example. First, some limit must be placed on all variables. Otherwise it would be impossible to cover the entire surface. In the mapping example it was the continental boundaries. Second, for each independent variable a number of specific points that are uniformly spaced and cover its whole range are chosen. The objective function is then obtained at each possible combination of these points. If enough points have been taken, the area within which the optimum can occur has been greatly reduced. This remaining area can then be explored further, using another factorial design to more closely pinpoint the best combination of independent variables.

The number of points used in a factorial design can be reduced by using statistical techniques.[2] Even so, unless the region that is being investigated is very restricted, a factorial design requires so many points to cover a surface adequately that it is impractical for most design problems. It provides a lot of information about the general nature of the surface, but in most design problems all that is desired is the optimum conditions. These can usually be obtained more efficiently using sequential procedures.

## Sequential Procedures

These methods move one step at a time. A strategy that delineates the particular method dictates how to proceed once the results of the most recent tests are available. The one-at-a-time method is an example of an algebraic sequential procedure. The goal of the algebraic sequential methods is to find and follow a path to the summit. The geometric sequential methods attempt to isolate the area in which the maximum may exist.

For all these procedures it will be assumed that the surface is *unimodal*. If we are looking for a maximum, this means that there is only one peak; if we start rising toward that peak from any starting point, as long as we never go downward we can reach the zenith. If we were climbing a unimodal mountain, we could never reach any point other than the summit from which the only direction is down. A level portion is permissible. If a minimum is being sought, the definition of unimodal is similar; the surface has only one valley, and the lowest point can be found following any path provided no upward steps are taken.

A unimodal surface has been chosen because we have no way of dealing sequentially with a surface that has two or more peaks or valleys. The only reasonable approach is to start the sequential procedure at a number of widely disparate points and to determine whether the paths converse toward the same optimum. If two or more different peaks or valleys are indicated, the investigator must find the optimum for each possible peak or valley and then select the best. There is no way of knowing that all peaks or valleys have been explored except to map the whole surface finely using some factorial design.

When the algebraic methods are used, care must be taken that the constraints are obeyed. This usually means following a boundary until the search leaves the vicinity of the constraints. This should be kept in mind while reading about the various procedures.

**Steepest Ascent**

To understand the method of steepest accent, consider a hiker who wishes to reach the summit of a volcanic island (assumed unimodal) by taking the shortest path. The shortest path is the steepest. Suppose our hiker is a mathematician, and rather than use his eyesight he decides to determine mathematically the best direction in which to proceed. Further, he knows that if he goes 20 ft to the north of his current position he will rise 20 ft and if he goes 20 ft to the east he will fall 10 ft. If the surface were a plane, he could approximate it by the following equation:

$$A = C_0 + C_1 N + C_2 E \tag{3}$$

where  $A$ = elevation at a specific point
       $N$ = number of feet the point is north of a base point
       $E$ = number of feet the point is east of a base point
       $C_0, C_1$, and $C_2$ are constants

Consider any two points on the plane. The difference in elevation between them is

$$\Delta A = C_1 \Delta N + C_2 \Delta E \tag{4}$$

where  $\Delta A = A_2 - A_1$ = difference in elevation between points 2 and 1
       $\Delta N = N_2 - N_1$ = difference in the northerly direction between points 2 and 1
       $\Delta E = E_2 - E_1$ = difference in the easterly direction between points 2 and 1

From the data given, when the climber goes 20 ft directly north the elevation increases 20 ft:

$$20 = C_1\, 20$$

or

$$C_1 = 1.0$$

Similarly, when he goes east 20 ft his elevation decreases 10 ft:

$$-10 = C_2\, 20$$

or

$$C_2 = -0.5$$

The result is

$$\Delta A = 1.0 \Delta N - 0.5 \Delta E \tag{5}$$

To travel in the steepest direction he should go 2 ft north for every foot he goes west. The higher the coefficient in front of a variable, the greater the change that will occur from increasing that variable a unit step. When the constraints allow it, the variables having the highest coefficients are the ones that are changed the most.

In this example a linearization was performed. For most mountains this is a gross simplification, and at best its validity can only be assumed in a region close to the experimental points. Anyone who has ever climbed a hill can verify this. If the equation developed were assumed valid everywhere, the hiker could theoretically climb to a point beyond the moon by following the path given. Thus, after he has gone a short distance the hiker should determine whether he is still on the best path.

One strategy that has often been used is to proceed along the path of steepest ascent until a maximum is reached. Then another search is made. A path of steepest ascent is determined and followed until another maximum is reached. This is continued until the climber thinks he is in the vicinity of the global maximum. To aid in reaching the maximum, the technique of using three points to estimate a quadratic surface, as was done previously, may be used.

When the response surface is elliptical, this general strategy seems to wander back and forth across the surface, especially in the vicinity of the global maximum.[1] The method of parallel tangents (PARTAN)[3] was developed to help speed up the search. After two times through the above procedure a line is drawn between the original starting point (the first base point, $b_0$) and the best point obtained (the new base point, $b_1$). A search for a maximum is made on the continuation of this line. Note that this line is not a path of steepest ascent. Then the total procedure is repeated: a search is made to find the path of steepest ascent and this is followed to a maximum; then another search is made and the resulting path of steepest ascent followed until a peak is reached. A line is drawn between this peak, the best point obtained so far (the new base point, $b_2$), and the previous base point ($b_1$), and a maximum along this line is obtained. This pattern is repeated until the vicinity of the global maximum is apparently reached.

One problem with this or any other method using gradients is that the "best path" obtained is dependent on the units used. If different units are used a different path will be indicated. To illustrate this, suppose it is desired to improve the yield (y) of a plug flow reactor when the feed rates and compositions are constant. At the usual operating conditions of 50 psia and 500°K a yield of 60 lb/hr is obtained. In what order should the pressure (P) and the temperature (T) be changed? To reduce costs, it is desirable to minimize the number of experiments performed, hence the method of steepest ascent is to be used. When a test is performed at 50 psia and 510°K, the yield is found to be 60 lb/hr. When another experiment is run at 60 psia and 500°K, the yield is again 60 lb/hr. If the surface is linearized it can be expressed as:

$$y = C_0 + C_1 P + C_2 T \tag{6}$$

or
$$\Delta y = C_1 \Delta P + C_2 \Delta T \tag{7}$$

where $y$ = yield
$P$ = pressure
$T$ = temperature
$\Delta P = P_2 - P_1$
$\Delta T = T_2 - T_1$
$\Delta y = y_2 - y_1$
$C_0, C_1$, and $C_2$ are constants

If the pressure is expressed in psia and the temperature in °K, the constants can be determined and Equation 7 becomes

$$\Delta y = \Delta P + \Delta T \tag{8}$$

The path of steepest ascent is the one for which the pressure increases one psia for each degree Kelvin the temperature changes. If the pressure units were not psia but atmospheres, then Equation 7 would become

$$\Delta y = 14.7 \, \Delta P' + \Delta T \tag{9}$$

This equation says that the steepest path will be followed if the temperature is increased 1°K while the pressure is increased 14.7 atmospheres or 216 psia. If the pressure had been measured in mm Hg, the equivalent equation would have indicated that the pressure should increase 0.000374 psi for each degree Kelvin the temperature is increased. Figure 14-2 shows the results.

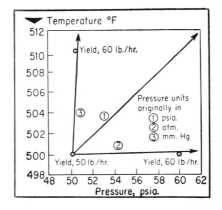

Figure 14-2   Three different paths of steepest ascent which result from using the same data but different units.
Courtesy of Baasel, W.D.: "Exploring Response Surfaces to Establish Optimum Conditions," *Chemical Engineering*, Oct. 25, 1965, p. 149.

The reason for this is that the coefficients which determine the direction of steepest ascent are the changes in the objective function per unit change in the independent variables. If there are two independent variables and only the units of one are changed, then only one of the coefficients will change. Since it is the numerical value of the coefficient which determines the direction of steepest ascent, a change in only one coefficient means a change in that direction. If the units of all the independent variables are the same, there is no problem, since if the units of one were changed they would all be changed. In the volcano-climbing example, if the height and distance were in meters Equation 5 would remain unchanged, since all the coefficients are dimensionless.

All this is the same as saying that when the units of the independent variables are different we must assume some equivalence for them. In the example given, is

1° Kelvin equivalent to 1 psia, 1 atm, 1 mm Hg, or some other pressure unit? No definite answer can be given. Since this is always true, the method of steepest ascent can only tell whether a variable should be increased or decreased. It cannot tell how much one independent variable should be moved in relation to another one having different units.

The reader should therefore look critically at all gradient methods. Some very sophisticated ones are available that have worked well for some very complex systems.[1, 4-7]. They should not be rejected because of the limitations just noted. In fact, some schemes have been devised to avoid the problem of choosing the "right units." It would be wise, however, to realize that making a large number of tedious calculations to obtain the exact conditions for the next experimental point may not be very reasonable.

### Direct Search

The direct search methods[8] use many of the basic ideas developed so far. They suppose that if a step in a given direction is good a larger one in the same direction will be better. Conversely, if a step results in a worse response, in the future a smaller step should be made in the opposite direction. The method follows.

Suppose there are independent variables. After a starting point ($b_0$) is picked a move is made sequentially in each of the $n$ orthogonal (perpendicular) directions over a preselected distance. After each move a test is made to see if the new point is better than the previous one. The best of the two points is retained and the next step is made from there. After n steps have been made, a straight line is drawn from the original starting point through the "best" point obtained so far. This line is extended a prechosen distance and a test is made. If this is better than all previous points it becomes the new base, $b_1$. If it is not, the previous best point is made the new base. The procedure is then repeated with two changes. First, if the previous step in one of the directions resulted in a failure (a point having a poorer value of the objective function), then the next time a change is made in that variable it is made in the opposite direction. Second, the point $b_1$ becomes the starting point of the new line. This method is independent of units but is dependent on the step size.

There are many variations of this method. To illustrate the procedure, a variation developed by Rosenbrock will be discussed. It is one of the best optimization methods known[6,7] when there is no experimental error. The method is also very useful for determining constants in kinetic and thermodynamic equations that are highly nonlinear. An example of this type of application is given in reference 9.

### Rosenbrock's Direct Search Procedure

This procedure[9,10] begins by electing a step size for each of the independent variables. These step sizes are all made unity by changing the units. A starting point is chosen and a unit step change is made sequentially in each of the variables. As before, if the objective function improves the move is considered success; the point is retained and the search continues from it. If the move is a failure the point is rejected and the next move is made from the previous point. After each variable has been changed once, a second round of changes is begun in the same way but with different step sizes. If the previous move made by changing the variable was a success, the next step will be $\alpha$ times the previous step size ($\alpha > 1$). If the previous move was a failure the next time that variable is changed the step will be in the opposite direction and its size will be changed by the factor $\beta$ ($0 < \beta \le 1$).

This can be generalize as follows for two independent variables.

If the $m^{th}$ move of the first variable is a success,

$$a_{m+1} = \alpha a_m \tag{10}$$

where $a_{m+1}$ = amount the first independent variable is changed
on the $(m+1)^{-th}$ move of the variable

$a_m$ = amount the first independent variable is changed
on the $m^{-th}$ move of the variable

$\alpha$ = constant ($\alpha > 1$)

If the $m^{-th}$ move is a failure (the yield is less than some previous yield)

$$a_{m+1} = -\beta a_m \tag{11}$$

$$\beta = \text{constant } (0 < \beta \le 1)$$

Similarly, if the $m^{-th}$ move of the second independent variable is a success

$$b_{m+1} = \alpha b_m \tag{12}$$

where $b_{m+1}$ = amount the second independent variable is changed on the
$(m+1)^{-th}$ move of the variable

$b_m$ = amount the second independent variable is changed on the
$m^{-th}$ move of the variable

If the $m^{-th}$ move is a failure then

$$b_{m+1} = -\beta b_m \tag{13}$$

This procedure is continued until there is a success and a failure connected with each of the independent variables. Then a new set of orthogonal directions is obtained. The first direction is obtained by connecting the initial point with the best point obtained. When there are many independent variables the Gram-Schmidt orthonormalization method should be used.[10] The whole procedure is then repeated, with the "best" point obtained so far becoming the new origin.

For an example, let us return to the plug flow reactor for which it was desired to obtain the optimum yield by varying the temperature and pressure. An initial step size must be chosen for each of the independent variables. For the temperature let it be 10°F, and for the pressure let it be 5 psia. To make these step sizes equal to unity, let the temperature units be A and those for the pressure be B. A is equivalent to 10°F and is zero at 0°F. B is equivalent to 5 psia and is zero for a perfect vacuum. These units are given along with the regular units as the axes in Figure 14-3.

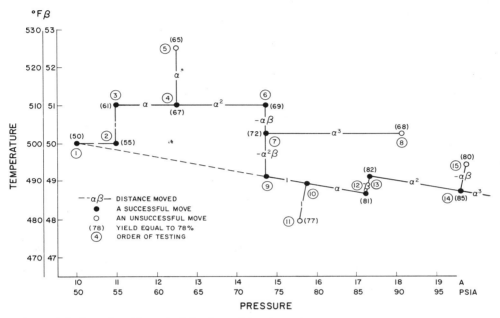

Figure 14-3   Rosenbrock's method for determining the optimum reactor conditions.

The starting point is P = 10A and T = 50B and the yield is 50 lb/hr. A unit step will now made in one of the two directions and the yield determined. Assume the pressure is increased and the yield is 55 lb/hr. This move was a success since the desired response, an increase in the yield, was obtained. This is shown on Figure 14-3. Next the temperature is increased and another success occurs.

Before we can proceed, $\alpha$ and $\beta$ must be selected. Let $\alpha = 1.5$ and $\beta = 0.5$. Since the initial steps were successful both step sizes are increased to $\alpha$.

However, this time when the temperature is increased a failure occurs. Therefore, the next time the temperature is changed it is decreased $\alpha\beta$ ($\alpha\beta = 0.75$). The second time the pressure was increased an increase in yield occurred. Therefore, the next time the pressure is changed it will be increased $\alpha^2$ ($\alpha^2 = 2.25$). By the time each variable has been changed 4 times (point $P_9$) every variable has had a success and a failure. Therefore a new set of directions is chosen. The first has the same direction as the vector through points $P_1$ and $P_9$. The second is perpendicular to that. A unit step is now made in each direction. The new points are easy to find geometrically using a draftsman's compass or a ruler and pencil. With more than two variables the use of a computer greatly facilitates matters.

### Geometric Methods

All the previous solutions for continuous variables have involved moving in the direction of a maximum or minimum. They are like the game in which an item is hidden from one of the participants. Then as he seeks to find it the others tell him whether he is getting hotter (nearer) or colder (going away). The geometric procedure is like "Twenty Questions." In this game one of the participants tries to identify a given person, place, or thing by limiting the realm of possibilities.

In the geometric method[1,11] experimental results are used to minimize the region in which the optimum exists. The response is obtained for a number of points that are located very near one another. The number of points should be one greater than the number of independent variables. From the results a surface (this is a line when there are two independent variables) representing a constant value of the response is constructed. This method hypothesizes that on one side of this surface will be all the points that yield a better response, and therefore the optimum must lie on that side of the surface.

Again, the example of a plug flow reactor will be used to illustrate the procedure. The response is, as usual, unknown until experiments are performed at a given point. However, so that the reader can easily follow what is being done the response surface is shown in Figure 14-4. Assume point A is the starting point. Since there are two independent variables (T and P) two more responses in the vicinity of A must be obtained (points B and C). As in previous examples, three points are often chosen by merely changing in sequence each of the independent variables from the conditions at point A. From these three points a linear estimate of the surface is obtained of the form

$$y = C_0 + C_1 T + C_2 P \tag{14}$$

Then $y$ in Equation 14 is set equal to the yield at point $A$ and $T$ is obtained.

$$T = \frac{y_A}{C_1} - \frac{C_0}{C_1} - \frac{C_2}{C_1} P = b + mP \tag{15}$$

where $y_A$ = yield at point $A$

$b = \dfrac{y_A}{C_1} - \dfrac{C_0}{C_1}$ = intercept of a line having a constant yield $y_A$

$m = -\dfrac{C_2}{C_1}$ = slope of a line having a constant yield $y_A$

The line given by equation 15 would represent a given yield for the whole surface if that surface were linear. It goes through point A and is plotted on Figure 14-4. As can be seen, it is only an approximation of the true constant yield curve. Again, if this surface were linear, all the points below the line would have a yield of less than 15% (region 1) and the maximum yield could not fall in this region. Even though the surface is not linear, this assumption will still be made.

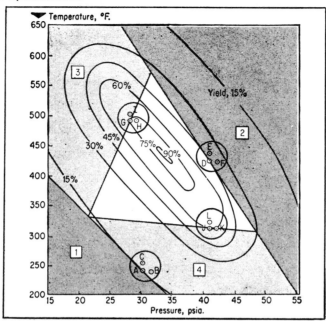

Figure 14-4   A graphical method of finding the optimum.
Courtesy of Baasel, W.D., "Exploring Response Surfaces to Establish Optimum Conditions," *Chemical Engineering*, Oct. 25, 1965, p. 148.

A second area to be tested is now chosen. Points D, E, and F are evaluated and region 2 is eliminated. This is followed by the elimination of regions 3 and 4. The region of the maximum is now given by the unshaded areas of Figure 14-4. This procedure can be continued indefinitely.

Until the region is bounded, the selections of the areas in which to perform the tests must be based on engineering judgment. Once it is bounded the next point may be chosen as either the midpoint, the minimax point, the center of the volume (area), or the centroid. The midpoint is the easiest to calculate and the worst theoretically. Its coordinates are one-half the sum of the highest and lowest values in the bounded region for each of the independent variables. The minimax is the hardest to calculate but the best choice theoretically. Conceptually, the minimax point is determined by computing the smallest possible region that could be eliminated for each point in the bounded area. The point that gives the largest minimum is the minimax point. The methods for obtaining the center of the volume (area) and centroid are given in most analytic geometry books.

The Geometric method works only for *strongly unimodal surfaces*. For such a surface, the straight path connecting any point with the optimum will have no dips in it. Figure 14-5 is an example of a surface that is unimodal, but not strongly. If a straight line is drawn from point A to the summit, the yield decreases and then increases. This verifies that the surface is not strongly unimodal. If point A were picked as the starting point of a geometric procedure, the shaded region would be eliminated. This is the region that contains the maximum.

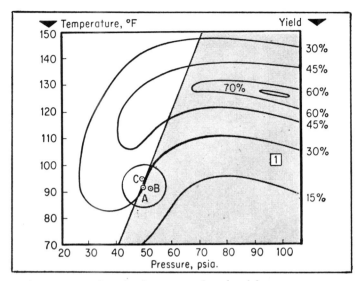

Figure 14-5    A response surface that is not strongly unimodal.
              Courtesy: Baasel, W.D., "Exploring Response Surfaces to Establish Optimum Condi-
              tions," Chemical Engineering, Oct. 25, 1965, pg. 149.

The geometric methods also work poorly when there is experimental error. They
are, however, independent of the scale or units used. They are not used very often
by design engineers.

### Evolutionary Operations

All the algebraic and geometric methods for optimization presented so far work
when either there is no experimental error or it is smaller than the usual absolute
differences obtained when the objective functions for two neighboring points are
subtracted. When this is not the case, the direct search and gradient methods can
cause one to go in circles, and the geometric method may cause the region containing the maximum to be eliminated from further consideration.

Box,[12,13,2] using the method of steepest accent, developed a method called
evolutionary operations (EVOP), that is applicable when experimental error is
significant. The basis for this method is that if the response of a system at given
operating conditions is measured enough times, the mean of the responses obtained
will closely approximate the true population mean. This is true even when there are
large experimental errors. To determine the direction in which to proceed, a sample
response is obtained at two levels (different values) for each of the independent
variables. The order in which these points are taken should be randomized and each
point should be sampled a number of times. Box uses a statistical technique called
*analysis of variance* to determine when the difference in response at the two levels
is significant. Once a difference is established, the procedure does not differ
radically from the first gradient method presented. For details, consult the refer-

ences given. EVOP can determine small differences in the response even when the experimental error is large if enough tests are made.

This method is especially valuable in an operating plant where a large amount of bad product cannot be tolerated. The engineers can direct the operators to make small changes that do not vary the product quality greatly, and obtain enough information to determine the direction in which they should proceed to improve the system. However, getting the operators to maintain the independent variables constant during the runs may be difficult when all they seem to be doing is going in circles sampling and resampling the same points. After all, the movement toward better process conditions is slow and may not even be apparent to the operators. If the yield of a 1,000,000,000 lb/yr plant is increased 1%, this can result in an extra 10,000,000 lb/yr of product at no increase in production costs. There may even be a reduction in costs, since the load on the waste treatment facilities may be reduced.

AMOCO estimated that by using EVOP it had increased its annual income by $800,000 in two years.[14] This was brought about by increasing the recovery of products and by decreasing operating costs. In one case a de-ethanizer limited the amount of butane that could be recovered from an oil absorption plant in Kansas. It was experimentally determined that because the column flooded at 263 gal/min (0.996 m³/min), the maximum throughput was 257 gal/min (0.973 m³/min). When the effect of changing the column pressure and the feed-preheat temperature was investigated using EVOP, it was found that the flow rate could be increased to 306 gal/min (1.16 m³/min). The result was that AMOCO did not have to spend an estimated $110,000 for a new fractionating column.

### Simplex Method

The simplex method [15,16] gets its name from a geometric figure. A simplex is a triangle in two dimensions and a tetrahedron in three dimensions. In $n$ dimensions the simplex is a figure consisting of $n + 1$ points all connected together by lines. If any point in a simplex is reflected through the centroid of the plane consisting of all the other points to a point on the other side of plane, such that both points are equidistant from that plane, a new simplex is formed. For an equilateral triangle this is equivalent to drawing a line ADE from one of the vertices (A) through the center (D) of the line joining the other two vertices (B and C). The new point (E) that will replace the vertex A is on the line through A and D but on the opposite side of line BC from A and is the same distance from the line BC as A is. The result is another equilateral triangle.

What has been described above is in essence the simplex procedure. This is illustrated in Figure 14-6. The point that is reflected is the one that has the worst response. Thus, the procedure merely says: determine the worst point. Get rid of it by creating a new simplex. Then repeat the procedure until the region of the optimum is reached.

Sometimes, upon reflection the new point still turns out to be the lowest. This means that if the above procedure is followed that point would merely be reflected back to its previous conditions. This procedure would not result in any forward

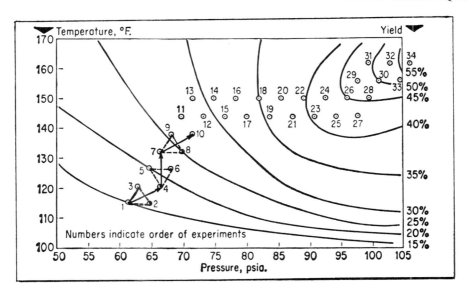

Figure 14-6    The simplex method for finding the optimum.
           Courtesy: Baasel, W.D., ''Exploring Response Surfaces to Establish Opimum Condi-
           tions,'' Chemical Engineering, Oct. 25, 1965, pg. 151.

progress. One alternative under these circumstances would be to keep the best of
those two points and to reflect the second worst point in the simplex instead of the
worst. Another procedure would be to reduce the step size. A point on the line
between the original point and its reflection would be taken such that it is closer to
the plane constructed from the other points. A new simplex should be created
around the best point with the distances between the points cut in half.

A variation[6] that speeds up the search can be used when the reflection point is
superior to all the other points. Here the response of a point located on an extension
of the line formed by the original point and its reflection is obtained. If it gives a
better response than the reflection, the extension point is made the starting point of
a new simplex. If the extension point does not give a better response, the reflection
point is retained and the simplex procedure is continued, with the extension point
being ignored.

The simplex procedure can be used when there is experimental error. Under
these circumstances one point may be so much higher than any of the surrounding
points that it never becomes eliminated. The result could be that the system rotates
forever around this point. To prevent this, the age (number of reflections made after
its determination) of each point is determined after each reflection. If the age of any
point is greater than a certain number, that point is reevaluated. If it is still higher
than any of the surrounding points, the step size should be reduced or the region
should be checked to determine if this is a true optimum.

This method is easy to use and easier to calculate than EVOP. It can easily be
used with a qualitative or quantitative response. The only decision that needs to be

made is which point gives the worst response. Even if the second worst is chosen by mistake, it will not slow the search down appreciably.

## END GAME

When one seems to be going nowhere with a search procedure, it is time to stop and evaluate what is happening. Usually the first thing to do is to reduce the size of each step and continue the search procedure. If it is still bogged down, then this point may be the maximum (minimum) or may be on a rising (falling) ridge or a saddle. A rising ridge is a narrow rising path that falls off steeply in all directions but one. This one direction may even curve. Often in a search procedure one is not riding the ridge but is taking points alternately on either side of it. This can cause it to appear that a maximum has been reached. A saddle surface looks like the usual horse saddle. The response surface in the region appears to be a valley when approached from some directions (the front and back of the horse saddle) and a peak when approached from others (the sides of the saddle). It can thus appear as a maximum or a minimum.

To determine the nature of the surface the investigator should run a second-order factorial design[2] of the region. This is a fancy way of saying that enough data should be obtained to approximate a second-order surface. By taking derivatives it can be determined if a maximum or minimum has been reached, and a prediction can be made as to where this might be. If there is not a maximum or minimum, the direction in which a search should be continued can be determined.

## ALGEBRAIC OBJECTIVE FUNCTIONS

In all the problems discussed, the response surface was unknown and could be approximated only by making some tests. If everything in the objective function is known and can be expressed algebraically and the variables are continuous, a number of other techniques such as *linear programming*[17,18,19] can be used. These will not be discussed here because this is usually not the case for plant designs.

## OPTIMIZING OPTIMIZATIONS

The methods just presented can be used for any number of variables. However, optimizing all the possible variables of a plant in one massive optimization is a Herculean task. The usual approach is to reduce the number of variables to those that strongly affect the performance index. For instance, in the polystyrene example the cost of electricity is almost insignificant and can be ignored. However, the amount of water added to the reactor may be very important. An optimization is made for the major variables. Then the effects of the minor variables are considered either in groups or separately.

Another approach called *dynamic programming*[20,21] can be used if the process contains no recycle streams and no feedback. This procedure is based on the concept that the last step in the process can be optimized  for  each of the possible combinations of inputs to that stage by the methods just presented. Then the last two stages can be optimized together for all possible inputs. This optimization is simplified, however, because the output from the second-to-the-last phase is fed to the last stage, where the optimum conditions for these feed conditions have already been determined. This effectively reduces the number of variables involved in the optimization to those involved in the second-to-last stage. This procedure is continued until the whole process has been optimized.

Rudd[21] has presented an adaptation of this method that covers systems involving feedback. Its development here is beyond the scope of this book. The reader should consult reference 21 for a description of this method.

## OPTIMIZATION AND PROCESS DESIGN

The amount of time allowed for optimizing a process design is usually inadequate. Most process engineering departments are kept busy working on rush projects. This means that the more sophisticated optimization techniques are not used except in special instances. There is also another reason not to use them. When tests are made in the laboratory or in pilot plants, all the conditions cannot be the same as in a full-scale plant. Hence, the optimum may occur at a different point in the full-sized plant and in the laboratory or pilot plant. To illustrate this, consider how the dosages for various drugs are determined. Before new drugs are released to the public, they must be thoroughly tested using animals. These animals may be likened to the engineers' pilot plants. Yet no one would expect a man to respond in exactly the same way as a chimpanzee. Hence, before the drug is released for general consumption it must be tested on human "guinea pigs." No one would be confident of the results if the optimum dosage for chimpanzees were scaled up only on the basis of the weight difference between them and humans.

Chemical plants may be simpler than biological systems, but there still are a number of things that cannot be perfectly scaled up. For instance, in Chapter 5, a reactor was sized. There it was shown that for a given ratio of radius to height, no two cylindrical vessels having different radii could have the same ratio of area to volume. Since the area determines the rate of heat removal and the volume the rate of heat generation, the optimal conditions at which the reactor should run often depend on the size of the equipment. This indicates that attempting to find the exact optimum, even when time and money are available is not usually worth the effort unless the tests are preformed on the actual process equipment in the very plant one wishes to optimize. This is, of course, impossible in a design situation. For this reason the process engineer is generally content merely to find the vicinity of the maximum. He will leave it to the plant manager to find the best operating conditions.

## PROBLEMS

Figure 14-1 gives the yield of a chemical reactor versus temperature and pressure. Starting at point A and using the one-at-a-time procedure, obtain using a quadratic estimate the optimum pressure at a temperature of 250°F if the following data are obtained:

$$P = 29 \text{ psia} \quad y = 15\%$$
$$P = 35 \text{ psia} \quad y = 20.2\%$$
$$P = 41 \text{ psia} \quad y = 24\%$$

Compare your results with the answers given in the chapter.

Problem 2.

Repeat Problem 1 if:

$$P = 29 \text{ psia} \quad y = 15\%$$
$$P = 35 \text{ psia} \quad y = 19\%$$
$$P = 41 \text{ psia} \quad y = 24\%$$

Discuss your results.

Problem 3.

It is desired to determine the optimum values of the constants in the van der Waals equation:

$$\left( P + \frac{a}{V^2} \right) (V - b) = RT$$

where  $P$ = pressure, atm
$V$ = volume, 1/mole
$T$ = temperature, °K
$R$ = gas constant, 1 atm/°K mole
$b$ = constant, $cm^3$/mole
$a$ = constant, $1^2$ atm/mole²

The following data* have been obtained for a specific number of moles of methane.

| $P$ atm | $V$ liters | $T$ °C |
|---|---|---|
| 1.0 | 1.0 | 0 |
| 1.0 | 0.741 | -70 |
| 1.0 | 1.7363 | 200 |
| 100 | 0.282 | -70 |
| 100 | 0.7845 | 0 |
| 100 | 1.7357 | 200 |
| 1,000 | 1.7656 | -70 |
| 1,000 | 2.000 | 0 |
| 1,000 | 2.7861 | 200 |

*Source of data: Perry, J.H. (ed.): *Chemical Engineer's Handbook,* Ed. 4, McGraw-Hill, New York, 1963, Section 3 p. 104.

Using the following methods, obtain the optimum values of $a$ and $b$:

(a)   one-at-a-time
(b)   steepest ascent
(c)   Rosenbrock's
(d)   geometric
(e)   simplex

$$\begin{array}{cc} k_1 & k_2 \\ A \rightarrow B \rightarrow C \end{array}$$

For this system, if only $A$ is charged to the kettle

$$C_B = \frac{C_{A0} k_1}{(k_2 - k_1)} \, e^{-k_1 t} - e^{-k_2 t}$$

where   $C_B$   = concentration of $B$, mole fraction

$C_{A0}$   = 1.0 = initial concentration of $A$, mole fraction

$k_1$ and $k_2$ = reaction rate constants, min$^{-1}$

$t$       = time, min

The following data have been obtained for a given run:

| $t$ | 0 | 1.0 | 2.0 | 3.0 | 4.0 | 5.0 | 6.0 | 7.0 | 8.0 | 9.0 | 10.0 |
|---|---|---|---|---|---|---|---|---|---|---|---|
| $C_B$ | 0 | 0.086 | 0.148 | 0.192 | 0.220 | 0.239 | 0.248 | 0.250 | 0.247 | 0.241 | 0.233 |

Determine the optimum values for $k_1$ and $k_2$ using the following methods:

(a)   one-at-a-time
(b)   steepest ascent
(c)   direct search
(d)   simplex
(e)   geometric

## REFERENCES

1. Wilde, D.J.: *Optimum Seeking Methods*, Prentice-Hall, Englewood Cliffs, 1964, p. 130.
2. Davies, O.L.: *Design and Analysis of Industrial Experiments*, Hafner, New York, 1954.
3. Shah, B.V., Buehler, R.J., Kempthorne, O.: *The Method of Parallel Tangents (PARTAN) for Finding an Optimum*, Technical Report No. 2, Office of Naval Research, Contract NONR 530(05), Statistical Laboratory, Iowa State University, Ames, Apr. 1961 (revised Aug. 1962).

4. Davidon, C.W.: *Variable Metric Method for Minimization,* Argonne National Laboratory, Argonne ANL-5990, 1959.
5. Buehler, R.J., Shah, B.V., Kempthorne, O.: "Some Algorithms for Minimizing an Observable Function," *Journal of the Society for Industrial and Applied Mathematics,* Mar. 1964, p. 74.
6. Barneson, R.A., Brannock, N.F., Moore, J.G., Morris, C.: "Picking Optimum Methods," *Chemical Engineering,* July 27, 1970, p. 132.
7. Perry, R.H., Singer, E.: "Practical Guidelines for Process Optimization," *Chemical Engineering,* Feb. 28, 1968, p. 163.
8. Hooke, R., Jeeves, T.A.: "Direct Search Solution of Numerical and Statistical Problems," *Journal of the Association for Computing Machinery,* Apr. 1961, p. 212.
9. Baasel, W.D.: "Exploring Response Surfaces to Establish Optimum Conditions," *Chemical Engineering,* Oct. 25, 1965, p. 147.
10. Rosenbrock, H.H.: "An Automatic Method for Finding the Greatest or Least Value of a Function," *Computer Journal,* Oct. 1960, p. 175.
11. Wilde, D.J.: "Optimization by the Method of Contour Tangents," *AIChE Journal,* Mar. 1963, p. 183.
12. Box, G.E.P.: "The Exploration and Exploitation of Response Surfaces: Some General Considerations and Examples," *Biometrics,* Mar. 1954, pp. 16-60.
13. Box, G.E.P., Hunter, J.S.: "Condensed Calculations for Evolutionary Operation Programs," *Technometrics,* Feb. 1959, p. 77.
14. Kelley, P.E.: "EVOP Technique Improves Operation of AMOCO's Gas Producing Plants," *Oil and Gas Journal,* Oct. 29, 1973, p. 94.
15. Spendley, W., Hext, G.R., Hemsworth, F.M.: "Sequential Application of Simplex Design in Optimization and Evolutionary Operation," *Technometrics,* Nov. 1962, p. 441.
16. Carpenter, B.H., Sweeney, H.C.: "Process Improvement with 'Simplex' Self Directing Evolutionary Operation," *Chemical Engineering,* July 5, 1965, p. 117.
17. Iscol, L.: "How to Solve Optimization Problems," *Chemical Engineering,* Feb. 19, 1962, p. 107.
18. Boas, A.H.: "Optimization via Linear and Dynamic Programming," *Chemical Engineering,* April 1, 1963, p. 85.
19. Wilde, D.J., Beightler, C.S.: *Foundations of Optimization,* Prentice-Hall, Englewood Cliffs, 1967.
20. Menhauser, G.H.: *Introduction to Dynamic Programming,* Wiley, New York, 1966.
21. Rudd, D.F., Watson, C.C.: *Strategy of Process Engineering,* Wiley, New York, 1968.

**Additional References**

Wilde, D.J., Beightler, C.S.: *Foundations of Optimization,* Prentice Hall, Englewood Cliffs, 1967.
Wilde, D.J.: Optimum Seeking Methods, Prentice Hall, Englewood Cliffs, 1964.
Box, G.E.P., Draper, N.R.: *Evolutionary Operations,* Wiley, New York, 1969.

# Chapter 15

# Digital Computers and Process Engineering

The use of computers for design began in earnest in the 1960s. It was given great impetus by the introduction of the third generation of computers around 1964. These were faster, cheaper machines that had large memories. Each year since then, there has been an increase in computer use for process and plant design. This trend can be expected to continue.

When used properly the computer is a very effective tool; when it is used incorrectly, the results can be disastrous. Many lay people believe that computers can think. They believe that all you need to do is to ask a computer a question and it will provide the correct answer. They have blind faith in its results. If the computer output gives an answer, it must be right. Every company using computers has probably been burned more than once by this very attitude, which has persisted even in highly intelligent people who should have known better.

Descartes based his thesis for the existence of God on the premise that man could not conceive of anything greater than himself unless that thing existed. The correctness of this has been debated for years. However, one statement, that is beyond debate is that computers cannot be greater than the men who build and program them. A computer can do only what a man has told it to do. It cannot be any more accurate than the information that has been supplied to it. It cannot do anything that it has not been told to do.

A computer can do only three things: add, subtract, and decide whether some value is positive, negative, or zero. The last capacity allows the computer to decide which of two alternatives is best when some quantitative objective function has been selected. The ability to add and subtract permits multiplication and division, plus the approximation of integration and differentiation.

Man can do all these things without a computer. The advantages a computer has are its ability to perform calculations very rapidly and accurately and its capability of storing large amounts of data and recovering them quickly. Assuming the computers are properly used, their disadvantages are the costs of preparing or buying the programs and the expenses associated with running and maintaining those programs.

## COMPUTER PROGRAMS

The use of computers for equipment design and for planning schedules (CPM, PERT) has already been discussed. A number of other situations where computers are useful follow. No attempt will be made to indicate how computer programs should be written. However, I shall attempt to show what must be considered both by the person who writes and the one who uses the program.

### Trial-and-Error Solutions

The obvious time to use computers is when some calculation is repeated over and over again. This can be in a trial-and-error calculation such as the calculation of the rate of return. As noted in Chapter 10, the best way to do this is to assume an interest rate, perform the calculations, and determine whether the net present value is zero. If it is not, another choice is made, and the net present value for this choice is calculated. This procedure is repeated until the desired answer is obtained.

Two other problems that fit this category are calculating the number of stages in a multicomponent distillation problem, and obtaining the material balance when complicated recycle operations occur.

The length of time needed to perform these calculations, and hence, the operating cost, depends on the starting point, and optimization procedure, and the accuracy desired. In the previous chapter it was stated that the best starting point should be one based on engineering judgments. This is still true, and many programs have a means for the engineer to insert the initial estimate. If he does not, the computer must use a built-in starting point. The previous chapter also gave the advantages and disadvantages of some optimization techniques, and should be helpful in choosing an efficient one.

The engineer must resist the temptation to obtain an answer accurate to five significant figures when one to two significant figures is adequate. He should also try to avoid exploring a region that is unimportant. For instance, the rate of return does not need to be closer than 1% to the exact answer, because the data used in obtaining it do not warrant a better answer. There is also no reason to obtain an answer even this precise if the rate of return is below a certain value. In most cases, it is probably adequate merely to indicate that it is below that value.

### Problems Done Frequently

Whenever the same series of calculations is repeated a number of times, even with different sets of data, the use of a computer should be considered. For instance, the calculations of the net present value is very straightforward. It can easily be done using tables, a calculator, and/or a slide rule. However, it can also be done on a computer, and this would relieve the engineer of the responsibility of repeatedly performing the calculations. This will give him some time to analyze and compare the results. Besides this, he can also obtain from the same data the payout period and the return on the investment. He could even combine this with the program for the rate of return, and obtain all the major economic indicators for the same effort previously required to obtain any single one.

Other examples of this type of program are those for equipment design—for instance, the detailed design of heat exchangers or fractionating columns.

### Time-Varying Information

The determination of the heat-transfer area required for a given situation is independent of when the calculation is made, but the cost of purchasing and installing that exchanger depends on when the order is placed. It is also affected by the number purchased, and whether it is a standard or a specialty item.

To understand the intricacies of this problem, consider a costing program. The purchase price at a given time for each piece of equipment may be stored by standard type and size. For instance, the cost of each specific standard centrifugal pump could be stored individually. Alternately, a general equation could be developed that gives the approximate cost of any standard centrifugal pump manufactured of carbon steel as a function of flow rate times the pressure drop. Generally equations are most frequently used because they require less memory space. Some constraints must be placed on these equations, since they do not apply when nonstandard items are specified. This applies to unusually large or small items as well as those that are specially designed and constructed. Somehow, the computer program must recognize when these items occur and, in these cases signal the engineer to provide the cost data.

As new information is obtained, not only must the equations be updated, but all the factors used in determining bare module cost factors must be updated (see Chapter 9). After all, new methods for manufacturing, assembling, and installing equipment can be expected to occur, and these will affect the installed cost. Some simple means to update must be provided for all programs needing information that varies with time. One way to do this is to provide, besides the program, a data bank in which all the cost information is stored. Then changes can be made in the data bank without affecting the program.

Each time one of these changes is made, there is a chance for an error to occur. So after each change the program must be checked to see that it responds correctly under as many varied situations as possible. Most programs have special input card decks, or tapes, that have been prepared to perform these tests. These are merely a set of specific input data for which the responses have been hand-calculated. If the computer and hand-calculated results agree, the program is considered correct. Of course, when the data bank is altered the answers are altered. This means a corrected set of hand calculations must be made whenever the data bank is changed.

### Simulations

A simulation is a mathematical approximation of a system. The simulation of a reactor tells how the output changes with the changing input and the system variables. Simulations have at least as many constraints placed on them as cost estimates.

Some simulation equations are based on physical laws. Material and energy balances fall into this category. Others are based on generalizations. The determination of heat transfer coefficients from the equations referred to in Chapter 8 would fit this category. Still others are determined statistically from experimental data, and may apply only to the specific plant from which the data were taken. In all except the first case, great care must be taken when these equations are used to extrapolate data.

### Steady-State Plant Simulation

A plant simulation is the set of equations necessary to approximate the response of a chemical plant to various changes. A steady- state plant simulation is one that predicts the eventual outputs when the inputs and all the internal variables are held constant. It does not say how the outputs are reached. A dynamic plant simulation is one that predicts how the outputs of a plant will change when a known change in the input occurs. It gives the path the process follows in going from one steady state to another.

Most plant simulations have been steady-state simulations. This is to be expected, since just as a baby must learn to crawl before he can walk, so the simpler steady-state problems must be solved before the unsteady-state ones can be tackled. However, unsteady- state plant simulations are being attempted, and undoubtedly sometime in the future this will be a common tool for plant designers.

A plant simulation can be developed in two ways. One is to obtain very precise equations for each step in the process and then put these together. The second is to approximate them very crudely and, after the program is running, to improve each of the estimations gradually.

The result, in either case, is a large number of equations, usually nonlinear, that must be solved simultaneously. The next step is to devise a method of solving these equations that converges rapidly to the answer. Usually some optimization techniques must be employed. An improper choice of procedures can result in a program that takes so long to obtain an answer that it is too expensive to run. This is especially probable when there are a number of recycle streams that interact.

This procedure could be followed for each plant that is to be simulated, but it is so long, tedious, and expensive that a number of groups have tried to develop some general programs applicable to any plant. These are based on the concept that all plants consist of individual units connected together by flow lines. These units are distillation columns, pumps, heat exchangers, and so on. For each specific type of unit, the mathematical simulations are identical although the process variables and physical constants differ. This implies that the same simulation equations can be used for a number of similar items in the same, or different, plants. A comprehensive package of these programs can thus provide all the equations necessary to determine the output of all the equipment.

The next thing that is needed is a program that keeps track of all the process and utility streams, and determines the order in which the individual equipment calculations will be performed. This is sometimes referred to as the *executive program*. The user of this system has merely to put into computer language the flow diagram, which identifies the units (areas of heat exchangers, number of trays in a distillation column) and their interrelations, and to list the operating characteristics of each unit (the pressures, temperatures, exit compositions), the input variables to the plant (not the individual units), the physical properties of substances that are not included in the program, and any known kinetic information. To speed up the optimization, the engineer is also often asked to provide an educated-guess starting point and the sequence in which the calculations should be performed.

The accuracy of the results emanating from the program depends on how well the individual simulation equations predict the real outputs. When better estimating techniques become available, the old equations should be replaced by more accurate or faster ones. This means the equations should be continuously updated, like the cost data in a previous example.

Crowe et al. have written a book entitled *Chemical Plant Simulation* [2] that gives the details of the steady-state simulation of a contact sulfuric acid plant. It uses an executive program named PACER. This and many other such programs as COPS, Flowsim, GPFS, and PDA are for sale. [3]

### Dynamic Plant Simulations

Even though I know of no generalized dynamic programs, their potential importance should not be underestimated. The most optimized plant that can be constructed from steady-state considerations would probably be a nightmare to run, if it could be run. Every process must be designed so that it can be started up and can quickly recover from unplanned upsets. A steady-state design does not consider either of these situations. As a result, the plant may take a long time to return to the desired conditions after a disturbance occurs. It is even possible that the desired steady-state conditions may be unstable. This means that it would be almost impossible to reach those conditions. For instance, a pendulum has two steady-state conditions. One is hanging directly down. The other is one hundred and eighty degrees away, standing straight up in the air. This second position is unstable. It is almost impossible to attain and once attained, the slightest breeze or vibration will cause it to leave that position. A steady-state program could not anticipate that.

### Process Design

A process design is generally at least as difficult to program as a simulation. For simulations the physical equipment is fixed and the output is predicted as a function of the input and the process variables (temperature, pressure, and the like). For a process design the equipment is not specified, but must be chosen subject to constraints given by the scope. In this case, general outputs are specified. One of the things the designer will want to determine is the savings that might result when the internal outputs are changed. For the polystyrene example, an engineer might try to determine whether it would be cheaper to remove more water in the centrifuge and less in the dryer, or vice versa. He also might want to know if a flash dryer should be used instead of a rotary dryer.

The procedure for programming a design problem does not differ greatly from the plant simulation example. The major difference is that instead of simulation equations that determine the output from the input conditions, design equations must be used that size the equipment needed to perform the desired changes. There is no reason why the executive programs mentioned previously cannot be adapted for this purpose.

## SENSITIVITY

Every major plant design, cost, and simulation program should have a means of determining how each variable affects the final results. This is called sensitivity. It is important because it reveals what variables have the greatest effect on the result. These are the ones that must be investigated thoroughly if an optimal design is to be obtained. In fact, a detailed evaluation of the variables that cause only minor variations in the final result is usually not worth the effort. Often those variables can be assumed to be constant during an optimization procedure. This greatly increases the efficiency of an optimization procedure.

## PROGRAM SOURCES

The first place to look for computer programs is in the computer center of the company or university with which the engineer is affiliated. Perhaps they do not have the program he wants, but often his employer may be a member of a computer-user's organization, where one is available. These are program-sharing groups that are sponsored by computer manufacturers. They charge only a nominal fee to cover the costs of duplicating a program card deck or tape and reproducing the documentation. Alternatively, if the company has access to a time-sharing computer, it probably can obtain a number of programs developed by the owners of the time-shared computer. Another inexpensive source of programs is the various governmental agencies. For instance, the Bureau of Mines offers a program for calculating the capital and operating costs of plants.[4] For these items the monthly catalog of government documents should be consulted.

In addition to these, a number of programs are available for sale or lease. Some of these programs have been developed to make money. One of the areas where this is done is marketing. This is because one of the most important variables in determining whether the new plant will be built is the prospect for selling the product once it is on the market. To estimate this, all current and future uses and sources of the product, as well as those for any competing substances, must be evaluated. The price trends, possible technological changes, and the probability of new plants being built by competitors must also be considered. In 1970 Battelle spent over 15 months and $140,000 to build a marketing model for sulfur and its derivatives. It expected to sell this for $23,000 to a number of organizations.[5]   Other programs sell for over $100,000.[6]

A brief description of some programs that might be useful to chemical engineers is given in references 3 and 7. Another more extensive list of programs is given in the *International Programs Quarterly*. This publication allows subscribers to list programs they have for rent or sale, or ones they desire. Reference 8 lists a number of periodicals and organizations that provide information on available computer programs.

## EVALUATION OF COMPUTER PROGRAMS

Anyone who has tried to use in one computer a program written for a different computer knows the problems that can develop in getting it to run and give correct answers. For this reason, no program should be considered unless it is written in a language that is acceptable by the computer on which it will be used. For engineers, this is usually a specific kind of Fortran or Algol. Avoid programs written in assembler languages, since these languages vary for different models of the same computer family. If the language is acceptable, the user should determine whether the storage requirements of his computer are adequate to handle the program. Trying to reduce the size of a program to make it fit in one's machine can be difficult. The next step is to evaluate the documentation that has been provided. The documentation should give the logical procedure that was used in preparing the program. It should list all the equations that were used, all approximations that were made, the method of solution, and any optimization procedures used. From this the evaluator should be able to tell whether those methods are acceptable. He must be certain that a program is not so accurate that it costs too much to run. Conversely, he must be sure it is accurate enough.

Each programmer will build certain constraints into each program. Often these will not be explicitly stated, but must be inferred from the equation he uses and the method of solution. For instance, a general equation may be used to evaluate a heat transfer coefficient. In the documentation it may be noted that the equation applies only to turbulent flow, but it may neglect to say it also does not work for liquid metals. When the program was written, the programmer probably never considered that he would ever be asked to solve a problem involving liquid metals, so this constraint was never mentioned. To be fully general, both constraints should be mentioned, and a test should be made within the program to see that they are met. Here the evaluator should be wary because not all constraints mentioned may be incorporated in the program. This can be checked by reading the printout of the program, or by testing the program directly.

The documentation also gives the input and output formats. On a complex computer program the user will rarely be completely satisfied with these. Often the user would like to be able to transmit the output directly to management or operators so they can make certain decisions. There will be some format that is best for insuring that these results can be interpreted quickly and easily. This "best" format undoubtedly will change as new procedures are adopted. The evaluator should check to see that both the input and output formats are flexible or can be made flexible. The evaluator should also determine whether the data bank can be easily updated.

If the user is considering purchasing the program, he should determine what provisions the seller has made, not only for updating the data bank, but for updating the program itself when errors are found or when there are pertinent technological changes. An agreement to correct errors is especially important on complex programs because for these programs it is difficult to check all the various possibilities. For this reason, errors in even the most widely used programs may still be encountered.

For programs that are to be purchased, the user should determine whether the supplier will assist him in training people to use the program and in initially getting any bugs out of it that may appear when it runs on the user's computer. This can save a lot of time and money because the supplier has men familiar with the program, and the user does not.

Finally, the evaluator should arrange to test the program. At this time the constraints, the accuracy of the answer, and the time required to obtain a solution should be checked.

If a program has passed all these hurdles, it is probably a good idea to purchase it rather than spend the time and effort to develop a similar one. Even if an acceptable one has not been found, in the process of evaluating other people's work some useful ideas will probably have been acquired. If these, along with the ideas presented in this section, are adopted, a very useful program should be developed.

## References

1 Guthrie, K.M.: "Pump and Valve Cost," *Chemical Engineering,* Oct. 11, 1971, p. 151.

2 Crowe, C.M., Hamielec, A.E., Hoffman, T.W., Johnson, A.I., Shannon, P.T., Woods, D.R.: *Chemical Plant Simulation*; Prentice-Hall, Englewood Cliffs, 1971.

3 Hughson, R.V., Steymann, E.H.: "Computer Programs for Chemical Engineers," Part 2 *Chemical Engineering,* Sept. 17, 1973, p. 127.

4 Johnson, P.W., Peters, F.A.: *A Computer Program for Calculating Capital and Operating Costs*, Department of Interior, U. S. Government Printing Office, Washington, D.C., I28.27 no. 8426, 1969.

5 "Computers in Marketing, They're Coming in Force and Bringing a New Way of Life," *Chemical Week,* Feb. 25, 1970, p. 50.

6 "Technological News Letter," *Chemical Week,* Mar. 10, 1971, p. 32.

7 Hughson, R.V., Steymann, E.H.: "Computer Programs for Chemical Engineers—1973, Part I," *Chemical Engineering,* Aug. 20, 1973, p. 121.

8 Kravitz, S., Meyers, M.: "Where and How to Obtain Computer Programs," *Chemical Engineering* Mar. 23, 1970, p. 140.

### Additional References

Hendry, J.E., Hughes, R.R.: "Generating Separation Process Flowsheets," *Chemical Engineering Progress*, June 1972, p. 71

Horowitz, J., Hullender, W.C.: "Computer Graphics and Scale Models," *Chemical Engineering Progress*, June 1972, p. 45.

Leesley, M.E.: "Process Plant Design by Computer," *Process Technology*, Nov. 1973, p. 403.

Kehat, E., Shacham, M.: "Chemical Process Simulation Programs—1," *Process Technology*, Jan./Feb. 1973, p. 35.

Bresler, S.A., Kuo, M.T.: "Cost Estimating by Computer," *Chemical Engineering*, May 29, 1972, p. 84.

Enyedy, G., Jr.: "Cost Data for Major Equipment," *Chemical Engineering Progress*, May 1971, p. 73.

Harris, R.E.: "Distillation Designs Using FLOWTRAN," *Chemical Engineering Progress*, Oct. 1972, p. 56.

Robins, D.L., Mattia, M.M.: "Computer Program Helps Design Stacks for Curbing Air Pollution," *Chemical Engineering*, Jan. 29, 1968, p. 119.

Brown, I.D.: "Computer-Aided Pipe Sketching," *Chemical Engineering Progress*, Oct. 1971, p. 41.

# Chapter 16

# Pollution and Its Abatement

Pollution is the release of something undesirable to the environment. By "undesirable" is meant something that is either harmful or unpleasant to some person, place, or thing. The industrial fumes that are destroying the stone artwork on the castles and cathedrals in Europe and the smog in the Los Angeles area are examples.

The concept of pollution is not new. The Chinese *Record of Rites of the Elder Tai*, which was written in the first century BC, warned against man polluting his own environment.[1] The Romans subjected themselves to lead poisoning by using lead vessels for wines and medicines. This resulted in a high incidence of stillbirths, deformities, and brain damage.[2] In the Middle Ages a king needed a number of residences because of the stench that developed after he and his court lived in any one place a short time. They did not have flush toilets. Coleridge (1772-1834) wrote the following:[3]

### Cologne

In Köln, a town of monks and bones,
And pavements fanged with murderous stones,
And rags, and hags, and hideous wenches,
I counted two-and-seventy stenches,
All well-defined, and separate stinks!
Ye nymphs that reign o'er sewers and sinks,
That river Rhine, it is well known,
Doth wash your city of Cologne;
But tell me, nymphs, what power divine
Shall henceforth wash the river Rhine?

When horses were still the chief mode of transportation in New York City, it is estimated they daily deposited 2,500,000 lb (1,100,000 kg) of manure and 60,000 gallons (200 m³) of urine on the streets.[4] Think of what it might be like today if the internal combustion engine had not been invented.

## WHAT IS POLLUTION? WHEN IS IT BAD?

While pollution has been around a long time, there are many areas of disagreement as to whether something is a pollutant or not. To some people rock music is noise pollution. Others enjoy it and play it at top volume. Men and women often use perfumes cosmetically. Yet sometimes they use an amount or scent that is disagreeable to some others; to the wearer it is pleasant, while to those offended by it, it is a pollutant. To a person gliding around a mountain lake in a canoe, the roar of an outboard motor may be noise pollution. The motorboat operator, however, feels it is a quick way to his favorite fishing spot on the other side of the lake. He may even enjoy hearing the 12-hp motor purr. DDT was felt to be a godsend that would forever rid man of malaria and insect pests, until it began destroying higher forms of life and accumulating in the adipose tissues of man and other living organisms.[5] Then a battle began (which has not yet ended) concerning whether the good it can do outweights the bad effects.

Sometimes pollution is acceptable in small amounts. When Columbus landed in the New World there were an estimated 15,000 Indians living in the area that is now Ohio (1970 population 10,652,017).[6] Did these Indians pollute the environment? They made open fires, something that is now banned in all municipal areas, and they urinated and defecated at large. Still, since they were so few in number, the cleansing actions of the air, water and biosphere could rapidly assimilate their wastes and no accumulating harm was done. In the strict sense of the word "pollution," they did "dirty" the streams and anyone drinking the water shortly afterwards could contract the diseases whose infective agents were present in their wastes. However, with a little care this was unlikely, and their pollution was not considered bad.

Some industrialists still take the same stand. They argue that as long as the river or air can eventually assimilate the waste they discharge they should not be required to do any purification. One highly praised method of avoiding installing water-pollution-control equipment is to inject oxygen directly into the stream.[7] This increases the rate at which the "bugs" can assimilate organic wastes. Others have argued that certain rivers should be designated industrial sewers. This would keep the costs of production down and would only be a minor inconvenience, since swimmers and fishermen could use other streams.

Contrariwise, conservationists have argued that no foreign substances should be discharged into the air and waters. These are natural resources and should be kept "pure." The loggers who are using horses in the Bull Run Reservoir area near Portland, Ore., are required to diaper their horses to protect the water quality. What about wild animals such as deer? They, like all the other plants and animals, excrete wastes. Often it seems that conservationists consider that it is only man, his domestic animals, factories, and machines that pollute. Purity implies wastes only from "natural sources."

The Great Smoky Mountains are so named because of a haze that is usually present. This is caused by hydrocarbon emissions from trees. The concentrations are sometimes so high that the forest areas are in violation of the clean air standards set by the Environmental Protection Agency. Does this mean the trees should be chopped down?

There are many cases where even though the pollution is bad it is condoned. To many developing countries some smoke and water pollution is merely a visible sign of an increase in the standard of living and of a possible end to malnutrition and high infant mortality. The possible good effects outweigh for them the effects of pollution. Some bauxite mining operations in Jamaica are an example. Even in Pittsburgh, Pa., a number of years ago the smoke from the steel mills was a welcome sight because it meant there were jobs.

## DETERMINING POLLUTION STANDARDS

It is up to governments (and in a democracy that means the people) to decide how much, if any, pollution should be permitted. They in turn must rely upon scientists and engineers to tell them how various elements and compounds affect man and the environment, so reasonable laws can be enacted. Yet not enough is known about all the substances and their interrelations to anticipate many problems that may arise. Consider the following case, where copper from an overhead power line located in a polluted area was lethal to sheep.

In the Netherlands a number of dead sheep were found in a pasture. The cause of their demise was not readily apparent, so the sheep were tested extensively, and it was found that they died of copper poisoning. Upon further investigation it was found that the copper content of the soil under the high-voltage copper power line where they were grazing was twice as high as that just 100m (328 ft) windward. The highest concentration of the copper in the soil was only around 50 ppm.* Yet in sheep, which are more susceptible to copper poisoning than other domestic animals, this proved to be fatal.

Apparently the copper had somehow come from the power line. When other similar areas, under power lines were investigated, no substantial increase in the concentration of copper was evident. This situation was different from the others in that it was in an industrialized area. It is hypothesized that the erosion of the copper was much greater there because of the presence of sulfur dioxide in the air. This is known as a synergistic effect. By themselves neither the $SO_2$ nor the copper line would have caused the death of the sheep. They both had to be present.

As an example for setting standards, let us consider mercury. The dangers of mercury are well known. As a metal or combined with inorganic compounds it can affect the liver or kidneys, although ordinarily its concentration in nature is not high enough to do harm and in these forms it is not retained by the body. However, methyl mercury and other alkyl mercury compounds are in a different category. They attack the central nervous system and are retained in man for a much longer period of time. Their half-life in the human body is around 70 days.[8] The concentration of methyl mercury in the blood at which effects have been noted is 0.2 ppm.[8] This means that a 155 lb (70 kg) man should not take in an average of over 0.3 mg per day.[8]

The primary source for man of methyl mercury is fish. Fish have the ability to

*ppm = parts per million.

assimilate methyl mercury through their gills as well as from their food, so that their flesh may contain it at levels several thousand times higher than the surrounding waters. Pike have been caught that contained 3,000 times more mercury per gram than the water in which they were caught.[9] Some fish caught in Lake St. Clair (sometimes called the most-fished lake in North America), which was polluted by mercury discharges from American and Canadian industry, contained 7 ppm methyl mercury.[10] If fish containing this much mercury were part of a person's daily diet, it would not be long before he would show the effects of mercury poisoning.

To help prevent this, the U. S. Public Health Service has recommended a maximum limit of 0.5 ppm mercury in any food. If the fish are to have less than this level of methyl mercury and the concentration factor is 3,000, then the surrounding water in which the fish live should have less than 0.16 ppb (parts per billion). Currently the oceans have about 0.1 ppb, but it is not known whether this is in the form of organic or inorganic compounds.[8] It is also not known whether fish can convert inorganic mercury into methyl mercury.[8] However, a large number of microorganisms can do this, so possibly its usual form is unimportant.

Man can also obtain mercury from the water he drinks. In 1970 the suggested maximum allowable amount of mercury in drinking water was 5 ppb.[10] This posed a problem at that time because there was no simple quantitative method capable of determining concentrations that low. However, by the end of 1970 a new method that could measure down to 1 ppb was announced.[11]

Why have companies been allowed to discharge mercury into rivers and lakes? The reason is that the discovery of such high concentration levels was a surprise to nearly everyone. It was not until 1965 that a Swedish scientist discovered that inorganic mercury could be converted under natural conditions to methyl mercury.[10] Also, as has just been noted, it was not until very recently that a method for easily measuring mercury at very low levels was available. Once the danger was recognized in 1970, the industry responded very quickly. For instance, Dow Chemical reduced its discharge of mercury from 200 lb/day (90 kg/day) to 0.5 lb per day (0.25 kg/day) in less than 12 months.[12]

The mercury concentration standards set for food and water are based on the information just presented. The reader should, on the basis of this information, decide how much mercury a company should be allowed to discharge in its effluent and then consider the impact of this restriction on the various industries that use mercury. As he continues through this chapter he should keep in mind what zero pollution would mean in terms of pollution-abatement procedures.

**Parts per Billion**

Just how much is a part per billion? One ppb is approximately equivalent to traveling 1 ft (0.4m) on a journey from the earth to the moon. It is equivalent to 1 mill (0.1¢) in a professional's life earnings (assuming an average of $25,000/yr for 40 years). It is equivalent to approximately 2.2 sec out of an average person's life (70 yrs). This is very small, but it is at these concentrations that we must monitor pollution.

## Trace Elements

We are just beginning to understand the effects that trace elements and compounds may have on man and his environment. For most of these we do not know what the toxic levels in man and animals are. Not only are some very dangerous in very small amounts, like cadmium and mercury, but others are necessary.

In July, 1970, 14 trace elements were known to be essential to human health. One of these is cobalt. Yet at least one researcher suggests that the addition of small amounts of cobalt to stabilize beer foam may have resulted in the deaths of a number of people in Minneapolis and Omaha. He theorizes that cobalt was necessary to activate the toxicity of selenium, which is naturally present in those areas. This is another example of a synergistic affect. Separately neither would have been harmful, but together they could cause fatalities.

Selenium, which at present does not appear to be essential to man, apparently can be either beneficial or harmful to man, depending on very small differences in the concentrations.[13] Molybdenum can also be a boon or a detriment to health. It helps stabilize enamel and prevent caries in teeth, but also causes osteoporosis, a weakening of the bones.[13]

When a scientist tries to find out what should be the maximum and minimum levels for trace elements, he rapidly runs into another problem: the variability within each species. Each person and animal is unique. Physiologists note that each individual responds differently to various medicines. This poses problems to the doctor who must prescribe the dosage when he has never tried it on that individual previously. Diseases strike some persons and not others. When a cold or influenza epidemic hits a city, some people get sick and others do not. Similarly, the amount of a given trace element an individual absorbs from eating a given item is different for each person. It is dependent, among other things, on the size of the liver and stomach and the person's weight.

Another consideration is that the human body is highly adaptable. A person who rarely consumes alcohol requires less to make him tipsy than a regular drinker. A dope addict keeps requiring greater and greater amounts of the substance to get a high. Thus, over a period of time, the human body may adapt to a higher level of a given substance in the body.

## Pollution Regulations

To summarize what has been presented, the following should be known about each element and compound before pollution standards are set:
  1. At what concentration is it likely to be harmful
     to man or other living things?
  2. What plants or animals are likely to concentrate the material,
     and by how much?
  3. How will other compounds or elements interact with it,
     and will the results be harmful to man or other living things?
  4. Will it be retained, or merely pass through man?

There are at best a very few substances for which this information is available.

An attempt to determine this information for 60 different elements was begun in 1967 in Britain. The goal was to determine the normal levels of these elements in man and compare them with those in his air, food, and water to find out what effects one had on the other. Unfortunately, this project was curtailed in 1972.[14] Since this information is not known, laws regarding pollution are being made in partial ignorance. Hence, as more information becomes available these laws will be changed.

Presently there are laws governing the emission of certain known toxic and noxious substances like lead, mercury, cadmium, phenol, sulfur dioxide, and nitrous oxide. Other compounds are generally lumped together and some general measurement such as Ringlemann numbers or biochemical oxygen demand (BOD.) is specified. Certain standards have also been set for specific equipment used by certain industries, like sewage sludge incinerators, storage vessels for petroleum liquids, and waste gas incinerators, boilers, and process heaters used in petroleum refineries. These laws can be determined by contacting the federal Environmental Protection Agency as well as the various pollution control agencies in the individual states. Specific addresses are listed annually in the "Environmental Engineering" deskbook issue of *Chemical Engineering*. A summary of the air-pollution codes was given in the 1973 edition. Also in that issue was a brief description of some of the many periodicals concerned with the environment, which the reader may wish to consult.

## MEETING POLLUTION STANDARDS

It is the job of the process engineer to see that any new plant meets federal and state regulations. He may do this by designing a system to handle the wastes, or he may arrange for a municipality or a central processing facility to treat his wastes. In the latter case he must be careful, since if the plant processing his wastes goes on strike or is temporarily shut down he may be forced to shut down his own facilities to avoid stiff pollution fines.

### Designing a Pollution-Abatement System

The design of a pollution-abatement system should begin with the preliminary design of the plant. The best place to reduce pollution is at its source. Chapters 8 and 12 give some suggestions as to how this can be done. For instance, installing a final polishing filter following a regular filter may recover 80% to 90% of solids that would ordinarily be considered pollution. They might even be recycled. In any case, the pollution-abatement problem has been greatly reduced.

In evaluating the sources of pollution, every section of the plant must be scrutinized to see under what circumstances it could be the cause of pollution. For instance, wherever there is machinery there is a good chance for lubricating oil or grease to drip to the floor or ground and eventually enter the water supply. For this reason, the drainage from the areas where this could occur is sent through treatment facilities.

The engineer must also consider some things that are usually considered clean.

Most people feel that rainwater is pure water and the runoff from storms is not polluted. If true, this would mean that rainwater does not need to be sent through any pollution-abatement system. Some authorities have even proposed that all communities have two separate sewer systems. One, which would go to a treatment plant, would conduct all sanitary wastes. The other would handle only rainwater and would exit directly into a nearby river or lake. The problem is that particulate matter may settle out of the air by gravity or fall out when the air comes up against a solid object; the rain itself may scrub pollution out of the air; and there may have been spills or leaks that left pollutants on the ground. All of these things can be picked up by the runoff from a rainstorm. Some tests conducted after a long dry spell have indicated that the runoff from a storm may be more polluted than the water in a sanitary sewer. However, if a plant is carefully designed so that all the areas where pollution can potentially occur drain into the sewer that goes to treatment facilities, the remainder of the rainwater can be sent directly to a river or lake. It is important to minimize the amount of rainwater processed because it can substantially increase the volume of material that must be treated and it dilutes the other streams, making removal of pollutants more difficult.

Another stream that should rarely need to be sent through a waste-treatment facility is cooling water that is used once and then discarded. Sometimes, however, the tubes or shells of heat exchangers develop leaks and a contaminant may enter the water. To protect against this, these streams should be monitored to make certain that if a leak does occur it will be detected immediately. Then when a leak does occur the water should be sent through a pollution-abatement system until repairs can be made. Since this may pose quite an extra load on the system or the contaminated effluent may require special treatment before it is discharged, holding ponds may be built to provide temporary storage until it can be treated properly.

After all the sources of pollution have been pinpointed, the next step is to categorize the waste and to determine what is the best way to remove it from the effluent. This categorization should state whether it is a solid, liquid, or gas, its concentration, and the rate at which it is being produced. Next the desired purity of the effluent should be set. This should be based on projected regulations, not on current codes.

In 1972 Aaron Teller[15] suggested that where special information was unavailable gaseous emissions should be less than 2 ppm of a pollutant, with a concentration at the plant borders of less than 2 ppb. The particulate loading of the air should be kept to less than 0.02g/scf* (0.00055g/m³) and there should be no discharge of liquid wastes.

Once the standards are set and the pollution sources analyzed, the solution to the problem is no different than the design of any other chemical plant. The methods used may be any of those presented previously plus a number that are more esoteric. The one thing that is different, as noted, is that the pollutants are often present in very low concentrations. When this is true such standard methods as distillation, extraction, and crystallization are usually too expensive.

In general it is not a good idea to mix weak and strong solutions prior to treatment,

*scf = standard cubic feet

since many of the pollution-abatement schemes are essentially means for concentrating the pollutants before some final disposition is made. Similarly, streams should not be diluted before treatment. In some cases it may be wise to change the process to achieve a more concentrated stream. For instance, the advantages of direct cooling were given in Chapter 8. However, if the stream being cooled is polluted, direct cooling increases the amount of material that must be processed through pollution equipment and thereby increases its size and cost. In this case an economic analysis is required to determine whether direct cooling is best.

In evaluating a proposed pollution solution the system must be examined to see that a waste is removed and not merely transferred from one operation to another. Consider a system whereby polluted air is scrubbed by water to remove a heavy metal. The metal is then removed from the water by coagulation methods. The resultant slurry is then burned, which results in the heavy metal again entering the air, from which it is again removed by scrubbing. This system does not allow for the removal of the contaminant; hence it can only accumulate within the system until something fails. Everything that enters a pollution abatement system must be removed somewhere. Ideally it is converted into a harmless substance like water, nitrogen, or carbon dioxide or a salable substance like sulfuric acid, sulfur, or hydrochloric acid. But substances like mercury or cadmium cannot be converted into harmless material so they must either be recovered or be discharged to the surroundings in a way that can never harm the environment.

### Distant Disposal

In the past the most common method of disposing of a harmful substance was to discharge it far enough away from people that no one was directly affected. For air pollutants, this meant constructing a very high chimney. Hopefully all the undesirable components were highly dispersed before the effluent reached ground level. If they were still obnoxious when they reached ground level, the polluter hoped they came down in a different political jurisdiction that could not stop his operations.

In the chemical process industries these chimneys are usually between 200 and 400 ft high. They should be at least 2.5 times the height of the nearest building, and the gas leaving should have an exit velocity of at least 40 ft/sec (12 m/sec). The desired exit velocity should be higher if the plant is in an area of very high winds. The construction costs generally ranged from $200,000 to $400,000 in 1968. The details are given in reference 16.

When the waste stream was liquid or solid it was disposed of by pumping it into a deep underground well or dumping it into the ocean. The estimated costs to haul and dump in the ocean are given in Table 16-1.[16] The areas in which the geological structure is such that wastes may be discharged into deep wells is given in references 18 and 19. These wells range from several hundred to over 12,000 ft (60-3,700 m) deep.[16] Their costs[19] range from $20,000 to $1,500,000.

All of the distant disposal techniques have come under attack. In the future their use will be more closely controlled and may even be eliminated.

Table 16-1

Average Marine Disposal
Costs ($/ton) in 1968 for Liquid
and Solid Wastes

| Type of Waste | Pacific Coast | Atlantic Coast | Gulf Coast |
|---|---|---|---|
| Dredging spoils | 0.43 | 0.56 | 0.25 |
| Bulk industrial wastes | 1.00 | 1.80 | 2.30 |
| Containerized industrial wastes | 53. | 7.70 | 28. |
| Refuse | 15. | — | — |
| Sludge | — | 1.00 | — |
| Construction and demolition debris | — | 0.75 | — |

Source: Smith, D.D., Brown, R.P.: "Deep-Sea Disposal of Liquid and Solid Wastes," *Industrial Water Engineering,* Sept. 1970, p. 20.

## AIR POLLUTION ABATEMENT METHODS

There are two types of air pollution. One is the presence of particulate matter and the other is the presence of unwanted or too highly concentrated gases. These are usually treated as two different separation problems, although sometimes the same equipment can be used to remove both types of pollution.

### Particulate Removal

Particulate matter is categorized mainly by its size, which is usually given in microns ($\mu$; $1 \times 10^{-6}$m). The smaller the particles, the more difficult they are to remove. Yet these are the ones that can do some of the greatest damage. The ones below 2 $\mu$ can easily enter and injure the respiratory systems of men and animals. They affect the ability of the atmosphere to transmit radiation and to form rain, snow, clouds, and hail. They also soil and damage the various substances that they contact.[20] In this area of study there is still a lot to be learned. The characteristics of particles and particle dispersoids are given in Table 16-2. The most common devices for removing fine particles from air are electrostatic precipitators,[21] fabric filters[22] scrubbers,[23] and afterburners.[24]

In electrostatic precipitators the gas passes between highly charged electrodes. As the particles pass through with the gas they pick up a charge and are attracted to the oppositely charged electrode, where they remain until the electrode is cleaned by washing, vibration, or rapping (Fig. 16-1). The most-used cleaning technique is washing. This results in dirty water that must be treated to remove what has been added.

The cleaning action of a fabric filter is based on the assumption that the air will pass through the fabric while the particles are retained (Fig. 16-2). The gases must be cooled to 180°F (82°C) if cotton bags are used and 550°F (288°C) if fiberglass is

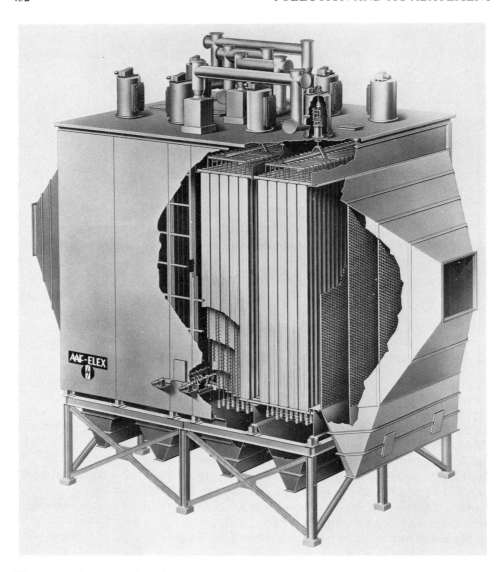

Figure 16-1   A cutaway view of an electrostatic precipitator. The gas flows in one side and out the other. The particles in the incoming air are attracted to the charged plates. The plates are periodically rapped. This causes the particles to fall to the bottom where they are collected. Courtesy of the American Air Filter Company.

used.[25] These filters may be cleaned by shaking or by blowing air back through them. The yield is a powder, which in some cases may be recycled or sold and in others must be treated as solid waste.

Scrubbers depend on the absorption of the particles in a liquid stream that runs

**ON STREAM AIR FLOW**                    **REVERSE STREAM CLEANING ACTION**

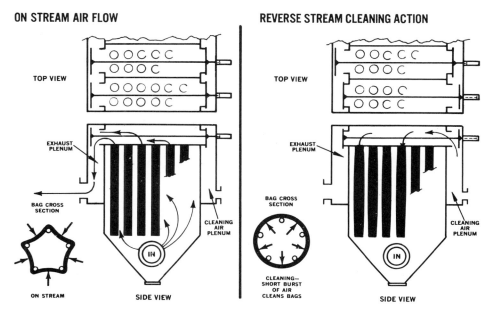

Figure 16-2    A cloth filter. The dirty air enters in the side near the bottom as shown in the figure on the left. The large particles drop to the bottom. The small particles are carried upward and deposited on the exterior surface of bag (shown in black).The air passes through the bag and exits through the exhaust plenum. The bags are cleaned periodically, but only a few at a time. When this occurs the exhaust plenum is closed as shown in the figure on the right. A burst of air is admitted to the inside of the bags; this knocks the small particles off the sides and they settle to the bottom where they are discharged through a rotary valve. Courtesy of the Dustex Division of American Precision Industries, Inc.

countercurrent to the gas stream. The liquid must then be cleaned before it can be discharged from the plant or reused.

Afterburners may be of the flame, thermal, or catalytic type. In each case the object is to cause a chemical reaction which will result in an acceptable product, such as water and carbon dioxide. This is not possible, hence this is an undesirable method, if heavy metals, sulfides, halogens, or phosphates are present. The costs associated with this method are given in reference 22.

The efficiencies of electrostatic precipitators, bag filters, and scrubbers are given in Figure 16-3. The costs for installing and operating these devices are given in reference 20. This source also describes some exotic methods, such as the use of thermophoretic or diffusiophoretic forces and sonic agglomeration, that have not yet been commercialized.

Some other devices that can be used when the main obnoxious material is large particles are gravity settlers, cyclones, and inertial devices. Gravity settlers rely on gravity to allow the particles to fall into collecting plates. These should not be used on particles smaller than $40\mu$.[25] Cyclones (see Fig. 8-10) depend on centrifugal

Table 16-2

CHARACTERISTICS OF PARTICLES AND PARTICLE DISPERSOIDS

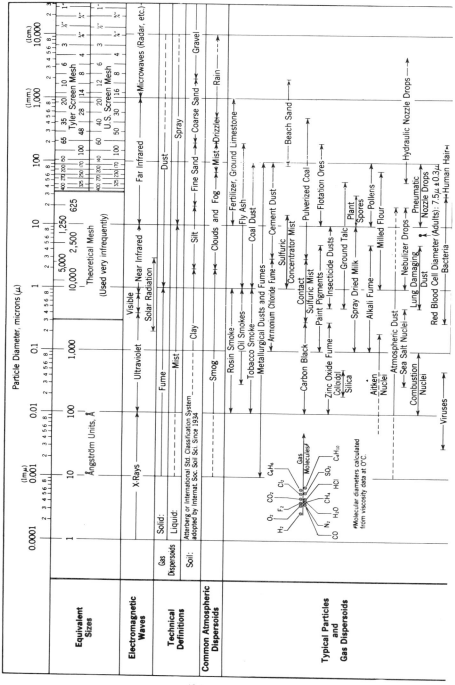

Table 16-2 (*continued*)

**Methods for Particle Size Analysis**

- Sieving
- Electroformed Sieves
- Microscope
- Ultramicroscope+
- Electron Microscope
- Centrifuge
- Ultracentrifuge
- Elutriation
- Sedimentation
- Impingers
- Turbidimetry++
- X-Ray Diffraction+
- Permeability+
- Adsorption+
- Scanners
- Light Scattering++
- Nuclei Counter
- Electrical Conductivity
- Visible to Eye
- Machine Tools (Micrometers, Calipers, etc.)

+ Furnishes average particle diameter but no size distribution.
++ Size distribution may be obtained by special calibration.

**Types of Gas Cleaning Equipment**

- Ultrasonics (very limited industrial application)
- Settling Chambers
- Centrifugal Separators
- Liquid Scubbers
- Cloth Collectors
- Packed Beds
- Common Air Filters
- Impingement Separators
- High Efficiency Air Filters
- Thermal Precipitation (used only for sampling)
- Electrical Precipitators
- Mechanical Separators

**Terminal Gravitational Settling** [for spheres, sp. gr. 2.0]

| | Reynolds Number | Settling Velocity, cm/sec. |
|---|---|---|
| In Air at 25°C, 1 atm. | | |
| In Water at 25°C | | |

**Particle Diffusion Coefficient,* cm²/sec.**

| | |
|---|---|
| In Air at 25°C, 1 atm. | |
| In Water at 25°C | |

Particle Diameter, microns (μ)

0.0001  0.001 (1mμ)  0.01  0.1  1  10  100  1,000 (1mm.)  10,000 (1cm.)

*Stokes-Cunningham factor included in values given for air but not included for water

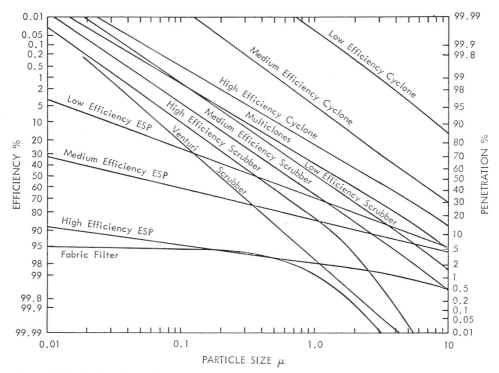

Figure 16-3   The fractional efficiency of various devices for removing various size particles from air.
Vandegrift, the originator of this figure, cautions that in the region below 0.5 microns the
data are rather sparse. ESP stands for electrostatic precipitator.
Source: Vandegrift, A.E., Midwest Research Institute.

forces to force the particles to the sides of a vessel. They then fall to the bottom and
are collected. This works for particles larger than 10 $\mu$.[15] The inertial devices force
the gas to go through some sharp changes in direction. The particles, which cannot
change direction as fast as the air molecules, collide with the barriers and fall into a
collector. This is not recommended when the particles are smaller than 20 $\mu$.[15] The
capital and operating costs for these items can be obtained from references 15 and
26.

**Gas Removal**

The removal of gases from air generally involves either absorption or adsorption.
The absorption may be either into a liquid or solid. Often a reactant is used that
forms a complex with the gas being removed. The reactant must then be regenerated and the pollutant removed in some concentrated form. The use of a reactant
usually greatly increases the efficiency of the removal process.[27]

As an example, consider the dry absorption of $SO_2$ by magnesium dioxide. The
solid magnesium dioxide is brought into contact with $SO_2$ and magnesium sulfate is

recovered along with the excess magnesium dioxide. The magnesium sulfate is then reacted with ammonia and oxygen or air to regenerate the magnesium dioxide and produce ammonium sulfate for sale.[28]

A liquid absorption process for the removal of $SO_2$ involves the absorption of the $SO_2$ into a solution of ammonia and water with resultant formation of ammonium sulfide. The liquid is then sent to an oxidizing unit to form ammonium sulfate, which can be sold as a by-product or reacted with milk of line to regenerate the ammonia and produce gypsum.[28]

In all the absorption systems any good type of contacting system may be used. For liquid systems this would include packed towers, spray towers, and wet cyclones. Reference 15 gives a list of these devices together with their power requirements and their advantages and disadvantages.

The adsorption systems involve the adsorption of the pollutant on the surface of a solid. The solid can then be regenerated by passing hot gases such as steam through the system. A concentrated pollutant is then recovered; hopefully it can be converted into a by-product or fuel. The most common adsorbents are activated carbon, silica gel, alumina, and molecular sieves.[29]

The absorption and adsorption processes must be designed specifically for each waste-gas system. Most are relatively new processes, and generalized cost data are not available.

To find out more about the equipment for air, water, and solid pollution abatement, the reader may wish to write to the various manufacturers. These are listed annually in the "Environmental Engineering" deskbook issue of *Chemical Engineering*.

## WATER POLLUTION ABATEMENT METHODS

The treatment of polluted waters usually begins with some type of physical treatment such as screening and sedimentation or filtration, followed by chemical and/or biological treatment. The traditional municipal treatment plant uses mainly physical and biological treatment, although it may include a clarification step. Because this procedure has been well developed, most industrial treatment has been performed in the same way, with some extra chemical steps like adsorption or ion exchange added to remove specific items. Recently, however, a complete chemical system has been developed that totally avoids biological treatment. The first municipal plant to adopt this philosophy started up at Rosemount, Minn., in 1973. One of the reasons for shifting away from biological systems is that they often fail to produce a pure enough effluent. This means the water must be further treated with chemical methods. Under these circumstances, which are occurring more frequently as water-quality standards increase, total use of chemical treatment is probably the most economical method. Another advantage is that biological systems can be severely harmed, if not destroyed, by poisons such as heavy metals— their recovery from such upsets can take anywhere from 7 to 21 days—while most chemical systems are not harmed by these poisons. The third advantage of chemical systems is that they can absorb sudden surges and overloads with much smaller effluent changes than biological systems.[30]

### Basic Physical Treatment (Primary Treatment)

The first step in most water-treatment operations is screening. Its purpose is to remove large articles like tree branches, dead animals, shoes, and bolts from the water. The first of these screens usually has openings that range from 1.5 to 6 in (4-15 cm). This is followed by another with openings from 0.25 to 1.5 in (0.6-4 cm). The flow through these screens should be around 2 ft/sec (0.6 m/sec). [31] The solids are raked off and ground up before further treatment.

The next step is to remove any suspended solids remaining, using either sedimentation or floation techniques. Sedimentation is the gravity separation of heavy particles. For it to work efficiently the fluid must be essentially stationary. The flow through a settling tank is usually between 500 and 2,000 gal/ft$^2$ of surface area per day (6-25 m/day); the lower rates are more common in industrial situations. The depth of the tank is between 4 and 14 ft (1.2-4.3 m). [32] The solids that collect on the bottom of the sedimentation tank are removed by rakes or plows that pull them toward a central exit or they are withdrawn through holes in the bottom by a very slow water flow rate. The result is a sludge that must undergo further purification. An alternative to a settling tank is a large lagoon. This must be periodically drained and the solids removed, or it will eventually fill up with solids.

In flotation systems air is bubbled through the water. As the air passes through the liquid it attaches itself to the particles that are less than $100\mu$, and together they float to the surface. A skimmer then pushes them into a trough from which they are collected for further treatment. The skimmer also removes any oils or greases that may have accumulated on the surface. It is regularly installed with most sedimentation as well as flotation devices, because oils and greases can adversely affect biological and chemical processes. Flotation units are smaller and generally give a more concentrated sludge than do sedimentation tanks.

While primary treatment is always used in municipal plants, because no one knows what might end up in a city sewer, it may be eliminated from industrial plants [33] when the suspended solids are less than 125 mg/l and oils and greases are less than 50 mg/l.

### Chemical Treatment

Chemical treatment is what the name implies—the addition of a foreign substance to effect the removal of unwanted substances. This includes such operations as neutralization, coagulation, ion exchange, and electrodialysis. These, along with the advanced physical systems, have been referred to at times as tertiary treatment or advanced treatment processes.

There are many more processes than will be discussed here. Only those systems that are currently in industrial use will be considered. This will be true for advanced physical and biological systems as well.

#### Neutralization

Before any stream can be discharged into a lake or river, it must be at a pH between 6.5 and 8.5. Even if this were not a requirement there would be at least two reasons to neutralize the waste. The first is to minimize corrosion and the use of expensive construction materials in future treatment operations. The second is to prevent damage to the microorganisms when biological treatment systems are used. These systems are very adversely affected by any large changes in the pH of the incoming water.[34]

The obvious chemicals to use are any available waste acids or bases. When these are not present or inadequate, the best acidic choice is concentrated sulfuric acid (66° Baume). It can be stored at this concentration in a carbon-steel vessel. It is less corrosive, less costly, and less likely to produce atmospheric pollution than its nearest competitor, hydrochloric acid.

Limestone is the cheapest basic product (0.2¢/lb in 1968; 0.4¢/kg). However, it can become coated with calcium sulfate, which can almost stop neutralization from occurring. Some kind of scouring may be necessary to prevent this. Lime may be used, but it also can become coated and rendered ineffective. Soda ash and caustic soda are other alternatives, but they cost 8 to 10 times as much as limestone.[35] Sometimes limestone or lime is used to raise the pH to between 4.0 and 5.0, and the more expensive but more easily controllable soda ash or caustic soda is used to obtain the desired final pH. The only equipment needed is storage tanks for the acids and bases and mixing tanks, equipped with the proper controls.

### Coagulation

This process, also known as clarification, involves the addition of a chemical to neutralize the charge on any colloids so they can agglomerate, and then the addition of some more of the same or other compounds to aid the formation of a floc that will settle and in the process entrap small particles. The most common coagulants are alum, lime, ferric chloride, ferric sulfate, aluminate, and polyelectrolytes. Polyelectrolytes are much more expensive than the other compounds, but there are some claims that under certain circumstances they can be 20 times more effective.[36] A list of the ideal concentrations and the properties of the coagulants is given in reference 35. A list of polyelectrolytes along with their properties and methods for preparing them is given in reference 37. Coagulants can be most helpful when particles are below 50$\mu$ in size. These particles will settle out, but so slowly that the sedimentation equipment would need to be extremely large and hence very expensive.[37]

The traditional clarification step does not involve the recycling of the floc. A more recent high-rate process speeds up the settling by recycling some of the floc.[30] It has also been determined that gentle mixing after coagulation brings about further agglomeration, which increases the settling rate. This can increase the surface rates given previously for sedimentation tanks by 2 to 4 times.[30] The designs of sedimentation and coagulation tanks are similar. The solids collected require further treatment. Since nearly all the exiting streams that contain concentrated pollutants will require more processing, the need for this will not be reiterated for each operation discussed.

### Precipitation

The removal of dissolved inorganic compounds is usually accomplished by precipitation. This step is based on the common ion effect. When a salt dissolves in water, it forms two ions. The salt will continue to dissolve in water until the product of the ionic concentrations is less than the solubility constant. The concentration of one of the ions can be increased by adding another compound that dissociates into that ion plus another. When the concentration of that ion is increased enough, the first compound will begin to precipitate out. If this is done judiciously an undesired ion can be replaced by one that is acceptable, or sometimes two compounds can be precipitated out simultaneously. Fluoride and iron can be removed by using calcium hydroxide, sulfide by using copper sulfate, and phosphates by using either aluminum sulfate or iron chloride.[35] Coagulants are often added to speed the settling of the precipitates.

### Chemical Oxidation and Reduction

Chemical oxidation and reduction are still other ways of getting rid of unwanted compounds. The most common oxidant is chlorine. It may be added as chlorine gas, hypochlorites, or chlorine dioxide. Chlorine gas is usually used because it is the cheapest. The major disadvantages of using chlorine in any form is that it may react with organic compounds to form potentially toxic, long-lived chlorinated hydrocarbons, and any unreacted chlorine that is discharged into a stream may destroy beneficial organisms. The other major oxidant is ozone. This has seen a limited use, because ozone has a short half-life and therefore it must be generated at the plant site. Because of its short half-life (20 min in water) ozone does not have any of the disadvantages to life downstream that chlorine does. Ozone's other advantage is it requires only a 5 min contact time where chlorine requires 30 min. [38] It is more expensive than chlorine but competitive with sodium hypochlorite. Both chlorine and ozone kill or inactivate pathogenic organisms as well as eliminating odors and color. They may both be used to oxidize cyanides to nitrogen and carbon dioxide and to treat phenolic wastes.[35,38]

Permanganates are occasionally used for oxidizing organic compounds, dissolved gases, and metal ions. Like dichromate, hydrogen peroxide, and other occasionally used oxidants, they are very expensive.

The need for reducing agents is much less than for oxidizing agents. The most common agents are ferrous chloride or sulfate, sodium metabisulfite, hydrogen sulfide, and sulfur dioxide.[35]

### Ion Exchange

This method is used to remove ionic species such as heavy metals, phosphates, or nitrates. It is the reversible exchange of ionic species between a resin and the liquid. For example, a cation resin will exchange positive ions such as hydrogen ions for copper ions that are in solution. Similarly, certain anion resins might replace phosphate ions with hydroxyl ions.

The most common method of contacting is to pass the liquid to be purified through a fixed bed of ion-exchange resin. Then periodically this flow must be stopped and the ion-exchange resin regenerated. Cation exchangers are usually regenerated by using a concentrated acid solution, and anion exchanges by using a concentrated base. In the regeneration step the resin is returned to its original state and the ions that had been substituted on the resin go back into solution. The net result is a solution that is much more concentrated than the feed to the system. For many substances like heavy metals, this is usually the first step in a recovery process. For others it is merely a means of concentration before ultimate disposal occurs.

Continuous ion-exchange units are also available in which ion exchange and resin regeneration both occur constantly. In this case both the resin bed and the fluid being purified are in motion. Continuous units are usually more economical than batch units when the rate of ion removal exceeds 1 lb/min (0.5 kg/min).[39]

Because of its expense, ion exchange is mainly useful for removing substances that are present in very low concentrations. For a discussion of various resins and some design data, see references 40 and 41. Some resins can be made reasonably selective, while others are general. In the latter case it is often desirable to remove many of the ions present in the water prior to the ion-exchange step, so a product may be obtained that can easily be recovered and purified.

### Electrodialysis
Electrodialysis is another method of separating ions, a membrane is used that selectively passes anions or cations. The transfer is accomplished by the induction of an electromotive driving force that causes the permeable ions to be transferred across the membrane from a solution of low concentration to one of higher concentration. See references 42, 43, and 44 for the description of equipment and situations where this method is used.

### Advanced Physical Methods
These methods are any physical method that has not been mentioned under the section on basic physical methods. They differ from the chemical methods in that no chemical change takes place.

### Adsorption
Adsorption is one of the key steps in most physical/chemical pollution-abatement systems. Its main role is to remove dissolved organics, thus eliminating the need for biological treatment. Adsorption has already been mentioned in connection with the purification of gases. Nearly everything said there applies equally well to liquid systems, except the material on regeneration. In a liquid system regeneration is more of a problem. To speed up desorption heat is often used, along with a purge gas or liquid. To simplify things the purge fluid may also be the heating medium. The system must then be cooled before it can be reused for adsorption. This regeneration procedure is often quite lengthy; sometimes three columns are run in parallel. One is performing adsorption, another is being used for thermal regeneration, and the final one is being cooled.[45] One disadvantage to thermal regeneration, especially when activated carbon is involved, is that the loss of the adsorbent is rather high.[46]

Another method that may work is to change the pH. This works with weak organic acids and bases. In this case the material is adsorbed at some optimum pH. To desorb it the pH is changed and the adsorbed material is removed in a more highly concentrated form. Dow has used this method to remove acetic acid and phenol from a by-product NaC1 stream.[46]

Another regeneration method, which is not currently used industrially, is to use a solvent to absorb the compounds adsorbed by the carbon. The solvent would then be purified and reused. The recovered material could then be further purified or prepared for disposal. Before the bed could be reused following regeneration, the solvent would have to be removed. This could be done by using a purge gas. The solvent could then be recovered from the purge stream by sending it through a condenser.[46]

The last regeneration system that will be mentioned is in some ways the simplest. The adsorbed material is burned off under controlled conditions.

Care must be used in selecting the adsorbent. Some are also good catalysts, and under the right conditions oxidation or fires could occur. Also, when there are unsaturated compounds, polymerization could occur. When that happens regeneration is usually impossible.[27]

Fixed-bed systems are the most common, but some countercurrent fluidized beds are in use. Flow diagrams are given in reference 47. The superficial velocities of gases in fixed beds should be about 1 ft/sec (0.3 m/sec) and those for liquids about 1 ft/min (0.3 m/min).[48] See references 48 and 49 for more design information.

### Filtration

Filtration is a standard unit operation that has been known for thousands of years. A liquid containing solids is made to pass through granular solids or a porous septum. The liquid passes through and, if it is an efficient system, most of the solids are retained on the filter medium. After a sufficient amount of solid material has collected this is removed and the filter can then be used again. The most common granular solid used is sand. The porous septum may be cloth, fiber glass, steel mesh, or tightly wound coils.

Filters are used only at specific places in the purification of water. Sand filters are used as a final polishing step or prior to a membrane or ion-exchange process, and vacuum rotary-drum filters are used for dewatering sludges.

The simplest sand filter is merely a bed of sand through which the water moves at very slow rates. A high-speed version has been developed that uses layers of sand of different sizes (Fig. 16-4). The water passes through the coarsest layers first.[50] Sometimes the layers may contain compounds other than sand, such as coal or garnet. When space is a problem a combination unit involving a plastic filter cloth and sand can be used.[51]

Granular solid filters (see Fig. 16-4) are cleaned by backwashing with water. The backwashing is usually preceded by an air scouring to assure better cleaning.[30] The backflow water rate should be fast enough to fluidize the bed.

A vacuum rotary-drum filter consists of a porous septum that surrounds an empty rotating drum. The bottom of the drum is immersed in the sludge liquor. A vacuum pulled on the inside of the drum causes the water to enter the drum through the filter medium. The solids that cannot pass through are retained on the surface of the medium. As the drum rotates, the solids at the surface of the septum are lifted out of

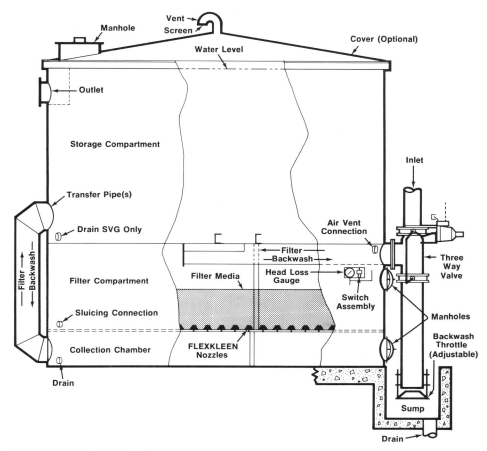

Figure 16-4  A media (sand) filter.
The liquid material to be filtered enters at the point marked "inlet" and passes into the filter above the filter media. It passes down through the media, where the solids are retained, into the collection chamber. The purified liquid passes up through the transfer pipe into a storage compartment, which is totally separated from the filtering section, and leaves at the point marked outlet. To clean the filter media, the feed is shut off and the valve to the sump opened, purified water is then forced up through the media and the solids which have been collected are carried into the sump.
Courtesy of EIMCO, the Processing Machinery Division of Envirotech Corp.

the fluid and air is drawn through them, because a lower pressure exists inside the drum. When the solid cake that has been formed reaches the discharge point, an air blowback pushes the solids off while a scraper or some other device breaks up the solid cake and helps direct it into a collecting vessel.

One problem that can develop is that the filter becomes clogged because of fine particles from the sludge that are caught in the pores, precipitation that has occurred during the dewatering stage, or a thin nonporous layer that has formed on the surface. One way to counteract this clogging is to use a continuous-belt drum filter

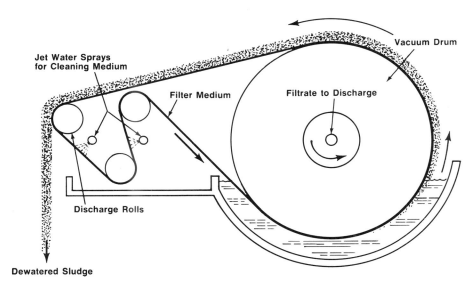

Figure 16-5   A continuous belt drum filter. This works on the same principal as the vacuum drum filter.
The drum is immersed in the liquid to be filtered. A vacuum is pulled in the drum to create
the pressure drop necessary to cause the fluid to flow through the filter media into the
drum. The solids are retained on the surface of the filter medium. After the belt and the
adhered solids leave the fluid, the vacuum continues and most of the liquid in the solids is
removed. The solids are then discharged and both sides of the belt are cleaned.
Courtesy of EIMCO, the Processing Machinery Division of Envirotech Corp.

(Fig. 16-5). Here the filter medium is attached to a continuous belt instead of
directly to the drum. The belt, after leaving the rotary drum, makes some abrupt
changes in direction, which loosens and discharges the cake. The septum is then
washed from both sides with water[52] to remove any remaining material. This wash
often is charged to the filter.

Another method, which is even more successful in preventing binding of the
septum, is the use of a precoat.[53] Before filtration is begun a coating of 2-6 in (5-15
cm) of diatomaceous earth or perlite filter aid is deposited on the surface of the
septum. During filtration operations the scraper is set so that it slowly removes the
precoat and, of course, with it the materials that would have plugged the filter. Since
the precoat causes a considerable pressure drop, the rate of filtration is slowed
down. Flow rates may vary from 2 to 50 gal/hr/ft² (0.025-0.60 m/hr). The precoat
material costs around 3 or 4¢/lb and is used at the rate of 10-15 lb/1,000 gal of feed
(1,200 to 1,800 kg/m³).

The rates of filtration are strongly affected by the compounds present and their
concentrations. Before being filtered or centrifuged, the sludges being dewatered
should contain between 5 and 15% dry solids. Since it takes energy to perform this
dewatering, the more concentrated the sludge, the less expensive this step will
be.[53] A list of rotary drum sizes is given in reference 54.

### Reverse Osmosis

Reverse osmosis is a membrane system that allows the passage of water but impedes the passage of dissolved salts and other molecules. The driving force is a large pressure gradient, 600-1,000 psi (42-70 kg/cm²) across the membrane. The membrane most commonly used in the early 1970s is made of cellulose acetate and has very small pores (around 5 Angstroms). The only way a compound can pass through is to progressively hydrogen-bond to the acetate molecules. Thus, only compounds that can hydrogen-bond can be passed. Cellulose acetate is very sensitive to temperature and loses its selectivity above 85°F (30°C).[55] New membranes are being designed, and some having a different selectivity and different characteristics will undoubtedly be commercially available soon.

Reverse osmosis systems are subject to fouling due to precipitation onto the surface, particulate binding, and biological growths. To minimize the fouling, some compounds like iron salts that often pose problems should be removed prior to treatment. In other cases the formation of precipitates can be minimized by controlling the pH. Biocides can be used to prevent the growth of microorganisms, and filtration can be used to remove small particles. Still, no matter how hard one tries, fouling will occur. The membrane may be cleaned by washing with water, followed by a solution of citric acid to remove hardness scales and one containing enzyme detergents to remove microorganisms.[55] The costs of operating a reverse osmosis system are given in references 55 and 44.

### Ultrafiltration

The third membrane process that has been used successfully in water purification is ultrafiltration. As with reverse osmosis, the driving force is pressure. However, in ultrafiltration the separation is merely based on the size of the molecules. Here the passage of molecules having molecular weights above 100 can be deterred. The pressure differences are usually between 20 and 50 psi (1.4-3.5 kg/cm²).

This method is used mainly to remove high-molecular-weight materials such as proteins, colloids, viruses, and bacteria. The same types of problem encountered with the use of reverse osmosis membranes are encountered here, and the proposed solutions are the same.

### Foam Fractionation

This is very similar to the flotation procedure described under basic physical treatments. In the case of foam fractionation, not only are the pollutants raised to the surface where they can be skimmed off, but a froth, like beer foam, is produced in which the pollutants become concentrated. The key to the process is the adsorption of the pollutants onto the surface-active agents that cause the froth to form. Sometimes a surfactant is added so that non-surface-active components can be removed.

In some cases the air is dissolved in the water being treated rather than being bubbled through. This is done at a pressure of around 45 psig (4.2 kg/cm²), and the whole stream or a portion of it may be pressurized.[56] When the pressure is reduced to atmospheric, bubbles of around 50 $\mu$ are formed. For a pressurized system the flow rate in the flotator should be around 3 gal/min/ft$^2$ of surface area (0.04 m/min). This is an excellent way to remove oils or suspended materials.[57]

**Biological Treatment**

Biological treatment systems use microorganisms to remove the soluble organic waste. The organisms metabolize (eat) the organic matter and in the process convert it into insoluble cellular matter (they multiply) plus carbon dioxide and water (some organic matter is burned to supply energy). The cellular matter can then be removed with the aid of the flocculating agents discussed previously. Biological treatment is sometimes called secondary treatment.

The purpose of the system is to bring the water into contact with the organisms. Then if they find the diet (soluble organic compounds) to their liking they will feast upon it. But just like man, they have other needs besides food.

Aerobic bacteria, the kind most often used, require oxygen. In lagoons and ponds the oxygen is mainly supplied by algae that live there. To obtain oxygen at a faster rate, trickling filters may be used. Here water trickles over a bed of rocks or other media upon which the microorganisms live. Since there is only a thin layer of liquid, the major means of oxygen transport is by diffusion. In most other systems, either a sparging system, which can deliver air or pure oxygen, or an aeration system, where the liquid is agitated in the presence of air, is used to provide a more abundant supply of oxygen.

Another need the organisms have in common with man is some inorganic materials. They need small quantities of nitrogen, phosphorous, and sulfur, plus trace quantities of iron, calcium, magnesium, manganese, zinc, boron, potassium, and cobalt.[34] These are generally present in most municipal waters but may be absent from certain industrial waste streams. If this is so, they must be added.

The "bugs," as the microorganisms are frequently called, can also get sick and die if they are not treated properly. They dislike rapid changes in types of food (organic compounds), pH, and temperature; a pH below 6.5 or above 9.0, a salt concentration in excess of 5,000 mg/1; and the presence in any more than trace quantities (10 mg/1) of heavy metals.[34] If they are killed, the only way to obtain a new group of residents for the waste-treatment system is to grow them, and this takes time.

Bugs can be trained by deprivation methods to eat compounds they ordinarily shun. Dow developed a strain of microorganisms that would eat phenol. However, the bugs, when fed more easily digested organics than phenol, often lost their adaptation and would rather die than eat it again.

Besides the aerobic microorganisms there are also anaerobic ones. These exist and multiply where no dissolved oxygen is present. Saprophytic bacteria produce organic acids and alcohols. The methane bacteria will then convert these compounds into cells plus carbon dioxide and methane. The methane may be recovered and burned as fuel. If any sulfur is present it will eventually be converted to $H_2S$.

### BOD and COD

The efficiency of an aerobic biological system is the percentage of the soluble organics that can be converted into $CO_2$ and $H_2O$ or insoluble biological solids. This is equivalent to the percentage of the biological food that is eaten by the bugs. The *biochemical oxygen demand* (BOD) is the total amount of oxygen it would take a biological system to convert all the biological food available in the water to carbon dioxide and water.

Microorganisms are capable of converting between 30 and 70% of their biological food into insoluble material, the rest of the material being converted to carbon dioxide and water. This means that even for a 100% efficient system the effluent still contains a lot of organic material that could, by other than biological means, be converted to carbon dioxide and water. Some of this is material the organisms cannot digest, and the rest is solid material that the organisms have made. The *chemical oxygen demand* (COD) is the theoretical amount of oxygen it would take to totally oxidize the organic matter present.

Traditionally, wastes have been classified by their BOD and treatment systems by the fraction of the BOD that was removed. This is because the ultimate disposal was into a nearby river or lake, and if the BOD was high the bugs in the stream or lake might use up all the dissolved oxygen when eating the waste. This would cause all the fish and other animals that live in the water to die of asphyxiation. The purpose of a secondary treatment plant was to prevent this from happening. As we gain more knowledge, COD removal may become as important or more important than BOD considerations.

### Stabilization Ponds

These are ponds where the water to be biologically purified is charged and the action takes place *au-naturel*. The average residence time of the fluid ranges from 4 to 20 days. The loading of the organic material should be between 4 and 10 lb BOD/acre-ft/ day (15-37 kg/hectare m/day). This is obviously a very slow process, although if the average residence time is long enough up to 90% of BOD can be removed. If the pond is deeper than 5 ft (1.5 m) and the amount of waste charged is high, all the oxygen may be depleted in certain areas of the pond. Anaerobic bacteria may therefore be thriving and methane and hydrogen sulfide odors may be present.[34]

### Aerated Lagoons

These are similar to stabilization ponds except that oxygen is added by mechanical surface aerators. This cuts the residence time of the water by 80%. It also means that the depth of the lagoons can be increased to 18 ft (5.5 m) and aerobic conditions can be maintained. The surface aerator power level is usually between 0.008 and 0.06 hp/1,000 gal (0.0002-0.016 hp/m³). This is not enough power to keep all the solids in solution. To do that would require 0.05-0.1 hp/1,000 gal (0.013-0.026 hp/m³).[33]

### Trickling Filters

In a trickling filter the water slowly drains through a bed of solids upon which the microorganisms are growing. When rocks are used the bed is between 5 and 10 ft high (1.5-3 m). The water may be recycled between 1 and 10 times. The total water flow rate through the system is between 10,000,000 and 40,000,000 gal/acre/day (100,000-400,000 $m^3$/hectare/day). The BOD loading should be between 0.015 and 3.00 lb BOD/$yd^3$ of media per day (0.009-1.8 kg/$m^3$/day).

Recently, by using synthetic media that have a large surface area, the possible depth of the bed has been increased by a factor of 4 and the BOD loadings have been tripled.[34] This has increased the possible water rates by a factor of 10.

A trickling filter is not used when highly pure water is desired. Usually only between 40% and 70% of the BOD is removed even when synthetic media are used.

### Activated Sludge

The activated sludge unit consists of a mixing area, a reaction area, and a separation area. The entering feed is first mixed with recycled sludge containing microorganisms. This is then charged to a thoroughly agitated container into which oxygen or air is fed. This may be considered a continuous-stirred tank reaction. The stream leaving enters a clarifier from which purified water and sludge are withdrawn. Only a portion of the sludge is recycled, the remainder being further treated before disposal.

The purpose of recycling the sludge is to maintain the composition of the bugs, feed, and oxygen at a prechosen level. This level may be chosen to produce a high reaction rate or a high BOD removal rate. They do not occur at the same conditions. The high reaction rate occurs when a large supply of food and oxygen is present. Under these conditions, the bugs can gorge themselves. However, since the BOD in the aeration basin is approximately the same as that in the exiting fluid, the effluent will contain a high amount of BOD. Conversely, a low concentration of BOD in the basin and the exiting stream would mean a lower reaction rate and a higher residence time for the fluid.

Under the high-rate conditions, the fluid remains approximately 2 hours in the aeration basin, and the BOD of feed is reduced between 50 and 70%. When maximum conversion is desired, over 95% of the BOD can be removed. This is termed an *extended activated sludge process,* and the average residence time is at least 18 hours. The conventional conditions are a compromise, with 90% of the BOD being removed and the fluid being detained around 5 hours.[34] Table 16-3 gives the design information for these systems.

The difference in the amount of solid matter produced per unit of BOD removed can be explained from the reaction conditions. When there is a large amount of organic matter, as in the high-rate process, there is plenty of food for everyone and all the bugs thrive. Under the conditions that exist in the extended system, many microorganisms do not get enough food and die. They are then metabolized by their neighbors. The result is that larger a percentage of the BOD that is removed from the incoming stream is converted to carbon dioxide and water. It should be noted that as the amount of BOD in the effluent decreases, the difficulty of separating the solids from the effluent also increases.

There have been some arguments over whether pure oxygen or air should be used as a source of oxygen. Generally, unless there are high concentrations of organic wastes, it does not seem economically justifiable to use pure oxygen.[58] Pure oxygen should be used only for high-rate systems.

Table 16-3

Design Conditions for Activated Sludge Units

| | Type of System | | |
| | Extended | Conventional | High Rate |
|---|---|---|---|
| BOD. Loading<br>lb BOD in feed/lb<br>of suspended solids<br>per day in recycling | 0.03-0.1 | 0.3-1.2 | 1.5-4.0 |
| Oxygen Needs<br>lb oxygen/lb BOD<br>in feed | 1.3-1.8 | 0.7-1.3 | 0.45-0.65 |
| Waste Sludge Produced<br>lb solids/lb BOD<br>removed | 0.1-0.2 | 0.33-0.55 | 0.65-0.85 |
| Conc. of Suspended Solids<br>in recycle: mg/l | 5,000-7,000 | 2,000-4,000 | 1,000 |

Source:   Lesperance, T.W.: "Biological Treatment," *Chemical Engineering,* Oct. 14, 1968, p. 89.

### Anaerobic Methods

These methods are used when the organic concentration of the feed exceeds 1% of the total[32] and sometimes when it is less. This may occur in the preparation of the sludge for ultimate disposal or in various food-processing industries. The reaction rates are lower and the systems are more sensitive to toxic material than for aerobic systems. The total detention time ranges from 4 to 60 days.

The process is conducted in totally enclosed equipment, often in two stages, and the methane produced is usually burned to supply energy. This process will become more desirable as the price of natural gas increases.

Anaerobic digesters reduce BOD levels between 40 and 70%. They can also remove nitrates.[59] When two digesters are used, the first is agitated and the second is not. Less power is required than for activated sludge units.

**Treatment of Biological Sludges**

The sludges from biological treatment systems have usually been dewatered on a vacuum filter, or occasionally in a centrifuge, and then sold, given away, or spread over land. In the last case, if the fields over which the sludge is spread are nearby, filtration may not be necessary. Milwaukee filters and dries its sludge before selling it as fertilizer. The sludge has a high nitrogen content because of brewery wastes. This makes it highly desirable. Most municipal sludge is not of this quality. It has been estimated that preparation of the sludge to form compost costs around $8/ton.[60] A flow diagram is given in reference 60. The problem with producing compost for sale is that the market is small.

Industrial treatment plants may not be able to use their biological sludges as fertilizer or spread them on nearby fields if certain substances are present. They may be forced to dry and burn them instead. In some cases, as when heavy metals are present, they may have no recourse but to recover them or use deep-well or ocean-dumping methods. Often anaerobic digestors are used to reduce the amount of sludge and hence the size of subsequent equipment or the cost of ultimate disposal.

Sometimes these wastes are nontoxic and can be deposited in a sanitary landfill. This generally costs between $4 and $5 per ton.[61] Landfill operations can, however, be expected to increase in price as land becomes scarcer, pollution laws become tougher, and maybe even disposal taxes are levied.

The multiple-hearth incinerator (Fig. 16-6) can accept sludges containing between 60 and 75% water. The operating costs run between $0.50 and $5.00 per ton of dry solids, with total costs between $8 and $14 per ton. Design information is given in reference 62. When the sludges contain more water, fluidized-bed incinerators are sometimes used. Their operating costs run between $11 and $21 per ton of dry solids and capital costs are $15/ton.[60] See reference 63 for more details. All incinerators must have the proper air-pollution abatement devices attached.

Sometimes, instead of incinerators, so-called oxidation processes can be used. In this context oxidation differs from combustion in that no flame is present and the temperatures are much lower. A high-pressure process (1,750 psi or 123 kg/cm$^2$) that operates at a temperature of 525°F (275°C) has total costs around $33/ton of dry solids.[62] Others operate at a pressure of around 600 psi (42 kg/cm$^2$). They can treat sludges containing up to 99% water.

The Porteous process changes the physical characteristics of the sludge by directly contacting it with steam for between 30 and 45 minutes at pressures between 180 and 230 psi (12.5-16 kg/cm$^2$). The result after dewatering is a sterile sludge that is much more compact and easy to handle. This conditioning can reduce dewatering, incineration, and other processing costs.

## Combinations of the Various Chemical, Physical, and Biological Processes

The process engineer dealing with pollution abatement must decide which of the processes that have been described will be used and how these will be arranged. The 1970 costs for a number of systems are given in Table 16-4 along with the percent

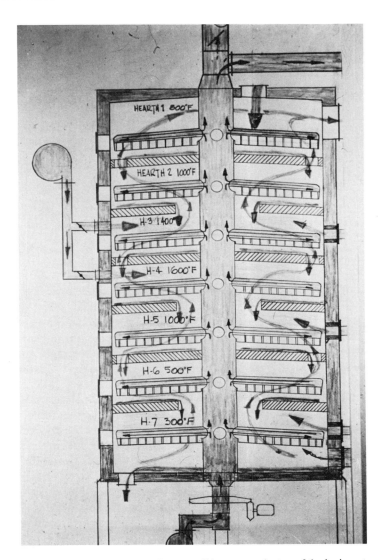

Figure 16-6 Multiple-hearth incinerator. The wet solids enter at the top of the incinerator onto the top hearth. A rake-like device plows the sludge across the top to drop holes where they descend to the hearth below. The agitation of the sludge exposes the maximum amount of surface to the hot gases, which promotes drying.
Courtesy of the North American Manufacturing Company.

removal of pollutants. Other costs and treatment efficiencies are given in references 33 and 64. The traditional municipal waste treatment plant has involved primary treatment followed by biological treatment, sometimes coagulation, and chlorination. As has been indicated, this series of treatments does not give a pure

Table 16-4

Cost and Efficiency Comparison

| Process | Cost in cents/1,000 Gal. For Various Plant Sizes, in mgd.[a] | | | | | Cumulative Percent Removals | | | | |
|---|---|---|---|---|---|---|---|---|---|---|
| | 1 | 3 | 10 | 30 | 100 | BOD | COD | S.S.[b] | P | N |
| Primary treatment | 4.4 | 3.2 | 2.4 | 2.0 | 1.7 | 35 | 35 | 50 | 5 | 5 |
| Secondary treatment | 5.5 | 4.0 | 3.2 | 2.9 | 2.7 | 80-90 | 50-70 | 80-90 | 25-45 | 30-40 |
| Sludge handling | 10.8 | 7.9 | 6.2 | 5.2 | 4.5 | – | – | – | – | – |
| Disinfection (Cl$_2$) | 0.8 | 0.7 | 0.6 | 0.6 | 0.6 | – | – | – | – | – |
| Sand filtration | 7.5 | 5.2 | 3.5 | 2.5 | 1.6 | 85-95 | – | >95 | – | – |
| Activated carbon[c] | 16.0 | 12.0 | 8.0 | 5.8 | 4.1 | >95 | >95 | – | – | – |
| Ammonia stripping[d] | 3.3 | 2.0 | 1.6 | 1.5 | 1.4 | – | – | – | – | 90 |
| Electrodialysis[e] | 20.0 | 17.0 | 14.0 | 11.0 | 9.0 | – | – | – | >95 | >95 |

[a]Millions of gallons per day; [b]Suspended solids; [c]Granular carbon; [d]Not including pH adjustment; [e]Not including brine disposal.

Source:   Smith, C.V., Di Gregorio, D.: "Advance Wastewater Treatment," *Chemical Engineering*, Apr. 27, 1970, p. 73.

enough effluent in many cases, and adsorption or some other processes have been added.

The competitive physical/chemical system that is being installed at Rosemount, Minn., consists of primary treatment followed by coagulation, sand filtration, activated carbon adsorption, another filtration step, ion exchange, and oxidation. This can produce a highly purified water at less cost than a system involving primary, secondary, and tertiary treatment.[30]

An industrial treatment system may require some chemical pretreatment before biological treatment; even some physical treatment may be desirable. Also, since the concentrations of the pollutants are usually greater and more predictable than those in municipal wastes, the engineer can design a more specific system than is possible for municipal treatment plants.[64] In this case all possibilities must be carefully evaluated by the process engineer.

Finally, usually a couple of large holding ponds or lagoons are constructed to even out loads and to provide for storage when major upsets occur. These may be constructed of any impervious material like clay, bentonite, wood, concrete, or metal. They may also be lined with a synthetic material like nylon, rubber, polyvinyl chloride, or polyethylene[65] to prevent any seeping from occurring. The costs for such ponds are given in reference 65.

## CONCENTRATED LIQUID
## AND SOLID WASTE TREATMENT PROCEDURES

Concentrated wastes can result from bad product being made, unsalable by-products, contamination of products, laboratory wastes, and previously mentioned pollution-abatement steps that concentrate the pollutants. Whatever their source, if recovery is impossible, they must be eliminated. The most common means are incineration or pyrolysis followed by landfill operations and/or compacting. As mentioned in Chapter 8, heat recovery methods should be used wherever possible. This is especially true for all incinerators and pyrolysis equipment.

## Incineration

This has been mentioned as a means for treating sludges. The furnaces mentioned there could be adapted to process other wastes.

The cheapest incinerator is an open-pit type. With proper design, particulate emissions have been reduced to 0.25 g/ft³ (7 mg/m³). However, there is no way to clean the exit gases, so materials that form obnoxious or harmful products cannot be charged if pollution standards are to be met. Dupont has developed a special design in which the air is supplied through high velocity nozzles. The air jets produce a cylindrical rolling flame that results in higher burning temperatures and more thorough mixing of air and fuel.

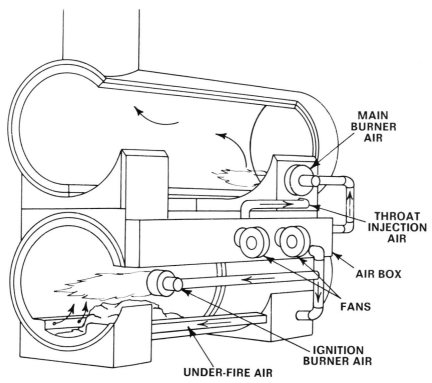

Figure 16-7    A two-chamber incinerator. The raw waste is charged to the lower incinerator- Here it is burned with a limited air supply in order to develop a high temperature which will gasify most of the solid waste material. Between the chambers the gases pass a restriction (throat) where air is injected. They then enter the second chamber and the flame of the main burner.
Courtesy of the Trane Company.

Multiple-chamber incinerators (Fig. 16-7) provide for different regions so that after the primary ignition chamber the gases will spend enough time at a high enough temperature, under turbulent conditions in the presence of oxygen, to ensure nearly complete combustion. The passages are tortuous to permit settling of the entrained fly ash. They are usually followed by a wet scrubber plus any other air-pollution abatement devices that may be desirable. Reference 61 gives the costs for these incinerators and wet scrubbers.

Rotary kilns can be used when there are low-ash liquid and solid wastes to be burned. However, they have a problem in that it is difficult to supply them with enough air, and the effluent is often smoky unless an afterburner is used. Also, the refractory material must be replaced yearly.[61]

### Pyrolysis

Pyrolysis is the decomposition of organic compounds in the absence of a flame and air, at temperatures often as high as 3,000°F (1,650°C). The result is a number of different, simpler organic compounds. The type and amount of each depend on the temperature and the time the material is at the elevated temperatures. The resultant products might be sold as a synthetic gas or may contain valuable compounds that can be separated out and sold. Even when these alternatives are not feasible, the gas leaving the pyrolysis units has two advantages over the usual material charged to the burners. First, it is a gas, and second, it contains very little ash. This means it will pose fewer pollution and operating problems if it is burned to produce energy. This energy can be used to provide the high temperatures required in the pyrolysis unit. At large throughputs the capital and operating expenses for these units can be less than for an incinerator.[61]

### Ultimate Disposal

After combustion or pyrolysis the waste can sometimes be used as a filler in making concrete roads or blocks. It can be compacted and disposed of as landfill, put in the ocean, or deposited in some underground mine. Care must be taken that any leaching that may occur after final disposal will not introduce any hazardous or noxious substances into the air or the water.

### References

1. Orleans, L.A., Suttmeier, R.P.: "The Mao Ethic and Environmental Quality," *Science* **170**: 1173, Dec. 11, 1970.
2. Taiganides, E.P.: "Everything You Always Wanted to Know about Ecology but Were Afraid to Ask," banquet address given April 14, 1972, to the North Central Section of ASEE, Cleveland.
3. "War against Water Pollution Gets More Lively," *Chemical Engineering,* May 23, 1966, p. 88.
4. "Do You Know That...," *Civil Engineering,* Apr. 1972, p. 29.
5. Harrison, H.L., Loucks, O.L., Mitchell, J.W., Parkhurst, D.F., Tracy, CoR., Watts, D.G., Vannacone, V.J.,Jr.: "Systems Studies of DDT Transport," *Science* **170**: 503, Oct. 30, 1970.
6. Shepard, P., McKinley, D. (eds.): *The Subversive Science,* Houghton Mifflin, Boston, 1969, p. 84.
7. "For Rivers: Breathing Room," *Chemical Week,* June 21, 1969, p. 131.

8. Hammond, A.S.: "Mercury in the Environment: Natural and Human Factors," *Science* **171**:789, Feb. 26, 1971.

9. "Mercury's Turn as Villain", *Chemical Engineering*, July 27, 1970, p. 84.

10. "Tighter Limits on Mercury Discharge," *Chemical Week*, July 22, 1970, p. 35.

11. April, R.W., Hume, D.N.: "Environmental Mercury: Rapid Determination in Water at Nanogram Levels," *Science* **170**:849, Nov. 30, 1970.

12. Rosenzweig, M.D.: "Paring Mercury Pollution," *Chemical Engineering*, Feb. 22, 1971, p. 70.

13. Maugh, T.H.,II: "Trace Elements: a Growing Appreciation of Their Effects on Man," *Science* **181**:253, July 20, 1973.

14. "Elements-in-man Research Runs Out of Gold", *Chemical Week*, Jan. 26, 1972. p. 39.

15. Teller, Aaron J.: "Air Pollution Control," *Chemical Engineering*, May 8, 1972, p. 93.

16. Carleton-Jones, D.S.: "Tall Chimneys," *Chemical Engineering*, Oct. 14, 1968, p. 166.

17. "Ocean Pollution and Marine Waste Disposal," *Chemical Engineering*, Feb. 8, 1971, p. 60.

18. Sheldrick, M.G.: "Deepwell Disposal: Are Safeguards Being Ignored," *Chemical Engineering*, Apr. 7, 1969, p. 74.

19. Talbot, J.S.: "Deep Wells," *Chemical Engineering*, Oct. 14, 1968, p. 108.

20. Vandegrift, A.E., Shannon, L.J., Gorman, P.G.: "Controlling Fine Particles," *Chemical Engineering*, June 18, 1973, p. 107.

21. Sickles, R.W.: "Electrostatic Precipitation," *Chemical Engineering*, Oct. 14, 1968, p. 156.

22. Munson, J.S.: "Dry Mechanical Collectors," *Chemical Engineering*, Oct. 14, 1968, p. 147.

23. Imperato, N.F.: "Gas Scrubbers," *Chemical Engineering*, Oct. 14, 1968, p. 147.

24. Brewer, G.L.: "Fume Incineration," *Chemical Engineering*, Oct. 14, 1968, p. 160.

25. Vandenhoeck, P.: "Cooling Hot Gases before Baghouse Filtration," *Chemical Engineering*, May 1, 1972, p. 67.

26. Alonso, J.R.F.: "Estimating the Costs of Gas-Cleaning Plants," *Chemical Engineering*, Dec. 13, 1971, p. 87.

27. Fair, J.R., Crocker, B.B., Null, H.R.: "Trace-Quantity Engineering," *Chemical Engineering*, Aug. 7, 1962, p. 60.

28. Maurin, P.G., Jonakin, J.: "Removing Sulfur Dioxides from Stacks," *Chemical Engineering*, Apr. 27, 1970, p. 173.

29. Rickles, R.N.: "Waste Recovery and Pollution Abatement," *Chemical Engineering*, Sept. 27, 1965, p. 133.

30. Larkman, D.: "Physical/Chemical Treatment," *Chemical Engineering*, June 18, 1973, p. 87.

31. Geinopolos, A., Katz, W.J.: "Primary Treatment," *Chemical Engineering*, Oct. 14, 1968, p. 79.

32. Gurnham, C.F.: "Control of Water Pollution," *Chemical Engineering*, June 10, 1963, p. 191.

33. Eckenfelder, W.W., Jr., Ford, D.L.: "Economics of Waste Water Treatment," *Chemical Engineering*, Aug. 25, 1969, p. 109.

34. Lesperance, T.W.: "Biological Treatment," *Chemical Engineering*, Oct. 14, 1968, p. 89.

35. Kemmer, F.N., Odland, K.: "Chemical Treatment," *Chemical Engineering*, Oct. 14, 1968, p. 83.

36. "US Cleanup Funds: Boost for Polyelectrolytes," *Chemical Week*, Mar. 3, 1971, p. 55.

37. Fitzgerald, C.L., Clemens, M.M., Reilly, P.B., Jr.: "Coagulants for Waste Water Treatment," *Chemical Engineering Progress*, Jan. 1970, p. 36.

38. "$O_2$ & $O_3$—Rx for Pollution," *Chemical Engineering*, Feb. 23, 1970, p. 46.

39. "Ion Exchange: Steady Does it," *Chemical Week*, Aug. 24, 1968, p. 31.

40. Michalson, A.W.: "Ion Exchange," *Chemical Engineering*, Mar. 18, 1963, p. 163.

41. Applebaum, S.B.: *Demineralization by Ion Exchange*, Academic Press, New York, 1968.

42. Lacey, R.E.: "Membrane Separation Processes," *Chemical Engineering*, Sept. 4, 1972, p. 56.

43. Wilson, J. (ed.): *Demineralization by Electrodialysis*, Butterworths, London, 1960.

44. Lacey, R.E., Loeb, S. (eds.): *Industrial Processing with Membranes*, Wiley, New York, 1972.

45. Lukchis, G.M.: "Adsorption Systems, Part III: Adsorption Regneration," *Chemical Engineering*, Aug. 6, 1973, p. 83.

46. Fox, R.D.: "Pollution Control at the Source," *Chemical Engineering*, Aug. 6, 1973, p. 72.

47. Smith, C.V., DiGregorio, D.: "Advance Wastewater Treatment," *Chemical Engineering*, Apr. 27, 1970, p. 71.

48. Lukchis, G.M.: "Adsorption Systems, Part II: Equipment Design," *Chemical Engineering*, July 9, 1973, p. 83.
49. Cooper, J.C., Hager, D.G.: "Water Reclamation with Activated Carbon", *Chemical Engineering Progress*, Oct. 1966, p. 85.
50. Hayes, R.C.: "Advanced Water Treatment via Filter," *Chemical Engineering Progress*, June 1969, p. 81.
51. "Sand Filter Saves Space," *Chemical Engineering*, Sept. 21, 1970, p. 112.
52. "Water Pollution Control," *Chemical Engineering*, June 21, 1971, p. 65.
53. Dahlstrom, D.: "Sludge Dewatering," *Chemical Engineering*, Oct. 14, 1968, p. 103.
54. Perry, J.H. (ed.): *Chemical Engineers' Handbook*, McGraw-Hill, New York, 1963, Section 19 p.76.
55. Kaup, E.C.: "Design Factors in Reverse Osmosis," *Chemical Engineering*, Apr. 2, 1973, p. 46.
56. Shell, G.L., Boyd, J.L., Dalstrom, D.A.: "Upgrading Waste Treatment Plants," *Chemical Engineering*, June 21, 1971, p. 97.
57. Brunner, C.A., Stephan, D.G.: "Foam Fractionation," *Industrial and Engineering Chemistry*, May 1965, p. 40.
58. Characklis, W.G., Busch, A.W.: "Industrial Waste Water Systems," *Chemical Engineering*, May 8, 1972, p. 61.
59. Eliasen, R., Tchobanoglous, G.: "Advanced Treatment Processes," *Chemical Engineering*, Oct. 14, 1968, p. 95.
60. 'Solid Waste Treatment," *Chemical Engineering*, June 21, 1971, p. 155.
61. Witt, P.A., Jr.: "Solid Waste Disposal," *Chemical Engineering*, May 8, 1972, p. 109.
62. Sebastian, F.P., Cardinal, P.J.: "Solid Waste Disposal," *Chemical Engineering*, Oct. 14, 1968, p. 112.
63. Sohr, W.H., Ott, R., Albertson, O.E.: "Fluid and Sewage Sludge Combustion," *Water Works and Waste Engineering*, Sept. 1965, p. 90.
64. Sawyer, G.A.: "New Trends in Wastewater Treatment and Recycle," *Chemical Engineering*, July 24, 1972, p. 121.
65. Kumar, J., Jedlicka, J.: "Selecting and Installing Synthetic Pond-Linings," *Chemical Engineering*, Feb. 5, 1973, p. 67.

## Additional References

"Environmental Engineering," deskbook issues of *Chemical Engineering*, Oct. 14, 1968; Apr. 27, 1970; June 21, 1971; May 8, 1972; June 18, 1973.

Bond, R.G., Straub, C.P.: *CRC Handbook of Environmental Control*, 4 vols., CRC Press, Cleveland, 1973.

Lund, H.F. (ed.): *Industrial Pollution Control Handbook*, McGraw-Hill, New York, 1971.

The following publications of the Environmental Protection Agency:

*Estimating Costs and Manpower Requirements for Conventional Wastewater Treatment Facilities*, 17090 DAN 10/71, Oct. 1971.

*Capital and Operating Costs of Pollution Control Equipment Modules*. Vol. II: *Data Manual*, R5-73-023b, July, 1973.

*Projected Wastewater Treatment Costs in the Organic Chemical Industry*, 12020 GND 07/71, July 1971.

*Preliminary Investigational Requirements—Petrochemical and Refinery Waste Treatment Facilities*, 12020 EID 03/71, Mar. 1971.

*Inorganic Chemicals Industrial Profile*, 12020 EJI 07/71, July 1971.

*State of the Art Review: Water Pollution Control Benefits and Costs'* vol. I, 600573-008a, Oct. 1973.

*Cost Analysis of Water Pollution Control: An Annotated Bibliography*, R5-73-017, Apr. 1973.

All publications, U.S. Government Printing Office, Washington, D.C.

Vandegrift, A.E., Shannon, L.J.: *Particulate Pollutant System Study*. vol. II, *Handbook of Emission Properties*, NTIS, National Technical Information Service, Commerce Department PB 203522, Springfield, Va., 1971.

Vervalin, C.H.: "Contact These Sources for Environmental Information," *Hydrocarbon Processing*, Oct. 1973, p. 71.

Alsentzer, H.A.: "Treatment of Industrial Wastes at Regional Facilities," *Chemical Engineering Progress*, Aug. 1972, p. 73.

Monaghan, C.A.: "Environmental Management: A Planned, Multi-plant Approach," *Hydrocarbon Processing*, Oct. 1972, p. 173.

Beychok, M.R.: "Wastewater Treatment," *Hydrocarbon Processing*, Dec. 1971, p. 109.

Thompson, C.S., Stock, J., Mehta, P.L.: "Cost and Operating Factors for Treatment of Oily Waste Water," *Oil and Gas Journal*, Nov. 20, 1972, p. 53.

Thomson, S.J.: "Data Improves Separator Design," *Hydrocarbon Processing*, Oct. 1973, p. 81.

Raynor, R.C., Porter, E.F.: "Thickeners and Clarifiers", *Chemical Engineering*, June 20, 1966, p. 198.

Keith, F.W., Jr., Moll, R.T.: "Matching a Dewatering Centrifuge to Waste Sludge," *Chemical Engineering Progress*, Sept. 1971, p. 55.

Ferrel, J.F., Ford, D.L.: "Select Aerators Carefully," *Hydrocarbon Processing*, Oct. 1972, p. 101.

Nogaj, R.J.: "Selecting Wastewater Aeration Equipment," *Chemical Engineering*, Apr. 17, 1972, p. 95.

Newkirk, R.W., Schroeder, P.J.: "Reverse Osmosis Can Help with Wastes," *Hydrocarbon Processing*, Oct. 1972, p. 103.

Leitner, G.F.: "Reverse Osmosis for Water Recovery and Reuse," *Chemical Engineering Progress*, June 1973, p. 83.

Nusbaum, I., Cruver, R.E., Sleigh, J.H., Jr.: "Reverse Osmosis—New Solutions and New Problems," *Chemical Engineering Progress*, Jan. 1972, p. 69.

Albertson, O.E., Vaughn, D.R.: "Handling of Solid Wastes," *Chemical Engineering Progress*, Sept. 1971, p. 49.

Kent, G.R.: "Make Profit from Flares," *Hydrocarbon Processing*, Oct. 1972, p. 121.

# Appendix A

## Conversion Factors

**Length**

| | | |
|---|---|---|
| 1 foot | = | 12 inches  =  0.3048 meters |
| 1 inch | = | 2.54 centimeters |
| 1 mile | = | 5,280 feet  =  1.609 kilometers |
| 1 meter | = | 100 centimeters  =  3.281 feet  = 39.37 inches |
| 1 micron | = | 0.0001 centimeter |

**Area**

| | | |
|---|---|---|
| 1 square mile | = | 640 acres  = 259 hectares |
| 1 hectare | = | 10,000 square meters  = 2.471 acres |

**Volume**

| | | |
|---|---|---|
| 1 cubic foot | = | 1,728 cubic inches  = 28.32 liters |
| 1 U.S. gallon | = | 0.833 Imperial gallons = 3.785 liters = 231 cubic inches |
| 1 U.S. barrel  (petroleum) | = | 42 U.S. gallons = 35 Imperial gallons |
| 1 liter | = | 1,000 cubic centimeters = 0.2642 U.S. gallons |

**Mass**

| | | |
|---|---|---|
| 1 pound | = | 0.454 kilograms |
| 1 U.S. short ton | = | 2,000 pounds = 907 kilograms |
| 1 U.S. long ton | = | 2,240 pounds |
| 1 metric ton | = | 1,000 kilograms = 1.102 U.S. short tons = |
| 1 kilogram | = | 1,000 grams  = 2.205 pounds |
| 1 grain | = | 64.8 milligrams |

**Pressure**

| | | |
|---|---|---|
| 1 atmosphere | = | 760 millimeters of mercury |
| | = | 29.92 inches of mercury |
| | = | 14.696 pounds per square inch |
| | = | 1.033 kilograms per square centimeter |
| | = | 1,013 millibars |
| 1 atmosphere (metric) | = | 1 kilogram per square centimeter |
| | = | 10,000 kilograms per square meter |
| | = | 10 meters head of water |
| | = | 14.22 pounds per square inch |
| 1 bar | = | 1.02 kilogram per square centimeter |
| 1 pound per square inch | = | 2.036 inches of mercury |
| | = | 2.309 feet of water |
| | = | 0.0703 kilograms per square centimeter |
| 1 foot of water | = | 0.433 pounds per square inch |

**Miscellaneous**

| | | |
|---|---|---|
| 1 British Thermal Unit (BTU) | = | 0.2520 kilocalories |
| | = | 778.17 foot pounds |
| | = | 107.6 kilogram meters |
| 1 kilocalorie | = | 3,088 foot pounds |
| | = | 427 kilogram meters |
| | = | 3.968 BTU |
| 1 BTU per pound | = | 0.556 kilocalories per kilogram |
| 1 foot pound | = | 0.1383 kilogram meter |
| | = | 1.3558 joules |
| 1 kilowatt | = | 738 foot pounds per second |
| | = | 102 kilogram meters per second |
| | = | 1.341 horsepower |
| | = | 1.360 metric horsepower |
| 1 horsepower | = | 33,000 foot pounds per minute |
| | = | 550 foot pounds per second |
| | = | 76.04 kilogram meters per second |
| | = | 0.746 kilowatt |
| | = | 1.014 metric horsepower |
| 1 metric horsepower | = | 32,550 foot pounds per minute |
| | = | 542 foot pounds per second |
| | = | 75 kilogram meters per second |
| | = | 0.986 horsepower |
| | = | 0.735 kilowatt |
| 1 kilowatt hour | = | 3,412.14 BTU |
| | = | 860 kilocalories |
| 1 horsepower hour | = | 2,545.1 BTU |

Appendix B

Cost Data for Process Equipment
Offsite Facilities, Site Development,
and Structures

The data in this section are in my opinion the best available. They were developed from data presented by Kenneth M. Guthrie and first appeared in the March 24, 1969 and Jan. 13, 1969, issues of *Chemical Engineering*. I am very indebted to Mr. Guthrie and *Chemical Engineering* for allowing me to reproduce this material. All these data are based on mid 1968 costs. Examples 9-6, 9-7, and 9-8, along with the capital cost estimation of a 150,000,000 lb/yr polystrene plant, show how to use the figures and tables.

Table B-1

Bare Module Factor for Figures B-1 to B-7

| Base dollar magnitude $100,000 | Single Unit | Multiple Units | | | | |
|---|---|---|---|---|---|---|
| | | Up to 2 | 2 to 4 | 4 to 6 | 6 to 8 | 8 to 10 |
| Process Furnaces | | 2.27 | 2.19 | 2.16 | 2.15 | 2.13 |
| Direct fired heaters | | 2.23 | 2.16 | 2.13 | 2.12 | 2.10 |
| Shell-and-tube exchangers | 3.39 | 3.29 | 3.18 | 3.14 | 3.12 | 3.09 |
| Air coolers | 2.54 | 2.31 | 2.20 | 2.18 | 2.16 | 2.14 |
| Pressure vessels—vertical | 4.34 | 4.23 | 4.12 | 4.07 | 4.06 | 4.02 |
| Pressure vessels—horizontal | 3.29 | 3.18 | 3.06 | 3.01 | 2.99 | 2.96 |
| Centrifugal pumps and drivers | 3.48 | 3.38 | 3.28 | 3.24 | 3.23 | 3.20 |
| Reciprocating pumps and drivers | 3.48 | 3.38 | 3.28 | 3.24 | 3.23 | 3.20 |

Source:   Guthrie, K.M.: "Capital Cost Estimating," *Chemical Engineering*, Mar. 24, 1969, p. 114.

Table B-2
Bare Module Factors and Unit Costs for Various Items (mid-1968)

| | Units | Unit Cost ($) | Size Exponent | Bare Module Factor |
|---|---|---|---|---|
| Double Pipe Heat Exchanger . . . . . | Ft$^2$ | 266.00 | 0.07 | 1.83 |
| Centrifugal Gas Compressor | | | | |
| And motor . . . . . . . . . . . . . . | | | | |
| With Turbine . . . . . . . . . . . . . | BHP | 527.00 | 0.82 | 3.21 - 2.93 |
| Reciprocating Gas Compressor | BHP | 606.00 | 0.82 | 3.21 - 2.93 |
| Run by Steam . . . . . . . . . . . . | | | | |
| And Motor . . . . . . . . . . . . . . | BHP | 564.00 | 0.82 | 3.21 - 2.93 |
| And Gas Engine . . . . . . . . . . | BHP | 679.00 | 0.82 | 3.21 - 2.93 |
| Packaged Boiler Units | BHP | 959.00 | 0.82 | 3.21 - 2.93 |
| < 250 psi saturated steam . . . . . | Lb/hr | 29.20 | 0.70 | 1.83 |
| Field-Erected Boiler Units | | | 0.80 | 1.96 |
| < 400 psi saturated steam . . . . . | LB/hr | 54.60 | 0.80 | 1.46 |
| Power Generating Facilities . . . . . | Kva | 520.00 | 0.80 | 1.75 |
| Cooling Tower | | | 0.75 | |
| Cooling Range 15 °F . . . . . . . | Gpm | 721.00 | 0.60 | 1.75 |
| Cooling Range 20 °F . . . . . . . | Gpm | 1,118.00 | 0.60 | 1.75 |
| Cooling Range 25 °F . . . . . . . | Gpm | 1,406.00 | 0.60 | 1.75 |
| Vertical Storage Tank$^a$ | | | | |
| < 1500 gallons . . . . . . . . . . . | Gal | 275.00 | 0.28 | 1.96 |
| 1500 - 40,000 . . . . . . . . . . . . | Gal | 409.00 | 0.30 | 1.96 |
| 40,000 - 5,000,000 . . . . . . . . | Gal | 11.80 | 0.63 | 2.52 |
| Horizontal Pressure Storage Vessel | | | | |
| < 150 psi . . . . . . . . . . . . . . | Gal | 16.10 | 0.65 | 2.20 |
| 150-200 . . . . . . . . . . . . . . | Gal | 18.50 | 0.65 | 2.20 |
| 200-250 . . . . . . . . . . . . . . | Gal | 21.30 | 0.65 | 2.20 |
| Spherical Pressure Storage Vessel | | | | |
| < 30 psi . . . . . . . . . . . . . . | Gal | 20.50 | 0.70 | 2.32 |
| 30-50 . . . . . . . . . . . . . . . . | Gal | 22.20 | 0.70 | 2.32 |
| 50-75 . . . . . . . . . . . . . . . . | Gal | 24.40 | 0.70 | 2.32 |
| 75-100 . . . . . . . . . . . . . . | Gal | 25.60 | 0.70 | 2.32 |
| 100-125 . . . . . . . . . . . . . . | Gal | 29.10 | 0.70 | 2.32 |
| 125-200 . . . . . . . . . . . . . . | Gal | 32.60 | 0.70 | 2.32 |
| Mechanical Refrigeration | | | | |
| (evaporation temperature) | | | | |
| 40 °F . . . . . . . . . . . . . . . . | Tons | 1,509.00 | 0.70 | 1.42 |
| 20 °F . . . . . . . . . . . . . . . . | Tons | 2,943.00 | 0.70 | 1.42 |
| 0 °F . . . . . . . . . . . . . . . . | Tons | 3,395.00 | 0.70 | 1.42 |
| -20 °F . . . . . . . . . . . . . . . . | Tons | 5,961.00 | 0.70 | 1.42 |
| -40 °F . . . . . . . . . . . . . . . . | Tons | 6,850.00 | 0.70 | 1.42 |

$^a$For aluminum multiply by 1.4; for rubber-lined, multiply by 1.48; for lead-lined, multiply by 1.55; for stainless steel, multiply by 3.20; for glass-lined (10,000 gal), multiply by 4.25.

Table B-3
Additional Bare Module Factors and Unit Cost Data

| | Unit | Unit @ Cost, $ | Size Exponent | Bare Module Factor |
|---|---|---|---|---|
| Agitators | | | | |
| Propellers .............. | Hp. | 350 | 0.50 | 2.09 |
| Turbine ............... | Hp. | 750 | 0.30 | 2.09 |
| Air compressors (cap.) | | | | |
| 125 psig. (cap.) ........... | Cu. ft./min. | 2,900 | 0.28 | 2.07 |
| Air conditioners | | | | |
| Window vent ............. | Ea. | 300 | —— | 1.45 |
| Floor-mounted .......... | Ea. | 200 | —— | 1.45 |
| Rooftop 10 ton ........... | Ea. | 3,800 | —— | 1.55 |
| 20 ............... | Ea. | 6,500 | —— | 1.55 |
| 30 ............... | Ea. | 8,100 | —— | 1.55 |
| Air dryers (cap.) | Cu. ft./min. | 200 | 0.56 | 2.25 |
| Bagging machines (cap.) | | | | |
| Weight ............... | Bags/min. | 3,300 | 0.80 | 1.87 |
| Volume .............. | Bags/min. | 1,000 | 0.80 | 1.87 |
| Blenders (cap.) | Cu. ft./min | 850 | 0.52 | 2.08 |
| Blowers * fans (cap.) | Cu. ft./min | 7 | 0.68 | 2.05 |
| Boilers (industrial) | | | | |
| 15 psig ................ | Lb/hr. | 400 | 0.50 | 1.94 |
| 150 ................ | Lb/hr. | 440 | 0.50 | 1.94 |
| 300 ................ | Lb./hr. | 500 | 0.50 | 1.94 |
| 600 ................ | Lb./hr. | 560 | 0.50 | 1.94 |
| Centrifuges | | | | |
| Horizontal basket ......... | Dia., in. | 140 | 1.25 | 2.03 |
| Vertical basket .......... | Dia., in. | 310 | 1.00 | 2.03 |
| Soild bowl (SS) ........... | Hp | 1,900 | 0.73 | 2.06 |
| Sharples (SS) ............. | Hp | 5,200 | 0.68 | 2.06 |
| Conveyors (length) | | | | |
| Belt, † 18 in. wide ........ | Ft. | 450 | 0.65 | 2.18 |
| 24 ............... | Ft. | 540 | 0.65 | 2.18 |
| 36 ............... | Ft. | 620 | 0.65 | 2.12 |
| 42 ............... | Ft. | 700 | 0.65 | 2.12 |
| 48 ............... | Ft. | 750 | 0.65 | 2.12 |
| Bucket (height) ........... | | | | |
| 30 tons/hr. (8 in. × 5 in.)... | Ft. | 220 | 0.65 | 2.37 |
| 75 tons/hr. (14 in. × 7 in.).. | Ft. | 400 | 0.83 | 2.37 |
| 120 tons/hr. (15 in. × 8 in.). | Ft. | 500 | 0.83 | 2.37 |
| Roller, 12 in. wide ......... | Ft. | 7 | 0.90 | 2.18 |
| 15 ............... | Ft. | 8 | 0.90 | 2.18 |
| 18 ............... | Ft. | 9 | 0.90 | 2.13 |
| 20 ............... | Ft. | 10 | 0.90 | 2.13 |
| Screw, 6 in. dia. ........... | Ft. | 230 | 0.90 | 2.05 |
| 12 ............... | Ft. | 270 | 0.80 | 2.05 |
| 14 ............... | Ft. | 290 | 0.75 | 2.05 |
| 16 ............... | Ft. | 300 | 0.60 | 2.05 |
| Vibrating, 12 in. wide ....... | Ft. | 80 | 0.80 | 2.12 |
| 18 ............... | Ft. | 110 | 0.80 | 2.12 |
| 24 ............... | Ft. | 120 | 0.90 | 2.06 |
| 36 ............... | Ft. | 150 | 0.90 | 2.06 |
| Cranes (cap.) | | | | |
| Span 10 ft. ............. | Tons | 1,800 | 0.60 | 1.29 |
| 20 ............... | Tons | 2,400 | 0.60 | 1.29 |
| 30 ............... | Tons | 3,800 | 0.60 | 1.29 |
| 40 ............... | Tons | 4,800 | 0.60 | 1.29 |
| 50 ............... | Tons | 6,300 | 0.60 | 1.29 |
| 100 ............... | Tons | 8,500 | 0.60 | 1.29 |
| Crushers (cap.) | | | | |
| Cone ............... | Tons/hr. | 750 | 0.85 | 2.03 |
| Gyratory ............. | Tons/hr. | 55 | 1.20 | 2.03 |
| Jaw ................. | Lb./hr. | 520 | 0.35 | 2.05 |
| Pulverizers .............. | | | | |

## Table B-3 (*Continued*)

| | | | | |
|---|---|---|---|---|
| Crystallizers (cap.) | Tons/day | 5,500 | 0.65 | 2.26 |
| Growth | Tons/day | 7,900 | 0.55 | 2.26 |
| Forced circulation | Gal. | 170 | 0.70 | 2.06 |
| Batch | | | | |
| Dryers (area) | Sq. ft. | 3,000 | 0.45 | 2.24 |
| Drum | Sq. ft. | 1,900 | 0.38 | 2.24 |
| Pan | Sq. ft. | 3,100 | 0.45 | 2.24 |
| Rotary vacuum | | | | |
| Ductwork | Lin./ft. | 5.42 | 0.55 | 1.29 |
| (Shop fabricated and field erected) | Lin./ft. | 8.00 | 0.55 | 1.29 |
| Aluminum | Lin./ft. | 15.12 | 0.55 | 1.29 |
| Galvanized | | | | |
| Stainless | | | | |
| Dust collectors (cap.) | | | | |
| Cyclones | Cu. ft./min. | 3 | 0.80 | 2.18 |
| Cloth filter | Cu. ft./min. | 25 | 0.68 | 2.18 |
| Precipitators | Cu. ft./min. | 390 | 0.75 | 2.18 |
| Ejectors (cap.) | | | | |
| 4 in. Hg suction | Lb./hr. | 2,000 | 0.79 | 1.42 |
| 6 | Lb./hr. | 200 | 0.67 | 1.42 |
| 10 | Lb./hr. | 200 | 0.55 | 1.42 |
| 4-stage barometric | | | | |
| 2.5 mm Hg suction | Lb./hr. | 2,500 | 0.45 | 1.45 |
| 5.0 | Lb./hr. | 1,400 | 0.48 | 1.45 |
| 10.0 | Lb./hr. | 900 | 0.53 | 1.45 |
| 20.00 | Lb./hr. | 700 | 0.54 | 1.45 |
| 5-stage barometric | | | | |
| 0.5-mm. Hg suction | Lb./hr. | 4,200 | 0.50 | 1.49 |
| 0.8 | Lb./hr. | 3,200 | 0.50 | 1.49 |
| 1.0 | Lb./hr. | 2,800 | 0.48 | 1.49 |
| 1.4 | Lb./hr. | 2,500 | 0.49 | 1.49 |
| Elevators (height) | | | | |
| Freight 3,000 lb | Ft. | 3,600 | 0.32 | 1.29 |
| 5,000 | Ft. | 4,000 | 0.32 | 1.29 |
| 10,000 | Ft. | 5,400 | 0.32 | 1.29 |
| Passenger 3,500 lb | Ft. | 3,900 | 0.48 | 1.29 |
| Evaporators | | | | |
| Forced circulation | Sq. ft. | 6,000 | 0.70 | 2.45 |
| Vertical tube | Sq. ft. | 1,200 | 0.53 | 2.45 |
| Horizontal tube | Sq. ft. | 800 | 0.53 | 2.45 |
| Jacketed vessel (glasslined) | Gal. | 1,000 | 0.50 | 2.25 |
| Filters (effective area) | | | | |
| Plates and press | Sq. ft. | 330 | 0.58 | 2.31 |
| Pressure leaf-wet | Sq. ft. | 410 | 0.58 | 2.31 |
| dry | Sq. ft. | 1,500 | 0.53 | 2.31 |
| Rotary drum | Sq. ft. | 1,400 | 0.63 | 2.06 |
| Rotary disk | Sq. ft. | 1,000 | 0.78 | 2.06 |
| Flakers (effective area) | | | | |
| Drum | Sq. ft. | 1,300 | 0.64 | 2.05 |
| Generator sets (portable) | | | | |
| 10 kw | Ea. | 1,500 | ------ | 1.29 |
| 15 kw | Ea. | 2,000 | ------ | 1.29 |
| 25 kw | Ea. | 3,000 | ------ | 1.29 |
| 50 kw | Ea. | 5,000 | ------ | 1.29 |
| 100 kw | Ea. | 7,000 | ------ | 1.29 |
| Hydraulic presses (plate area) | | | | |
| 100 psig | Sq. ft. | 2,500 | 0.95 | 2.24 |
| 300 | Sq. ft. | 3,600 | 0.95 | 2.24 |
| 500 | Sq. ft. | 5,000 | 0.95 | 2.24 |
| 1,000 | Sq. ft. | 6,200 | 0.95 | 2.24 |
| Mills (cap.) | | | | |
| Ball | Tons/hr. | 550 | 0.65 | 2.20 |
| Roller | Tons/hr. | 5,000 | 0.65 | 2.20 |
| Hammer | Tons/hr. | 500 | 0.85 | 2.20 |

| | | | | |
|---|---|---|---|---|
| Screens (surface) | | | | |
| vibrating single . . . . . . . . . . . | Sq. ft. | 900 | 0.58 | 1.70 |
| double . . . . . . . . . . . | Sq. ft. | 1,100 | 0.58 | 1.70 |
| Stacks (height) | | | | |
| 24 in. (CS) . . . . . . . . . . . . . | Lin./ft. | 25.83 | 1.00 | 1.60 |
| 36 in. (CS) . . . . . . . . . . . . . | Lin./ft. | 58.20 | 1.00 | 1.60 |
| 48 in. (CS) . . . . . . . . . . . . . | Lin./ft. | 78.25 | 1.00 | 1.60 |
| Tank heaters (area) | | | | |
| Stream coil* . . . . . . . . . . . . . | Sq. ft. | 94.12 | 0.32 | 1.61 |
| Immersion . . . . . . . . . . . . . . | Kw. | 18.75 | 0.85 | 1.55 |
| Weigh Scales | | | | |
| Portable beam . . . . . . . . . . . | Ea. | 250 | ----- | 1.29 |
| dial . . . . . . . . . . . . . . . | Ea. | 1,500 | ----- | 1.29 |
| Truck 20 ton . . . . . . . . . . . . | Ea. | 4,000 | ----- | 1.39 |
| 50 . . . . . . . . . . . . . . . | Ea. | 7,200 | ----- | 1.39 |
| 75 . . . . . . . . . . . . . . . | Ea. | 8,500 | ----- | 1.39 |

*a* All unit costs are based on mid-1968. These are not general unit costs.
† For enclosed conveyors walkway multiply by 2.10
\* Stainless factor is 2.4
Source: Guthrie, K.M.: "Capital Cost Estimating," *Chemical Engineering*, Mar. 24, 1969, p. 114.

## Table B-4

## Direct Costs for site development*

| | Unit | Field Min. | Installation, $ Norm | Max. | | Unit | Field Min. | Installation, $ Norm. | Max. |
|---|---|---|---|---|---|---|---|---|---|
| Dewatering and drainage | | | | | 4-in.-thk. | | | | |
| Pumping sys- | | | | | gravel | Sq. yd. | 0.55 | 0.87 | 1.19 |
| tem (rented) | Day | 25 | 32 | 40 | 6-in.-thk | | | | |
| Wellpoint de- | | | | | gravel | Sq. yd. | 1.00 | 1.38 | 1.76 |
| watering sys- | | | | | | | | | |
| tem | Month | 6,500 | 7,500 | 8,500 | Parking lots | | | | |
| Drainage | | | | | Black-top | | | | |
| trench | Lin. ft. | 0.75 | 0.85 | 0.95 | surface | Sq. yd. | 5.30 | 6.25 | 7.54 |
| | | | | | Sewer facilities | | | | |
| Fencing | | | | | Asbestos Ce- | | | | |
| Complete | | | | | ment pipe | | | | |
| fence (light) | Lin. ft. | 1.34 | 1.88 | 2.42 | (general) | Lin. ft. | 4.55 | 4.85 | 5.15 |
| Complete | | | | | Concrete pipe | | | | |
| fence (heavy) | Lin. ft. | 1.51 | 2.13 | 2.75 | (reinforced) | | | | |
| Chain link | Lin. ft. | 5.48 | 5.93 | 6.38 | 18 in. dia. | Lin. ft. | 5.65 | 5.80 | 5.95 |
| Gates 6 ft. | | | | | 36 in. | Lin. ft. | 14.75 | 15.96 | 17.23 |
| (light) | Ea. | 55.50 | 67.50 | 79.50 | 72 in. | Lin. ft. | 50.33 | 52.18 | 54.03 |
| (heavy) | Ea. | 69.25 | 82.50 | 95.75 | Vitrified clay | | | | |
| (chain) | Ea. | 105.65 | 128.00 | 150.65 | piping | | | | |
| Corner posts | Ea. | 31.50 | 32.00 | 32.50 | 18 in. dia. | Lin. ft. | 7.55 | 7.80 | 8.45 |
| | | | | | 24 in. | Lin. ft. | 15.15 | 16.10 | 17.05 |
| Fire protection | | | | | 36 in. | Lin. ft. | 33.95 | 36.20 | 38.95 |
| Pumps | | | | | Septic tank | | | | |
| Firehouse | Allowance | 100,000 | 150,000 | 200,000 | (45,000 gal.) | Ea. | ----- | 7,500 | ----- |
| Firetrucks (2) | | | | | | | | | |

## Table B-4 (*Continued*)

| Land surveys and fees | | | | |
|---|---|---|---|---|
| General surveys and fees | % total cost | 4.0 | 9.0 | 14.0 |
| Soil tests | Ea. | 300 | 400 | 500 |
| **Landscaping** | | | | |
| General | Sq. yd. | 1.50 | 1.70 | 1.90 |
| **Piling** | | | | |
| Wood (untreated) | Lin. ft. | 1.70 | 2.15 | 2.60 |
| Wood (creosoted) | Lin. ft. | 2.15 | 2.60 | 2.90 |
| Concrete: precast | Lin. ft. | 6.75 | 7.00 | 7.25 |
| cast in place | Lin. ft. | 4.75 | 6.62 | 8.50 |
| Steel pipe (concrete filled) | Lin. ft. | 7.50 | 9.50 | 11.50 |
| Steel section | Lin. ft. | 7.40 | 8.50 | 9.50 |
| Sheet piling, steel | Sq. ft. | 1.45 | 2.60 | 3.75 |
| wood | Sq. ft. | 1.25 | 1.75 | 2.25 |
| Pile driver setup | Ea. | 6,800 | 7,500 | 8,200 |
| **Roads, walkways, paving** | | | | |
| Paving | | | | |
| 4-in.-thk. reinf., 6-in. subbase | Sq. yd. | 6.35 | 7.87 | 8.39 |
| 6-in.-thk reinf., 6-in. subbase | Sq. yd. | 7.61 | 9.37 | 10.13 |
| 2-in.-thk. asphalt top, existing base | Sq. yd. | 2.37 | 3.12 | 3.87 |
| 2-in.-thk. asphalt top, 4-in. sub-base | Sq. yd. | 3.58 | 4.68 | 5.78 |
| 3-in.-thk. asphalt top, 12-in. sub-base | Sq. yd. | 6.37 | 7.62 | 8.87 |
| Gravel surface 2-in.-thk. gravel | Sq. yd. | 0.33 | 0.58 | 0.83 |

| Site clearing, excavation and grading | | | | |
|---|---|---|---|---|
| Site preparation | | | | |
| Machine cuts | Cu. yd. | 0.50 | 0.56 | 0.63 |
| Clearing and grubbing | Sq. yd. | 0.13 | 0.15 | 0.18 |
| General grading | Sq. yd. | 0.43 | 0.44 | 0.48 |
| Final leveling | Sq. yd. | 0.25 | 0.31 | 0.38 |
| **Foundation excavation** | | | | |
| Machine excavation | Cu. yd. | 1.50 | 1.63 | 1.75 |
| Machine plus hand trim | Cu. yd. | 2.50 | 3.44 | 3.75 |
| Hand work | Cu. yd. | 7.56 | 10.00 | 12.50 |
| **Trench excavation** | | | | |
| Machine 3½ ft. deep × 2 ft. wide | Lin. ft. | 0.38 | 0.44 | 0.50 |
| Machine 4 ft. × 3 ft. | Lin. ft. | 0.56 | 0.63 | 0.68 |
| Machine 4½ ft × 4 ft. | Lin. ft. | 1.12 | 1.13 | 1.25 |
| Machine 5 ft. × 5 ft. | Lin. ft. | 1.38 | 1.50 | 1.63 |
| Hand labor | Cu. yd. | 8.75 | 10.12 | 13.75 |
| **Trench shoring** | | | | |
| Sheeting | Sq. ft. | 1.25 | 1.52 | 1.75 |
| **Trench and foundation backfill** | | | | |
| Machine plus hand trim | Cu. yd. | 1.44 | 1.56 | 1.68 |
| Hand labor only | Cu. yd. | 5.79 | 6.25 | 6.75 |
| **Miscellaneous materials** | | | | |
| Sand | Cu. yd. | 3.05 | 4.80 | 5.55 |
| Gravel | Cu. yd. | 1.50 | 2.25 | 3.00 |
| Dirt fill | Cu. yd. | 1.30 | 2.15 | 3.00 |
| Crushed stone | Cu. yd. | 2.55 | 4.37 | 5.19 |

\* To obtain bare module costs multiply by 1.29 for solids operations and 1.34 for others.

Source:   Guthrie, K.M.: "Capital Cost Estimating," *Chemical Engineering*, Mar. 24, 1969, p. 114.

## Table B-5

## Unit costs for offsite facilities*

| | Field Installation, $ | | | | Field Installation, $ | | |
| | Unit | Min. | Norm | Max. | | Unit | Min. | Norm | Max. |
|---|---|---|---|---|---|---|---|---|---|
| Air systems | | | | | Pallet truck | | | | |
| Instrument air | | | | | Hydraulic 4,000 | | | | |
| Compression facilities, | | | | | lb. | Ea. | | 930 | |
| air dryer, air receiver, | | | | | Electric 4,000 | | | | |
| and distribu- | | | | | lb. | Ea. | | 3,600 | |
| tion | $M** | 18.75 | 43.75 | 62.50 | Payloaders | | | | |
| Plant air | | | | | 2 cu. yd. (gas) | Ea. | | 21,000 | |
| Compression facilities, | | | | | 4 cu. yd. | | | | |
| air receiver and distri- | | | | | (gas) | Ea. | | 33,700 | |
| bution | $M | 12.50 | 31.25 | 50.85 | 2 cu. yd. | | | | |
| | | | | | (diesel) | Ea. | | 22,900 | |
| Blowdown and flare | | | | | 4 cu. yd. | | | | |
| For general purposes | | | | | (diesel) | Ea. | | 36,500 | |
| (Including flare lines, | | | | | Tank trailers | | | | |
| blowdown drum and dis- | | | | | Carbon steel | Ea. | | 14,800 | |
| posal pit) | $M | 81.75 | 102.58 | 187.52 | Aluminum | Ea. | | 21,600 | |
| | | | | | Stainless | Ea. | | 36,500 | |
| Cooling tower and CW dis- | | | | | Tractors | | | | |
| tribution | | | | | Gasoline | Ea. | | 12,500 | |
| Use 1.15 design factor | | | | | Diesel | Ea. | | 27,500 | |
| on estimated through- | | | | | Tractor shovel | | | | |
| put. | | | | | 2-cu. yd. | | | | |
| Cooling tower costs | | See Table B-2 | | | bucket | Ea. | | 26,300 | |
| Distribution systems | | | | | 3-cu. yd. | | | | |
| For general | | | | | bucket | Ea. | | 28,700 | |
| purposes | Gpm. | 12.58 | 36.50 | 43.25 | 4-cu. yd. | | | | |
| | | | | | bucket | Ea. | | 35,600 | |
| River intake installation | | | | | Automotive shipping | | | | |
| For general | | | | | facilities | | | | |
| purposes | Gpm. | 8.22 | 16.25 | 24.37 | One outlet per | | | | |
| | | | | | 2,000 bbl. | | | 9,800 | |
| Fireloop and hydrants | | | | | Docks and wharves | | | | |
| For general | | | | | Light construction | | | | |
| purposes | $M | 12.58 | 22.50 | 40.24 | 2-in. deck | Sq. ft. | 5.15 | 5.63 | 6.25 |
| | | | | | 3-in. | Sq. ft. | 6.25 | 6.87 | 7.50 |
| Fuel systems | | | | | Medium construction | | | | |
| Fuel oil (includes | | | | | 3-in. | Sq. ft. | 8.75 | 9.38 | 10.12 |
| pumps, storage, piping, | | | | | 4-in. | Sq. ft. | 10.15 | 11.25 | 12.50 |
| controls and distribu- | | | | | Heavy construction | | | | |
| tion) | $M | 6.25 | 25.12 | 43.75 | 4-in. | Sq. ft. | 12.50 | 15.62 | 18.75 |
| Fuel gas (includes re- | | | | | concrete | Sq. ft. | 17.50 | 21.25 | 25.25 |
| ceiver, piping, controls | | | | | | | | | |
| and distribution) | $M | 12.50 | 37.52 | 62.58 | | | | | |

| | | | | |
|---|---|---|---|---|
| **General water systems** | | | | |
| Treated water | | | | |
| Filtered and | | | | |
| softened | Gal. | 0.15 | 0.23 | 0.30 |
| Distilled | Gal. | 0.65 | 0.92 | 1.20 |
| Drinking & service water | | | | |
| General | | | | |
| facilities | $M | 2.50 | 5.40 | 7.58 |
| **Power generation and distribution** | | | | |
| Use 1.10 design factor on estimated consumption. | | | | |
| Generating | | | | |
| facilities | Kw. | See Table B-2 | | |
| Electrical distribution | | | | |
| For general | | | | |
| purposes | Kw. | 87.5 | 93.75 | 98.75 |
| Main transformer stations | | | | |
| Three phase, 60 cycle | | | | |
| Capacity | | | | |
| 3,000 kva. | Kva. | 33.0 | 37.0 | 44.0 |
| 5,000 | Kva. | 20.0 | 23.0 | 26.0 |
| 10,000 | Kva. | 13.0 | 14.0 | 16.0 |
| 20,000 | Kva. | 10.0 | 12.0 | 13.0 |
| Secondary transformer stations | | | | |
| 4,200/575 v. | | | | |
| 600 kva | Kva. | 30.1 | 33.8 | 42.3 |
| 1,000 | Kva. | 20.1 | 25.2 | 31.5 |
| 1,500 | Kva. | 15.6 | 19.5 | 24.3 |
| 2,000 | Kva. | 14.8 | 18.5 | 23.2 |
| 13,200/575 v. | | | | |
| 600 kva | Kva. | 28.2 | 35.3 | 44.2 |
| 1,000 | Kva. | 21.2 | 26.5 | 33.1 |
| 1,500 | Kva. | 16.6 | 20.8 | 26.2 |
| 2,000 | Kva. | 15.4 | 19.3 | 24.1 |
| **Receiving, shipping, storage** | | | | |
| Automotive | | | | |
| Forklift trucks | | | | |
| 3,000 lb | Ea. | | 7,800 | |
| 5,000 lb | Ea. | | 11,000 | |
| 10,000 lb | Ea. | | 16,200 | |

| | | | | |
|---|---|---|---|---|
| **Dredging** | | | | |
| General | | | | |
| operations | Cu. yd. | 4.32 | 10.81 | 17.28 |
| Tankage | | See Table B-2 | | |
| General | | | | |
| **Railroad** | | | | |
| Straight track | | | | |
| (railroad sliding) | Lin. ft. | | 26.25 | |
| Turnout | Ea. | | 2,800 | |
| Bumper | Ea. | | 790 | |
| Blinker and gate | Ea. | | 9,300 | |
| Grade and ballast | Lin. ft. | | 6.25 | |
| Locomotives (battery) | | | | |
| 9 ton | Ea. | | 35,000 | |
| 12 ton | Ea. | | 41,800 | |
| Locomotives (diesel) | | | | |
| 1½ tons | Ea. | | 11,000 | |
| 3 tons | Ea. | | 14,000 | |
| Tank car (10,000 gal.) | Ea. | | 10,800 | |
| Railroad shipping facilities | | | | |
| One outlet per 2,000 bbl | Ea. | | 4,800 | |
| **Steam generation and distribution** | | | | |
| Use 1.10 design factor on estimated consumption. | | | | |
| Package boilers (up to 150,000 lb./hr.) | Lb./hr. | See Table B-2 | | |
| Field erected (above 150,000 lb./hr.) | Lb./hr. | See Table B-2 | | |
| Steam distribution | | | | |
| For general purposes | Lb./hr. | 0.94 | 1.52 | 1.68 |
| **Yard lighting and communications** | | | | |
| For general purposes | $M | 18.75 | 52.25 | 93.75 |
| **Yard transfer lines and Pumps** | | | | |
| For general purposes | $M | 17.25 | 31.25 | 56.25 |

* Multiply by 1.29 to obtain the bare module factor.
** When $M appears in Unit Column it means the costs should be multiplied by $1,000.
Source: Guthrie, K.M.: "Capital Cost Estimating," *Chemical Engineering*, Mar. 24, 1969, p. 114.

Table B-6

## Bare Module Costs of Single Story Building Shells

| Building Type | Base Height (ft.) | Shell Cost ($/ft² of Floor Area) | | |
|---|---|---|---|---|
| | | Low | Medium | High |
| Administration offices | 10 | 5.54 | 8.80 | 12.38 |
| Cafeterias | 12 | 2.87 | 7.60 | 11.00 |
| Compressor houses (with bridge crane) | 20 | 3.92 | 6.14 | 7.35 |
| Control house (equipped) | 10 | 4.61 | 7.60 | 12.18 |
| Garages | 15 | 2.36 | 3.76 | 5.75 |
| Maintenance shops | 20 | 3.24 | 5.31 | 6.40 |
| Laboratories and medical | 10 | 6.91 | 9.96 | 13.20 |
| Process buildings | 20 | 3.35 | 5.30 | 9.54 |
| Warehouses | 20 | 2.98 | 4.64 | 6.27 |

Source: Guthrie, K.M.: "Capital Cost Estimating," *Chemical Engineering*, Mar. 24, 1969, p. 114.

Table B-7

## Adjustment Factors for Multiple-Story Buildings, $F_n$

| Number of Stories (Multiple of Base Height | $F_n$ |
|---|---|
| 1 | 1.0 |
| 2 | 1.4 |
| 3 | 1.9 |
| 4 | 2.5 |

Source: Guthrie, K.M.: "Capital Cost Estimating," *Chemical Engineering*, Mar. 24, 1969, p. 114.

Table B-8

## Bare Module

## Costs of Building Services and Equipment

| Building Services | Cost Range $/Ft² of Effective Floor Area | | |
|---|---|---|---|
| | Low | Average | High |
| Air conditioning | 4.87 | 6.89 | 9.10 |
| Lighting and electrical | | | |
| Process buildings, cafeterias, offices, laboratories, medical, | 1.95 | 2.27 | 2.60 |
| control, compressor houses. | 2.92 | 3.25 | 3.58 |
| Warehouses, Maintenance shops | 0.91 | 1.17 | 1.43 |
| Heating and ventilating | 1.30 | 1.95 | 2.60 |
| Plumbing (general) | 1.57 | 2.21 | 2.83 |
| Fire prevention equipment (includes alarms, extinguishers, and sprinkler systems) | 1.17 | 1.43 | 1.69 |

Source: Guthrie, K.M.: "Capital Cost Estimating," *Chemical Engineering*, Mar. 24, 1969, p. 114.

Table B-9

Bare Module

Cost of Furniture and Equipment

| Furniture and Equipment | Cost Range, $/ft^2$ of Effective Floor Area | | |
|---|---|---|---|
| | Low | Medium | High |
| Laboratory equipment | 10.40 | 20.80 | 32.50 |
| Office equipment | 3.90 | 6.50 | 9.10 |
| Shop equipment | 5.20 | 7.80 | 10.40 |
| Cafeteria equipment | 4.55 | 5.85 | 7.15 |

Source:   Guthrie, K.M.: "Capital Cost Estimating," *Chemical Engineering,* Mar. 24, 1969, p. 114.

Table B-10

Bare Module Cost for

Steel Structures

| Height, ft | Bare Module Cost $/Cubic ft^3$ | | |
|---|---|---|---|
| | Light | Medium | Heavy |
| Up to 10 | 0.06 | 0.25 | 0.51 |
| 10 to 20 | 0.13 | 0.37 | 0.66 |
| 20 to 30 | 0.19 | 0.45 | 0.79 |
| 30 to 40 | 0.23 | 0.51 | 0.89 |
| 40 to 50 | 0.25 | 0.56 | 0.96 |
| 50 to 60 | — | 0.61 | 1.02 |
| 60 to 70 | — | 0.66 | 1.09 |
| 70 to 80 | — | 0.69 | 1.14 |
| 80 to 90 | — | 0.74 | 1.18 |
| 90 to 100 | — | 0.76 | 1.21 |
| 100 to 200 | — | 1.02 | 1.80 |
| 200 to 300 | — | — | 2.03 |

Source:   Guthrie, K.M., "Capital Cost Estimating," *Chemical Engineering,* Mar. 24, 1969, p. 114.

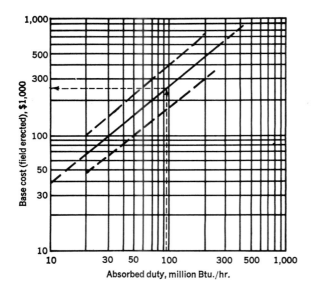

**Required**
Furnace type
Absorbed heat duty, Btu./hr.
Design pressure, psig.
Radiant tube material

**Basis of chart**
Process heater
Box or "A-frame" construction
Carbon steel tubes
Design pressure, 500 psi.
Field erected

**Time base**
Mid-1968

**Exponent**
Size exponent 0.85
$Cost_2 = cost_1 (size_2/size_1)^{*}$

**Included**
Complete field erection
Subcontractor indirects

**Process Furnace Cost, $** = [Base cost$(F_d + F_m + F_p)$]Index
**Pyrolysis or Reformer Furnace Cost, $** = [Base cost $(F_d + F_p)$] Index
**Adjustment factors**

| Design Type | $F_d$ | Radiant Tube Material | $F_m$* | Design Pressure, Psi. | $F_p$* |
|---|---|---|---|---|---|
| Process heater | 1.00 | Carbon steel | 0.00 | Up to  500 | 0.00 |
| Pyrolysis | 1.10 | Chrome/moly | 0.35 | 1,000 | 0.10 |
| Reformer (without catalyst) | 1.35 | Stainless | 0.75 | 1,500 | 0.15 |
| | | | | 2,000 | 0.25 |
| | | | | 2,500 | 0.40 |
| | | | | 3,000 | 0.60 |

*If these factors are used individually, add 1.00 to the above values.

Figure B-1   Process furnaces.
Source: Guthrie, K.M.: "Capital Cost Estimating," *Chemical Engineering,* Mar. 24, 1969, p. 114.

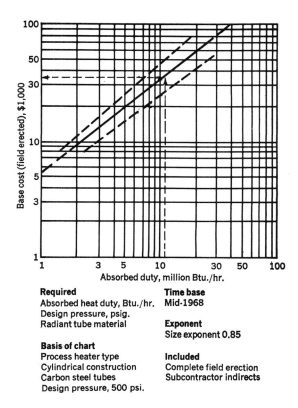

**Required**
Absorbed heat duty, Btu./hr.
Design pressure, psig.
Radiant tube material

**Time base**
Mid-1968

**Exponent**
Size exponent 0.85

**Basis of chart**
Process heater type
Cylindrical construction
Carbon steel tubes
Design pressure, 500 psi.

**Included**
Complete field erection
Subcontractor indirects

**Fired Heater Cost, \$** $= [\text{Base cost}(F_d + F_m + F_p)]\text{Index}$

**Adjustment factors**

| Design Type | $F_d$ | Radiant Tube Material | $F_m{}^*$ | Design Pressure, Psi. | | $F_p{}^*$ |
|---|---|---|---|---|---|---|
| Cylindrical | 1.00 | Carbon steel | 0.00 | Up to | 500 | 0.00 |
| Dowtherm | 1.33 | Chrome/moly | 0.45 | | 1,000 | 0.15 |
| | | Stainless | 0.50 | | 1,500 | 0.20 |

*If these factors are used individually, add 1.00 to the above values.

Figure B-2 Directed fired heaters.
Source: Guthrie, K.M.: "Capital Cost Estimating," *Chemical Engineering,* Mar. 24, 1969, p. 114.

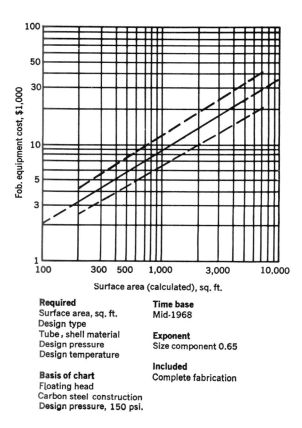

Surface area (calculated), sq. ft.

**Required**
Surface area, sq. ft.
Design type
Tube, shell material
Design pressure
Design temperature

**Basis of chart**
Floating head
Carbon steel construction
Design pressure, 150 psi.

**Time base**
Mid-1968

**Exponent**
Size component 0.65

**Included**
Complete fabrication

**Exchanger Cost, $** = [Base cost $(F_d + F_p) \times F_m$] Index

**Adjustment factors**

| Design Type | $F_d$ | Design Pressure, Psi. | $F_p$* | |
|---|---|---|---|---|
| Kettle, reboiler | 1.35 | Up to 150 | 0.00 | *If these factors are |
| Floating head | 1.00 | 300 | 0.10 | used individually, |
| U tube | 0.85 | 400 | 0.25 | add 1.00 to these |
| Fixed tube sheet | 0.80 | 800 | 0.52 | values. |
| | | 1,000 | 0.55 | |

**Shell/Tube Materials, $F_m$**

| Surface Area, Sq. Ft. | CS/ CS | CS/ Brass | CS/ Mo | CS/ SS | SS/ SS | CS/ Monel | Monel/ Monel | CS/ Ti | Ti/ Ti |
|---|---|---|---|---|---|---|---|---|---|
| Up to 100 | 1.00 | 1.05 | 1.60 | 1.54 | 2.50 | 2.00 | 3.20 | 4.10 | 10.28 |
| 100 to 500 | 1.00 | 1.10 | 1.75 | 1.78 | 3.10 | 2.30 | 3.50 | 5.20 | 10.60 |
| 500 to 1,000 | 1.00 | 1.15 | 1.82 | 2.25 | 3.26 | 2.50 | 3.65 | 6.15 | 10.75 |
| 1,000 to 5,000 | 1.00 | 1.30 | 2.15 | 2.81 | 3.75 | 3.10 | 4.25 | 8.95 | 13.05 |
| 5,000 to 10,000 | 1.00 | 1.52 | 2.50 | 3.52 | 4.50 | 3.75 | 4.95 | 11.10 | 16.60 |

Figure B-3    Shell-and tube exchangers.
Source: Guthrie, K.M.: "Capital Cost Estimating," *Chemical Engineering,* Mar. 24, 1969,
p. 114.

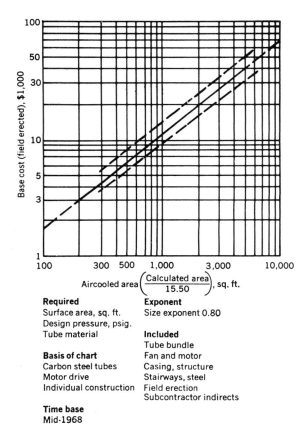

Aircooled area $\left(\dfrac{\text{Calculated area}}{15.50}\right)$, sq. ft.

**Required**
Surface area, sq. ft.
Design pressure, psig.
Tube material

**Basis of chart**
Carbon steel tubes
Motor drive
Individual construction

**Time base**
Mid-1968

**Exponent**
Size exponent 0.80

**Included**
Tube bundle
Fan and motor
Casing, structure
Stairways, steel
Field erection
Subcontractor indirects

Air Cooler Cost, $ = [Base cost$(F_p + F_t + F_m)$]Index

**Adjustment factors**

| Pressure Rating, Psi. | $F_p$ | Tube Length, Ft. | $F_t$* | Tube Material | $F_m$* |
|---|---|---|---|---|---|
| 150 | 1.00 | 16 | 0.00 | Carbon steel | 0.00 |
| 250 | 1.05 | 20 | 0.05 | Aluminum | 0.50 |
| 500 | 1.10 | 24 | 0.10 | Stainless | 1.85 |
| 1,000 | 1.15 | 30 | 0.15 | Monel | 2.20 |

*If these factors are used individually, add 1.00 to the above values.

Figure B-4   Air coolers.
Source: Guthrie, K.M.: "Capital Cost Estimating," *Chemical Engineering,* Mar. 24, 1969, p. 114.

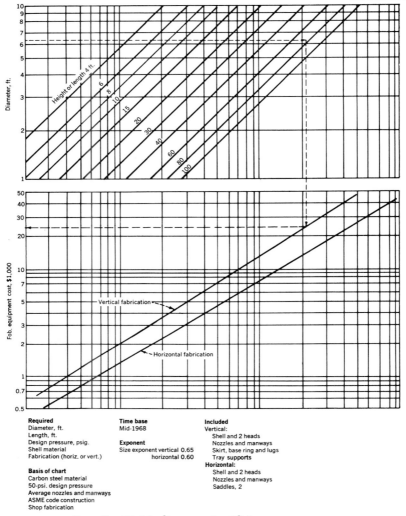

Required
Diameter, ft.
Length, ft.
Design pressure, psig.
Shell material
Fabrication (horiz. or vert.)

Basis of chart
Carbon steel material
50-psi. design pressure
Average nozzles and manways
ASME code construction
Shop fabrication

Time base
Mid-1968

Exponent
Size exponent vertical 0.65
                 horizontal 0.60

Included
Vertical:
    Shell and 2 heads
    Nozzles and manways
    Skirt, base ring and lugs
    Tray supports
Horizontal:
    Shell and 2 heads
    Nozzles and manways
    Saddles, 2

**Process Vessel Cost, \$** = [Base cost $\times F_m \times F_p$]Index

**Adjustment factors**

| Shell Material | $F_m$ Clad | $F_m$ Solid | Pressure Factor Psi. | $F_p$ |
|---|---|---|---|---|
| Carbon steel | 1.00 | 1.00 | Up to    50 | 1.00 |
| Stainless 316 | 2.25 | 3.67 | 100 | 1.05 |
| Monel | 3.89 | 6.34 | 200 | 1.15 |
| Titanium | 4.23 | 7.89 | 300 | 1.20 |
|  |  |  | 400 | 1.35 |
|  |  |  | 500 | 1.45 |
|  |  |  | 600 | 1.60 |
|  |  |  | 700 | 1.80 |
|  |  |  | 800 | 1.90 |
|  |  |  | 900 | 2.30 |
|  |  |  | 1,000 | 2.50 |

Figure B-5    Pressure vessels, vertical and horizontal.
Source: Guthrie, K.M.: "Capital Cost Estimating," *Chemical Engineering*, Mar. 24, 1969, p. 114.

CAPITAL COSTS . . .

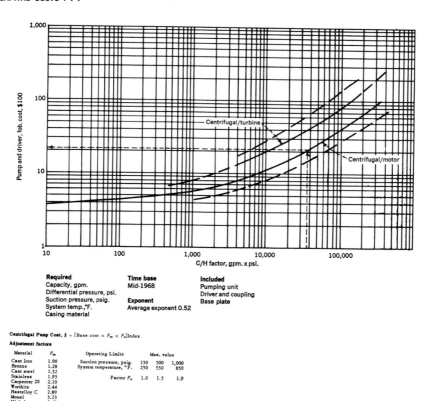

| Required | Time base | Included |
|----------|-----------|----------|
| Capacity, gpm. | Mid-1968 | Pumping unit |
| Differential pressure, psi. | | Driver and coupling |
| Suction pressure, psig. | **Exponent** | Base plate |
| System temp.,°F. | Average exponent 0.52 | |
| Casing material | | |

**Centrifugal Pump Cost, $ = [Base cost × $F_m$ × $F_o$]Index**

**Adjustment factors**

| Material | $F_m$ | Operating Limits | Max. value | | |
|----------|-------|------------------|------|------|------|
| Cast iron | 1.00 | Suction pressure, psig. | 150 | 500 | 1,000 |
| Bronze | 1.28 | System temperature, °F. | 250 | 550 | 850 |
| Cast steel | 1.32 | | | | |
| Stainless | 1.93 | Factor $F_o$ | 1.0 | 1.5 | 1.9 |
| Carpenter 20 | 2.10 | | | | |
| Worthite | 2.44 | | | | |
| Hastelloy C | 2.89 | | | | |
| Monel | 3.23 | | | | |
| Nickel | 3.48 | | | | |
| Titanium | 8.98 | | | | |

Figure B-6    Centrifugal pumps and drivers.
Source: Guthrie, K.M.: "Capital Cost Estimating," *Chemical Engineering,* Mar. 24, 1969, p. 114.

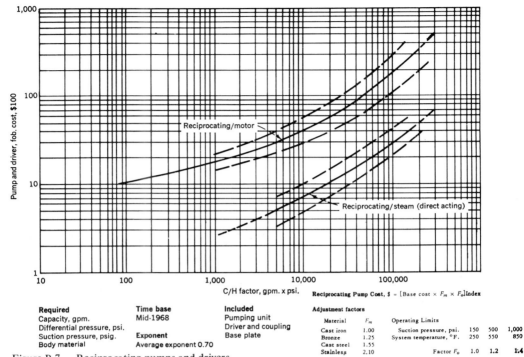

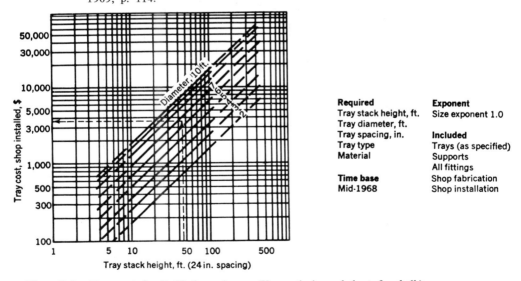

Figure B-7    Reciprocating pumps and drivers.
                Source: Guthrie, K.M.: "Capital Cost Estimating," *Chemical Engineering,* Mar. 24,
                1969, p. 114.

Figure B-8    Tray costs for distillation columns. (Use vertical vessel charts for shell.)
                Source: Guthrie, K.M.: "Capital Cost Estimating," *Chemical Engineering,* Mar. 24, 1969,
                p. 114.

1. Total equipment: $E=$(Source)X(Index)

  Source=Published data, personal files or vendor quotations

2. Direct material: $(E+M)=(E)$ $(F_r)$ $(F_m)$X(Index)

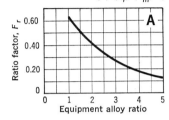 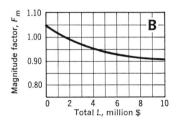

3. Direct field labor: $L=(E)$ $(F_r)$ $(F_m)$X(Index)

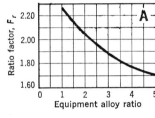 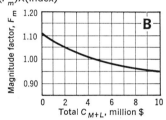

4. Direct material-and-labor cost: $C_{M+L}=(E)$ $(F_r)$ $(F_m)$X(Index)

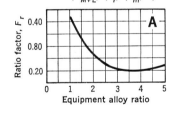 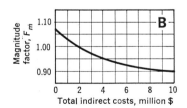

5. Indirect costs$=(C_{M+L})$ $(F_r)$ $(F_m)$X(Index)

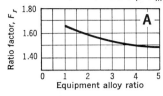 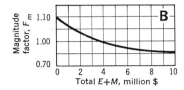

6. Total bare-module cost$=(E)$ $(F_r)$ $(F_m)$X(Index)

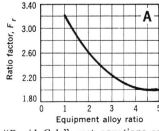 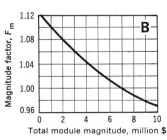

Figure B-9    "Rapid Calc" cost equations and chart.
        Source: Guthrie, K.M.: " 'Rapid Calc' Charts," *Chemical Engineering*, Jan. 13, 1969, p. 138.

# PHYSICAL PROPERTIES AND CONSTANTS OF PURE VINYL CHLORIDE

The following table contains the latest information available on the properties of vinyl chloride. Some of these properties were determined by Dow while others are calculations, estimations or determinations from the most reliable sources available.

| Property | Ref.* | Temp. °C | |
|---|---|---|---|
| 1. Molecular Weight | (1) | | 62.501 |
| 2. Boiling Point at 760 mm, °C | (2) | | —13.80 |
| 3. Melting Point, °C | (3) | | —153.69 |
| 4. Odor, uninhibited<br>inhibited | (4) | | Faint, sweet odor<br>Faint phenolic odor due to inhibitor |
| 5. Color | (4) | | Colorless |
| 6. Flash Point, Cleveland Open Cup | (1) | —78°(—108°F) | |
| 7. Explosive Limits, Vol. % in air | (5) | | 3.6 - 26.4 |
| 8. Liquid Density, g/ml | (6) (3) | 100°C<br>25°C<br>—20°C<br>—25°C<br>—30°C | 0.746<br>0.9013<br>0.9834<br>0.9918<br>0.9999 |

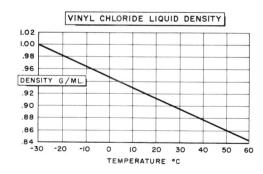

**VINYL CHLORIDE LIQUID DENSITY**

Data Courtesy of Dow Chemical Company.

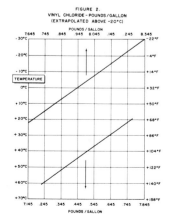

FIGURE 2.
VINYL CHLORIDE - POUNDS / GALLON
(EXTRAPOLATED ABOVE -20°C)

| Property | Ref.* | Temp. °C | |
|---|---|---|---|
| 9. Density Change with Temperature, g/ml/°C (between —30°C & —20°C) | (3) | | 0.00164 (approx.) |
| 10. Liquid Viscosity, cps | (3) | 100°C | 0.100 |
| | | 25°C | 0.185 |
| | | —10°C | 0.248 |
| | | —20°C | 0.274 |
| | | —30°C | 0.303 |
| | | —40°C | 0.340 |
| 11. Vapor Viscosity, cps | (7) | 100°C | 0.0138 |
| | | 25°C | 0.0108 |
| 12. Surface Tension, dynes/cm | (3) | 100°C | 5.4 |
| | | 25°C | 16.0 |
| 13. Refractive Index, D line | (8) | 25°C | 1.3642 |
| 14. Vapor Pressure, mm Hg | (2) | 100°C | 16,617 |
| | | 25°C | 2,943 |
| | | 0°C | 1,293 |
| | | —25°C | 470 |
| | | —50°C | 130 |
| | | —75°C | 24.4 |
| | | —100°C | 2.5 |
| 15. Change in Boiling Point with Pressure dt/dP °C/mm Hg | (3) | 25°C | 0.01355 |
| | | —13.37°C | 0.03423 |

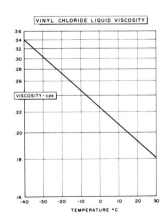

VINYL CHLORIDE LIQUID VISCOSITY

VISCOSITY - cps

TEMPERATURE °C

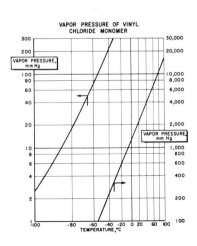

VAPOR PRESSURE OF VINYL CHLORIDE MONOMER

VAPOR PRESSURE, mm Hg

VAPOR PRESSURE, mm Hg

TEMPERATURE, °C

| | | | |
|---|---|---|---|
| 16. Heat Capacity (Liquid), Cp, cal./mole/deg. | (3) | 100°C<br>25°C | 26.25<br>20.56 |
| 17. Heat Capacity (Vapor), Cp, cal./mole/deg. | (9) | 100°C<br>25°C | 14.88<br>12.82 |
| 18. Latent Heat of Vaporization<br>▲ Hv, cal/mole | (3) | 100°C<br>25°C<br>—13.8°C | 3,310<br>4,710<br>5,250 |
| 19. Latent Heat of Fusion,<br>▲ Hm, cal/mole | (3) | | 1,172 |
| 20. Heat of Formation<br>▲ Hf, 298°K, kcal/mole | (3) (10) | | 8.480 |
| 21. Free Energy of Formation,<br>▲ Ff 298°K, kcal/mole | (3) | | 12.386 |
| 22. Heat of Polymerization,<br>▲ Hp, kcal/mole | (3) | | —25.3 ± 0.5 |
| 23. Critical Temperature, $t_c$, °C | (11) | | 14/ |
| 24. Critical Pressure, $P_c$, atmospheres | (11) | | 56 |
| 25. Critical Volume, $v_c$, cc/mole | (11) | | 179 |
| 26. Critical Compressibility (PV/RT), Z | (11) | | 0.29 |
| 27. Volumetric Shrinkage upon Polymerization, approximate, % | (3) | | 35 |
| 28. Dielectric Constant at Frequency of $10^6$ Hz | (3) | 17.2°C<br>—21°C | 6.26<br>7.05 |
| 29. Dissipation Factor at Frequency of $10^6$ Hz | (3) | —21°C | 0.0011 |
| 30. Solubility — Soluble in $CCl_4$, ether, ethanol and most organic solvents | | | |
| 31. Solubility of water in Vinyl Chloride, % | (1) | | 0.11 |

## REFERENCES

(1) R. R. Dreisbach; American Chemical Society Advances in Chemistry Series No. 22 (1959)

(2) R. A. McDonald, S. A. Shrader, D. R. Stull; J. Chem. Eng. Data 4, 311 (1959)

Antoine Equation for calculating vapor pressures of vinyl chloride. Log. Pmm = 6.85978-892.23/(238.04+t) (t=°C)

(3) Unpublished Data The Dow Chemical Company

(4) M. C. A. Chemical Safety Data Sheet SD-56 (1954); Properties and Essential Information for Safe Handling and Use of Vinyl Chloride

(5) T. Numano, T. Kitagawa; Kogyo, Kagaku Zasshi 65, 182-4 (1962)

(6) M. C. A. Res. Proj.; Selected Values of Properties of Chemical Compounds, June 20, 1956

(7) P. Zaloudik; Chem. Prumysl 12, 81-3 (1962)

(8) A. W. Francis; J. Chem. Eng. Data 5, 534-5 (1960)

(9) K. A. Kobe, R. H. Harrison; Petroleum Refiner 30 No. 11,151 (1951)

(10) Average from

J. R. Lachner, E. Emery, E. Bohmfalk, J. D. Park; J. Phys. Chem. 60, 492-5 (1956) and

J. R. Lachner, H. B. Gottlieb, J. D. Park; Trans. Faraday Soc. 58, 2348-51 (1962)

(11) Calculated by Lydersen's Method; A. L. Lydersen; U. of Wisconsin. Eng. Exp. Station Report No. 3, April 1955

# Index